Methods in Plant Ecology

Methods in Plant Ecology

EDITED BY
P. D. MOORE & S. B. CHAPMAN

SECOND EDITION

BLACKWELL SCIENTIFIC PUBLICATIONS
OXFORD LONDON EDINBURGH
BOSTON PALO ALTO MELBOURNE

© 1976, 1986 by
Blackwell Scientific Publications
Editorial offices:
Osney Mead, Oxford, OX2 0EL
8 John Street, London, WC1N 2ES
23 Ainslie Place, Edinburgh, EH3 6AJ
52 Beacon Street, Boston
 Massachusetts 02108, USA
667 Lytton Avenue, Palo Alto
 California 94301, USA
107 Barry Street, Carlton
 Victoria 3053, Australia

First published 1976
Second edition 1986

Printed and bound by The Alden Press,
Osney Mead, Oxford

DISTRIBUTORS

USA and Canada
 Blackwell Scientific Publications Inc
 P O Box 50009, Palo Alto
 California 94303

Australia
 Blackwell Scientific Publications
 (Australia) Pty Ltd
 107 Barry Street
 Carlton, Victoria 3053

British Library
Cataloguing in Publication Data

Methods in plant ecology.—2nd ed.
 1. Botany—Ecology
 I. Moore, Peter D. II. Chapman, S.B.
 581.5 QK901

ISBN 0-632-00989-6
ISBN 0-632-00996-9 Pbk

Contents

Chapter 5 continued

6 Chemical Analysis, 285

S. E. ALLEN, H. M. GRIMSHAW and A. P. ROWLAND

List of Contributors

S. E. ALLEN *Institute of Terrestrial Ecology, Merlewood Research Station, Grange over Sands, Cumbria LA11 6JU*

D. F. BALL *Institute of Terrestrial Ecology, Bangor Research Station, Penrhos Road, Bangor, Gwynedd LL57 2LQ*

P. BANNISTER *Department of Botany, University of Otago, Box 56, Dunedin, New Zealand*

R. BHADRESA *Field Studies Council, Flatford Mill Field Centre, East Bergholt, Colchester, Essex CO7 6UL*

S. B. CHAPMAN *Institute of Terrestrial Ecology, Furzebrook Research Station, near Wareham, Dorset BH20 5AS*

F. B. GOLDSMITH *Department of Botany and Microbiology, University College, University of London, Gower Street, London WC1E 6BT*

H. M. GRIMSHAW *Institute of Terrestrial Ecology, Merlewood Research Station, Grange over Sands, Cumbria LA11 6JU*

C. M. HARRISON *Department of Geography, University College, University of London, 26 Bedford Way, London WC1H 0AP*

M. J. HUTCHINGS *School of Biological Sciences, University of Sussex, Biology Building, Falmer, Brighton, Sussex BN1 9QG*

P. D. MOORE *Department of Biology, King's College, University of London, 68 Half Moon Lane, London SE24 9JF*

A. J. MORTON *Imperial College Field Station, University of London, Silwood Park, Ascot, Berkshire SL5 7PY*

S. D. PRINCE *School of Biological Sciences, Queen Mary College, University of London, Mile End Road, London E1 4NS*

D. ROBINSON *Department of Soil Fertility, Macaulay Institute for Soil Research, Craigiebuckler, Aberdeen AB9 2QJ*

I. H. RORISON *Unit of Comparative Plant Ecology, Department of Botany, University of Sheffield, Sheffield S10 2TN*

A. P. ROWLAND *Institute of Terrestrial Ecology, Merlewood Research Station, Grange over Sands, Cumbria LA11 6JU*

Preface

The last century has seen the gestation, birth and rapid evolution of the discipline of ecology. The very word has moved from obscurity into limelight, perhaps even into cliché, in the development of the English language over the past few decades. Concern for the environment and the recognition that man is not merely a factor influencing his surroundings, but also suffers the consequences of environmental mismanagement, has led to changing attitudes on the part of scientists and society. Ecology is no longer merely quantitative natural history, it is becoming an increasingly rigorous discipline as more specific (and more general) questions are being asked of it by concerned members of society.

The demands made upon ecologists as a consequence of problems ranging from pesticide misuse to acid rain, from nuclear winter projections to carbon dioxide accumulation, have placed the ecologist in an increasingly difficult dilemma. He now has the technical capacity to construct elegant analogue models of ecological systems, but often he lacks the raw data to feed into them. Some 10 years ago, when the first edition of this book was being prepared, the situation was quite different. Practical field and laboratory methods in ecology were developing rapidly, influenced by such concerted international efforts as the International Biological Programme, but the limitation on ecological advance lay in the area of data logging and analysis. The rapid development of computer science and, in particular, the current universal availability of quite sophisticated microcomputers, has broken this particular log jam and has left us in a position where more extensive and more accurate field and laboratory data are the major requirements for further advancement of the science. It is an awareness of this need which has stimulated the preparation of a second edition of this book.

Inevitably, the development of new ideas and methods during the past 10 years has not been evenly distributed over the entire field of plant ecology. Certain areas, such as production ecology and chemical analysis of ecological materials, had seen a rapid development in the 1970s under the patronage of IBP. In certain subjects, such as soil study, the chapters contained in the first edition have profoundly influenced more recent work. The subsequent developments in these areas have been significant, but perhaps not as dramatic and spectacular as, say, that of vegetation analysis, which has benefited specifically from the computer revolution, so its coverage is consequently expanded here. The new, numerical power in the hands of the ecologists has also led us to introduce a chapter on data handling, for so much of one's experimental design in field and laboratory depends upon an early appreciation of the analytical processes available and the careful choice of the most appropriate methods for specific problems.

One area of ecology which has emerged and flourished over the past 10 years is the study of plant populations. It has clearly become necessary to include in this book a section relating to the practical aspects of such study.

Another development area is that of physiological ecology. There has been a very healthy and productive movement of ideas across the barrier which once existed between laboratory-based plant physiology and field-based plant ecology, which has resulted in benefits to both subjects. Perhaps it is no longer even possible to regard them as separate. The consequence, as far as this book is concerned, is an expansion of the space devoted to this area of ecology; it now occupies two chapters (covering, respectively, nutrient and water relations) rather than the former one.

Another barrier which is becoming increasingly blurred is that between plant and animal ecology. A new section illustrating the opportunities presented in this twilight zone covers faecal analysis and exclosure studies as an aid to observing the consequences of plant/animal interactions. Perhaps, in future editions, we shall see further expansion of this theme into such areas as the chemistry and significance of secondary plant products, and pollination biology.

The study of environmental history has developed along two new paths. First, it has become a more integrated discipline, using evidence from a wide range of biological and chemical sources and, second, it has moved beyond the stage of regional vegetation and climatic reconstruction to one of local historical development of sites, studied by means of small sedimentary basins and soils. This aspect, it is felt, makes the subject more valuable to general ecologists who are interested in local and geologically recent events in the field locations of their studies.

All of the authors have been under pressure to confine their contributions to the minimum of space in order to keep down costs and thus make this compendium available to a greater number of ecologists. In many of the areas covered there have been important, useful and easily available publications produced recently which it would be wasteful to reprint or paraphrase here. So, wherever such methodological accounts are readily available, authors have been encouraged to refer to them rather than reproduce them. This book remains, however, a practical manual rather than a reference list, and important methods will be found here, discussed in detail and containing hints, warnings, advice and encouragement to those who would otherwise become lost in the current morass of published material.

We have aimed the book at a fairly broad spectrum of ecologists, from the specialist to the generalist, from the professional research worker to the undergraduate project student, from the field conservationist to the laboratory technician. It is hoped that specialists may find within the chapter dealing with their specific field some new ideas, information and guidance. It is also hoped that such specialists will look beyond their own field towards the problems encountered by others and how they have been tackled. In this way the greatest possible benefit will be gained from the book, and perhaps it may contribute to an even more rapid development of methods in plant ecology over the next decade.

Many people have contributed to the compilation of this volume and have helped authors with individual chapters. We extend to them all our thanks. Specific thanks must be given to Bob Campbell of Blackwell Scientific Publications Ltd, Oxford, for his boundless energy, enthusiasm and optimism, and also to Penny Baker for her laborious and careful preparation of the text for publication.

P.D. Moore
S.B. Chapman

1 Production ecology and nutrient budgets

S.B. CHAPMAN

1 Introduction

An ecologist can find himself involved in the study of biological production and nutrient budgets for one of a number of reasons. He may wish to use the estimates of primary production for a comparison of sites within a particular type of ecosystem, or to use them as the basis for comparing very different types of ecosystem (Westlake, 1973). The study of primary production, nutrient budgets and energy flow are important in attempting to understand the function of natural communities, but it should be remembered that they represent only one particular approach to the problem and that other viewpoints in ecology may be just as important in helping to obtain a more complete analysis and understanding of ecosystems and ecological processes.

One particularly important feature of production ecology is the way that it provides a strong and unifying link between a number of different aspects of the subject. The distinctions between types of ecologists working in this field tend to break down, and an individual engaged in a production study may often wonder whether he is a botanist, a zoologist or even a pedologist.

This chapter is divided into three main sections; the first provides an introduction to some of the more important definitions and concepts that relate to production ecology, the second describes some of the methods that are available for the estimation of primary production and associated processes, and the third deals with methods that are relevant to the study of nutrient budgets.

2 The ecosystem

2.1 The ecosystem concept

In a paper presented in 1935, Sir Arthur Tansley dealt with a number of terminological and conceptual problems that beset ecologists of the time. He rejected such contemporary terms as 'complex organisms' and 'biotic community', and introduced the term ECOSYSTEM in the following terms:

> 'Though the organisms may claim our primary interest, when we are trying to think fundamentally we cannot separate them from their special environment, with which they form one physical system.'

The ecosystem has since been defined by many authors as a functional unit that includes the biotic components (organisms including man) and the abiotic components (environmental physico-chemical) of a specified area (Fig. 1.1). While it is generally recognized that the ecosystem includes inter-relationships between biotic and abiotic factors it is often forgotten that the definition includes 'a specified area'. Tansley stated, 'ecosystems are of the most various kinds and sizes. They form one category of the multitudinous

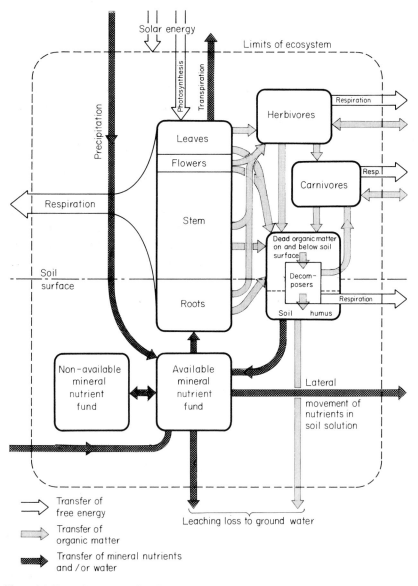

Figure 1.1 General representative diagram of an ecosystem.

physical systems of the universe, which range from the universe as a whole down to the atom'. Unfortunately, many people now use the term ecosystem in a general way without defining the limits of the system to which they refer. Whenever the term ecosystem is used in reference to an ecological study the physical and biological limits of the system should be made clear. It should be remembered that ecosystems are inter-related functional units and that any separation is an artificial division for the purposes of simplification and investigation. The historical development of the ecosystem concept has been described in detail by Major (1969).

2.2   Ecosystem modelling

In recent years most ecologists have become familiar with the terms 'ecosystem modelling' and 'systems analysis' even if they are not sure of their full meaning or implications. A model is no more than an abstraction that serves to describe or simulate all or part of some process or situation. Models can take a number of forms: mathematical models, word models, box models and flow diagrams. It is inevitable that modelling, the systems approach, and the ability of computers to handle large amounts of complex data, have a great deal to offer ecology in the future. It should be emphasized that systems analysis is not the only way of studying an ecosystem, but that a particularly important feature of the approach is the way in which it emphasizes the need to define and to quantify the basic components of the system.

A detailed account of systems analysis and modelling cannot be attempted here but Smith (1970), Jeffers (1972, 1978), Reichle *et al.* (1973), de Wit & Goudriaan (1974), Hall & Day (1977), Clark & Roswall (1981) and Smith (1982) are references that will be of interest to those requiring an introduction to the subject.

3   Production, decomposition and accumulation

3.1   Definitions, concepts and units

For the successful measurement of production, and the associated processes of decomposition and accumulation, it is necessary for a number of basic terms to be defined and for the relationships between them to be clearly understood.

3.1.1   Production

It would be most satisfactory if gross primary production was the fundamental estimate upon which other estimates of production could be based or against which comparisons could be made. If this were the case then a number of problems such as root production would possibly seem less intractable. Unfortunately, the measurement of gross production involves the use of sophisticated and expensive apparatus that does not lend itself readily to the degree of replication generally required in ecological studies, and reasonable proximity to a laboratory is often an important requirement of such techniques. A good case can be made that, as far as some components of the ecosystem are concerned, it is only net primary production that is important. In the case of some herbivores it may only be one particular fraction of the net primary production that is of interest. In a great deal of the ecological literature primary production is taken to be synonymous with net production. It is not intended to deal with methods for the estimation of gross primary production but workers requiring information upon this subject should refer to a specialized text such as that by Coombs & Hall (1982).

PRODUCTION is the weight or biomass of organic matter assimilated by an organism or community over a given period of time.

PRIMARY PRODUCTION is the production of organic matter by photosynthesis and SECONDARY PRODUCTION the subsequent conversion of that organic matter by heterotrophic organisms. Primary production can be expressed in two ways:

(a) GROSS PRIMARY PRODUCTION, the total amount of organic matter produced (including that lost in respiration) over a given period of time.

(b) NET PRIMARY PRODUCTION, the amount of organic matter incorporated by a plant or an area of vegetation (gross primary production minus the loss due to respiration) over a given period of time.

It is net primary production that is generally the concern of the plant ecologist and it is often further qualified by reference to some particular part of the plant or vegetation (aerial, root or seed production, etc.).

BIOMASS or STANDING CROP is the weight of organic matter per unit area present in some particular component of the ecosystem at a particular instant of time. Biomass is generally expressed in terms of dry weight and on occasion may be given in terms of ash free dry weight (see Section 4.4.1).

The relationship between biomass, time and production, has been given in a standard form in the handbooks produced for the International Biological Programme (Newbould, 1967; Miller & Hughes, 1968).

B_1 = Biomass of a plant community at time t_1.

B_2 = Biomass of a plant community at time $t_2 (= t_1 + \Delta t)$.

ΔB = Change in biomass during the period $t_1 - t_2$.

L = Plant losses by death and shedding during $t_1 - t_2$.

G = Plant losses by grazing etc. during $t_1 - t_2$.

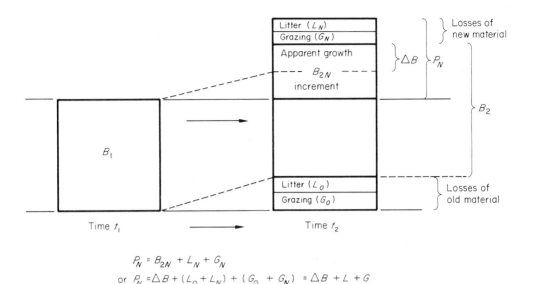

$$P_N = B_{2N} + L_N + G_N$$

$$\text{or } P_N = \Delta B + (L_O + L_N) + (G_O + G_N) = \Delta B + L + G$$

Figure 1.2 The relationship between production, apparent growth increment and change in biomass (based on Newbould, 1967).

P_N = Net production by the community during $t_1 - t_2$.
In terms of these symbols:

$$P_N = \Delta B + L + G,$$

so that if ΔB, L and G can be estimated satisfactorily P_N can be calculated. The use of this relationship requires estimates of biomass to be made at least twice, and generally with at least a year between the determinations.

There is an alternative approach (Fig. 1.2; Newbould, 1967): if the components of production can be recognized at the end of the growing season it should be possible to obtain an estimate of production from a single visit to the site.

$$P_N = P_{\text{flowers}} + P_{\text{green}} + P_{\text{wood}} + P_{\text{roots}}$$

By the end of the growing season some of the current year's production may well have been lost by grazing, or by death and loss as litter, so that the apparent growth increment will be an underestimate of production.

The addition of the total grazing and litter losses to the apparent growth increment will result in overestimation of the net production as part of the losses will have been from previous years' production.

3.1.2 Decomposition

DECOMPOSITION is the process by which organic matter is physically broken down and converted to simpler chemical substances, resulting in the production of carbon dioxide, water and the liberation of energy.

Decomposition represents a loss of energy and material from the eco-system as well as the transformation and movement of organic matter within the system. The decomposition of organic matter is of fundamental import-ance in the release of plant nutrients from the litter and soil organic matter, making them available for uptake and further plant growth.

3.1.3   Accumulation

ACCUMULATION is the rate of change in weight of some part of the ecosystem as a result of production and decomposition.

The accumulation of litter has been described and examined against the background of a simple mathematical model by Jenny *et al.* (1949) and by Olson (1963).

If B = Biomass or weight of organic matter.

B_{ss} = Biomass or weight of organic matter present under steady state conditions.

P = Production (input to the system).

k = Instantaneous fractional loss rate.

The rate of change of biomass in the system over some discrete time interval (t), such as a day or year can be expressed as:

$$\frac{\Delta B}{\delta t} = \text{input} - \text{losses (for that time interval)}.$$

If the input to the system remains constant, the instantaneous rate of change of weight (rate when the limits of ΔB and Δt approach zero) will be:

$$\frac{dB}{dt} = P - kB.$$

Under steady state conditions it follows that inputs to the system must equal losses,

$$P = kB_{ss},$$

so that if these assumptions are valid and two of these parameters can be measured then the third can be calculated. If at all possible production and the rate of decomposition should be measured independently.

If constant rates of production and decomposition are assumed (for some objections to this simple model see Section 3.4.6) a number of further relationships can be derived. Where sources of input to the system are removed, such as in litter bag experiments (Section 3.4.2) or in the case of soil organic matter under fallow conditions it can be shown that:

$$B = B_0 a e^{-kt}$$

where B = the amount of organic matter remaining after time t.

B_0 the initial weight of organic matter present at time t_0.

This expression for exponential decay is the same as that used to calculate the amount of a radioactive isotope remaining after a given period of time.

If the input to the system (production) is now assumed to remain at some value (P) instead of being zero, an expression can be derived that will provide an estimate of the weight of organic matter that will have accumulated after any period of time.

$$B = B_{ss}(1 - a e^{-kt})$$

or

$$B = \frac{P}{k}(1 - a e^{-kt})$$

where a is a constant.

This accumulation curve, often referred to as the MONO-MOLECULAR growth curve (Fig. 1.3), shows that with constant production and decomposition the weight of accumulated organic matter increases with time to a steady state value (B_{ss}). It is possible to calculate the time required to approach steady state conditions for any particular combination of rates of production and decomposition:

$$\frac{0{\cdot}6931}{k} = \text{time required to reach 50\% steady state biomass;}$$

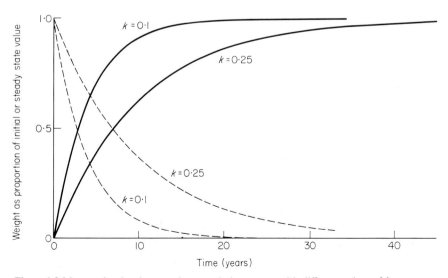

Figure 1.3 Monomolecular decay and accumulation curves with different values of k.

$$\frac{3}{k} = \text{time required to reach 95\% steady state biomass.}$$

In these expressions the constant k has been defined as the instantaneous fractional loss rate, but in many ecological investigations it is the fractional loss rate over a definite period of time, such as a month or year, (k') that is measured or estimated. When the loss is expressed as a fraction of the original weight the relationships between the two loss rates are:

$$k' = 1 - e^{-kt}$$

$$k = -\log_e (1 - k')$$

When the time interval is short (e.g. loss per day), or where decomposition is slow there are only small differences between the values of k and k', but when litter decomposition is measured over a whole year the differences between the two decay rates can be very important.

In this discussion it has been assumed that production remains constant with time and in many situations this is obviously not the case. The papers referred to by Jenny *et al.* and Olson discuss the application of these expressions to forest conditions where litter production is seasonal. Despite the assumptions of a simple exponential model it has much to offer in many ecological problems and demonstrates the basic relationships between production, decomposition and accumulation.

The monomolecular growth curve is just one of a series of curves that have been used to describe biological growth, or accumulation (Fig. 1.4). The monomolecular curve, so called because of its relevance to first-order chemical reactions, assumes that the rate of growth at any time will be proportional

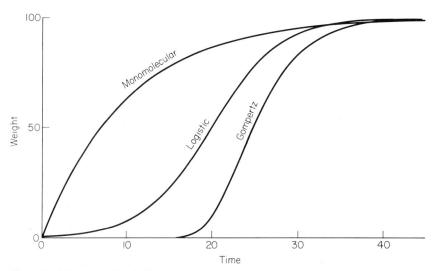

Figure 1.4 Monomolecular, logistic and Gompertz growth curves.

to the difference between biomass (B) and the maximum weight that will be attained or accumulated (B_{ss}) under steady state conditions.

$$\frac{dB}{dt} = k(B_{ss} - B)$$

In the LOGISTIC growth curve it is assumed that the rate of growth at any particular time is proportional to the weight at that time, as well as to a function of that weight in relation to the maximum weight that will be attained.

$$\frac{dB}{dt} = kB\left(\frac{1 - B}{B_{ss}}\right)$$

from which can be derived:

$$B = \frac{B_{ss}}{1 + e^{a-kt}}$$

The time taken to reach 50% of the steady state biomass is given by

$$t = a/k$$

The GOMPERTZ growth curve, developed by actuaries in the analysis of human mortality, is described by the equation:

$$B = B_{ss}\,e^{(-e^{a-kt})}$$

with the growth rate given by:

$$\frac{dB}{dt} = kB\log_e\left(\frac{B_{ss}}{B}\right)$$

When growth curves are fitted to ecological data they can be useful for comparative and descriptive purposes, but it is important not to attach too great a degree of biological significance to the constants obtained in the process.

The monomolecular, logistic and Gompertz curves are all related and can be expressed as a function described by Richards (1959).

$$B^{(1-m)} = B_{ss}^{(1-m)}(1 - a\,e^{-kt})$$

where m is a constant that controls the location of the point of inflexion. In terms of the specific curves described, when m = 0 the curve is equivalent to the monomolecular, when m $\approx$ 1 then the curve is a Gompertz and when m = 2 then the function is equivalent to that of a logistic curve. Further details of growth curves and growth analysis can be found in Hunt (1978, 1982) and Evans (1972).

3.1.4 Units

Various combinations of metric units have been, and are used to indicate the magnitude of dry matter production and associated processes. Whilst there would appear to be a move towards standardization and the use of grammes per square metre ($g\,m^{-2}$) as the basic unit for the study of primary production there are several points worth considering before choosing a set of units to express a particular set of results. There is a considerable advantage in using either kilogrammes per hectare ($kg\,ha^{-1}$), or tonnes per hectare (tonnes ha^{-1}) when the weights of organic matter involved are large.

$$1\,g\,m^{-2} = 10\,kg\,ha^{-1} = 0\cdot01\ tonnes\,ha^{-1}$$

A further point, apparently not thought to be important by many workers, is the scale at which a particular investigation is being conducted. For example, if several changes within a relatively small area of heathland are being considered (Barclay-Estrup, 1970) it might be appropriate to use the square metre as the basic unit upon which to express the results (i.e. $g\,m^{-2}$), but if it is an overall area of heathland that is under consideration (Chapman, 1967) then it might be more appropriate to use the hectare as the basic unit (i.e. $kg\,ha^{-1}$). Even if some such convention is not accepted or found to be practical, it is still important that the scale of the investigation is made clear to avoid confusion when the results from different investigations are compared.

The use of other terms and units such as dry weight, calories and joules are dealt with in later sections of this chapter.

3.2   Measurement of the above-ground standing crop

It was hoped that one of the results of the International Biological Programme would be to achieve some degree of standardization of the techniques used in ecological research. When different ecosystems, such as forest, grassland, bog and tundra are considered, a number of different techniques are required; but the definitions and concepts can be standardized so that results from different investigations can be compared. The most important advice is for an individual worker to think about his own particular research problem in the terms described in Section 3.1.

3.2.1 Direct harvest methods

Individual plant method

This approach is most suitable where separate plants occur at low densities and where only a few different species are present within the sampling area. Estimates of biomass per individual plant, are combined with estimates of species density (see Chapter 9) to obtain overall figures for the standing crop.

Harvested quadrat method

In vegetation such as grassland or heathland it is not possible to differentiate between individual plants and the vegetation must be sampled by means of random quadrats. The location and numbers of quadrats required for a particular degree of accuracy are discussed in Chapter 9. It has been suggested by Milner & Hughes (1968) that a standard error in the order of 10% of the mean is an acceptable level of accuracy for ecosystem studies.

The size and shape of quadrats are discussed in Chapter 9, but an important factor in the measurement of standing crop is the amount of plant material that a particular size of quadrat will provide. Chapman (1967) found that a quadrat 50 × 50 cm provided sufficient sample from younger stands of heathland but not so much from older stands that would cause practical difficulties in handling the samples for analysis.

Any technique used for sampling the above-ground standing crop must provide accurate and reproducible results, preferably with the minimum of effort on the part of the observer. If possible the results obtained should be independent of the actual person involved in the sampling process. This is important, especially when a sampling programme extends over a long period and a number of different people will be involved in the work.

The particular type of cutting device chosen will depend upon the individual study, it may be hand shears, secateurs, scissors or some form of mechanical device. Where the vegetation sample needs to be sorted into separate species it may be most convenient to do this at the cutting stage. Although this means a longer period in the field it will often save time, especially where species of grass are involved. Milner & Hughes suggest that point quadrats (Warren Wilson, 1960, 1963) can be used to determine the relative proportions of the different species in the vegetation, and that by calibrating the results against cut samples they can be applied to other harvested material. Heady & Van Dyne (1965) have developed a laboratory point sampling technique which determines the proportion of each species presented by examining chopped vegetation on a tray through a binocular microscope. Neither of these methods provide samples of individual plant species for chemical analysis, this can only be done by means of direct sorting. Any standing dead material must also be sorted out and separated from the harvested material. Particularly relevent is the work of Williamson (1976) which deals with the problems of plant death in the estimation of grassland production.

When vegetation enclosed by a quadrat is to be sampled for the weight of standing crop it is important to minimize the edge effects and to collect only vegetation contained within vertical extensions of the quadrat boundaries. The weight of standing crop is not necessarily the same as the weight of vegetation rooted within the quadrat, although in some cases they may be almost the same.

Satoo (1970) has listed three different ways in which harvesting techniques can be applied to woodland and forest conditions.

(a) Clear felling, normally used for the standardization of other indirect methods. An estimate of the standing crop (W) of an area is given by:

$$W = \Sigma w \tag{1}$$

where w = weight of each individual tree.

(b) Trees of mean cross-sectional area are harvested and the mean value of the sample trees ($\bar{w}$) are either multiplied by the number of trees per unit area (N), or the sum of the sample trees Σw is multiplied by the ratio of the basal area of all the trees in the area (G) to the sum of the cross-sectional area of the sample trees (g).

$$W = NW \tag{2}$$

or

$$W = \Sigma w \left(\frac{G}{\Sigma g} \right) \tag{3}$$

Satoo points out that it is often difficult to obtain sufficient 'average trees', and that a possible alternative is to stratify the trees contained within the sample area into different size classes.

(c) Another variation of this last method is to sample trees of different size classes at random and to calculate the total standing crop by means of equation (3).

In the majority of cases where changes in the standing crop are being measured it will be possible to distinguish and use the same limits for sampling, e.g. soil surface, at the beginning and the end of the observational period; but situations do occur where this can be difficult unless special methods are adopted. An example is the case of a sand dune where sand may be added to or lost from the soil surface during an experimental period. A similar situation exists upon a growing peat surface (Forrest, 1971), and the problem is most relevant where it is the *Sphagnum* upon the bog surface whose growth is being measured. Clymo (1970) has reviewed and tested a number of methods that are suitable for estimating the growth and production of *Sphagnum*. The growth of *Sphagnum* is predominantly apical and the difficulty is to establish some reference point against which the increase in standing crop can be measured.

Clymo has divided the methods that can be used into four groups:
(a) The use of innate time markers. These time markers include cyclic fluctuations in the arrangement of branches and the use of ^{14}C-dated peat profiles.
(b) The use of reference markers outside the plant. These markers include wires placed upon the *Sphagnum* carpet ('the cranked wire method') enabling relocation and measurement of samples with minimum disturbance, and the use of thread tied around the stems.
(c) The use of plants cut to known lengths. A method based upon an increase in length is probably more reliable than one based directly upon an increase in weight as material is likely to be lost from the stems during the observational period.
(d) Direct estimates of change in weight. Clymo has developed a method that is based upon weighing the plant under water at the beginning and at the end of the growth period. After the second underwater weighing the plant is dried and re-weighed in air so that the specific gravity (d) can be calculated.

$$d = \frac{D_h}{D_h - W_h}$$

D_h = dry weight at harvest,
W_h = weight under water at harvest.

If the specific gravity remains constant, and measurements suggest that this is a reasonable assumption, then:

$$D_s = \frac{W_s d}{(d - 1)}$$

$$G = d_h - D_{ss}$$

$$= D_h(1 - W_s/W_h)$$

where D_s = dry weight at start,
$\quad\quad W_s$ = weight under water at start,
$\quad\quad G$ = growth (increase in weight).

As long as care is taken to remove gas bubbles, by evacuating the plants, and corrections are made to allow for the effects of surface tension on the balance at the air–water interface, it is claimed that increases in weight in the order of 2 mg can be measured.

3.2.2 Indirect estimation of the standing crop

In situations such as forest or woodland the complete harvesting of a series of sample areas, or even a single area, will not be possible normally. Under such circumstances alternative methods are needed to estimate the above-ground standing crop. If some readily measured parameter, such as stem diameter or tree height, can be correlated with biomass of harvested samples then the relationship obtained can be used to obtain estimates of the standing crop in other similar areas of vegetation. Such correlations have been used extensively in the study of woodlands and forests and have been reviewed by Newbould (1967) and Satoo (1970). Statistical relationships used to estimate standing crop are based either upon an assumption as to the shape or form of the plant, or upon a direct correlation with weight.

The simplest assumption is that a log or stem is cylindrical in form with a mean cross-sectional area that can be based upon measurements taken at the mid-point or from the two ends of the log. Newbould (1967) suggests that the square root mean (D_m) of the end diameters (D_1 and D_2) should be taken.

$$D_m = \sqrt{\left(\frac{D_1^2 + D_2^2}{2}\right)}$$

It has been assumed that the volume of a tree trunk approximates to a parabaloid of rotation, the volume of which is given by:

$$V_p = \frac{\pi r^2 h}{2}$$

where r = radius at breast height,
    h = tree height.

Whittaker & Woodwell (1968) found that the parabolic volume (V_p) was often an overestimate of the true volume for shrubs (where the radius is measured at the base), while in the case of small trees it was an underestimate because the radius at breast height was small in relation to the true basal radius. For larger trees they found that the estimates of volume were close to the true values.

Where slices of the stem are available or increment cores can be taken with a Pressler type borer, the radial wood increment can be measured. The mean annual increment should be estimated from the previous 5–10 years' growth and a series of measurements made from different points around the circumference of the stem.

Measurements of radial increment can be combined with estimates of basal area to provide basal area increments (A_i).

$$A_i = \pi[r^2 - (r - i)^2]$$

where r = radius at base,
    i = mean annual increment.

An estimate of the annual volume increment (V_i) can be obtained from half the basal increment times the height of the tree,

$$V_i = 0.5(A_i h)$$

According to Newbould this estimate is often an underestimate, the true value is often somewhere between 1·0 and 1·5 times the estimated value. Whittaker & Woodwell (1968) found close correlations between the estimated volume increment and wood growth in their studies.

Estimates of rate of wood volume increment can be based upon assumptions of linear or exponential growth.

$$\Delta V = \frac{(v - v')}{n} \qquad \text{Linear growth}$$

$$\Delta V = V(1 - e^{-r}) \qquad \text{Exponential growth}$$

where V = volume at time of felling,
V' = volume n years ago,

$$r = \frac{1}{n} \log_e (V/V')$$

To convert estimates of volume to weight it is necessary to determine the specific gravity. This can be done from samples obtained by destructive sampling, but in many cases it will be necessary to make use of increment cores. When this is done, precautions must be taken to avoid compression errors (Stage, 1963; Walters & Bruckmann, 1964).

The choice of a suitable regression model for predictive purposes should be based upon known or reasonably assumed relationships. A model that has been widely employed is based upon the law of allometric growth:

$$\log_e Y = a + b \log_e X$$

where Y = weight of the standing crop, or some component of the standing crop or production,
X = some readily measured parameter of the standing crop, a + b are constants.

This type of expression has been used to relate measurements such as basal diameter, or diameter at breast height (DBH) to weight, volume or production (Kittredge, 1944; Ovington & Madgwick, 1959; Whittaker & Woodwell, 1968; Satoo, 1970; Andersson, 1970, and others).

In some cases a more satisfactory estimate can be obtained by including measurements of both diameter and height in the regression:

$$\log_e Y = a + b \log_e (d^2h)$$

Once a series of suitable relationships have been established, the weight of standing crop can be calculated from the relevant dimensions of individuals present in the sample area. If the estimate of total standing crop (W) is obtained from an allometric regression, by taking the sum of the antilogs of predicted values for individuals, it will be biassed and an underestimate of the true value. Mountford & Bunce (1973) suggest multiplication by a factor $e^{s^2/2}$ to correct for this bias:

$$W = e^{s^2/2} \sum_{i=1}^{N} e^{a+b\mathbf{x}}$$

where s^2 = estimated variance about the regression line.

Mountford and Bunce also discuss the calculation of confidence limits for estimates derived from allometric relationships. Corrections for bias in regression estimates after logarithmic transformation have also been discussed by Beauchamp & Olson (1973).

A logarithmic regression cannot be used in situations where the weight of some component, such as fruiting bodies or dead wood, may be absent. In such cases Whittaker & Woodwell (1968) have suggested the use of regressions of the type:

$$W_d = a + bD^3$$

where W_d = weight of dead wood or fruit,

D = branch basal diameter.

Regression techniques can also be of value in non-woodland types of vegetation. Leaf length or tussock diameter might be correlated with standing crop in grassland (Scott, 1961; Mark, 1965) and used to obtain estimates of standing crop between harvests. Bliss (1966) has used such a combination of regression and clipping techniques in the study of an alpine ecosystem in the US.

3.2.3   Crop meters

Electrical capacitance is a function of the surface area of the capacitor plates, their arrangement and the nature of the di-electric material between them. If a suitable apparatus is placed on the ground so that the vegetation lies between an arrangement of electrodes the resulting capacitance will depend upon the weight and moisture content of the standing crop. Once calibrated, such an instrument can provide rapid and non-destructive estimates of the weight of standing crop. The method was first used by Fletcher & Robinson (1956) and other workers have developed the method (Alcock, 1964; Hyde & Lawrence, 1964; Johns et al., 1965). The use of such an instrument for grassland is described by Alcock & Lovett (1967). In practice, subsamples of the vegetation must be taken to correct for moisture content, difficulties may arise if used on very wet ground and in some cases the results obtained may not be sufficiently accurate to justify use of the apparatus. An alternative approach to the non-destructive estimation of standing crop is the use of beta-ray attenuation (Mott *et al.*, 1965).

3.2.4   Frequency of sampling

The frequency of sampling required in a production study will depend very much upon the particular investigation. Measurements of the standing crop of an area of woodland made over a period of 3–10 years will allow the rate of change to be calculated which, combined with litter production and grazing losses, will provide an estimate of net aerial production.

On lowland heathlands where the weight of standing crop varies with age, the sampling programme must cover the whole range of different aged stands of vegetation, and these must be sampled at suitable intervals to build up a growth curve for the standing crop. It is therefore necssary to establish the age of a number of stands of vegetation by ring counts from cut stems or increment borings, or from historical records (see Chapter 10). It is always advisable to obtain a number of estimates of age from as many independent sources as possible. In some investigations a single harvest at the end of the growing season will not be sufficient; for example, a study of the ground flora in a woodland will require a very different sampling programme from that of the tree species. The phenology of the individual plant species must be considered when designing any sampling programme (Lieth, 1970).

3.2.5   Treatment of samples

All primary production data are expressed in terms of dry weight so that harvested samples must be dried in an oven at temperatures somewhere between 80° C and 105° C. The exact temperature will depend upon circumstances, but it is important to dry the material quickly to minimize the loss of weight of organic matter by decomposition. Forced draught ovens are available which allow temperature to be reached quickly and to be maintained throughout the oven (Grassland Research Institute Staff, 1961). When for some reason the plant material is not dried at 105° C it is suggested that subsamples should be dried at this temperature and a factor used to convert the weight obtained to equivalent dry weights at 105° C.

Having dried and weighed the samples they must then be milled or ground for any subsequent analysis (see Chapter 6).

3.3   Litter production

Production of litter by the above-ground vegetation represents a major component of the net primary production, and its measurement is important whether it be in relation to primary production, or for consideration of other relationships within the ecosystem. Bray & Gorham (1964) have examined the production of litter within forest ecosystems and their review should be consulted by anyone concerned with litter production. The mineralization and release of plant nutrients within the litter layer are processes in which the soil fauna play an important part, and it is easy to see that the soil zoologist and the production botanist have many problems in common. It has been shown that production, decomposition and accumulation are inter-related (Section 3.1.3) but for the sake of convenience the methods involved in the estimation of litter production will be dealt with in this section and those for litter decomposition and accumulation in Section 3.4.

Litter production can be defined as the weight of dead material (of both plant and animal origin) that reaches unit area of the soil surface within a standard period of time. All material that dies does not immediately fall to the ground and in some types of vegetation, such as grassland and other tussocky plant communities, the litter may contain only a small proportion of the dead material. It was to distinguish between the different locations of dead organic matter in the ecosystem that the term 'standing dead' was introduced (Odum, 1960; Gore & Olson, 1967; Forrest, 1971).

3.3.1   Direct estimation of litter production

Different types of litter trap have been described in the literature, but as has been stated previously the design of a piece of apparatus for a particular project must depend upon the special requirements of the individual problem. Whatever the local conditions a litter trap must fulfil a number of basic requirements.

(a) It must intercept the litter fall before it reaches the ground with as little aerodynamic disturbance as possible.

(b) It must retain material once it has been trapped.

(c) It should be designed or placed so that litter already on the soil surface cannot enter the trap.

(d) Water must be allowed to drain from the trap without loss of litter (especially fine litter material).

(e) The size and number of traps must provide an estimate of the required degree of accuracy.

A selection of litter traps is shown in Fig. 1.5.

One difficulty arising in the use of litter traps, especially in open or exposed situations, is the failure to trap or the subsequent loss of trapped material from the traps. Chapman et al. (1975a) have found a good correlation between the numbers of Calluna seed capsules in litter traps and the

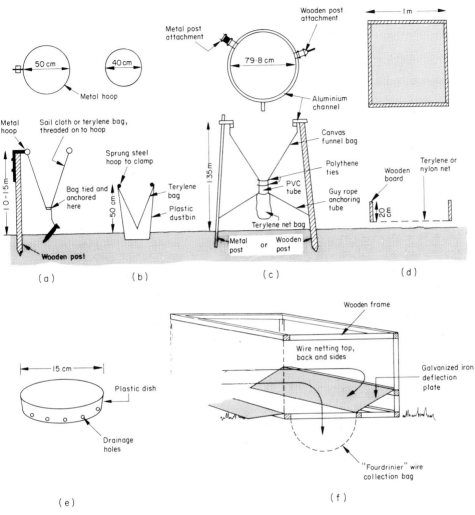

Figure 1.5 A selection of litter traps suitable for use in different types of vegetation: (a)–(d) Newbould (1967); (e) after Chapman (1975a); (f) after Woodell (pers. comm.).

numbers produced per unit area of older and denser heathland vegetation, but in younger and more open stands the numbers of capsules trapped were significantly lower than the numbers produced by the vegetation. Wherever possible some similar check upon the efficiency of a litter trap should be made.

In some cases the amount of litter being added to the soil surface each year will include a 'blow in' and a 'blow off' component; and it is important to consider whether any such lateral movement of material within the eco-system is likely to be important. S.R.J. Woodell (pers. comm.) has used suitable traps (Fig. 4.5f) to study the movement of litter in a narrow band of old beechwood in Oxfordshire, and has been able to measure the amounts of litter entering and leaving the wood as a result of wind action.

The length of time between emptying traps should be kept as short as possible. Weekly collections are desirable, especially in damp and humid areas or where the breakdown and leaching of litter may be appreciable between trapping and collection. The phenology of the particular plant species present will be important when deciding upon the actual timing and frequency of emptying the traps. It is suggested that estimates of litter production should be based upon observations carried out over a period of at least 3 years. Sykes & Bunce (1970) have demonstrated the fluctuations in litter fall within a deciduous woodland over such a 3-year period. Medwecka-Kornás (1971) has suggested that a suitable level of accuracy for estimates of litter production would be to obtain 95% confidence intervals equal to about 5% of the mean. The number of traps used will vary, and an initial sampling period may well be necessary to establish the actual number required, but in most cases twenty traps at each sampling site will be sufficient.

Under some types of vegetation it may be necessary to make independent estimates of some components of the litter and then combine the results for an overall estimate of litter production. In woodland ecosystems components of the litter such as branches, or even whole trees, will not be measured by trapping; and it will be necesary to record the appearance of these larger items in suitably sized plots (Healey & Swift, 1971). As they appear they must be removed, or labelled so that they can be recognized at a later date.

When litter material is required for chemical analysis it will be necessary to prevent chemical contamination and fouling by birds (see Section 4.2.1). If the traps are not made of suitably inert materials the frames or supports must be painted with bitumastic or some other suitable paint.

3.3.2 Indirect estimate of litter production

It will soon become apparent that it is difficult or almost impossible to trap litter under some types of vegetation, and that in some situations the estimates obtained from litter traps will not be very reliable; younger stands of heathland have already been given as an example. In such cases it is necessary to obtain estimates of litter production by some indirect method.

As leaves senesce and die, considerable amounts of material are translocated to other parts of the plant before they fall as litter. It is therefore extremely important to assess the magnitude of this translocation if estimates of the standing crop of leaves are to be used to predict potential litter production. An estimate of the weight of material lost from a leaf before it falls as litter can be obtained from a comparison of the dry weight/surface area ratios of live and dead leaves. Such comparisons assume that no changes in surface area take place but this assumption can be checked against a series of measurements on individual marked leaves. In some cases it may be more satisfactory to compare dry weight/leaf length ratios to obtain an estimate of the weight lost by reabsorption.

In cases where the green matter produced in a single season is not all shed before the next growing season, it is necessary to develop a model that takes into account the life expectancy of the green material, and to include this in the estimation of potential litter. Estimates of litter production by *Calluna* heathland can be checked by such an indirect method (Chapman *et al.*, 1975a). Much of the green matter produced by *Calluna* is in the form of lateral short shoots, which either drop as litter at the end of the growing season or persist to grow and increase in length during the next season. The growth increments can be recognized, so that the age of an individual shoot can be determined, and a partial estimate of the potential litter production obtained from the difference between the weight of green shoot material present in one year and the weight of the 'old short shoots' remaining at the end of the following year. This difference in weight of green material must be corrected for loss of weight due to translocation (approximately 20%) and added to the weight of the flower and woody components of the litter production. Bray & Gorham (1964) and Burrows (1972) report similar losses of about 20% in the dry weight of leaves from a variety of plant species before litter fall. If it can be assumed that the production of green matter is constant from year to year then the annual loss of green matter can be estimated from the difference between the total weight of green shoots and the weight of old green shoots present at the end of the growing season. If measurements of the amount of shoot material that have been consumed by grazing animals have been made they should be incorporated in the calculation of litter production.

3.4   Decomposition and accumulation of litter

The rate of accumulation of litter upon the surface of the soil is the result of the interaction between litter production and the rate of litter disappearance.

Litter disappears from the ecosystem as a result of combined losses due to decomposition, mineralization, leaching, animal consumption, wind transport and in some cases harvest by man (Medwecka-Kornás, 1971; Anderson, 1973b; Anderson & McFadyen, 1976; Swift *et al.*, 1979).

In some cases the disappearance of litter can be equated with decomposition, but this is not necessarily so and it may be necessary to apply a correction to the rate of disappearance to obtain an estimate of the rate of litter decomposition. As the accumulated litter is the result of a series of annual inputs combined with the progressive decomposition of the litter layer it can be divided into a number of horizons depending upon the degree of decomposition or humification.

3.4.1   Quadrat methods for estimating litter disappearance

A direct approach to the measurement of litter disappearance has been used by Wiegert & Evans (1964) in their study of old field systems in Michigan. They developed a technique where a series of 'paired plots' were selected, and in which individual pairs of plots were assumed to be identical in terms of weight and composition of the litter present. At the start of the experiment (t_0) all dead vegetation present on one of the plots was removed and weighed (W_0). To prevent any more dead material being added to the plots during the observation period all the green matter was removed from both the plots. After a suitable period of time (t_1) the litter on the second plot was sampled and weighted (W_1). The rate of disappearance of litter (r) was calculated assuming an exponential rate:

$$r = \frac{\log_e (W_0/W_1)}{t_1 - t_0}$$

This basic method has been modified by Lomnicki et al. (1968) where they have avoided the assumption that decay remains constant despite the fact that the live vegetation has been removed from the plots. At the beginning of the experiment (t_0) they sampled the dead material (W_0) from one of the plots but did not remove any of the live vegetation. At the end of their observational period (t_1) they sampled the dead material from both of the plots. The weight of dead material collected from the previously sampled plot they designated (h) and that from the unsampled plot (g). It follows that:

$$g = W_0 + h - (W_0 - W_1),$$

and

$$W_1 = g - h.$$

This modification of the basic method of Wiegert & Evans assumes that the litter produced during the observational period ($t_1 - t_0$) is not affected by the removal of the dead material from the plot. Lomnicki and his associates compared their techniques with the original method and found good agreement between the two procedures, but claim that their modification of the method is more convenient to use. Their parameter (h), representing the weight of litter produced during the observation period, can be combined with the change in standing crop (*b*) to provide an estimate of net aerial production (P_n).

$$P_n = h + (b_1 - b_0)$$

To obtain this estimate of net production, (*h*) should be measured over suitably short periods and the results summed to provide an annual estimate. This method can be applied to a variety of vegetation types, especially those where standing dead is an important component of the ecosystem.

3.4.2   Measurement of disappearance from contained litter samples

If samples of litter, either freshly produced or collected from the litter layer, are contained in some way that makes it possible to retrieve them at a later date then the rate of disappearance of the litter can be measured directly. Commercially available nylon hairnets have been employed for this purpose by Bocock & Gilbert (1957), Bocock *et al.* (1960) and Bocock (1964). Nylon net bags, made from ballroom dresses, were used by Shanks & Olson (1961) to demonstrate that the rate of loss of weight by leaves on the soil surface was dependent upon the plant species involved, their chemical composition and the prevailing climatic conditions.

Edwards & Heath (1963) have shown that the size of the mesh used in nylon bag experiments is important because it controls the types of organism that can enter the bags and participate in the decomposition process. In their experiments they assessed the rate of decomposition by estimating the area of leaf discs (2·5 cm in diameter) that had been lost after a given period of time. Results of the effect of mesh size upon the rate of breakdown in litter bag experiments have also been reported by Bocock (1964) and Anderson (1973b).

The choice of a particular size of mesh for litter bag experiments must take into account the type of litter under investigation. In an experiment on the breakdown of *Calluna* litter by Chapman (1967) it was found that a mesh larger than 2 mm could not be used as it was unable to retain the relatively small sized heather shoots, but that this size of mesh probably did not exclude any important component of the heathland soil fauna.

Where the object of the investigation is to study the loss of weight of even-aged or 'fresh litter' then plant material derived from current year's growth is used to fill the litter bags. An alternative approach is to fill the bags with material collected from the accumulated litter on the soil surface, when the sample will be composed of organic matter of different ages. The results will provide an integrated estimate of the rate of loss of organic matter from the whole litter layer.

In some cases significant weights of animal material may enter the litter bag and unless removed will affect the results obtained. Nylon litter bags have in fact been used by a number of workers to study the invasion of plant litter by soil organisms (Crossley & Hoglund, 1962).

The rate of disappearance of litter from mesh bags is estimated from random samples taken at intervals throughout the experiment and dried in an oven. It is important to make sure that all the bags contain the same dry weight of litter at the start of the experiment and, as the litter bags need to be sampled several times throughout the experiment, sufficiently large numbers of the bags must be set out at the beginning of the exercise. It is possible that the method used by Clymo (1970) to estimate the initial dry weights of shoots of *Sphagnum* in growth experiments (Section 3.2.1) might be used to obtain estimates of the initial dry weights of litter bags.

3.4.3   The measurement of litter disappearance from labelled or tagged samples

An alternative to the use of litter bags is to mark or label individual fragments or items of litter so that they can be retrieved at a later date. An example is the work of Frankland (1966) where she studied the decay of bracken (*Pteridium aquilinum*) by labelling individual petioles with plastic labels. Latter & Cragg (1967) marked 200 individual leaves of *Juncus squarrosus* with paint in a study of decomposition that also used litter bags. Hayes (1965a, b) attached coniferous leaf litter to lengths of nylon thread to enable him to relocate individual leaves, and compared results with those obtained from litter bags. He found lower rates of disappearance from the litter bags as fragments were not so easily lost from the experimental system.

Murphy (1962) used a paint containing the radioisotope tantalum-182 to label leaves while they were still on trees, and was able to relocate individual leaves for up to 12 months after they had fallen as litter. He measured decomposition directly from the loss of weight.

Olson & Crossley (1963) combined the use of nylon litter bags with the application of radioactive isotopes in a study of forest litter decomposition. They introduced radioactive tracers into the trees, producing radioactive leaves and subsequently radioactive litter. The litter was placed in nylon mesh bags 10 cm square that were then contained in plastic sandwich boxes with holes in the sides and glass fibre mesh at the base. The boxes were placed on the forest floor and counts of the radioactivity of the bags enabled the amount of radioactive material that had been transferred to the forest floor to be calculated.

3.4.4   Respirometry and litter decomposition

It has already been shown that the loss of weight from a litter bag, or from the litter layer itself cannot necessarily be equated with the mineralization and loss of carbon from the ecosystem in the form of carbon dioxide. It would seem that the obvious approach to this problem would be to measure the uptake of oxygen, or the evolution of carbon dioxide by respirometry. Parkinson & Coups (1963) measured the respiration of soil and litter placed in specially designed respirometer flasks. Howard (1967) has described an 'experimental tube' that can be placed upright in a box in the field and protected from rain, but kept moist by watering with distilled water when necessary (about 5 ml every 2 weeks). These tubes can be returned to the laboratory at intervals and connected to a Gilson respirometer in a flask of the type illustrated, and at the end of the experiment the contents of the tubes can be dried, weighted, ashed and the weight of organic matter lost during the experimental period calculated.

If sufficient attention is paid to the moisture content and aeration of the experimental material (Ross & Boyd, 1970) it would seem that the main disadvantages of respirometry as a method for the study of litter decomposition are disturbance and the isolation of the litter from the natural litter layer. If these limitations are recognized, respirometry can provide useful estimates of carbon losses from the litter layer. The direct estimation of the amount of carbon dioxide evolved from the soil and litter layer can be measured by other methods but these are described in Section 3.5.7.

3.4.5   Comparative studies of litter decomposition

In many cases the object of a decomposition study is to investigate litter decomposition at one particular type of site, but as the rate of decomposition is dependent upon both the type of organic matter and upon other site factors, an alternative approach is to investigate the decomposition of a standard substrate at a number of different sites. Golley (1960) buried cellulose sheet (5 × 5 cm) held in aluminium frames, and then estimated the area of the sheet remaining at monthly intervals. Latter, Cragg & Heal (1967) buried cellulose film and strips of unbleached calico in the soil to study decomposition, and Went (1959) used cellophane for the same purpose. Other organic substrates that have been used as standard substrates include filter paper, cotton or linen cords (bootlaces) and wood pulp cellulose.

Benefield (1971) has suggested a different approach to the comparative study of decomposition in soils and litter that is dependent upon making comparisons of cellulase activity in soils from different sites. The method involves incubating a known amount of standard cellulose powder with a soil sample and then determining colorimetrically the glucose formed as a result of enzymic hydrolysis.

3.4.6 Measurement of litter accumulation

The weight of organic matter that accumulates upon the surface of the soil varies greatly under different types of vegetation. Accumulation is minimal where decomposition is rapid, while at the other extreme an organic layer several metres deep can develop on peatlands. In the case of peatland sites, dating techniques (Chapter 10) are available and the rate of accumulation of organic matter with time can be investigated (Durno, 1961; Clymo, 1965).

Direct estimation of the weight of accumulated litter can be difficult when there is no well defined boundary between the soil and litter layers, but in most cases it is possible to define practical limits that make reproducible sampling possible. It is often most convenient to combine sampling of the litter layer with that of the standing crop, especially where the litter is to be sampled by quadrats.

The accumulation of litter on the soil surface can also be sampled with a corer or tubular sampling tool. Capstick (1962) used samplers that had different cross-sectional areas (1.0, 2.63, 8.5 and 41.5 cm^2) to sample the litter layer in a forest ecosystem, and found that a 1.0 cm^2 sampler cut through the litter layer easily and with little disturbance of the experimental area. The results from this size of sampler showed only a low degree of scatter and agreed well with estimates derived from larger samples. He found inconsistencies when trying to follow changes in the weight of litter after litter fall, but these were just as likely to occur with samplers of a larger size. Frankland *et al.* (1963) used a corer 81 cm^2 in cross-sectional area, to sample the litter layer from woodlands in the Lake District, but found it difficult to demonstrate any significant differences between their sampling sites. Before using a corer to sample the litter layer it is important to carry out a trial sampling programme to establish the size and number of samples that may be required; or whether such a method is even suitable for the investigation about to be undertaken.

The litter layer will often contain considerable amounts of mineral soil and to make the estimates of litter accumulation comparable with those obtained for litter production and above-ground standing crop it is advisable to determine the loss-on-ignition (LOI) of the litter samples and to express the results on an ash free, or a standard (e.g. 5%) ash weight basis.

Root growth within the litter layer can be a problem when one is trying to estimate the weight of litter that has accumulated at some sites. Chapman *et al.* (1975b) found it to be an important factor when dealing with the litter layer from older stands of *Calluna* heathland. In such cases some estimate of the root content of the litter layer and its contribution to the accumulation of litter must be made.

Jenny *et al.* (1949) and Olson (1963) have described and discussed litter accumulation within forest ecosystems by means of a simple exponential

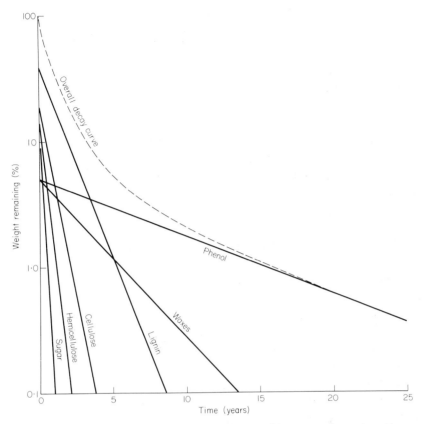

Figure 1.6 The relationship between the overall decomposition curve of organic matter and the exponential decay curves of the individual constituents. (Based on data from Minderman, 1968.)

model (see Section 3.1.3). While the assumptions made in this model make it convenient for the purposes of computation, and may be reasonable in some situations, the processes of decomposition and accumulation cannot always be described adequately by such a simple model. Minderman (1968) has stated that 'it is difficult to give a quantitative gravimetric picture of this phenomenon', and has shown that if the different chemical constituents of the litter decay exponentially but at different rates, then the overall decay curve for the litter will be curved when plotted on a semi-logarithmic basis (Fig. 1.6). This departure from the predictions of the simple exponential model is most important where decomposition is rapid, but is also dependent upon the chemical composition of the litter being studied. The effect of the assumptions, that are implied when the simple exponential model is used to make predictions, is discussed in relation to roots and soil organic matter in Section 3.5.6.

3.5   Roots and soil organic matter

The importance of root production need not be emphasized, comparisons of the relative productivity of different ecosystems or studies of their nutrient budgets that are based solely upon above-ground data are so incomplete that one may well doubt the validity of many of the conclusions that are drawn from them.

The same principles apply to the estimation of root production as to the measurement of above-ground production except that the practical problems are very much greater. The information required for the direct estimation of root production is the change of weight of root and the amount of root that has died in a given period of time. When some part of the above-ground standing crop dies it generally falls to the surface of the soil as litter, or can be identified as standing dead; in either case it can be measured. When part of the root system dies it is retained, and is often difficult or almost impossible to distinguish from the living roots. This factor combined with the difficulty of sampling root systems will help to explain the relative lack of data available about root production. In 1964 Olson stated that 'the kinetics of underground production and loss rates remain as one of the most challenging ecological and agricultural problems during the half century to come'. Despite the fact that much of that half century has now passed it is probably almost as true a statement as it was in 1964.

The problems involved in the estimation of root production have been described by Newbould (1967), Milner & Hughes (1968), Ghilarov et al. (1968) and Head (1971). The relationship between root production and net productivity has been reviewed by Bray (1963). Kononova (1961) has dealt with many aspects of soil organic matter in his book entitled *Soil Organic Matter*. As the problems involved in the estimation of root production are so closely linked to those of soil organic matter they are considered jointly in this section.

3.5.1   Estimation of root biomass and examination of root systems

The estimation of root biomass is entirely dependent upon the sampling and extraction of roots from soil samples or from the soil profile, both of which are reviewed by Schuurman & Goedewaagen (1971) and Lieth (1968). The methods of sampling fall into three main categories, the exposed soil face, soil cores and blocks, and the excavation or exposure of the root system *in situ*.

The use of soil pits involves considerable effort and labour, and is only of use where few replicates are required or plenty of labour is available. The exposed soil face can be sampled as an intact monolith by using a specially made container (Chapter 5), or use can be made of a 'pinboard' (Ashby,

1962; Schuster, 1964; Sheikh & Rutter, 1969; Schuurman & Goedewaagen, 1971). The pinboard consists of a baseboard of suitable size with wires or nails protruding from one side at closely and regularly spaced intervals. The board is pressed against the face of the soil pit so that the nails are pushed into the soil and hold the root system in place when the soil particles are washed away. If necessary a car jack can be used to press the board into place. At times it may be possible to use jets of water from a hose and motor-driven pump to expose the root system of an area of vegetation or an individual plant by direct washing in the field.

In the majority of investigations it will probably be found that some form of soil corer or auger will be the most convenient way of obtaining soil and root samples. A great variety of corers and augers are described in the literature, and as is usually the case when a whole range of variations in a particular technique exists, it means that no single corer will be suitable for all situations. Corers vary from relatively simple tubular devices to complex power-operated augers (Kelly *et al.*, 1948; Welbank & Williams, 1968; Schuurman & Goedewaagen, 1971). The type of corer chosen for a particular project will depend upon the soil type, and the diameter and depth of core required.

Root growth has been studied by the installation of inspection windows against the sides of soil pits (Ovington & Murray, 1968; Rogers & Head, 1968). Periodic observation or the use of time-lapse photography has enabled the periodicity of root growth to be studied and the life history of individual roots to be followed. While a great deal of information has been obtained in this way it is very difficult to express the results in terms of root production.

Roots can also be studied *in situ* by the preparation of thin sections of soil, and a number of techniques have been described for the preparation of soil sections (see Chapter 5). Not many of the methods described were intended, or have been used, for the examination of roots, but Sheikh & Rutter (1969) have used soil sections to investigate the distribution of roots in relation to the size of the pore spaces in the soil. The methods of Haarlov & Weis-Fogh (1953), Minderman (1956), and that of Anderson & Healey (1970) have much to recommend them for the purpose of examining roots. These methods use gelatine or agar for embedding, so the soil does not have to be dried as when being impregnated with resin, and the roots and soil animals can be seen with great clarity. Some difficulty may be encountered when cutting sections of soil embedded in gelatine if mineral grains are present in any quantity, but this problem was overcome by Minderman where he used hydrofluoric acid to dissolve away sand grains between the stages of embedding and sectioning his soil samples.

The length of larger roots can be measured from photographs or shadow-grams prepared from the extracted roots. The measurements can be made with a wheel-type measurer. Newman (1966) has described a method of measuring the length of fine roots. The roots contained in a transparent flat bottomed dish are placed over a base marked with a series of randomly placed straight lines. It can be shown that whatever the shape of the root, an estimate of its length is given by:

$$R = \frac{\pi N A}{2H}$$

where R = total length of root sample,
   N = number of intersections between roots and straight lines,
   A = area of rectangle inside which lines are drawn,
   H = total length of the straight lines.

3.5.2   Extraction of roots from soil samples

Once the soil samples have been collected it remains to separate and extract the roots. Pinboard samples are immersed and allowed to soak in water. The water level is then maintained just below the surface of the sample and sprinkling is carried out either mechanically or by hand to wash the soil material away, leaving the roots in place between the nails or pins.

Samples obtained by corer or auger are generally divided into subsamples representing different depths or soil horizons. The relative ease or difficulty encountered in extracting the roots will depend very much upon the particular type of soil. Pre-soaking the sample in aqueous sodium pyrophosphate (270 g per 100 litres) or 1% sodium hexametaphosphate may help to disperse some soils. Dahlman & Kucera (1965) followed such a treatment by soaking in 0·8% sodium hypochlorite to assist in separating the clay and humus fractions from the roots. The effects of any such pretreatment, as well as soaking and washing in water should be considered if root samples are required for any subsequent chemical analysis.

Separation of roots from the remainder of the soil can be done by one of two main methods. The sample can be placed over a nest of sieves of different mesh sizes and washed, or it can be subjected to some sort of flotation or elutriation technique. In the first case, the coarser soil material will be retained and have to be sorted by hand while the roots will be collected upon a number of different sized sieves. With flotation methods the soil organic matter, roots and finer mineral fractions will be separated from and leave behind the larger and denser mineral fractions.

Small samples can be placed in a litre beaker or graduated cylinder, and allowed to soak. After stirring and dispersion the roots, organic matter and finer soil material can be separated by decantation. The roots can then be

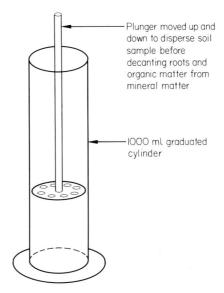

Figure 1.7 Simple apparatus for separating samples of soil and roots.

separated by washing over a nest of sieves. A simple piece of apparatus for separating roots by this method is illustrated in Fig. 1.7.

The flotation method of root separation can be mechanized to make it suitable for use with larger numbers of samples, and make the process more reproducible by giving the samples a uniform treatment. Apparatus has been described by McKell *et al.* (1961) and Cahoon & Morton (1961), both of which are shown in Fig. 1.8, and operate by jets of water entering the bottom of the apparatus, creating a vortex that carries the roots, organic matter and less dense fractions upwards and out into a nest of sieves. It has been

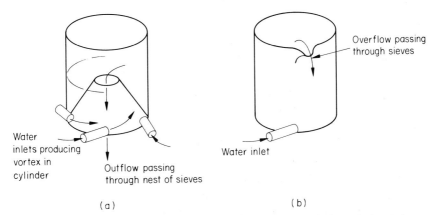

Figure 1.8 Apparatus for extracting roots from soil samples by elutriation: (a) after Cahoon & Morton (1961); (b) after McKell *et al.* (1961).

suggested by Chapman (1970) that ultrasonic cleaners might be of use in 'cleaning' the root fractions collected on the nest of sieves.

Other flotation methods, such as the one described by Salt & Hollick (1944) for the extraction of soil fauna, may sometimes be useful for the extraction of root material. The flotation is carried out in a concentrated solution of magnesium sulphate (specific gravity c. 1·2), sodium chloride, potassium bromide or zinc chloride (see Southwood, 1966) and as air is bubbled through the solution from the bottom of the container (the Ladell can) the surplus liquid and organic matter pass over the lip and are filtered through bolting silk. The animals can then be separated from the other organic matter by differential wetting in xylene or benzene.

Froth flotation has been used to extract subfossil seeds from soils in archaeological studies (Charles, 1972). A suitable chemical (the collector) is chosen for which the surface of the material to be extracted has a preferential adsorption so rendering it water repellent and aerophilic. When air is passed through a suspension of mixed solids those particles with aerophilic surfaces will be raised into a stabilized froth from which they can be separated. In the case of the subfossil seeds, cresol or kerosene was used as the collector and pine oil (terpineol) as the frothing agent.

Having separated the roots by one means or another they can be sorted into different size classes, dried and weighed. However carefully the roots have been washed they will still contain a certain amount of mineral soil contamination and it will be necessary to ash the dried root samples and to express the results on an ash-free dry weight basis, or to correct them to some standard ash weight content (i.e. 5%, Newbould, 1967).

3.5.3 Separation of live and dead roots

One of the greatest problems in calculating root production from root biomass data is that of distinguishing live from dead roots. Jacques & Schwass (1956), and Aimi & Fujimaki (1958) have tried to use tetrazolium salts to distinguish live roots but the success of such a method depends upon the individual root system and must by its nature be tedious to apply, especially to very fine roots. A number of workers have based their separations of live from dead roots upon visual assessments, but this is very difficult especially with the fine roots. Kononova (1961) describes a flotation method (ascribed to Pankova) for the separation of roots in which it is claimed that living roots can be separated from dead roots by repeated elutriation. Sator & Bommer (1971) tried both chemical and enzymatic methods to differentiate living from dead but reported little success. They were of the opinion that radioactive labelling techniques were the most promising.

Radioisotopes have been used in root studies to label live roots, to estimate rates of decay and turnover times, and to follow the transfer of materials from one part of the plant to another.

The root system can be labelled by the incorporation of ^{14}C into the plant through the leaves, either by the foliar application of a substance such as urea (Yamaguchi & Crafts, 1958; Nielson, 1964) or by the photosynthetic assimilation of $^{14}CO_2$ (Ueno et al., 1967; Dahlman & Kucera, 1968; Ellern et al., 1970). The assimilation takes place within some form of enclosure covering the vegetation or an individual leaf, and after a suitable period the roots can be sampled and their activity measured either by counting or by autoradiography.

An alternative technique is to inject an isotope into the soil at selected depths, and to establish the pattern of root activity by sampling the aboveground vegetation for radioactivity. Boggie et al. (1958) used ^{32}P as a tracer to investigate the rooting depths of plants growing on blanket peat by this method. They describe an improved technique for placing the tracer at different depths in a later paper (Boggie & Knight, 1962).

The root extension of forest trees (Hough et al., 1965; Ferrill & Woods, 1966) and the rooting pattern of forest understorey species (Nimlos et al., 1968) have been studied using iodine-131 as the trace. When using labelled iodine it has been reported that prior sterilization of the soil with methylbromide around the point of application increases the rate of absorption of the isotope (Woods et al., 1962).

Phosphorus-32 and calcium-45 have been injected into the stumps of tree species (Woods & Brock, 1964; Woods, 1970) to demonstrate the transfer of labelled material from the roots of one species to another in the ecosystem. It was concluded that root exudates or mutually shared mycorrhizal fungi were probably the important factors in the transfer rather than root grafts.

The problem of root exudates and their contribution to the total net primary production is generally recognized, but in most cases it is then conveniently forgotten because of the practical difficulties that remain to be solved. The use of radioactive tracers will undoubtedly figure prominently in the eventual solution of the problem. To date, most of the work on root exudates would appear to have been confined to studies carried out under sterile conditions in the laboratory and the extension of the work to field conditions is a major step.

Dahlman & Kucera (1968) have used $^{14}CO_2$ to label the root system of grassland vegetation, and by following the reduction in the radioactivity of the roots over a period of about 2 years they have obtained estimates of a root system turnover time of just over 4 years. Caldwell & Camp (1974) have followed the dilution of $^{14}C/^{12}C$ ratios in structural carbon of root systems to estimate turnover times and root production.

Radioactive tracers will be found to be of greater use in the study of roots of some types of vegetation than in others and they do not provide a universal panacea to all root production problems. It must also be remembered that radioactive materials can only be used in accordance with local regulations.

3.5.4   Relationship between root biomass and production

The relationship between biomass and production that applies to the above-ground component of the vegetation applies equally to the root system but it is more difficult to apply.

$$P = \Delta B + L + G$$

Other expressions relating root production to biomass have been used or suggested, such as:

$$\text{root production} = \frac{\text{max. root biomass}}{\text{turnover time}}$$

The turnover time is the time required for the decomposition of a weight of organic matter equal to the weight of the root biomass (the reciprocal of the decomposition rate, K) so that the expression is the same as:

$$L = kX_{ss}$$

and therefore assumes steady state conditions.

Remezov $et\ al.$ (1963) have suggested a slight modification of this type of expression to calculate the annual loss of roots.

$$R = W_a + \frac{w - w_a}{n}$$

where R = annual loss of roots,
$\quad W_a$ = weight of roots of annual species,
$\quad W$ = total weight of roots,
$\quad n$ = average length of life of roots of perennial species.

Newbould (1967) suggests that:

$$\frac{\text{above-ground production}}{\text{above-ground biomass}} = K \times \frac{\text{below-ground production}}{\text{below-ground biomass}}$$

but points out that as few accurate estimates are available for root production it is difficult to estimate the value of K, but he suggests that the further assumption that K is equal to unity might be made until better estimates are available.

Dahlman & Kucera (1965) measured the annual increment of the root system of prairie vegetation by taking the differences between the maximum and minimum values that were obtained for the root biomass during the year. The periods of greatest difference were April–July in the A_1 horizon, July–January in the A_2 horizon and July–October in the B_2 horizon. By making the assumption that these increments could be equated with root production, they calculated that approximately 25% of the root system would be replaced each year. As production and decomposition are simultaneous processes their estimates of root production must therefore be minimum values. In some other ecosystems, such as *Calluna* heathland (S.B. Chapman, unpublished), it has not proved possible to demonstrate significant seasonal variations in the root biomass. The differences are either small compared with the variance of the mean or production and decomposition proceed in such a way that the net root biomass remains relatively constant.

3.5.5   Accumulation of organic matter in seral sites

There are obvious advantages to the ecologist if an ecosystem can be assumed to exist in a steady state condition, but in many cases this is clearly not true. Although the investigation may be complicated because the ecosystem is changing with time it can be an advantage when studying some aspects.

Where distinct and dateable stages in a succession exist, such as the series of glacial moraines studied by Crocker & Major (1955) and Crocker & Dickson (1957), or the sand dune systems described by Salisbury (1922, 1925), Burges & Drover (1953), Olson (1958) and Wilson (1960), it should be possible to measure the rate of certain parameters and to obtain estimates of the rates of input and loss from the system. One way to achieve this might be to fit accumulation curves of the type described by Olson (1963), but estimates obtained in this way would be approximate only and subject to the assumptions of constant rates of input and decomposition. Objections to the assumption of constant decay rates have been discussed in Section 3.4.6, and if a monomolecular accumulation curve is fitted to the hypothetical data given by Minderman (1968) it will be found to produce an underestimate of the true input to the system.

3.5.6   Measurement of soil respiration and carbon dioxide under field conditions

Root and litter production are the main sources of organic matter input to the soil. The total metabolism of the soil has been reviewed by Macfadyen (1971), and the evolution of carbon dioxide is one method for its measurement suggested by a number of workers. Where it has been measured it is generally found that the production of carbon dioxide is in excess of the amount that can be attributed to litter production and the decomposition of organic matter (e.g. Wanner, 1970).

The carbon sources for the decomposer cycle are mainly litter and root production. If the soil ecosystem can be assumed to approximate to a steady state condition then the annual carbon dioxide production will be proportional to litter and root production, plus root respiration. In such an argument it is important to consider soil respiration on an annual basis so that differences due to the translocation of organic matter to and from the roots are largely cancelled out. If it were possible to subtract the amount of carbon dioxide due to root respiration from the total amount of carbon dioxide produced annually in the soil, it would provide some form of estimate of the input of organic carbon to the soil. Estimates of root respiration derived from excised roots (Crapo & Coleman, 1972) have obvious disadvantages, but might be used to make such calculations in the first instance. Kucera & Kirkham (1971) have suggested an alternative approach: if estimates of total soil respiration can be obtained from a series of soils that contain different combinations of weight of root and organic matter, then it should be possible to apportion the total carbon dioxide to its various sources (Chapman, 1979).

A number of methods have been used to measure soil respiration under field conditions (Howard, 1966; Witkamp, 1966; Kosonen, 1968; Brown & Macfadyen, 1969; Witkamp & Frank, 1969; Wanner, 1970; Kucera & Kirkham, 1971; Chapman, 1971, and others). These can be classified according to the methods used to collect the gas samples, and to estimate the carbon dioxide produced (Bowman, 1968). In the majority of cases some form of container ('open box' Witkamp, 1966) is placed over the soil to collect the carbon dioxide, preferably having been set into the soil some time prior to taking measurements and then left in place so that the effects of disturbance are minimized. The container is covered with an airtight lid while actual measurements are being made.

The carbon dioxide evolved has been measured by means of infra-red gas analysers (Witkamp, 1966; Reiners, 1968; Witkamp & Frank, 1969), by absorption in soda-lime (Howard, 1966), in barium hydroxide (Witkamp, 1969), or by absorption in potassium or sodium hydroxide (Witkamp, 1966; Brown & Macfadyen, 1969; Chapman, 1971). Where the carbon dioxide has been absorbed in potassium hydroxide it has generally been by diffusion (Conway, 1950), and determined by titration.

Titration of the carbon dioxide evolved has the disadvantage of providing only one determination at the end of the observation period, and the rate is therefore calculated from this one single reading. Chapman (1971) has described a soil respirometer operating on the same principles as those just described but in which the carbon dioxide absorbed can be determined from

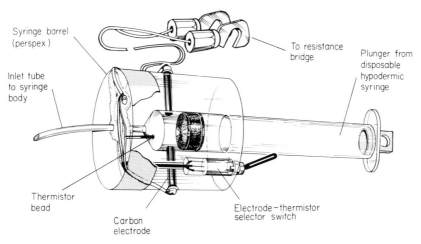

Syringe barrel
(perspex)

Inlet tube
to syringe
body

To resistance
bridge

Plunger from
disposable
hypodermic
syringe

Thermistor
bead

Carbon
electrode

Electrode – thermistor
selector switch

Figure 1.9 A syringe conductivity cell for use with a simple soil respirometer.

readings of the conductivity of the electrolyte taken at intervals throughout the period of measurement.

A modification of this conductiometric method is to build the electrode system into a syringe (Fig. 1.9) so that simpler containers for the electrolyte can be used. If the lids of commercially obtainable plastic pots are used to cover the respirometer cylinders a great number of units can be used at any one time. The electrolyte is drawn up into the syringe and the cell resistance and temperature of the solution measured at the end of the experimental period.

There are situations other than those already described where the ecologist will need to measure the concentration of carbon dioxide in the field. A selection of methods are available but it is up to the individual to assess their suitability for his own particular use.

Macfadyen (1970), in a paper describing a number of field techniques, has described an electrolytic syringe based upon the method of Köpf (1952). When a direct electric current is passed through a solution of sodium chloride, from a silver anode to a platinum cathode, sodium hydroxide will be produced.

$$NaCl \rightarrow Na^+ + Cl^-$$

$$Na^+ + H_2O \rightarrow NaOH + \tfrac{1}{2}H_2$$

$$Cl^- + Ag^+ \rightarrow AgCl$$

If a gas sample is shaken up and dissolved in the syringe an equivalent amount of sodium hydroxide can then be generated. Equivalence is indicated by the production of a pink colour from phenolphthalein contained in the electrolyte. The amount of sodium hydroxide generated can be calculated from the product of the current and the time.

$$1 \, \text{mA per min} \equiv 0.01445 \, \text{ml } CO_2 \text{ at } 10°C$$

A current of 1 mA passed for 1 minute is equivalent to 0.3% CO_2 with a 5 ml air sample and, as the end-point can only be timed to within about 1.5 seconds, the expected accuracy will not be greater than about 0.01% CO_2, but this error is independent of concentration.

Macfadyen has used this syringe in combination with small thin-walled polythene bags (c. 15 × 120 mm) buried in the soil. The bags were provided with sealed tubes leading to the surface of the soil, allowing gas samples to be withdrawn and analysed at suitable intervals. Other methods of estimating the carbon dioxide concentration in the soil have been described by Rutter & Webster (1962), and Martin & Pigott (1965). Rutter and Webster used a probe incorporating a ceramic cup filled with distilled water, that was allowed to equilibrate with the soil solution. One problem was to find the time that was required to reach equilibrium; they employed four probes and measured the rate of change of pH and used it to indicate equilibrium. Martin and Pigott inserted sterilized pyrex tubes, with small 'windows' covered with polytetrafluorethylene membrane, into the soil. The tubes contained 0.01 M sodium bicarbonate and after 10 days the pH of the solution was measured (to within 0.05 pH units) and the carbon dioxide concentration estimated from the relationship with pH (Table 1.1).

This electrometric method for determining the concentration of carbon dioxide in the soil has been examined further by Lee & Woolhouse (1966), and they emphasize the necessity for calibrating individual experimental procedures. They also point out that a particular weakness of the method is the long equilibrium time—up to 3 weeks in some cases—required for the range of CO_2 concentrations that might be found in the soil.

Table 1.1 pH of 0.01 M sodium bicarbonate solution at different concentrations of CO_2. (From Martin & Pigott, 1965.)

%CO_2	pH
0	9.53
0.03	9.31
0.25	8.65
1.00	8.05
4.00	7.45
100.0	6.05

3.5.7 Temperature and soil respiration

The intensity of soil respiration is strongly correlated with soil temperature and shows both an annual and a diurnal cycle (Witkamp, 1966, 1969; Witkamp & Frank, 1969). For comparative purposes it is normal to convert observed levels of respiration to equivalent values at some standard temperature. To do this and to be able to use recordings of soil temperature to obtain estimates of the annual evolution of carbon dioxide from the soil it is necessary to establish a relationship between soil respiration and temperature for a particular ecosystem. The most generally used formula being

$$\log R = a + bT$$

where R = rate of respiration,
T = temperature ($^\circ$C),
a = constant,
b = temperature coefficient.

The temperature coefficient can be expressed as the ratio of the respiration rate at a given temperature and that at a second temperature 10°C lower. The symbol Q_{10} is used to represent this coefficient.

$$Q_1 = \text{antilog } (10.b)$$

In practice, temperature coefficients with values in the order of 2·0 are obtained. Objections can be made regarding the use of this relationship over wide ranges of temperatures, but within the ranges occurring in the soil it has been found satisfactory (Wiant, 1967b; Reiners, 1968; Witkamp, 1969; Anderson, 1973a). Other relationships between temperature and respiration have been proposed (Krogh, 1914) and some workers have obtained better correlations from log–log regressions (Kucera & Kirkham, 1971).

Soil respiration is also a function of soil moisture (Wiant, 1967a; Macfadyen, 1971; Froment, 1972; Anderson, 1973a) and other factors such as the carbon dioxide concentration of the soil atmosphere (Macfadyen, 1971).

As a result of the diurnal cycle of carbon dioxide evolution from the soil it is preferable to base estimates of the daily respiration upon measurements taken over 24 hour periods.

4 Nutrient and energy budgets

The cycling of nutrients, the sources of additional nutrients and the pathways by which they are lost are of great importance in attempting to analyse and understand the working of an ecosystem. The production of organic matter and the flow of energy are important processes but are both influenced by the availability of nutrients and water. The relative importance of the nutrient budget (the nutrient income and loss account) varies from one type of ecosystem to another. Where the source of nutrients is restricted to rainfall, as in the case of ombrotrophic peatland, the overall nutrient budget is obviously of great interest. In more nutrient-rich situations the nutrient budget may assume relatively less importance than the internal cycling of some particular element, but it can be very unwise to try and rank such factors in order of importance when they are so intimately related.

To compile a nutrient or energy budget for any ecosystem it is essential to have a reliable framework of organic matter production and turnover upon which to base the estimates. Calculations of the nutrient or energy contents of the different components of the ecosystem are based upon the combination of biomass and estimates of composition or calorific content.

While there is no hard and fast division between a nutrient budget and a nutrient cycle, the study of the budget often presents fewer practical problems and may be the initial approach to any such investigation. As the individual worker can, and must, define the limits of the ecosystem under investigation, he can also subdivide the overall system into smaller units, and it is the recombination of the budgets from these subsystems that produces the overall picture of nutrient or energy flow through the complete system.

When some of the inputs to an ecosystem have been measured the question will often arise as to whether they represent a total or true input, or whether part merely represents some degree of recycling within the limits of the ecosystem. An example is found with the estimation of the nutrient input by rainfall. If the rain samples are filtered they will be found to contain mineral particles, pollen grains and other organic debris. Some of this material will have originated within the ecosystem; it should not therefore be included in any estimate of nutrient input, and presents a very real practical problem.

References to relevant published work are given in the sections dealing with particular aspects of nutrient and energy flow, but general references providing valuable background information include Rodin & Bazilovich (1967), and papers included in Young (1968) and Reichle (1970).

4.1   Estimation of nutrient and calorific content

Where independent estimates of weight and chemical composition (or calorific content) are combined to calculate the nutrient (or calorific content) of some particular component of the ecosystem, the relationship between the means and the standard deviations of the means (standard errors) are:

$$\bar{N} = \bar{k}, \bar{w}$$

where $\bar{N}$ = mean nutrient or energy content,

$\bar{k}$ = mean composition of calorific value,
$\bar{w}$ = mean weight or biomass.

$$S_{\bar{N}}^2 = \bar{k}^2 S_{\bar{w}}^2 + \bar{w}^2 S_{\bar{k}}^2 + S_{\bar{w}}^2 S_{\bar{k}}^2$$

where $S_{\bar{N}}^2$ = variance of the mean nutrient content ($\bar{n}$),
$\quad\quad S_{\bar{k}}^2$ = variance of the mean composition ($\bar{k}$),
$\quad\quad S_{\bar{w}}^2$ = variance of the mean biomass ($\bar{w}$).

and where:

$$S_{\bar{x}}^2 = \frac{\Sigma x^2 - \dfrac{(\Sigma x)^2}{n}}{n(n-1)}$$

when x = single observation,
$\quad\quad n$ = number of observations.

The derivation and proof of this relationship between the means and variances of combined estimates are given by Colquhoun (1971) and have been used by Rawes & Welch (1969) and Gyllenberg (1969).

To obtain an estimate of the nutrient content of the soil it is necessary to combine measurements of composition (i.e. %N), bulk density and soil volume. Methods of measuring the density of soil are described in Chapter 5. A relationship between soil density and the loss-on-ignition for a wide range of soils has been shown by Jeffrey (1970) and Harrison & Bocock (1981) and in the absence of actual determinations of soil density this relationship can be used to estimate the nutrient content of soil horizons. The expression for predicting the bulk density of the soil (y) from the percentage loss-on-ignition data (x) is:

$$y = 1 \cdot 482 - 0 \cdot 6786 \log_{10} x$$

4.2   Nutrient inputs

There are a number of different sources of input of nutrients to an ecosystem, including precipitation and atmospheric fallout, the lateral movement of nutrients through the soil from adjacent areas, nitrogen fixation, faunal migration, the weathering of soil minerals and in some cases the application of fertilizers (see Fig. 4.1).

4.2.1   Precipitation and atmospheric fallout

A survey of the chemical content of precipitation in north-western Europe has been carried out by Egnér et al. (1955–1960). The chemical composition of rainfall on a daily basis has been related to the amount of precipitation, to wind direction, to wind velocity and to temperature by Gorham (1958) working in the Lake District of England.

Where it is intended to collect rain samples for chemical analysis it is recommended that the amount of rainfall be measured with a standard rain gauge, and that a separate gauge made of polythene or some other chemically inert material be used to collect samples for analysis. The plastic funnels should be fitted with gauze filters to prevent contamination by insects or other debris, and the collecting bottle kept within a dark container to prevent algal growth. Microbiological growth can affect the composition of the samples by nutrient uptake and assimilation, but this can be reduced by impregnating the walls of the collecting bottle with iodine (Heron, 1962). This can be done by placing a few crystals of iodine in the bottle and putting the stoppered bottle into a warm oven (60°C) for several hours. Fouling of the rain gauges by birds is a problem that is generally encountered, and can sometimes be reduced by fitting the gauge with a crown of spikes (Egnér *et al.*, 1956), or by a ring around the funnel to act as a 'decoy' perch. Some people seem to have found these precautions useful while others have

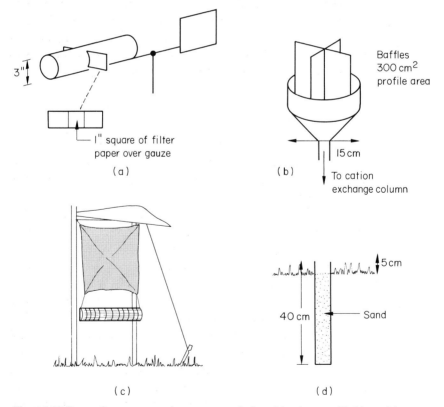

Figure 1.10 Types of apparatus used to measure salt deposition in coastal habitats: (a) Edwards & Claxton (1964); (b) Etherington (1967); (c) Randall (1970); (d) Malloch (1972).

reported only little benefit (Allen *et al.*, 1968). The gauges should be emptied as frequently as possible and the samples stored in a deep-freeze until the analyses are carried out.

Where rain samples are collected after passing through the tree canopy or after running down the trunks or stems of trees they are found to contain higher concentrations of dissolved nutrients than the incident rainfall (Madgwick & Ovington, 1959; Carlisle *et al.*, 1966). While some of this enrichment is due to the leaching of materials out of the plant tissue, part is due to the capture of airborne particles (aerosols) by the vegetation. These aerosol components arise from smoke, reactions between gases in the atmosphere, sea spray and mineral dust.

The capture of airborne particles by a woodland canopy has been studied by White & Turner (1970). Their sampling apparatus was mounted on a tower in and above the tree canopy, but they suggest that their results overestimate the nutrient income in the form of airborne particles, especially during the windier months of the year.

A different method has been used by Nihlgård (1970) in an attempt to assess the importance of aerosols in the enrichment of the precipitation passing through the woodland canopy. He arranged ten layers of plastic netting, one above another, in a woodland clearing with raingauges underneath. The concentration of nutrients in the water collected underneath the netting was found to be higher than in gauges collecting direct rain samples.

4.2.2   Salt spray

When working in coastal habitats it is important to consider both the direct effects of sea-spray upon the vegetation and also its importance as a source of plant nutrients.

Edwards & Claxton (1964) devised an instrument where a square of filter paper is held at right angles to the wind. Etherington (1967) in a study of potassium cycling and the production of a dune heath ecosystem used a set of plastic baffles over a funnel to intercept the salt spray. The drainage from the funnel was passed through a cation exchange column to absorb sodium and potassium. A 30-cm square of towelling material, weighted at the base and suspended beneath a plastic shelter was used by Randall (1970) to collect salt spray. The salt was extracted from the towelling by placing it in 300 ml of water and then estimated conductimetrically. Malloch (1972) buried sand-filled tubes vertically in the soil to estimate salt deposition on the Lizard Peninsula in Cornwall. The salt was extracted in 500 ml of deionized water and sodium determined by flame photometry after periods of 2–3 months. An advantage of this apparatus is that it is capable of being 'hidden from predation by tourists', a point worth considering when any apparatus must be left in a relatively public place.

4.2.3   Nitrogen fixation

The biological fixation of nitrogen has been studied by a number of different methods: kjeldahl analyses, ^{15}N-enrichment assayed by mass spectrometer, and by means of the radio-isotope ^{13}N, even with its great disadvantage of a 10-minute half-life. These methods are all far from ideal in that they are either insensitive, time-consuming or require complex apparatus such as a mass spectrometer.

With the discovery of the acetylene reduction technique by Dilworth (1966) and Schöllhorn & Burris (1967) these problems were largely overcome. The method has the advantage of being sensitive, capable of being used in the field and relatively inexpensive. The technique depends upon the fact that the enzyme nitrogenase is capable of reducing acetylene to ethylene and therefore uses the rate of acetylene reduction as an index of nitrogen fixation.

The use of the technique in the field has been described by Stewart *et al.* (1967) and by Hardy *et al.* (1968) and a review of the methods available is given by Bergersen (1980).

4.3   Nutrient losses

The principal losses of nutrients from an ecosystem are through leaching and the lateral movement of nutrients in the soil solution, by grazing, by the removal of harvested plant material, by the removal of nutrients in animal bodies, and by other processes such as fire, soil erosion and solifluction. It can be seen that a number of these losses of nutrients from one particular area represent an input to some other adjacent area or ecosystem.

4.3.1   Loss of nutrients in the soil solution

The methods available for the measurement of nutrient losses by leaching include direct sampling of the soil solution, the use of lysimeters and the analyses of drainage waters from catchment areas. In part these methods depend upon the combination of hydrological and analytical methods described in Chapter 6 (see also Chapter 4, p. 178).

When water is present in the soil in excess of field capacity it is free to drain through the soil and be lost from the ecosystem. Under these conditions the soil solution can be sampled by interception or by suction sampling. Parizek & Lane (1970) have reviewed a number of these techniques and describe a trench lysimeter where the drainage water is intercepted by metal pans inserted at intervals down the soil profile. The trench lysimeter described was used in conjunction with an irrigation experiment; in the absence of such an experimental treatment samples of water for analysis would only have been obtained in small quantities and at infrequent intervals. They considered the constructional problems involved and concluded that other procedures were preferable, especially where sampling was required at a number of different sites. In the same paper they describe the use of vacuum or suction lysimeters of the type previously used by Wagner (1962). These samplers consist of unglazed ceramic cups cemented to the end of plastic pipes (Fig. 1.11) that are connected to tubes to which a vacuum can be applied and any soil solution available collected and withdrawn for chemical analysis. This type of sampler will only extract water from the soil when it is present at tensions less than the suction that can be supplied by a vacuum pump. The amount of water sample that will be obtained is dependent upon both the water content and the 'water supplying power' of the soil. The results of any chemical analyses carried out on samples obtained from

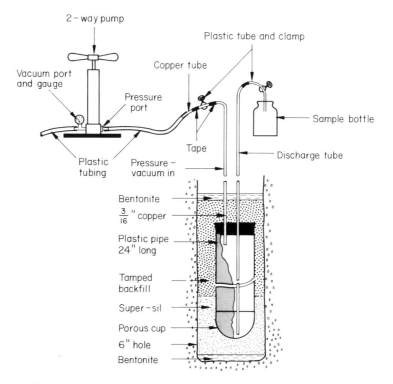

Figure 1.11 Cross-section of a typical pressure-vacuum lysimeter installation with the pressure pump attached to the lysimeter. The apparatus is ready to recover a water sample. (From Parizel & Lane, 1970).

suction lysimeters must be used in combination with soil moisture determinations to obtain estimates of nutrient flow through the soil profile.

Grover & Lamborn (1970) found that some types of ceramic cup used for sampling soil solution added calcium, sodium and potassium to solutions drawn through them and that they absorbed significant amounts of phosphorus from the soil solution. They found that leaching the cups in 1 N HCl reduced the contamination from all but calcium and reduced the loss of phosphorus to an acceptable level.

An investigation into the water and mineral budget of an area of Pennine moorland by Crisp (1966) provides an example of the estimation of nutrient losses from an ecosystem by the analysis of stream flow. Relationships between the concentrations of nutrients and the rate of run-off were established, enabling estimates of the quantities of nutrients lost in solution to be calculated from the volume of water leaving the catchment area.

It is important that the variation of nutrient concentration with stream flow be measured directly; alternatively, some device may be used which will collect water samples proportional in volume to the rate of run-off. If such a proportional sampler is used, the water samples collected at regular time intervals can provide an estimate of nutrients lost in solution. A simple proportional sampler of this type has been described by Eggink & Duvigneau (1963), consisting of a sampling chamber placed upstream from a weir and so shaped that the volume of water it contains at any particular time is proportional to a function of the head of water and, therefore, to the rate of flow over the weir. At intervals a valve in the base is opened and the sampling chamber fills to a level governed by the head of water flowing over the weir. The sample is transferred to a collecting chamber by means of compressed air, so providing a composite water sample representative of the total volume of run-off that has left the area during the sampling period.

In addition to the loss of nutrients in solution the movement of solid material in suspension may be important. Crisp obtained estimates of the amount of peat lost from his catchment area by using a siphon sampling system that incorporated a self-cleansing filter. Periodic samples of the filtrate were taken to allow a correction to be made for the loss of fine material through the filters. The relationship between the rate of flow through the siphons and the total flow in the stream was used to calculate the weight of solid material lost in suspension.

Perrin (1965) has examined the possibility of using drainage water analyses in the study of soil development, in particular the development of a chalk soil and the genesis of shallow brown soils on shales in central Wales.

4.3.2 Loss of nutrients by fire

In a number of habitats, such as savanna and heathland, the above-ground vegetation is burnt off at periodic intervals. Such fires are important for a number of reasons; the maintenance of a particular type of vegetation, the encouragement of new growth as potential grazing material or the mineralization and return of plant nutrients to the soil. After a fire the soil surface will be covered by a nutrient-rich layer of ash, but significant amounts of the nutrients contained in the above-ground vegetation and litter layer will have been lost from the ecosystem. Where the particular ecosystem is extensive many of the nutrients lost from one particular area will be distributed over a surrounding area of the same type of vegetation, but when the ecosystem is smaller or has become fragmented, such as heathland in lowland Britain (Moore, 1962), the loss of nutrients in a fire may represent a complete loss from the ecosystem.

The magnitude of these losses from heathland have been measured by placing cut vegetation over sheets of steel, burning and collecting the resulting ash (Chapman, 1967; Evans & Allen, 1971). Samples of plant material have been burnt under laboratory conditions (Allen, 1964; Evans & Allen, 1967) in attempts to control the conditions of the experiment rather more than is possible in the field procedure just described.

The actual losses of nutrients under field conditions depend very much upon local conditions at the time of the burn, such as wind, the weight and structure of the standing crop, and the temperature of the burn. The temperatures reached in natural fires have been measured by means of chemicals with different melting points (Beadle, 1940), by the use of heat sensitive paints (Whittaker, 1961) and by the use of thermocouples (Kenworthy, 1963). The distribution and changes in the nutrient content of heathland soils after fires have been examined by Allen *et al.* (1969) and by Hansen (1969).

4.4 Measurement of calorific values

4.4.1 Calorific values and calorimetry

In the establishment of a nutrient budget the result of chemical analyses are combined with biomass data to obtain estimates of nutrient content; in the construction of an energy budget the calorific values are combined with weights to provide estimates of calorific content.

The GROSS CALORIFIC VALUE is the number of heat units liberated when a unit weight of material is burnt in oxygen; and the residual materials are oxygen, carbon dioxide, sulphur dioxide, nitrogen, water and ash.

The basic unit of energy is the joule (J), but in the past the unit used to express the energy content of biological material has often been the calorie (c) or the kilocalorie (C) and results expressed in terms of kilocalories per gramme dry weight. In many cases it is more meaningful to calculate and express the results in terms of ash-free dry weight. Ecologists have often overlooked the fact that there are a number of differently defined types of calorie (15° calorie, steam calorie, international table calorie) each having a slightly different value and it is now universally accepted that the SI unit the joule (4·1840 joules = 1 thermochemical calorie) should be used to express all results in the study of ecological energetics (Lieth, 1968; Phillipson, 1971).

Wet combustion techniques (Ivlev, 1934), that depend upon conversion factors being applied to carbon, nitrogen and sometimes sulphur contents, have been used to obtain estimates of the calorific values of biological material. These methods are not generally as satisfactory or as convenient as the use of calorimetry.

In bomb calorimetry a sample of material is ignited and burnt inside a thick-walled stainless steel container (the bomb) that contains oxygen under pressure of up to 30 atmospheres. The increase in temperature of the bomb, as a result of the combustion of the sample, is measured and by comparison with the results obtained from the combustion of a standard material the calorific value of the sample can be calculated. The standard generally used is benzoic acid having a calorific value of 26 447 joules per gramme.

4.4.2 Preparation of samples

Collection and preparation of samples for the determination of calorific values may require greater care and precautions than are necessary when sampling for the estimation of the weight of standing crop. The sample must be dried in such a way that losses of volatile constituents are kept to a minimum. Freeze-drying and the use of vacuum ovens are recommended by some authors. When dried the material must be milled, homogenized and compressed into pellets. Lieth (1968) describes methods by which the sample material can be melted into wax or contained in gelatine capsules as alternatives to the use of pellets. In some cases, known amounts of benzoic acid have been added to the sample to aid combustion (Malone & Swartout, 1969), or to enable calorific values to be determined on small amounts of material (Richman, 1968).

4.4.3 Calorimeter corrections

Bomb calorimeters are of two main types, adiabatic and non-adiabatic. In the adiabatic type of calorimeter the bomb is surrounded by a water jacket whose temperature is controlled and remains the same as that of the bomb at all times. This means that no heat is transferred to or from the bomb and the calorific value (V) of a sample can be calculated from:

$$V = \frac{W\Delta t - \Sigma c}{G}$$

where W = the number of joules necessary to raise the temperature of the
water bath (and the bomb) by one degree centigrade,

Δt = the rise in temperature of the water bath,

Σc = the sum of necessary corrections (see below),

G = dry weight of the sample.

The micro-bomb calorimeter described by Phillipson (1964), suitable for small samples (5–100 mg dry weight) is an example of a non-adiabatic calorimeter. In this type of calorimeter heat will flow either in or out of the apparatus depending upon its temperature relative to that of the surroundings. This means that suitable corrections must be applied to the results obtained from such a calorimeter to allow for any heat exchange. The most convenient cooling correction (C) is the application of Dickinson's formula:

$$C = r_1(T_a - T_0) + r_2(T_n - T_a)$$

where r_1 = the pre-firing rate of temperature change, if the temperature is
rising r is negative,

r_2 = the post-firing cooling rate,

T_0 = time at temperature t_0,

T_n = time at temperature t_n,

T_a = time at temperature $t_0 + 0.60(t_n - t_0)$.

Where a chart recorder is used to record the temperature changes within the calorimeter the cooling correction can be applied by a graphical method direct on the chart (Fig. 1.12).

An alternative type of non-adiabatic calorimeter is the ballistic bomb calorimeter where larger samples of material are combusted. The time taken for the bomb to reach its maximum temperature is short (45–60 seconds) so that heat exchange is reduced and cooling corrections are unnecessary. This type of calorimeter is capable of producing results quickly and is convenient to use where large numbers of determinations are required and there is no shortage of sample material.

In addition to the cooling corrections described there are a number of additional corrections which may have to be applied in the calculation of calorific values. When electrical energy is used to ignite the sample there is an input of energy; this can be measured by means of blank determinations and subtracted from actual determinations of calorific value. If any of the firing wire is burnt during combustion, heat will be released, but when platinum wire is used this weight is small and the correction insignificant (418 J g^{-1}

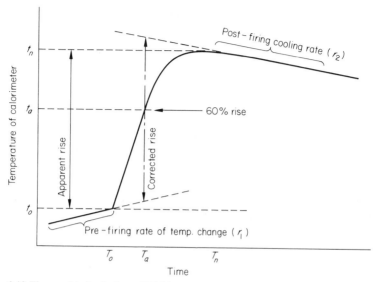

Figure 1.12 The graphical solution of Dickinson's cooling correction for non-adiabatic bomb calorimeters (for details see text).

platinum, $1402\,J\,g^{-1}$ nichrome). In the definition of calorific value it is assumed that the products of combustion will include nitrogen and sulphur. In practice these elements are oxidized to nitric and sulphuric acid which results in the evolution of extra heat. The washings obtained from the bomb can be titrated and used to estimate the quantities of these acids that have been formed, and the necessary corrections applied. With biological materials these corrections are small, generally less than 0·1% (Lieth, 1968), and are often omitted in ecological research.

5   Summary

This chapter has attempted to provide an introduction and background to the methods and concepts required for the investigation of primary production and nutrient circulation within a terrestrial ecosystem. It will be apparent that anyone hoping to find a single set of methods that can be applied to any set of ecological circumstances will be disappointed. It is, or should be, one of the attractions of research that the means of obtaining the relevant information may be one of the most difficult problems that has to be solved.

While it has not been possible to give a detailed account of results obtained from the methods described, the references supplied in the text should provide starting points for those interested in examining their own results against a more general background.

It is hoped that this chapter has already drawn attention to some of these problems and that it will encourage co-operation and liaison between botanists and zoologists in the study of natural communities.

6   References

AIMI R. and FUJIMAKI K. (1958) Cell-physiological studies on the function of root. 1. Chemical and enzymatic constitution associated with the structural differentiation of root in rice plant. *Proc. Crop Sci. Soc. Japan* **27**, 21–24.

ALCOCK M.B. (1964) An improved electronic instrument for estimation of pasture yield. *Nature, Lond.* **203**, 1309.

ALCOCK M.B. and LOVETT J.V. (1967) The electronic measurement of the yield of growing pasture. 1. A statistical assessment. *J. agric. Sci., Camb.* **68**, 27–38.

ALLEN, S.E. (1964) Chemical aspects of heather burning. *J. appl. Ecol.* **1**, 347–368.

ALLEN S.E., CARLISLE A., WHITE E.J. and EVANS C.C. (1968) The plant nutrient content of rainwater. *J. Ecol.* **56**, 497–504.

ALLEN S.E., EVANS C.C. and GRIMSHAW H.M. (1969) The distribution of mineral nutrients in soil after heather burning. *Oikos* **20**, 16–25.

ANDERSON J.M. (1973a) Carbon dioxide evolution from two temperate, deciduous woodland soils. *J. appl. Ecol.* **10**, 361–378.

ANDERSON J.M. (1973b) The breakdown and decomposition of Sweet Chestnut (*Castanea sativa* Mill.) and Beech (*Fagus sylvatica* L.) leaf litter in two deciduous woodland soils. 1. Breakdown, leaching and decomposition. *Oecologia* **12**, 251–274.

ANDERSON J.M. and HEALEY I.N. (1970) Improvements in the gelatine embedding technique for woodland soil and litter samples. *Pedobiologia* **10**, 108–120.

ANDERSON J.M. and McFADYEN A. (Eds) (1976) *The Role of Terrestrial and Aquatic Organisms in Decomposition Processes.* BES Symposium No. 17. Blackwell Scientific Publications, Oxford.

ANDERSSON F. (1970) Ecological studies in a Scanian woodland and meadow area, Southern Sweden. II. Plant biomass, primary production and turnover of organic matter. *Bot. Notiser* **123**, 8–51.

ASHBY W.C. (1962) Root growth in American Basswood. *Ecology* **43**, 336–339.

BARCLAY-ESTRUP P. (1970) The description and interpretation of cyclical processes in a heath community. II. Changes in biomass and shoot production during the *Calluna* cycle. *J. Ecol.* **58**, 243–249.

BEADLE N.C.W. (1940) Soil temperatures and their effect on the survival of vegetation. *J. Ecol.* **28**, 180–192.

BEAUCHAMP J.J. and OLSON J.S. (1973) Corrections for bias in regression estimates after logarithmic transformation. *Ecology* **54**, 1403–1407.

BENEFIELD C.B. (1971) A rapid method for measuring cellulase activity in soils. *Soil Biol. Biochem.* **3**, 325–329.

BERGERSEN F.J. (Ed.) (1980) *Methods for Evaluating Biological Nitrogen Fixation.* John Wiley and Son, New York.

BLISS, L.C. (1966) Plant productivity in alpine micro-environments on Mt. Washington, New Hampshire. *Ecol. Monogr.* **36**, 125–155.

BOCOCK K.L. (1964) Changes in the amounts of dry matter, nitrogen, carbon and energy in decomposing woodland leaf litter in relation to the activities of the soil fauna. *J. Ecol.* **52**, 273–284.

BOCOCK K.L. and GILBERT O.J.B. (1957) The disappearance of leaf litter under different woodland conditions. *Pl. Soil* **9**, 179–185.

BOCOCK K.L., GILBERT O.J.B., CAPSTICK C.K., TWINN D.C., WAID J.S. and WOODMAN M.J.

(1960) Changes in leaf litter when placed on the surface of soils with contrasting humus types. *J. Soil Sci.* **11**, 1–9.

BOGGIE R., HUNTER R.F. and KNIGHT A.H. (1958) Studies of the root development of plants in the field using radioactive tracers. *J. Ecol.* **46**, 621–640.

BOGGIE R. and KNIGHT A.H. (1962) An improved method for the placement of radioactive isotopes in the study of root systems of plants growing in deep peat. *J. Ecol.* **50**, 461–464.

BOWMAN G.E. (1968) The measurement of carbon dioxide concentration in the atmosphere. In *The Measurement of Environmental Factors in Terrestrial Ecology* (ed. R.M. Wadsworth), pp. 131–139. BES Symposium No 8. Blackwell Scientific Publications, Oxford.

BRAY J.R. (1963) Root production and the estimation of net productivity. *Can. J. Bot.* **41**, 65–72.

BRAY J.R. and GORHAM E. (1964) Litter production in forests of the world. *Adv. Ecol. Res.* **2**, 101–157.

BROWN A. and MACFADYEN A. (1969) Soil carbon dioxide output and small-scale vegetation pattern in a *Calluna* heath. *Oikos* **20**, 8–15.

BURGES A. and DROVER D.P. (1953) The rate of podsol development in the sands of Woy Woy district. *Aust. J. Bot.* **1**, 83–94.

BURROWS W.H. (1972) Productivity of an arid zone shrub (*Eremophila gilesii*) community in south-western Queensland. *Aust. J. Bot.*, **20**, 317–329.

CAHOON G.A. and MORTON E.S. (1961) An apparatus for the quantitative separation of plant roots from soil. *Proc. Am. Soc. hort. Sci.* **78**, 593–596.

CALDWELL M.M. and CAMP L.B. (1974) Below ground productivity of two cool desert communities. *Oecologia* **17**, 123–130.

CAPSTICK C.A. (1962) The use of small cylindrical samplers for estimating the weight of forest litter. In *Progress in Soil Zoology* (ed. P.W. Murphy), pp. 353–356. Butterworths, London.

CARLISLE A., BROWN A.H.F. and WHITE E.J. (1966) The organic matter and nutrient elements in the precipitation beneath a sessile oak (*Quercus petraea*) canopy. *J. Ecol.* **54**, 87–98.

CHAPMAN S.B. (1967) Nutrient budgets for a dry heath ecosystem in the south of England. *J. Ecol.* **55**, 677–689.

CHAPMAN S.B. (1970) The nutrient content of the soil and root system of a dry heath ecosystem. *J. Ecol.* **58**, 445–452.

CHAPMAN S.B. (1971) A simple conductimetric soil respirometer for field use. *Oikos* **22**, 348–353.

CHAPMAN S.B., HIBBLE J. and RAFAREL C.R. (1975a) Net aerial production by *Calluna vulgaris* on lowland heath in Britain. *J. Ecol.* **63**, 233–258.

CHAPMAN S.B., HIBBLE J. and RAFAREL C.R. (1975b) Litter accumulation under *Calluna vulgaris* on a lowland heathland in Britain. *J. Ecol.* **63**, 259–271.

CHAPMAN S.B. (1979) Some interrelationships between soil and root respiration in *Calluna* heathland in southern England. *J. Ecol.* **67**, 1–20.

CHARLES J.A. (1972) Physical science and archaeology. *Antiquity* **46**, 134–138.

CLARK F.E. and ROSWALL T. (Eds) (1981) *Terrestrial Nitrogen Cycles, Processes, Ecosystem Strategies and Management Impacts*. *Ecol. Bull.* (Stockholm) 33.

CLYMO R.S. (1965) Experiments on breakdown of *Sphagnum* in two bogs. *J. Ecol.* **53**, 747–758.

CLYMO R.S. (1970) The growth of Sphagnum: methods of measurement. *J. Ecol.* **58**, 13–49.

COLQUOHOUN D. (1971) *Lectures on Biostatistics*. Oxford.

CONWAY, E.J. (1950) *Microdiffusion Analysis and Volumetric Error*, 3rd edition. London, Crosby Lockwood.

CRAPO N.L. and COLEMAN D.C. (1972) Root distribution and respiration in a Carolina old field. *Oikos* **23**, 137–139.

CRISP D.T. (1966) Input and output of minerals for an area of Pennine moorland: the importance of precipitation, drainage, peat erosion and animals. *J. appl. Ecol.* **3**, 327–348.

CROCKER R.L. and DICKSON B.A. (1957) Soil development on the recessional moraines of the Herbert and Mendenhall glaciers, south-eastern Alaska. *J. Ecol.* **45**, 169–185.

CROCKER R.L. and MAJOR J. (1955) Soil development in relation to vegetation and surface age at Glacier Bay, Alaska. *J. Ecol.* **43**, 427–448.

COOMBS J. and HALL D.O. (Eds) (1982) *Techniques in Bioproductivity and Photosynthesis*. Pergamon, Oxford.

CROSSLEY D.A. JR. and HOGLUND M.P. (1962) A litter bag method for the study of micro-arthropods inhabiting leaf litter. *Ecology* **43**, 571–573.

DAHLMAN R.C. and KUCERA C.L. (1965) Root productivity and turnover in native prairie. *Ecology* **46**, 84–89.

DAHLMAN R.C. and KUCERA C.L. (1968) Tagging native grassland with carbon-14. *Ecology* **49**, 1199–1203.

DE WIT C.T. and GOUDRIAAN J. (1974) *Simulation of Ecological Processes*. Pudoc, Wageningen.

DILWORTH M.J. (1966) Acetylene reduction by nitrogen fixing preparations from *Clostridium pasteurianum*. *Biochim. biophys. Acta* **127**, 285–294.

DURNO S.E. (1961) Evidence regarding the rate of peat growth. *J. Ecol.* **49**, 347–351.

EDWARDS C.A. and HEATH G.W. (1963) The role of soil organisms in breakdown of leafy material. In *Soil Organisms* (eds J. Doeksen and J. van der Drift) pp. 76–84. North Holland Publishing Co., Amsterdam.

EDWARDS R.S. and CLAXTON S.M. (1964) The distribution of air-borne salt of marine origin in the Aberystwyth area. *J. appl. Ecol.* **1**, 253–263.

EGGINK H.J. and DUVIGNEAU H. (1963) Bemonstering van afvalwater naar hoeveelheid. (Sampling proportional to rate of sewage flow.) *Water* **47**, 69–71.

EGNÉR H., BRODIN G. and JOHANSSON O. (1956) Sampling technique and chemical examination of air and precipitation. *K. LantbrHögsk. Annlr.* **22**, 369–382.

EGNÉR H., ERIKSSON E. and BRODIN G. (1955–60) Current data on the composition of air and precipitation. *Tellus* 7–12.

ELLERN S.J., HARPER J.L. and SAGAR G.R. (1970) A comparative study of the distribution of the roots of *Avena fatua* and *A. strigosa* in mixed stands using a carbon-14 labelling technique. *J. Ecol.* **58**, 865–868.

ETHERINGTON J.R. (1967) Studies of nutrient cycling and productivity in oligotrophic ecosystems. 1. Soil potassium and wind blown sea-spray in a south Wales dune grassland. *J. Ecol.* **55**, 743–752.

EVANS C.C. and ALLEN S.E. (1971) Nutrient losses in smoke produced during heather burning. *Oikos* **22**, 149–154.

EVANS G.C. (1972) *The Quantitative Analysis of Plant Growth*. Blackwell Scientific Publications, Oxford.

FERRILL M.D. and WOODS F.W. (1966) Root extension in a long-leaf pine plantation. *Ecology* **47**, 97–102.

FLETCHER J.E. and ROBINSON M.E. (1956) A capacitance meter for estimating forage weight. *J. Range Mgmt.* **9**, 96–97.

FORREST G.I. (1971) Structure and production of north Pennine blanket bog vegetation. *J. Ecol.* **59**, 453–479.

FRANKLAND J.C. (1966) Succession of fungi on decaying petioles of *Pteridium aquilinum*. *J. Ecol.* **54**, 41–63.

FRANKLAND J.C., OVINGTON J.D. and MACRAE C. (1963) Spatial and seasonal variations in soil litter and ground vegetation in some Lake District woodlands. *J. Ecol.* **51**, 97–112.

FROMENT A. (1972) Soil respiration in a mixed Oak forest. *Oikos* **23**, 273–277.

GHILAROV M.S., KORDA V.A., NOVICHKOVA-IVANOVA L.N., RODIN L.E. and SVESHNIKOVA V.M. (Eds) (1968) *Methods of Productivity Studies in Root Systems and Rhizosphere Organisms*. USSR, Acad. Sciences, Leningrad.

GOLLEY F.B. (1960) An index to the rate of cellulose decomposition in the soil. *Ecology* **41**, 551–552.

GOLLEY F.B. and GENTRY J.B. (1965) A comparison of variety and standing crop of vegetation on a one-year and a twelve-year abandoned field. *Oikos* **15**, 185–199.

GORE A.J.P. and OLSON J.S. (1967) Preliminary models for accumulation of organic matter in an *Eriophorum/Calluna* ecosystem. *Aquilo Ser. Botanica* **6**, 297–313.

GORHAM E. (1958) The influence and importance of daily weather conditions in the supply of chloride, sulphate and other ions to freshwaters from atmospheric precipitation. *Phil. Trans. R. Soc. Ser. B.* **241**, 147–178.

GRASSLAND RESEARCH INSTITUTE STAFF (1961) *Research Techniques in use at the grassland*

Research Institute, Hurley. Bulletin 45. Commonwealth Bureau of Pastures and Field Crops, Hurley; Berks, England.

GROVER B.L. and LAMBORN R.E. (1970) Preparation of porous ceramic cups to be used for extraction of soil water having low solute concentrations. *Proc. Soil Sci. Soc. Am.* **34**, 706–708.

GYLLENBERG G. (1969) The energy flow through a *Chorthippus parallelus* (Zett.) (Orthoptera) population on a meadow in Tvärminne, Finland. *Acta Zool. fenn.* **123**, 1–74.

HAARLØV M. and WEIS-FOGH T. (1953) A microscopical technique studying the undisturbed texture of soils. *Oikos* **4**, 44–57.

HALL, C.A.S. and DAY J.W. (1977) *Ecosystem Modelling in Theory and Practice.* Wiley, New York.

HANSEN K. (1969) Edaphic conditions of Danish heath vegetation and the response to burning-off. *Bot. Tidssk.* **64**, 121–140.

HARDY R.W.F., HOLSTEIN R.D., JACKSON E.K. and BURNS R.C. (1968) The acetylene ethylene assay for N₂ fixation. Laboratory and field evaluation. *Plant Physiol.* **43**, 1185–1207.

HARRISON A.F. and BOCOCK K.L. (1981) Estimation of soil bulk density from loss-on-ignition values. *J. Appl. Ecol.* **8**, 919–927.

HAYES A.J. (1965a) Studies on the decomposition of coniferous leaf litter. 1. Physical and chemical changes. *J. Soil Sci.* **16**, 121–140.

HAYES A.J. (1965b) Studies on the decomposition of coniferous leaf litter. 2. Changes in external features and succession of micro fungi. *J. Soil Sci.* **16**, 242–257.

HEAD G.C. (1971) Plant roots. In *Methods of Study in Quantitative Soil Ecology: Population, Production and Energy Flow* (ed. J. Phillipson), IBP Handbook No. 18, pp. 14–23. Blackwell Scientific Publications, Oxford.

HEADY H.F. and VAN DYNE G.M. (1965) Prediction of weight composition from point samples of clipped herbage. *J. Range Mgmt.* **18**, 144–148.

HEALEY I.N. and SWIFT M.J. (1971) Aspects of the accumulation and decomposition of wood in the litter layer of a coppiced beech-oak woodland. In *VI Colloquium pedobiologiae*, pp. 417–430. Dijon 1970. Paris. Inst. National de la Recherche Agronomique.

HERON J. (1962) Determination of PO₄ in water after storage in polyethylene. *Limnol. Oceanogr.* **7**, 316–321.

HESKETH D.J. and JONES J.W. (Eds) (1980) *Predicting Photosynthesis for Ecosystem Models.* CRC Press Inc., Florida.

HOUGH, W.A., WOODS F.W. and McCORMACK M.L. (1965) Root extension of individual trees in surface soils of a natural Longleaf Pine–Turkey Oak stand. *For. Sci.* **11**, 223–242.

HOWARD P.J.A. (1966) A method for the estimation of carbon dioxide evolved from the surface of soil in the field. *Oikos* **17**, 267–271.

HOWARD P.J.A. (1967) A method for studying the respiration and decomposition of litter. In *Progress in Soil Biology* (eds O. Graff and J.E. Satchell), pp. 464–472. Braunschweig, Vieweg.

HUNT R. (1978) *Plant Growth Analysis.* Studies in Biology No. 96. Edward Arnold, London.

HUNT R. (1982) *Plant Growth Curves: An Introduction to the Functional Approach to Plant Growth Analysis.* Edward Arnold, London.

HYDE F.J. and LAWRENCE J.T. (1964) Electronic assessment of pasture growth. *Electron. Engng.* **36**, 666–670.

IVLEV V.G. (1934) Eine Mikromethode zur Bestimmung des Kaloriengehalts von Nahrstoffen. *Biochem. Z.* **275**, 49–55.

JACQUES W.H. and SCHWASS R.H. (1956) Root development in some common new Zealand pasture plants. *N.Z. J. Sci. Technol.* **37**, 569–583.

JEFFERS J.N.R. (1972) *Mathematical Models in Ecology.* BES Symposium No. 12. Blackwell Scientific Publications, Oxford.

JEFFERS J.N.R. (1978) *An Introduction to System Analysis: With Ecological Applications.* Edward Arnold, London.

JEFFREY D.W. (1970) A note on the use of ignition loss as a means for the approximate estimation of soil bulk density. *J. Ecol.* **58**, 297–299.

JENNY H., GESSEL S.P. and BINGHAM F.T. (1949) Comparative study of decomposition rates of organic matter in temperate and tropical regions. *Soil Sci.* **68**, 419–432.

JOHNS G.G., NICOL G.R. and WATKIN B.R. (1965) A modified capacitance probe for estimating pasture yield. *J. Br. Grassld. Soc.* **20**, 212–217.

KELLY O.J., HARDMAN J.A. and JENNINGS D.S. (1948) A soil sampling machine for obtaining two-, three-, and four-inch diameter cores of undisturbed soil to a depth of six feet. *Proc. Soil Sci. Soc. Am.* **12**, 85–87.

KENWORTHY J.B. (1963) Temperatures in heather burning. *Nature (Lond.)* **200**, 1226.

KITTREDGE J. (1944) Estimation of the amount of foliage of trees and stands. *J. For.* **42**, 905–912.

KONONOVA M.M. (1961) *Soil Organic Matter—Its Nature, its Role in Soil Formation and in Soil Fertility.* Pergamon, Oxford.

KÖPF H. (1952) Laufende Messungen der Bodenatmung im Freiland. *Landw. Forsch.* **4**, 186–194.

KOSONEN M. (1968) The relation between the carbon dioxide production in the soil and the vegetation of a dry meadow. *Oikos* **19**, 242–249.

KOSONEN M. (1969) Carbon dioxide production in relation to plant mass. *Oikos* **20**, 335–343.

KROGH A. (1914) The quantitative relation between temperature and standard metabolism in animals. *Int. Z. phys.-chem. Biol.* **1**, 491–508.

KUCERA C.L. and KIRKHAM D.R. (1971) Soil respiration studies in tallgrass prairie in Missouri. *Ecology* **52**, 912–915.

LATTER P.M. and CRAGG J.B. (1967) The decomposition of *Juncus squarrosus* leaves and micro-biological changes in the profile of *Juncus* moor. *J. Ecol.* **55**, 465–482.

LATTER P.M., CRAGG J.B. and HEAL O.W. (1967) Comparative studies on the microbiology of four moorland soils in the northern Pennines. *J. Ecol.* **55**, 445–464.

LEE J.A. and WOOLHOUSE H.W. (1966) A reappraisal of the electrometric method for the determination of the concentration of carbon dioxide in soil atmospheres. *New Phytol.* **65**, 325–330.

LIETH H. (1968) The measurement of calorific values of biological material and the determi-nation of ecological efficiency. In *Functioning of Terrestrial Ecosystems at the Primary Production Level* (ed. F.E. Eckardt), pp. 233–242. Proc. Copenhagen Symposium, UNESCO.

LIETH H. (1968) The determination of plant dry matter production with special emphasis on the underground parts. In *Functioning of Terrestrial Ecosystems at the Primary Production Level* (ed. F.E. Eckardt), pp. 179–184. Proc. Copenhagen Symposium, UNESCO.

LIETH H. (1970) Phenology in productivity studies. In *Analysis of Temperate Forest Ecosystems* (ed. D.E. Reichle), pp. 29–46. Springer-Verlag, Berlin.

LOMNICKI A., BANDOLA E. and JANKOWSKA K. (1968) Modification of the Wiegert-Evans method for estimation of net primary production. *Ecology* **49**, 147–149.

MACFADYEN A. (1970) Simple methods for measuring and maintaining the proportion of carbon dioxide in air, for use in ecological studies of soil respiration. *Soil Biol. Biochem.* **2**, 9–18.

MACFADYEN A. (1971) The soil and its total metabolism. In *Methods of Study in Quantitative Soil Ecology* (ed. J. Phillipson), pp. 1–13. IBP Handbook No. 18. Blackwell Scientific Publications, Oxford.

MCKELL C.M., WILSON A.M. and JONES B.M. (1961) A flotation method for easy separation of roots from soil samples. *Agron. J.* **53**, 56–57.

MADGWICK H.A.I. and OVINGTON J.D. (1959) The chemical composition of precipitation in adjacent forest and open plots. *Forestry* **32**, 14–22.

MAJOR J. (1969) Historical development of the ecosystem concept. In *The Ecosystem Concept in Natural Resource Management* (ed. G.M. van Dyne), pp. 9–22. Academic Press, New York.

MALLOCH A.J.C. (1972) Salt-spray deposition on the maritime cliffs of the Lizard peninsula. *J. Ecol.* **60**, 103–112.

MALONE C.R. and SWARTOUT M.B. (1969) Size, mass and caloric content of particulate organic matter in old-field and forest soils. *Ecology* **50**, 395–399.

MARK A.F. (1965) The environment and growth rate of narrow-leaved snow tussock, *Chionochloa rigida*, in Otago. *N.Z. J. Bot.* **3**, 73–103.

MARTIN M.H. and PIGOTT C.D. (1965) A simple method for measuring carbon dioxide in soils. *J. Ecol.* **53**, 153–155.

MEDWECKA-KORNÁS A. (1971) Plant litter. In *Methods of Study in Quantitative Soil Ecology* (ed. J. Phillipson), pp. 24–33. IBP Handbook No. 18. Blackwell Scientific Publications, Oxford.

MILNER C. and HUGHES R.E. (1968) *Methods for the Measurement of the Primary Production of Grassland.* IBP Handbook No. 6. Blackwell Scientific Publications, Oxford.

MINDERMAN G. (1956) The preparation of microtome section of unaltered soil for the study of soil organisms *in situ. Pl. Soil* **8**, 42–48.

MINDERMAN G. (1968) Addition, decomposition and accumulation of organic matter in forests. *J. Ecol.* **56**, 355–363.

MOTT G.O., BARNES R.F. and RHYKERD C.L. (1965) Estimating pasture yield *in situ* by beta-ray attenuation techniques. *Agron. J.* **57**, 512–513.

MOUNTFORD M.D. and BUNCE R.G.H. (1973) Regression sampling with allometrically related variables, with particular reference to production studies. *Forestry* **46**, 203–212.

MOORE N.W. (1962) The heaths of Dorset and their conservation. *J. Ecol.* **50**, 369–391.

MURPHY P.W. (1962) A radioisotope method for determination of rate of disappearance of leaf litter in woodland. In *Progress in Soil Zoology* (ed. P.W. Murphy), pp. 357–363. Butterworths, London.

NEWBOULD P.J. (1967) *Methods for Estimating the Primary Production of Forests.* IBP Handbook No. 2. Blackwell Scientific Publications, Oxford.

NEWMAN E.I. (1966) A method of estimating the total length of root in a sample. *J. appl. Ecol.* **3**, 139–146.

NEWTON J.D. (1923) Measurement of carbon dioxide evolved from the roots of various crop plants. *Scient. Agric.* **4**, 268–274.

NIELSON J.A. JR. (1964) Autoradiography for studying individual root systems in mixed herbaceous stands. *Ecology* **45**, 644–646.

NIHLGÁRD B. (1970) Precipitation, its chemical composition and effect on soil water in a beech and a spruce forest in south Sweden. *Oikos* **21**, 208–217.

NIMLOS T.J., VAN METER W.P. and DANIELS L.A. (1968) Rooting patterns of forest understory species as determined by radioiodine absorption. *Ecology* **49**, 1146–1151.

ODUM E.P. (1960) Organic production and turnover in old-field succession. *Ecology* **41**, 34–49.

OLSON J.S. (1958) Rates of succession and soil changes on southern Lake Michigan sand dunes. *Bot. Gaz.* **119**, 125–170.

OLSON J.S. (1963) Energy storage and the balance of producers and decomposers in ecological systems. *Ecology* **44**, 322–331.

OLSON J.S. (1964) Gross and net production of terrestrial vegetation. *J. Ecol.* **52** (Suppl.), 99–118.

OLSON J.S. and CROSSLEY D.A. JR (1963) Tracer studies of the breakdown of forest litter. In *Radioecology* (eds V. Schultz and A.W. Klements), pp. 411–416. Reinhold, New York.

OVINGTON J.D. and MADGWICK H.A.I. (1959) The growth and composition of natural stands of birch. 1. Dry matter production. *Pl. Soil.* **10**, 271–283.

OVINGTON J.D. and MURRAY G. (1968) Seasonal periodicity of root growth of birch trees. In *Methods of Productivity studies in Root Systems and Rhizosphere Organisms* (ed. M.S. Ghilarov), pp. 146–161. Leningrad, USSR Acad. Sciences.

PARIZEK R.P. and LANE B.E. (1970) Soil-water sampling using pan and deep pressure-vacuum lysimeters. *J. Hydrol.* **11**, 1–21.

PARKINSON D. and COUPS E. (1963) Microbial activity in a podzol. In *Soil Organisms* (eds. J. Doeksen and J. van der Drift), pp. 167–175. North Holland Publishing Co., Amsterdam.

PERRIN R.M.S. (1965) The use of drainage water analyses in soil studies. In *Experimental Pedology* (eds E.G. Hallsworth and D.V. Crawford), pp. 73–96. Butterworths, London.

PHILLIPSON J. (1964) A miniature bomb calorimeter for small biological samples. *Oikos* **15**, 130–139.

PHILLIPSON J. (1971) Other Arthropods. In *Methods of Study in Quantitative Soil Ecology* (ed. J. Phillipson), pp. 262–287. IBP Handbook No. 18. Blackwell Scientific Publications, Oxford.

RANDALL R.E. (1970) Salt measurement on the coasts of Barbados, West Indies. *Oikos* 21, 65–70.

RAWES M. and WELCH D. (1969) Upland productivity of vegetation and sheep at Moor House National Nature Reserve, Westmorland, England. *Oikos Suppl.* 11, 7–72.

REICHLE D.E. (ed) (1970) *Analysis of Temperate Forest Ecosystems.* Springer-Verlag, Berlin.

REICHLE D.E., O'NEILL R.V., KAYE K.V., SOLLINS P. and BOOTH R.S. (1973) Systems analysis as applied to modeling ecological processes. *Oikos* 24, 337–343.

REINERS W.A. (1968) Carbon dioxide evolution from the floor of three Minnesota forests. *Ecology* 49, 471–483.

REMEZOV H.P., RODIN L.E. and BAZILEVICH N.I. (1963) Instructions on methods of studying the biological cycle of ash elements and nitrogen in the above ground parts of plants in the main natural zones of the temperate belt (Russian). *Bot. Zh. S.S.S.R.* 48, 869–877.

RICHARDS F.J. (1959) A flexible growth function for empirical use. *J. Exp. Bot.* 10, 290–300.

RICHMAN S. (1968) The transformation of energy by *Daphnia pulex*. *Ecol. Monogr.* 28, 273–291.

RODIN L.E. and BAZILEVICH N.I. (1967) *Production and Mineral Cycling in Terrestrial Vegetation.* Oliver and Boyd, Edinburgh.

ROGERS W.S. and HEAD G.C. (1968) Studies of roots of fruit plants by observation panels and time-lapse photography. In *Methods of Productivity Studies in Root Systems and Rhizosphere Organisms* (eds. M.S. Ghilarov *et al.*), pp. 176–185. Leningrad, USSR Acad. Sciences.

ROSS D.J. and BOYD I.W. (1970) Influence of moisture and aeration on oxygen uptakes in Warburg respiratory experiments with litter and soil. *Pl. Soil* 33, 251–256.

RUTTER A.J. and WEBSTER J.R. (1962) Probes for sampling ground water for gas analysis. *J. Ecol.* 50, 615–618.

SALISBURY E.J. (1922) The soils of Blakeney Point: A study of soil reaction and succession in relation to plant covering. *Ann. Bot.* 36, 391–431.

SALISBURY E.J. (1925) Note on the edaphic succession in some sand dune soils with special reference to the time factor. *J. Ecol.* 13, 322–328.

SALT G. and HOLLICK F.S.J. (1944) Studies of wireworm populations. 1. A census of wireworms in pasture. *Ann. appl. Biol.* 31, 52–64.

SATOO T. (1970) A synthesis of studies by the harvest method: primary production relations in the temperate deciduous forests of Japan. In *Analysis of Temperate Forest Ecosystems* (ed. D.E. Reichle), pp. 55–72. Springer-Verlag, Berlin.

SATOR CH. and BOMMER D. (1971) Methodological studies to distinguish functional from nonfunctional roots of grassland plants. In *Integrated Experimental Ecology* (ed. H. Ellenberg), pp. 72–78. Springer-Verlag, Berlin.

SCHÖLLHORN R. and BURRIS R.H. (1967) Acetylene as a competitive inhibitor of nitrogen fixation. *Proc. natn. Acad. Sci. USA.* 58, 213–216.

SCHUSTER J.L. (1964) Root development of native plants under three grazing intensities. *Ecology* 45, 63–70.

SCHUURMAN J.J. and GOEDEWAAGEN M.A.J. (1971) *Methods for the Examination of Root Systems and Roots* (2nd edn). Centre for Agricultural Publications and Documentation, Wageningen.

SCOTT D. (1961) Methods of measuring growth in short tussocks. *N.Z. J. agric. Res.* 4, 282–285.

SHANKS R.E. and OLSON J.S. (1961) First year breakdown of leaf litter in southern Appalachian forests. *Science, N.Y.* 134, 194–195.

SHEIKH K.H. and RUTTER A.J. (1969) The responses of *Molinia caerulea* and *Erica tetralix* to soil aeration and related factors. 1. Root distribution in relation to soil porosity. *J. Ecol.* 57, 713–726.

SMITH F.E. (1970) Analysis of ecosystems. In *Analysis of Temperate Forest Ecosystems* (ed. D.E. Reichle), pp. 7–18. Springer-Verlag, Berlin.

SMITH O.L. (1982) *Soil Microbiology: a Model of Decomposition and Nutrient Cycling.* CRC Press Inc., Florida.

SOUTHWOOD T.R.E (1966) *Ecological Methods: with Particular Reference to the Study of Insect Populations.* Methuen, London.

STAGE A.R. (1963) Specific gravity and tree weight of single-tree samples of Grand Fir. *US For. Serv. Res. Pap. Intermt. For. Range Exp. Stn. No.* INT-4, pp. 11.

STEWART W.D.P., FITZGERALD G.P. and BURRIS R.H. (1967) *In situ* studies on N_2 fixation using the acetylene reduction technique. *Proc. Acad. Sci. USA* **58**, 2071–2078.

SWIFT M.J., HEAL O.W. and ANDERSON J.M. (1979) *Decomposition in Terrestrial Ecosystems* Studies in Ecology, Vol. 5. Blackwell Scientific Publications, Oxford.

SYKES J.M. and BUNCE R.G.H. (1970) Fluctuations in litter fall in a mixed deciduous woodland over a three-year period 1966–68. *Oikos* **21**, 326–329.

TANSLEY A.G. (1935) The use and abuse of vegetational concepts and terms. *Ecology* **16**, 284–307.

UENO M., YASHIHARA K. and OKADA T. (1967) Living root system distinguished by the use of Carbon-14. *Nature (Lond.)* **213**, 530–532.

WAGNER G.H. (1962) Use of porous ceramic pots to sample soil water within the profile. *Soil Sci.* **94**, 379–386.

WALTERS C.S. and BRUCKMANN G. (1964) A comparison of methods for determining volume of increment cores. *J. For.* **62**, 172–177.

WANNER H. (1970) Soil respiration, litter fall and productivity of tropical rain forest. *J. Ecol.* **58**, 543–547.

WARREN WILSON J. (1960) Inclined point quadrats. *New Phytol.* **59**, 1–8.

WARREN WILSON J. (1963) Estimation of foliage denseness and foliage angle by inclined point quadrats. *Aust. J. Bot.* **11**, 95–105.

WELBANK P.J. and WILLIAMS E.D. (1968) Root growth of a Barley crop estimated by sampling with portable powered soil-coring equipment. *J. appl. Ecol.* **5**, 477–482.

WENT J.C. (1959) Cellophane as a medium to study cellulose decomposition in forest soils. *Acta bot. neerl.* **8**, 490–491.

WESTLAKE D.F. (1963) Comparisons of plant productivity. *Biol. Rev.* **38**, 385–425.

WHITE E.J. and TURNER F. (1970) A method of estimating income of nutrients in a catch of airborne particles by a woodland canopy. *J. appl. Ecol.* **7**, 441–461.

WHITTAKER E. (1961) Temperatures in heath fires. *J. Ecol.* **49**, 709–716.

WHITTAKER R.H. and WOODWELL G.M. (1968) Dimension and production relations of trees and shrubs in the Brookhaven Forest, New York. *J. Ecol.* **56**, 1–25.

WIANT H.V. (1967a) Influence of moisture content on soil respiration. *J. For.* **65**, 902–903.

WIANT H.V. (1967b) Influence of temperature on the rate of soil respiration. *J. For.* **65**, 489–490.

WIEGERT R.G. and EVANS F.C. (1964) Primary production and the disappearance of dead vegetation on an old field. *Ecology* **45**, 49–63.

WILLIAMSON P. (1976) Above-ground primary production of chalk grassland allowing for leaf death. *J. Ecol.* **64**, 1059–1075.

WILSON K. (1960) The time factor in the development of dune soils at South Haven peninsula, Dorset. *J. Ecol.* **48**, 341–360.

WITKAMP M. (1966) Rates of carbon dioxide evolution from the forest floor. *Ecology* **47**, 492–494.

WITKAMP M. (1969) Cycles of temperature and carbon dioxide evolution from litter and soil. *Ecology* **50**, 922–924.

WITKAMP M. and FRANK M.L. (1969) Evolution of carbon dioxide from litter, humus and subsoil of a Pine stand. *Pedobiologia* **9**, 358–366.

WITKAMP M. and OLSON J.S. (1963) Breakdown of confined and non-confined Oak litter. *Oikos* **14**, 138–147.

WOODS F.W. (1970) Interspecific transfer of inorganic materials by root systems of woody plants. *J. appl. Ecol.* **7**, 481–486.

WOODS F.W. and BROCK K. (1964) Interspecific transfer of Ca-45 and P-32 by root systems. *Ecology* **45**, 886–889.

WOODS F.W., FERRILL M.D. and McCORMACK M.L. (1962) Methyl bromide for increasing Iodine-131 uptake by Pine trees. *Radiat. Bot.* **2**, 273–277.

YAMAGUCHI S. and CRAFTS A.S. (1958) Autoradiographic method for studying absorption and translocation of herbicides using Carbon-14 labelled compounds. *Hilgardia* **28**, 161–191.
YOUNG H.E. (Ed) (1968) *Symposium on Primary Production and Mineral Cycling in Natural Ecosystems.* University of Maine Press.

2 Faecal analysis and exclosure studies

R. BHADRESA

1 Faecal analysis

The identification of grazing preferences of herbivores can lead to a better understanding of the ecological balance and successional development of plant communities, especially where grazing animals are important consumers of primary productivity. This can directly aid in management and conservation of particular plant communities. One method by which such preferences can be studied, especially in mammalian herbivores, is faecal analysis and the aim of this chapter is to outline the basic methods employed in faecal analysis in order to obtain a qualitative and a quantitative assessment of the diet of herbivorous mammals. The advantages and the disadvantages are also considered.

Faecal analysis consists of the identification of plant fragments in faeces and is a useful technique for ascertaining the diet of herbivorous mammals especially since other methods, e.g. direct observation, analysis of stomach contents and/or fistulation, are often either impractical or inappropriate. The main advantage of faecal analysis is that the same population of herbivores can be continuously sampled without direct interference. The method has been successfully used by a number of workers to identify the food items of a range of different species of animals; for example, marsupials (Storr, 1961; Griffiths & Barker, 1966; Dunnet et al., 1973), rodents (Baumgartner & Martin, 1939; Brand, 1978), lagomorphs (Dusi, 1949; Hansen et al., 1969, 1973; Oldham, 1972; Williams et al., 1974; Bhadresa, 1977, 1981, 1982; Chapuis, 1979; Chapuis & Lefeuvre, 1980), wild ungulates (Stewart, 1967; Hansen & Reid, 1975; Hansen et al., 1977) and domestic ungulates (Martin, 1955, 1964; Hercus, 1960; Griffiths & Barker, 1966; Grant & Campbell, 1978). Various methods have thus appeared in the literature for preparing reference material and examining faecal samples.

Because of the uniqueness of each plant species' epidermal pattern and the retention, after digestion, of the cuticle in faeces, faecal analysis is useful in both qualitative and quantitative determinations of herbivore diets. The technique depends upon thorough observation of a reference collection of the cuticles of plant species found on a study area. A list of characteristics should be drawn up for each species and, on the basis of this list, a key can be constructed to help with identification of cuticles present in faecal samples. The collected faeces are subjected to a non-destructive separation of the epidermal fragments, which are then subsampled to allow small representative portions to be analysed under the microscope. Since it is useful to know not only what species are present in the faeces but also the proportions of the various species, a method is used to quantify these proportions. These proportions, however, might not reflect the actual proportions of plant species consumed because of differences in the surface area of epidermis per unit weight of particular species and, secondly, because of differential digestion of species consumed (Stewart, 1967). These, the major disadvantages of faecal analysis, however, can be effectively tackled by conducting controlled feeding trials on the herbivore concerned to ascertain relationships between the amounts of food plants consumed to proportions of cuticles of these plants in the faeces.

1.1   Methods of analysis

1.1.1   Making a reference collection of epidermal types

The leaf cuticle of potential food-plant species may be peeled off, scraped off with a blade or, in cases where the leaves are very small, gently macerated in 10% nitric acid. The epidermises may then be mounted in glycerol jelly on microscope slides. For making permanent stained mounts, the slow process of alcohol dehydration and final replacement by xylol and mounting in Canada Balsam or Euparol is necessary. Staining in safranin, acid fuchsin or gentian violet is conducted at the 70% alcohol stage.

The leaf epidermis is best studied by making drawings and/or taking photomicrographs and noting down the size and shape of various characteristics, e.g. epidermal cells, cell walls, stomata, trichomes and silica bodies. The stomatal patterns (amocytic, anisocytic, etc.), density of stomata and length to breadth ratios of epidermal cells can often be useful. The general works on plant anatomy by Metcalfe & Chalk (1950) and Esau (1977) are useful in this respect. Where possible, both the abaxial and adaxial surfaces of leaves should be observed. These characteristics can then be used to construct a key to aid the identification of epidermal fragments in faeces.

1.1.2   Collection of faecal samples

It is useful to know the time at which the faeces collected for analysis were deposited so that a comparison between the proportions of leaf epidermal fragments in faeces and the vegetation on offer at a particular time can be made. In the case of domestic herbivores this is simple because defecation can be observed, but for wild herbivorous mammals it is necessary, prior to collection, to clear plots of all previous faeces, thus ensuring the subsequent collection of fresh faeces only. The size of these plots will depend upon the size and habits of the herbivore concerned; for example, 4 × 4 m squares are quite adequate for rabbits but larger areas need to be cleared for larger mammals. To reduce edge effects and ensure collection of only those faeces deposited since the plots were cleared, it is advisable not to take faeces from within say a metre of the outer margin of the cleared plots. This method of collection is particularly applicable to periodic (weekly, monthly) collection of faeces for assessing seasonal changes in the diet of the herbivore. A knowledge of the colour changes associated with ageing and weathering of faeces can also be used for the estimation of the degree of freshness or otherwise of faeces, but this is less satisfactory. In this respect, a soil Munsell colour chart is useful in describing colour of faeces. Since areas are often inhabited by more than one species of herbivore, a knowledge of the characteristics of faeces of particular species is essential for accurate identification.

The collected faeces are dried and stored in a dry place until required for analysis. Faeces can survive for long periods in the dried state without deterioration. The faeces may also be stored in a formalin–acetic acid–alcohol mixture in the ratio 10 : 5 : 50 : 35 (of water) respectively.

1.1.3   Measurement of the size of faeces

Differences in the size of faeces may provide a useful way of comparing the diets of juvenile and adult animals and changes in, for example, the monthly frequencies of different-sized droppings may reflect changes in the age structure of the population of herbivores. Bhadresa (1981, 1982) ascertained the change in the size of the droppings with age in the rabbit (see Fig. 2.1) and used the relationship to distinguish between rabbit droppings belonging to juvenile and adult rabbits.

An indication of size may be obtained either by weighing or by estimating the volume of dried individual faeces. It is also useful to calculate the weight to volume ratio (or density) of faeces. Bhadresa (1982), for instance, has found that in the case of the European rabbit (*Oryctolagus cuniculus* L.), a fibrous and/or a monocotyledonous diet produces less dense faecal pellets while a dicotyledonous diet produces denser pellets. The density of faeces may thus provide a useful insight into the dietary habits of herbivores.

1.1.4   Preparation of faeces for analysis

Various techniques have been used in the separation of epidermal fragments in faecal samples, the simplest one being that described by Croker (1959). He dispersed the faeces in water and after subsampling, counted the epidermal fragments under a microscope. Zyznar & Urness (1969) recom-

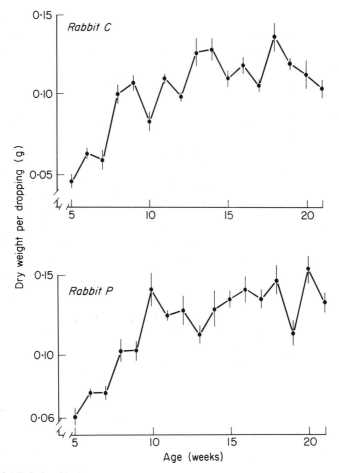

Figure 2.1 Relationship between dry weight per dropping and age, for two domestic New Zealand White rabbits; 95% confidence limits are presented.

mend soaking of the faeces in 10% sodium hydroxide to remove mucous coating to aid dispersal of the fragments. The method developed by Williams (1969) includes dispersing the faecal samples in 70% alcohol to extract the chlorophyll and placing them in boiling water and sodium hypochlorite. Stewart (1967), on the other hand, used nitric acid and potassium hydroxide for dispersing the epidermal fragments. The method described here in detail is a modification of Stewart's method and has been used extensively by the author with rabbit faeces. The method described below applies to small (*c.* 0·1 g each) droppings. The amounts of reagents used should be proportionately altered to accommodate larger or smaller faeces. For very large faeces, it is more practical to use smaller subsamples.

The dried faecal sample is inserted into a 50 ml conical flask and soaked in 2 ml distilled water, followed after 5 minutes by 2 ml concentrated nitric acid. The acid helps to loosen the faecal sample by dissolving away any mucous coating that might be present. The flask is warmed in a water bath set at 50–60°C for 5 minutes and dispersal of the fragments is achieved by gentle teasing with a glass rod; 10 ml of 1 M potassium hydroxide are then added to neutralize the acid and stop further reaction. The flask containing the fragments is then attached to a flask shaker and, using the wide end of a pasteur pipette, a 1 ml sample is removed while the flask is being shaken so that the subsample is representative of the entire faecal sample (for estimating absolute amounts of different epidermal fragments in a faecal sample, it would be necessary to remove a precise quantity, but for estimating relative proportions of the different epidermal types, the size of the subsample is not crucial). The subsample is placed in a crucible and, after a settling period of about 5 minutes, the supernatant is removed using a pasteur pipette. Distilled water (2–3 ml) is then added to wash the fragments; 2–3 drops are left with the sample while the rest is pipetted off. Using a spatula, the fragments are scraped off onto a labelled microscope slide, spread out to the size of a coverslip and allowed to dry. The drying may be assisted by the use of a slide-drying oven or plate, and after they have dried, they are mounted in glycerol jelly. If the density of fragments on the slide is very high and there is much overlapping of fragments, it is necessary to dilute the sample and remove a dilute sample for analysis.

1.1.5   Identification of the fragments in the faecal subsamples

Identification of the cuticular fragments in the droppings is carried out with the aid of illustrations, microphotographs and a key of the reference collection of cuticles. In most circumstances × 100 magnification is adequate for this purpose but, for critical assessments, a higher magnification is sometimes necessary. The majority of fragments on the slide can be identified down to the species level. However, in some cases it is not possible to do this and on the basis of, e.g. the epidermal pattern of random arrangment of isodiametric epidermal cells or long, narrow cells placed parallel to each other, fragments may be classified as belonging to a dicotyledonous or monocotyledonous species respectively. Similar criteria—namely, small cells with thick, smooth, straight cell walls—can be used to group the mosses. The faecal sample may yet contain fragments that cannot be categorized, including those that are either too deteriorated to show any cuticular pattern or those that are hardly digested, thus making them opaque. They also include fragments of fibrous and vascular plant tissues. This disadvantage, however, cannot be overcome and it must be assumed that the unidentifiable fragments represent similar proportions of species in the identifiable portion of the sample.

1.1.6   Quantification of the fragments in the subsamples

Basically, four methods are available for quantifying proportions of different cuticles in faeces (see Bhadresa, 1981).

(a) Counting the number of fragments in the entire subsample.

(b) Frequency counts in microscope fields, at a particular magnification.

(c) Direct estimation of surface areas of fragments.

(d) Estimation of surface areas of fragments using the point quadrat principle.

There is usually a great variation in the size of fragments in the faecal samples. Stewart (1967) and Bhadresa (1974) have shown that different species break down to different fragment sizes and for this reason, numerical counts of fragments would be biased towards species breaking down into smaller fragments. Frequency counts also suffer from similar considerations and although they indicate relative importance of different cuticles in the faeces, they do not reflect the true proportions of cuticles present. Direct estimation of surface areas of fragments is tedious and time-consuming since fragments are variously shaped (Stewart, 1967), and is therefore impractical in a study that requires analysis of large quantities of faeces. Estimation of surface areas of fragments using the point quadrat principle is considered the best method for assessing proportions of different cuticles in faecal samples. The method bears close similarities to point quadrats employed in vegetation analysis; however, because of the two-dimensional, static nature of the material for examination, faecal analysis using point quadrats is less tedious and more accurate than analysing vegetation in the same way.

The point quadrat is placed in the eyepiece and may take the form of a micrometer/crosswires or one made specially by marking a round coverslip with a grid of equidistant dots. Bhadresa (1982), for example, used one with nine dots placed in a square grid with dots 5 mm apart from each other. The size of the dots (0·5 mm) correlated with the size of the smallest identifiable fragment at × 100 magnification. When recording point-quadrat 'hits', the points (dots) are orientated parallel to the sides of the microscope slide. Hits on cuticles belonging to different species are recorded in straight line transects across the slide or by random location of microscopic fields. To reduce edge effects, a margin of > 2 mm of the coverslip is avoided at the time of recording and the distance between microscopic fields in the case of straight-lined transects, is kept such that there is no overlap between consecutive microscopic fields.

Tests should be carried out to determine the minimum number of hits required for estimating percentage cover of different cuticles; the number of hits required depending on the diversity of species in the sample. Running percentage means are recorded at, for example, 10-hits intervals (for a number of samples) until the values of the major components become stable.

The percentage means of species with sporadic occurrences may continue to vary; however, a compromise is needed when deciding upon the number of hits required, so that estimates of percentage cover of different cuticles give a reasonable representation of the rarer components and a good representation of the commoner ones.

The methods outlined above should enable accurate estimates to be made of the proportions of cuticles in faeces but these may not necessarily represent the actual proportions of food species consumed. The amount of cuticle of any particular species in the faeces depends on two major factors: (i) the surface area of epidermis per unit weight of species; (ii) the amount of cuticular material surviving digestion. In order to overcome these disadvantages, it is necessary to conduct controlled feeding trials on the herbivore concerned to ascertain a relationship between the amounts of food plant consumed to proportions of cuticles of these plants in the faeces.

1.2   Feeding trials

It must be realized that feeding trials are imperative in any serious study involving the use of faecal analysis. The quantitative results obtained become more meaningful only when the proportions of plant species in faeces can be related to the actual proportions consumed by the herbivore. In this way, when the population density of the herbivore and its daily food requirements are known, it is also possible to determine the amount of primary production of a particular species taken up by the herbivore. When dealing with a selective herbivore, such measures would be crucial in the management and conservation of the selected species or in controlling the spread of non-selected ones. In 'fragile' habitats (e.g. semi-deserts) used by man for his domestic herbivores, such measures would be pertinent in determining the true 'carrying-capacity' of the habitat. The feeding trials can also provide information on the survival of the epidermis of particular species, the time of throughput (the time between ingestion and appearance in faeces) and dry matter digestibilities of the different food plants for a particular species of herbivore.

1.2.1   Housing

Depending on its size, the herbivore may be caged indoors or in an outside enclosure where its only source of food is that supplied. It is important that external conditions be kept as constant as possible between feeding trials with different feed species. It is also necessary to conduct feeding trials on more than one animal concurrently to identify the variation between individuals.

1.2.2   Feeding experiments

Feeding experiments should be conducted by feeding only two food species at a time. A standard food species is usually adopted, and in each feeding experiment this is fed together with one of the other species. It is then possible to relate the proportions of all the species to the chosen standard and hence to each other. Choice of the standard species may be governed by abundance in the study area. Also, all the species selected for feeding experiments should, where possible, have clearly identifiable epidermal characteristics so that misidentifications do not affect the results. Duplications of the same feeding experiment are an important safeguard.

1.2.3   Collection of plant feeds

Depending on how long a feeding trial is to last, the amounts of feed collected should supply the full requirement of a particular feeding experiment, and preferably should be restricted to the current year's growth. Collection should be made separately for each species immediately prior to the feeding trial, the material being stored in black polythene bags and refrigerated until required. This retains the freshness of the collected material by reducing water loss for the duration of a feeding trial.

1.2.4   Details of a feeding experiment

The mammalian herbivore is kept on a manufacturer's diet of pellet food or on a species with a readily identified epidermal pattern between feeding trials. This staple diet is removed a day before the start of a feeding trial to ensure that the herbivore consumes all of the feed species supplied on the day after. Throughout the duration of a feeding trial, the amounts of feed supplied should be kept slightly lower than the normal requirement of the herbivore to avoid the problem of preferential feeding.

Equal amounts (in terms of fresh weight) of the two feeds—one the standard and the other the test feed—are supplied, e.g. each morning to each individual herbivore. It may not be necessary to collect faeces on the first day of the feed, but on subsequent days they can be collected at the same time as the feeds are given. Collection of faeces, however, should be continued for a minimum of 3 days after the test feed has been stopped and the herbivore is back on the staple diet.

A sample of all the feeds used in the feeding experiments should always be removed for the determination of dry weight. Dry weight of the collected droppings should also be determined. Thereafter, the faeces from different collections are stored until required for analysis. The same technique as described earlier should be used for faecal analysis.

The results derived from the feeding experiments may now be used to convert proportions of different cuticles in faeces collected in the wild to proportions consumed, in terms of both fresh and dry weights.

2   Exclosure studies

The main effect of grazing on the vegetation is to change the course of succession. Directly it can lead to an increase in unselected species and, by keeping the vegetation down, favour the spread of hemicryptophytes. Conversely, grazing can increase the competitive abilities of selected species (e.g. increased shoot and root growth) to the detriment of unselected ones. Vegetation may also be affected by trampling and faecal deposition. The sum effect of these various factors on the plant community can more readily be understood by erecting small exclosures designed to keep grazing animals out.

2.1   Methods

2.1.1   Siting of exclosures

Exclosures should be erected in fairly homogeneous areas, in terms of the vegetation, soil characteristics and topography. This is to ensure that future differences in the vegetation between grazed and ungrazed areas are mainly attributable to grazing. Similar-sized grazed plots should be marked out at the same time for comparative studies.

2.1.2   Design of exclosures

The size of the exclosures and the type of fencing used will depend upon the kind of herbivorous mammal from which the vegetation is being protected. Thus, for large mammals (e.g. deer, sheep, cows) barbed wire fences could be adequate, and the size of exclosures can vary in the range of 10–100 × 10–100 m squares or rectangles. Stakes for securing the barbed wire should be erected at 5 m intervals.

For smaller mammals, e.g. rabbits, exclosures need not be too large and 5 × 5 m square plots are adequate. The type of fencing used is, however, very important and wire-netting is recommended with a mesh size of 3–4 cm for rabbits and 0·5 cm for voles and mice. Many small mammals are habitual burrowers and therefore the wire-netting should be buried in the soil. An exclosure of this type is illustrated in Fig. 2.2.

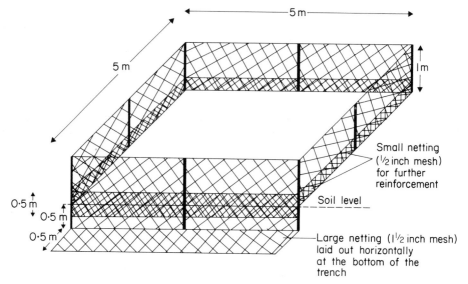

Figure 2.2 A rabbit exclosure showing the construction design with the front side uncovered.

2.1.3 Vegetation recording

Vegetation recording should be carried out at regular intervals and at similar times to assess long-term (e.g. annual) changes. Various methods may be used to record the vegetation and often a combination of methods may have to be used. It is worth remembering that to avoid edge effects such as rain splash and bird droppings, recording should not be carried out within a margin (of say 1 m) immediately inside and adjacent to the fencing of the exclosures. Various methods are described in Chapter 9.

3   References

Baumgartner L.L. and Martin A.C. (1939) Plant histology as an aid in squirrel food habits studies. *J. Wildl. Mgmt.* **3**, 266–268.

Bhadresa R. (1974) *Food preferences of rabbits*, Oryctolagus cuniculus (*L.*) *on Holkham sand-dunes*. M.Sc. thesis, King's College, University of London.

Bhadresa R. (1977) Food preferences of rabbits (*Oryctolagus cuniculus* L.) at Holkham sand dunes, Norfolk. *J. Appl. Ecol.* **14**, 287–291.

Bhadresa R. (1981). Identification of leaf epidermal fragments in rabbit faeces (with reference to heathland vegetation). Rogate Papers No. 4, King's College, London, pp. 27.

Bhadresa R. (1982) *Plant–rabbit interactions on a lowland heath.* Ph.D. thesis, King's College, University of London.

Brand M.R. (1978) A method to correct for differential digestibility in faecal analysis. *Amer. Midland Nat.* **40**, 228–233.

Chapuis J.L. (1979) *Le régime alimentaire du lapin de garenne*, Oryctolagus cuniculus (*L.*) *1758 dans deux habitats contrastés: une lande Bretonne et un domaine de l'Ile de France.* Le titre de Docteur en Trosieme Cycle, University of Rennes, France.

CHAPUIS J.L. and LEFEUVRE J.C. (1980) Evolution saisonniere due regime alimentaire du lapin de garenne, *Oryctolagus cuniculus* (L.), en lande: resultats de deux ans d'analyses. *Bull. Ecol.* **3**, 587–597.

CROKER B.H. (1959) A method of estimating the botanical composition of the diet of sheep. *N.Z. J. agric. Res.* **2**, 72–85.

DUNNET G.M., HARVIE A.E. and SMIT T.J. (1973) Estimating the proportions of various leaves in the diet of the opossum, *Trichosurus vulpecula* Kerr. by faecal analysis. *J. appl. Ecol.* **10**, 737–745.

DUSI J.L. (1949) Methods for the determination of food habits by plant microtechnique and histology and their application to Cottontail rabbit food habits, *J. Wildl. Mngmt.* **13**, 295–298.

ESAU K. (1977) *Anatomy of Seed Plants.* John Wiley, New York.

GRANT S.A. and CAMPBELL D.R. (1978) Seasonal variation in *in vitro* digestibility and structural carbohydrate content of some commonly grazed plants of blanket bog. *J. Br. Grassld. Soc.* **33**, 167–173.

GRIFFITHS M. and BARKER R. (1966) The plants eaten by sheep and by kangaroos grazing together in a paddock in South-eastern Queensland. *CSIRO Wildlf. Res.* **11**, 145–167.

HANSEN R.M., CLARK R.C. and LAWTHORN W. (1977) Foods of wild horses, deer and cattle in the Douglas Mountain Area, Colorado. *J. Range Mngmt.* **30**, 116–118.

HANSEN R.M. and FLINDERS J.T. (1969) Food habits of North American Hares. *Colo. State Univ. Range Sci. Dep. Sci.* Ser. No. 1. pp. 18.

HANSEN R.M., PEDEN D.G. and RICE R.W. (1973) Discerned fragments in faeces indicate diet overlap. *J. Range Mgmt.* **26**, 103–105.

HANSEN R.M. and REID L.O. (1975) Diet overlap of deer, elk and cattle in southern Colorado. *J. Range Mgmt.* **28**, 43–47.

HERCUS B.H. (1960) Plant cuticle as an aid to determining the diet of grazing animals. *Proc. Eighth Int. Grassld. Cong. (Reading)*, 443–447.

MARTIN D.J. (1955) Features of plant cuticle. *Trans. Proc. Bot. Soc. Edinb.*, **36**, 278–288.

MARTIN D.J. (1964) Analysis of sheep diet utilizing plant epidermal fragments in faeces samples. In *Grazing in Terrestrial and Marine Environments*, (ed. D.J. Crisp) BES Symposium No. 4, pp. 173–188. Blackwell Scientific Publications, Oxford.

METCALFE C.R. and CHALK L. (1950) *Anatomy of the Dicotyledons.* Oxford University Press, Oxford.

OLDHAM J. Study mentioned by Wooton R.J. (1972) Notes on the fauna of the Ynys las sand dunes, *Ynys las Nature Reserve Handbook*, NERC, Aberystwyth, pp. 71–81.

STEWART D.R.M. (1967) Analysis of plant epidermis in faeces—a technique for studying the food preferences of grazing herbivores. *J. appl. Ecol.*, **4**, 83–111.

STORR G.M. (1961) Microscopic analysis of faeces, a technique for ascertaining the diet of herbivorous mammals. *Aust. J. biol. Sci.* **14**, 157–164.

WILLIAMS O.B. (1969) An improved technique for identification of plant fragments in herbivore faeces. *J. Range Mgmt.* **22**, 51–52.

WILLIAMS O.B., WELLS T.C.E. and WELLS D.A. (1974) Grazing management of Woodwalton Fen—seasonal changes in the diet of cattle and rabbits. *J. appl. Ecol.* **11**, 499–516.

ZYZNAR E. and URNESS P.J. (1969) Qualitative identification of forage remnants in deer faeces. *J. Wildl. Mgmt.* **33**, 506–510.

3 Water relation and stress

P. BANNISTER

1 Introduction

Over the last decade the methodology of plant water relations has been well served, particularly by the comprehensive volume written by Slavík (1974), but also by the more general book on ecological methods by Kreeb (1977) which devotes much of its content to water relations and stress. The reader is referred to both texts as many methods are dealt with there in greater detail than can be encompassed in a single chapter.

Otherwise, there have been arguably few revolutionary advances in technique over the last 10 years, although the advent of microprocessors has allowed for more sophisticated control. This is particularly apparent in the design of porometers (Section 3.4.1) where the null-point method, with its control of flow rates, has tended to oust other designs; further developments are in the direction of porometers that measure photosynthesis as well as diffusive conductance. Other techniques have become dominant, such as the use of the pressure chamber to measure water potential in the field. Aspects such as these are consequently given a more detailed treatment in this chapter than well-established methods that have undergone little change.

This chapter attempts to cover all the major aspects of the measurement of water in physiological ecology; such as water in the environment; the uptake, transport and loss of water by plants; stomatal control; water content and water potential of tissues; and resistance to drought. The consideration of drought resistance is combined with resistance to other forms of stress, such as temperature, which are not otherwise covered in this volume.

1.1 Water content and water potential

There are two fundamental measures of water status that are commonly made, irrespective of whether water is in the atmosphere, soil, or plant. The first of these is water content, which is merely the amount of water present in a sample, and the second is a measure of the force with which water is held and which is usually expressed as water potential.

For example, atmospheric water content may be expressed as weight per unit volume (e.g. $g\,m^{-3}$) and that of plants and soil as water content per unit dry weight or per unit fresh weight as well as per unit volume. Often it is more convenient to express water content in relative terms; thus, relative humidity is the amount of water in the air expressed as a proportion of that at full saturation, whereas the relative water content of plant tissue is similarly the ratio of field water content to that when the tissue is fully saturated with water.

Although measures of water content indicate the amount of water present they do not necessarily give any indication of its status. Thus, a water content that may represent saturation in a sandy soil could indicate a severe water deficit in a more organic soil. The force with which water is held in a system and the direction of thermodynamic gradients are provided by the concept of water potential (Ψ) (Taylor & Slatyer, 1962; Slatyer, 1967). Water potential is essentially the difference between the chemical potential of water in the system (μ_w) and that of pure, free, water (μ_w^0). As earlier measures of water status had been expressed as equivalent pressures or tensions, water potential is expressed in similar terms by dividing the difference in chemical potential by the partial molal volume of water ($\bar{V}_w$). (This is essentially the volume occupied by one mole of water, which is the ratio of density to molecular weight.), viz.

$$\Psi = \frac{\mu_w - \mu_w^0}{\bar{V}_w} \tag{1}$$

An alternative expression relates the ratio of the water vapour pressure in the system (p) to that of saturated vapour (p_0) as follows

$$\Psi = \frac{RT}{\bar{V}_w} \ln\left(\frac{p}{p_0}\right) \tag{2}$$

where R is the gas constant, $8 \cdot 3143\,\mathrm{J\,mol^{-1}\,K^{-1}}$, and T is the absolute temperature (K).

If $\bar{V}_w$ is expressed in $\mathrm{m^3\,mol^{-1}}$ then the value for Ψ will be in pascals ($\mathrm{J\,m^{-3}} = \mathrm{N\,m^{-2}} = \mathrm{Pa}$) although the values obtained in biological systems are such that megapascals (MPa) or kilopascals (kPa) are more commonly used: (if $\bar{V}_w$ is in $\mathrm{cm^3\,mol^{-1}}$ then Ψ will be in MPa).

As the water potential of pure water is zero (e.g. in equation (2), at saturation $p = p_0$, thus $\ln (p/p_0) = 0$, therefore $\Psi = 0$); most systems have negative water potentials (although positive water potentials can occur in water subjected to increases in pressure or temperature). Water flows from regions of high potential to those of lower potential; normally this is from less negative to more negative potentials although it could be from positive to less positive or negative potentials.

The water potential of plant and soil systems is often subdivided into various components such as a pressure potential (ψ_p, usually positive), a solute potential (ψ_s, usually negative due to the lowering of water potential in the presence of solutes) and a matric potential (ψ_m, representing the force with which water is held in the matrix of the system, such as between soil particles or the fibrils of cell walls). Hence,

$$\Psi = \psi_s + \psi_p + \psi_m \tag{3}$$

Slatyer (1967) has used P to represent the pressure component, π for the solute component and τ for the matric component; the last two are expressed as positive values so that $\pi =$ osmotic pressure rather than the osmotic potential denoted by ψ_s. This usage differs from Taylor & Slatyer (1962) and is therefore confusing although it is sometimes convenient, as in the analysis of $P-V$ curves (Section 4.3). Hence,

$$\Psi = P - \pi - \tau$$

Previous measures of water tension, such as the suction force (*Saugkraft*) of Ursprung & Blum (1916) and the diffusion pressure deficit (DPD) of Meyer (1945) have usually been expressed as positive values (usually as pressures in atmospheres) so that

$$\text{DPD} = \pi + \tau - P$$

A further measure of tension, particularly in soils, has been the pF terminology of Schofield (1935), where tension is expressed as the common logarithm of the height of an equivalent water column in cm.

Because of the long history of terms such as DPD, Tinklin & Weatherley (1968) advocated the use of a water potential deficit ($\Delta\Psi = 0 - \Psi$) which would have a similar positive value. Current use favours Ψ rather than $\Delta\Psi$. Table 3.1 gives a list of conversion factors for various units that have been used to express water potential or its equivalent.

The measurement of water content and water potential in atmospheric, soil and plant systems is dealt with in greater detail in the following sections (Sections 2.1, 2.2, 4.2).

2   Water in the environment

2.1   Atmospheric water

Atmospheric water can exist as vapour, fall as precipitation or condense as fog or dew. All these states may be important to the ecologist; for example, changes in atmospheric moisture are known to have profound effects upon stomatal aperture, whereas rainfall and dewfall are the major inputs to the hydrological budget of a stand of vegetation. As major losses of water occur due to evaporative processes, evaporation is also considered in this section (Section 2.3).

Table 3.1 Numerical equivalences for various expressions of water potential

MPa $\equiv$ J g^{-1}	N m^{-2} = Pa	Bars	Atmospheres	Height of water (m)
1	10^6	10	9·87	101·7
10^{-6}	1	10^{-5}	$9·87 \times 10^{-6}$	$1·017 \times 10^{-4}$
0·1	10^5	1	0·987	10·17
0·1013	$1·1013 \times 10^5$	1·013	1	10.33
$9·83 \times 10^{-3}$	$9·83 \times 10^3$	0·0983	0·097	1

Conversions are at STP and assume 1 cm^3 of water weighs 1 g.

2.1.1   Atmospheric water vapour

This is usually measured as humidity which is commonly expressed either as a concentration (or density) of water in the air (e.g. as g m^{-3}) or as a partial pressure, often as millibars or mmHg but more properly in SI units as Pa or kPa. These measures are of *absolute humidity*: relative humidity (RH) is the ratio of the absolute humidity of the air to that of the air saturated with water vapour at the same temperature and is usually expressed as a percentage. Thus, in partial pressures,

$$\%\text{RH} = (p/p_0) \times 100 \tag{6}$$

The difference between the absolute humidity of the air and fully saturated air at the same temperature is the *atmospheric saturation deficit*; this has been often used as an indication of the potential evaporation from leaves or vegetation. Both the atmospheric saturation deficit and the relative humidity of an isolated volume of air change with temperature, because the water-holding capacity of air changes with temperature while the absolute humidity remains constant. The saturation deficit or the relative humidity will vary with temperature even when one or the other remains constant (Fig. 3.1). Consequently, the potential for evaporation may increase as temperature increases even when relative humidity remains the same. Atmospheric saturation deficits are thus more useful than relative humidities in most plant ecological work but even their use (when expressed in terms of concentration or partial pressures) has been criticized by Cowan (1977) as they are pressure dependent. He suggests the use of mole fraction gradient (see Section 3.4) which is independent of fluctuations in atmospheric pressure such as would occur with changes in altitude.

2.1.2   Measurement of atmospheric humidity

This is commonly calculated from the difference between a wet (T_w) and dry (T_d) temperature sensor. The sensors may be mercury-in-glass

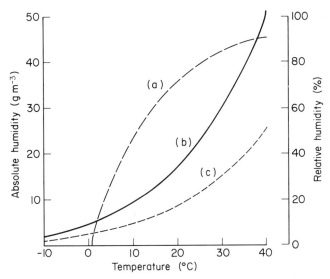

Figure 3.1 Relationships between absolute humidity, relative humidity, saturation deficit and temperature: (a) absolute humidity at saturation (or saturation deficit at 0% RH); (b) absolute humidity or saturation deficit at a constant 50% RH; (c) relative humidity at a constant saturation deficit of $5\,\mathrm{g\,m^{-3}}$.

thermometers, thermocouples, or thermistors and the instruments in which they occur are usually called psychrometers or hygrometers.

$$p_0 - p = \gamma(T_d - T_w) \tag{7}$$

where γ is the psychrometric constant which is sensitive to wind speed but less so at higher wind speeds. Accordingly, most hygrometers are subjected to wind, crudely in the whirling hygrometer (Fig. 3.2), but in a more controlled fashion in standard ventilated psychrometers such as the Assman-type. Where the sensors have a negligible heat capacity (as in thermocouple psychrometers) ventilation is not necessary (Fig. 3.2).

When a body of air is cooled, water (or ice) will condense from it at the temperature when its absolute humidity is equal to the saturation value. The temperature at which this occurs is the *dew point* and this principle is used in dewpoint hygrometers. In simple versions air is cooled until dew condenses on a mirror, but automatic versions have heating and cooling circuits which maintain the mirror at dewpoint temperature. The dewpoint is also employed in 'dewcel' hygrometers where the mirror is replaced by a layer of a salt such as lithium chloride and where there is a direct relationship between the dewpoint temperature and one at which a phase transition occurs in the salt (e.g. Holbo, 1981). Other hygrometers that use lithium chloride (or similar salts such as barium fluoride) rely on the change of electrical conductivity that occurs in the salt as ambient humidity alters (e.g. Jones & Wexler, 1960).

(a)

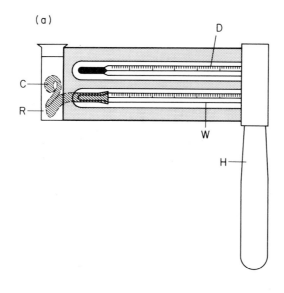

(b)

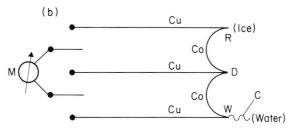

Figure 3.2 Measurement of atmospheric humidity. (a) Diagram of a hand-held whirling hygrometer: D, dry bulb thermometer; W, wet bulb thermometer; H, rotatable handle; C, cotton wick; R, water reservoir. (b) Circuit diagram for a thermocouple hygrometer Cu/Co, copper/constantan thermocouple wire; M, meter; R, reference junction; W, wet junction supplied by cotton wick (C).

Other methods of measuring water vapour include: (i) the use of infra-red gas analysers (Decker & Wien, 1960; Decker *et al.*, 1962), as the absorption of infra-red radiation is proportional to the amount of water in the air; (ii) the microwave hygrometer of Falk (1966) which uses the fact that the dielectric constant of air changes with humidity; (iii) the corona hygrometer of Andersson *et al.* (1954) which measures electrical discharge that is likewise dependent upon ambient humidity. The corona hygrometer is very sensitive to dust and other contaminants whereas the microwave hygrometer is not.

The properties of some materials change with fluctuations in humidity (e.g. human hair, cellophane, paper and polythene) and form the basis of some commonly used instruments for measuring humidity, such as the hair hygrograph (cf. Meteorological Office, London, 1981).

2.1.3 Precipitation

The measurement of precipitation is well-documented in meteorological literature (e.g. Meteorological Office, London, 1980) and also considered in the first edition of this book (Chapter 7.4, Painter 1976).

Rainwater is usually collected in rain gauges in which the amount of rainfall (mm) is calculated from the ratio of the volume collected to the area of the collecting orifice. Manual gauges merely consist of a cylinder containing a funnel leading to a collecting bottle which is emptied periodically into a measuring cylinder. Recording gauges empty themselves at intervals, usually corresponding to a given amount of rainfall. One of the problems of such gauges is that they cannot record any rain that is falling during the tipping cycle so that they are usually run in conjunction with storage gauges. The two most common recording gauges are the tipping bucket and tilting siphon gauges (Painter 1976). Commercial recording gauges may be expensive but a cheap and readily constructed plastic tipping bucket gauge has been recently described (Cornish & Green, 1982).

A major problem with the use of rain gauges lies with their placement rather than their design. This is particularly true of ecological use where gauges are placed on exposed hillsides, under trees, etc., where conditions are far from the meteorological ideal. Better estimates may be made where gauges are re-randomized around a site and Reiley *et al.* (1969) have described statistical procedures for making estimates of microclimate in such a system.

Although water may be splashed into a gauge, the major errors are losses. The gauge, particularly when placed at standard height (12 inches or 305 mm) presents an obstacle to airflow so that air is accelerated across the orifice and the catch is reduced. Drops may splash out of a gauge during heavy rain and the evaporation of drops that adhere to the catching surface represents another loss. Painter (1976) describes a gauge that is placed at ground level, surrounded by an anti-splash surface, and which may be more suitable for ecological use on exposed sites (Fig. 3.3). Rain gauges will also record other precipitation such as hail or sleet and be modified with shields to catch snow. The measurement of snow is fraught and gauges usually underestimate amounts (Painter, 1976).

2.1.4 Fog, dew and surface wetness

The ecologist is often interested in other than direct precipitation. Many upland regions are subject to fog and low cloud that is intercepted by plants (Mark & Holdsworth, 1979) and which usually protects them from water stress (Jane & Green, 1983a). Dewfall may be a major source of water

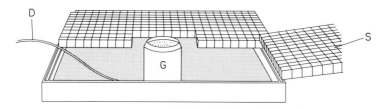

Figure 3.3 Non-splash rain gauge (Painter, 1976): G, rain gauge (flush with surface); S, non-splash surface (partially removed); D, drain tube.

for desert lichens (Lange & Matthes, 1981) and surface wetness is important in plant pathology (e.g. Weiss & Lukens, 1981).

Fog may be estimated by comparing the water trapped by an unshielded rain gauge and one equipped with a Grunow-type mesh screen that intercepts fog and other aerosols. Dew may be detected directly with a dew balance such as the Hiltner or Kessler type balances (Painter, 1976; Monteith, 1972). It may be argued that such gross measures of interception and dewfall have little direct relevance to plants which may differ from the measuring devices in their architecture, their ability to intercept moisture, and in their surface temperatures; all of which will influence dewfall. Lysimetric techniques (see Section 2.3) may be more appropriate (e.g. Jennings & Monteith, 1954; Collins, 1961) or techniques that involve the direct weighing of plant parts (e.g. Hirst, 1954). Various techniques for the detection of wetness on plant and other surfaces have been devised, and a number are listed by Slavík (1974); some more recent investigations are recorded by Weiss & Lukens (1981) and Weiss & Hagen (1983).

2.1.5 Evaporation

Evaporation is measured both by instruments (evaporimeters or atmometers) and by calculation from standard meteorological measurements (Meteorological Office, London, 1980).

Meteorological instruments include the Wild open-pan evaporimeters, the Livingston-type porous pot atmometers and the Piche evaporimeter. All suffer from problems, such as splashing from the open pan, fouling of the evaporation surface and even grazing of the filter paper discs used in Piche evaporimeters. Most seriously, evaporimeters are rarely directly comparable with a plant surface, particularly their surface temperatures which are generally lower than in most plants. Nevertheless, the Piche evaporimeter has often been considered the most suitable for ecological use (e.g. Eckardt, 1960; Slavík, 1974; Kreeb, 1977) and simple micro-Piche evaporimeters can be made from a filter paper disc and a glass capillary or a graduated pipette (e.g.

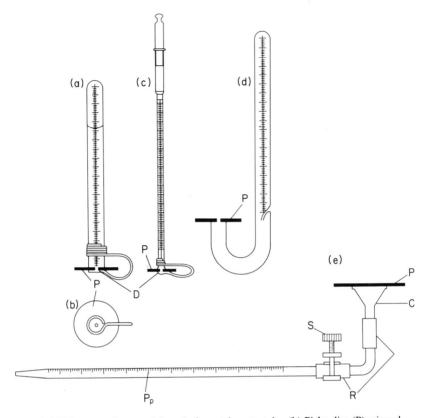

Figure 3.4 Piche evaporimeters: (a) made from a burette tube; (b) Piche disc (P), viewed from below and showing clamp and central hole (D); (c) Vertical micro-evaporimeter for short-term measurements; (d) U-shaped evaporimeter similar to that used by the French Meteorological Service; (e) Horizontal micro-evaporimeter made from a 1 ml pipette (Pp) as used by Bannister (1964b) (cf. Warren-Wilson, 1959); S, screw-clip; R, rubber tubing; C, porometer cup. (a–d after Kreeb, 1977).

Warren-Wilson, 1959, Fig. 3.4). Porous pot atmometers may be used in longer-term studies (e.g. Jessop, 1964) and can be modified for use in freezing conditions (Mark & Smith, 1962).

One of the most favoured methods is to compute potential evaporation from standard meteorological data by formula. The use of an atmospheric saturation deficit (see Section 2.1.1) is a first approximation which can be improved by taking account of the difference between air temperature and that of the evaporating surface. Furthermore, evaporation is a function of the total energy budget so that components such as radiation balance and wind have also to be taken into account. The Penman (1948) formula was the first of this type to come into general use. This and related formulae see both evaporation and transpiration as primarily physical processes, so that potential evapotranspiration from a crop surface is predicted by applying a simple coefficient (usually < 1) that is computed from daylength and season.

Monteith (1965) modified Penman's formula to take better account of changes in the diffusive conductance of plant surfaces and the derived equation is often termed the Penman–Monteith equation. A derivation is given by Jarvis (1981) and the resultant formula is increasingly used by plant ecologists to estimate evapotranspiration.

Transpiration (E_T) is a function of leaf diffusive conductance (g_w) to water and the vapour pressure difference between leaf and air ($p_L - p_A$); where p_L is the saturated vapour pressure at leaf temperature, c_P is the specific heat of dry air ($1006\,J\,kg^{-1}\,°C^{-1}$), ϱ is the density of dry air ($1\cdot22\,kg\,m^3$ at 15°C), γ is the psychrometric constant ($0\cdot655\,mg\,°C^{-1}$ at 15°C) and λ is the latent heat of evaporation of water at leaf temperature.

$$\lambda E_T = \frac{c_P\varrho}{\gamma}(p_L - p_A)g_w \tag{8}$$

Substitution of equation (8) into the Penman equation is simplified if the diffusive conductance to water (g_w) and heat (g_H) are considered equal. Then

$$\lambda E_T = \frac{sA + c_P\varrho Dg_A}{s + \gamma(1 + g_A/g_C)} \tag{9}$$

where A is available energy, s is the slope of the curve relating saturation vapour pressure to temperature, D is the *atmospheric* saturation deficit and g_C, g_A are canopy (stomatal) and boundary layer conductances respectively. The equation may be simplified to

$$E = aA + bD \tag{10}$$

where a and b are constants (Jarvis, 1981). This shows that, at constant conductance and temperature, transpiration is a simple function of available energy and atmospheric saturation deficit.

2.2   Soil moisture

2.2.1   Water content

This is measured simply by finding the difference in weight between fresh soil and the same sample oven-dried (usually at 105°C). Water contents are usually standardized by expression in terms of various soil characteristics, e.g. fresh weight, dry weight, pore space, volume of the fresh sample, saturated water content. All are fraught: fresh weight and volume vary with moisture content; contents per unit dry weight are misleading when bulk density changes between or within soils and, in soils of low bulk density (e.g. peats) may give rise to values $> 100\%$. For this reason volumetric water contents are often preferred for peats, but even then the commonly used method of determining a gravimetric water content and converting it to a volumetric measure by multiplying by bulk density may be misleading (Munro, 1982). Water content per unit saturated water content would provide a measure similar to relative humidity in the air or relative water content in plant tissues (Sections 2.1.1, 4.1) but as air is often trapped between pores in wetted soil full saturation may be difficult to achieve. Stewart & Adams (1968) have advocated expression per unit pore space as the best method for making comparisons between soils. Despite the alternatives, water content per unit dry weight of soil remains the most frequent expression.

Not all direct measures of soil moisture depend upon drying the soil (Slavík, 1974). If the particle density is known then soil air can be displaced in a pycnometer and water content calculated from weight changes (Aggarwal & Tripathi, 1975). This method is rapid and could be useful when dealing with a standard soil.

2.2.2   Available water and water status in the soil

Soil scientists are often concerned with the amount of soil moisture that is available to plants. The concept of plant-available water is related to the classification of soil water; which is usually subdivided into gravitational, capillary and hygroscopic water. Gravitational water drains away under the influence of gravity and thus is transient; when all this water is drained from the soil it is said to be at field capacity. This leaves capillary water which is trapped in the spaces between particles and is the major supply of water for plants: the plant wilts when the water potential of water in the capillaries is less than that in the plant, and the point at which the plant becomes incapable of recovery is the permanent wilting point. Plant-available water is usually taken as the difference in moisture content between field capacity and permanent wilting point. Some capillary water is still present from permanent wilting until the hygroscopic point is reached, beyond this (the point at which the soil appears air-dry) the hygroscopic water is in equilibrium, as vapour, with the soil air and is unavailable to plants. Field capacity and the permanent wilting point are often approximated by the water content retained at defined pressures (often 10 or 33 kPa for field capacity and 1·5 MPa for permanent wilting) but such values may differ from field measurements and be related to the physical properties of the soils (Ratcliff *et al.*, 1983). Conversely it may be possible to predict plant-available water from a knowledge of the soils' physical characteristics alone (Cassel *et al.*, 1983).

2.2.3   Determination of field capacity and permanent wilting point

The simplest method for the determination of field capacity is to measure the moisture content of prewetted blocks of soil from which moisture has ceased to drain (e.g. Salter & Haworth, 1961; Ratcliff *et al.*, 1983). Lysimeters (Section 2.3) could be used for this purpose. Alternatively, water can be extracted from soil by a known force, e.g. by equilibration on a suction plate (Richards, 1948), or by centrifugation. The soil water potential at field capacity is not constant and the force used may vary: the so-called moisture equivalent is determined at one-third of an atmosphere (approx. 33 kPa). Laboratory-determined values may differ from those determined in the field according to soil type (Ratcliff *et al.*, 1983).

Permanent wilting point has been determined in the field by observing the soil moisture content at which plants were killed (Ratcliff *et al.*, 1983) or in the laboratory by determining the permanent wilting of seedlings (e.g. Salter & Haworth, 1961; Salter & Williams, 1965; Slavík, 1974). Originally, the permanent wilting point was thought to be purely a soil characteristic (Briggs & Shantz, 1912) and therefore uninfluenced by the plants used to test it. This is not the case as the water potential of plants at permanent wilting can vary considerably but, as very small changes of moisture content around the permanent wilting point are reflected in large changes in water potential, the difference between plants is minimized when permanent wilting point is expressed in terms of moisture content. A 15 bar or 15 atmosphere equivalent (1·5 MPa) has often been used to approximate the permanent wilting point, usually by subjecting the sample to pressure and forcing the water through a semi-permeable membrane or porous plate (e.g. Waters, 1980). The pressure membrane technique was devised by Richards (1941) and is not only used for the determination of the permanent wilting point but also, in conjunction with the suction plate (Richards, 1948), for determining the relationship between water potential (strictly matric potential) over a wide range of water contents (Fig. 3.5).

Plant ecologists do not normally need to determine hygroscopic water but it can be done by equilibrating samples in atmospheres of known humidity (Croney *et al.*, 1952).

2.2.4   Determination of soil water potential

Water potential in the soil (ψ_{soil}) is the resultant of the force with which water is held in the soil matrix (matric potential, ψ_m) and the depression of water potential due to solutes in the soil solution (ψ_s).

$$\psi_{soil} = \psi_s + \psi_m \tag{11}$$

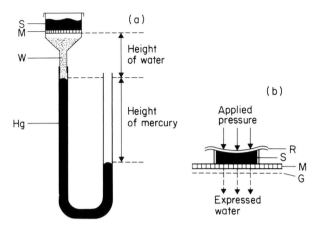

Figure 3.5 Determination of soil water content by (a) the suction plate method and (b) the pressure membrane. S, soil sample; M, membrane permeable to water; U, water; Hg, mercury; R, rubber sheet; G, metal grille. In (a) the applied tension is the equivalent height (in m of water) of the water and mercury columns and is, at equilibrium, a measure of the soil water (matric) potential. In (b) the samples are contained in a pressurized container and the applied pressure provides a measure of soil water (matric) potential. In both cases the soil water content can be determined at equilibrium.

Methods such as the suction plate and pressure plate measure only matric potential and ideally, the solute potential of the soil solution should be determined in addition for a full expression of soil water potential. Cryoscopic methods (see Section 4.2.4) are often used, the main problem being the extraction of a suitable sample of soil water. Various methods are discussed by Slavík (1974) but as the osmotic content of most soils is very low, ψ_{soil} can safely be approximated by ψ_m.

Many methods that are used to determine the water potential of plant tissue (Section 4.2) can be used to determine that of soil, although the principal ones are those that use vapour equilibration, particularly with psychrometers. Suitable psychrometers continue to be developed (e.g. Brown & Collins, 1980; Savage *et al.*, 1981a, b) but generally instruments of the type using thermoelectric cooling (e.g. Spanner, 1951; Monteith & Owen, 1958) are most suitable as they have a decreased temperature dependence (Slavík, 1974). As temperature fluctuations affect the vapour pressure deficit within the sample chamber (and hence the value of ψ), many soil psychrometers are designed to allow a rapid exchange of vapour between the soil and measuring chamber which may be made from porous pot (Rawlins & Dalton, 1967; Merill *et al.*, 1968; Rawlins, 1971) or wire mesh (Lang, 1968; Brown & Collins, 1980). Thermistors rather than thermocouples have been used (Kay & Low, 1970) whereas Richards (1965) has used a thermistor hygrometer which measures the temperature difference between a completely dry sensor and one in equilibrium with soil vapour. Further details of psychrometers are given in Section 4.2.1.

Soil water potential could also be determined by equilibrating samples above solutions of known water potential, although equilibration times for bulky soil samples would be much slower than for plant material. An alternative method is to calibrate the moisture content of filter paper against solutions of known water potential and allow these to equilibrate with soil. Eckardt (1960) and Al-Khafaf & Hanks (1974) describe such methods.

The relationship between water potential and water content characterizes soils in much the same way as it characterizes plants (Section 4.3), although it is influenced by the expresssion of water content that is used (Fig. 3.6). Soils with large pores will show larger changes in water content for a given change in water potential than those with small pores, and so will more readily lose their water to plants than closer textured soils.

2.2.5   Indirect methods of measuring soil moisture

Many properties of soil change with moisture content and can therefore be used as a measure. The principal ones considered here are changes in the electrical resistance of materials in moisture equilibrium with

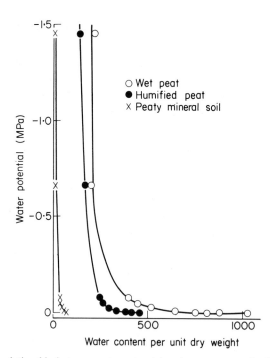

Figure 3.6 The relationship between water potential and water content in three different soils (from Bannister, 1976).

the soil, the attenuation of radiation by soil moisture, and the use of tensiometers. Techniques involving other soil properties are mentioned by Slavík (1974).

Moisture blocks

When blocks of gypsum, or any other porous material, are buried in the soil they will equilibrate with it; thus, changes in their weight can be used to assess changes in soil moisture (Davis & Slater, 1942; Rutter & Sands, 1958). The electrical resistance of the blocks will also vary with moisture content and this is the basis of the 'Bouyoucos block'. Bouyoucos (1953, 1954) used gypsum blocks, whereas Bouyoucos & Mick (1948) and Bouyoucos (1972) have used nylon or fibreglass blocks; these last types are more sensitive at high moisture contents and more resistant to corrosion but may be more sensitive to dissolved solutes.

Resistance of the blocks is read with a suitable meter, usually producing an alternating current that prevents polarization of the electrodes. Slavík (1974) gives circuitry for various meters; Hinson & Kitching (1964) describe a relatively simple meter and Kitching (1965) a more sophisticated type. More modern circuitry is given by Goltz et al. (1981) and provision may be made for automatic recording (Williams, 1980).

Calibration is problematical. A simple method is to place probes in a wire basket containing soil and to subject the whole to controlled desiccation (Slavík. 1974), but they may also be calibrated in a pressure membrane apparatus or equilibrated above solutions of known water potential (Slavík, 1974). Blocks with anomalous resistances are discarded. Tanner et al. (1949) have described a calibration that is based on the conductivity of blocks in distilled water.

Blocks must be inserted into the soil with the minimum of disturbance but with good contact between the block and the surrounding soil. The various problems associated with calibration—insertion in the soil, hysteresis during wetting and drying cycles, and the conductivity of the soil solution—have tended to discourage workers from using the blocks for anything more than an approximation of soil moisture status; although Taylor (1955) found that the standard error from blocks was only 12% higher than those from gravimetric determinations or the neutron probe.

Neutron probe and related techniques

In the neutron probe technique a source of fast neutrons is used, some of these are slowed by collision with hydrogen atoms and are counted (Stone et al., 1955; Visvalingham & Tandy, 1972). As water is the main source of hydrogen in the soil the probe can be calibrated against water content although other sources of hydrogen (e.g. organic matter) and other atoms (e.g. iron, boron and chlorine) may cause interference. The probe integrates over a relatively large volume of soil and thus eliminates some of the point-to-point heterogeneity that is inherent in most other methods. This integration makes the detection of small-scale pattern more difficult and hampers the use of the probe near the soil surface when its 'sphere of influence' cuts above the surface. Much contemporary effort has been put into rendering the method more suitable for use in the superficial layers of the soil (e.g. Hanna & Siam, 1980).

Other types of radiation are attenuated by soil moisture. Gamma rays have been used (e.g. Gurr, 1964; Oelsen, 1973). The technique needs expensive equipment and elaborate precautions and, like other nuclear techniques, may not be entirely suitable for routine ecological use. Like neutron scattering the technique is not entirely specific for water. Beta-rays have been used principally to determine the moisture content of plant tissues (e.g. Nakajama & Ehrler, 1964) but have been used to assess the moisture status of a nylon pad held in equilibrium with soil moisture (Newbould et al., 1968). Nuclear magnetic resonance is a further technique that has been used to assess soil moisture (Prebble & Currie, 1970).

Tensiometers

Tensiometers are commonly used at low water tensions (viz. high water potentials, usually $> -100\,kPa$). They consist of a water-filled porous pot connected to a manometer (Fig. 3.7); water in the porous pot is continuous with that in the soil so that the tension in the apparatus follows

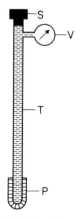

Figure 3.7 Soil tensiometer (diagrammatic). S, air-tight stopper; T, tube filled with water (no air); V, vacuum gauge (or manometer) to measure water tension; P, porous pot.

changes in tension in the soil water. The water must be air-free, otherwise the continuity is broken; this is one of the main disadvantages of the method as the tensiometers must be continually checked for leaks of air; another disadvantage is the slow response to changes in soil tension. Construction of tensiometers from readily available materials is discussed by various authors (Webster, 1966; Nelms & Spomer, 1980) whereas fast-response tensiometers have been described also (Klute & Peters, 1962; Leonard & Low, 1962; Rogers, 1974). Tensiometers may be linked to a recording system using suitable pressure transducers (Rice, 1969) but because of the constant attention needed, such a system may not be particularly advantageous.

The osmotic tensiometer of Peck & Rabbidge (1966) utilizes pressure changes in a solution of polyethylene glycol separated from the soil solution by a semi-permeable membrane and a ceramic plate. This tensiometer may be used over a much wider range of soil moisture tensions.

Further details of tensiometer design, placement and maintenance are given by Slavík (1974).

2.3 Water balance using lysimeters

In the water economy of an area of ground the inputs, precipitation plus condensation ($=P$), are balanced by the outputs, run-off (R), evapotranspiration (E) and changes in moisture content (S),

viz. $$P = S + R + E \qquad (12a)$$

or $$E = P - S - R \qquad (12b)$$

These equations are the basis of all lysimetric techniques. Precipitation can be measured directly using rain gauges (Section 2.1.2), run-off water can be collected (or, for the water balance of a whole catchment, stream flow may be measured) and changes in storage can be calculated from weight changes. Evapotranspiration can therefore be computed from equation (12b). Consequently, most lysimeters involve weighing a block of vegetated soil that is suspended in a hole but otherwise flush with the surrounding surface (e.g. Heatherley et al., 1980). A simple hydrostatic lysimeter that is suitable for ecological use has been described by Courtin & Bliss (1971). In this the container with soil and vegetation rests on a pressurized cushion; changes in pressure are used to estimate weight changes. Other lysimeters may be much larger and more sophisticated and have been used to examine the water relations of whole trees with a resolution of 150 g in a gross load of 2 tonnes (Gifford et al., 1982).

In long-term studies or in soils that are permanently wet the change in moisture content, S, may be ignored, hence

$$E(+S) = P - R \qquad (12c)$$

Non-weighing lysimeters have been used in upland sites where soils are at or near field capacity all the year round (e.g. Rowley, 1970). Microlysimeters suitable for use in permanently wet situations, and which could be adapted for ecological use, have been described by Tomar & O'Toole (1980).

3   Exchange of water with the environment

3.1   Uptake of water

There are no proven field methods for measuring the uptake of water by intact roots from the soil. However, lysimetric methods (Section 2.3) may give some estimate if all other components of the budget, including changes in soil moisture are known. Effectively such a method is equating transpiration with uptake (Section 3.3) and as a first approximation this is reasonably valid. Bulky plants, however, may have a considerable storage capacity (Waring et al., 1979) and this may contribute to the transpired water so that uptake may lag behind loss.

Otherwise, the most reliable methods of measuring uptake are potometric and usually involve working with artificially-reared plants. Suitable potometers are similar to those used for cut shoots, except that rooted plants may be used. They usually involve some apparatus recording the change in volume of a reservoir of water by following the retraction of a meniscus or the movement of a bubble in a glass capillary. One of the most simple potometer designs is to connect the plant by a flexible tubing to a graduated pipette or a burette (Fig. 3.8). The change in volume per unit time will give a measure of the rate of uptake.

A major problem in the use of potometers is in effecting an air-tight seal of the root or shoot into the apparatus. Evans & Vaughan (1966) describe detailed procedures for heat and pressure sealing, although silicone and other waterproof greases may be used where pressure differences are slight (e.g. Pettersson, 1966; Hu Ju & Kramer, 1967). Slavík (1974) gives details of various designs of potometer, including micropotometers that may be used with individual roots.

If a potometer containing a whole plant is weighed then both water uptake and loss can be calculated; Knight (1965) describes a very simple apparatus for this (Fig. 3.8). One drawback of potometers is that the resistance to water uptake at the root/soil interface is minimized so that rates of uptake may be spuriously high (Tinklin & Weatherley, 1966) although some potometers that use plants rooted in soil, have been devised (e.g. that of Lebedev & Solovev described by Slavík, 1974).

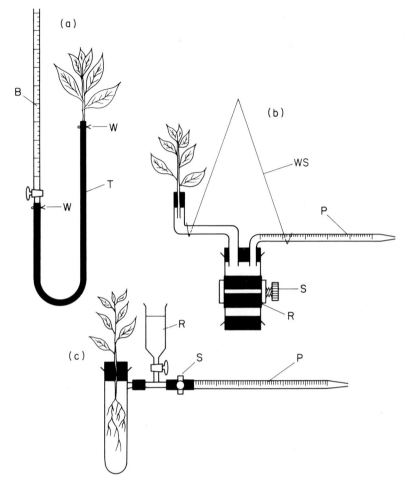

Figure 3.8 Simple potometers. (a) Potometer made from a burette (or pipette) (B) and a rubber tube (T) secured by wire (W). (b) Potometer to determine both uptake and loss (by weighing from a wire support, WS); P, pipette; S, screw clamp; R, rubber sleeve (after Knight (1965). (c) Potometer with reservoir (R) modified to take a whole plant.

3.2   Transport and storage of water

3.2.1   Water transport

Water transport through the stem may be measured by a variety of means. A marker such as a dye or a radioactive tracer may be introduced into the sap: most contemporary methods use radioactive tracers (Klemm & Klemm, 1964; Heine & Farr, 1973; Waring & Roberts, 1979). Direct injection may cause the fracture of the water column and the rapid transfer of the dye or tracer as the broken column retracts from the point of injection and the pathway taken by the marker may not be direct (Milburn, 1979a). If the marker can be introduced into the apoplastic water or without breaking columns (e.g. via a filled reservoir) then these problems may not apply. While it may be possible to trace a radioactive front without destroying the plant, dyes can be traced after destruction. Rates may be estimated by finding the distance that the marker moves in a known time.

The most commonly used method is the heat-pulse method originated by Huber (1932), refined by Huber & Schmidt (1937) and evaluated by Leyton (1970) and Stone & Shirazi (1975).

The technique uses a heat source to create a pulse and measures the time taken for the pulse to travel a small distance (say 2 cm) in the direction of flow. For rapid flow rates ($> 60\,\mathrm{cm\,h^{-1}}$) it is sufficient to measure the time taken for the pulse to reach the sensor, which is typically a thermocouple although thermistors have also been used (e.g. Swanson, 1962). If the two junctions of the thermocouple are arranged vertically then the lower will be heated first and there will be a positive output; but, as the pulse passes and the upper junction becomes warmer than the lower, then the output becomes negative (Fig. 3.9a). At low velocities, conduction of the pulse through the xylem walls becomes an important source of error and compensation methods are used with thermocouples situated both above and below the heater (Fig. 3.9b). Two arrangements are common: in the first the lower junction is placed nearer to the heater than the upper; in the second they are equidistant. In the first arrangement the lower junction is warmed soonest resulting in a negative deflection, changing to positive as the upper junction is warmed. The time taken for the negative deflection to be minimized is used in the calculation of pulse velocity and a correction is made for the speed of the pulse when there is no flow. In the second method the maximum (negative) deflection is used with a correction for the thermal diffusivity of wet wood with no water movement (Fig. 3.9c).

The velocity of movement of the pulse is not necessarily identical to the actual rate of movement of sap, as both convection within the sap and conduction by the stem tissue occur. Marshall (1958) relates the speed of the heat pulse to the sap flux and the density and specific heat of both sap and wet wood. Slavík (1974) and Kreeb (1977) give further details of technique and apparatus, and new variants and improvements continue to be made (e.g. Cohen et al., 1981; Sakuratani, 1981). Many of these provide for continuous recording and are more sensitive to low flow rates. Steady-state methods follow Vieweg & Ziegler (1960) (e.g. Saddler & Pitman, 1970; Allaway et al., 1981) and measure steady-state temperatures above and below a heater. The method of Redshaw & Meidner (1970) uses the cooling effect of the flux on a constantly heated wire.

Other methods have been proposed: Tyree & Zimmerman (1971) used the

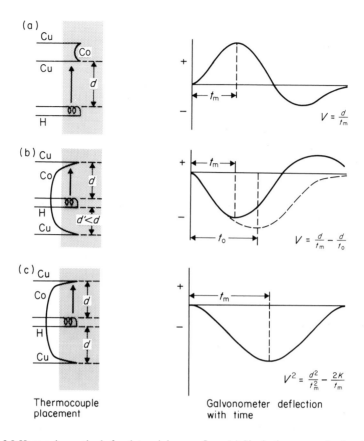

Figure 3.9 Heat-pulse methods for determining sap-flow. (a) Single thermocouple placed above heater (suitable for rapid rates of flow). (b) Compensation method with thermocouple junctions placed at unequal distances above and below the heater (suitable for low rates of flow). (c) Compensation method with thermocouple junctions placed equidistant above and below the heater. v, pulse velocity; d, distance between the heater and thermocouple junction; t_m, time of maximum absolute deflection with no sap-flow; k, a constant related to thermal diffusivity with no sap-flow; H, heater; Cu/Co, copper/constantan thermocouple wire.

difference in current when a known voltage was applied across a section of sapwood both against and with the direction of sapflow; whereas Sheriff (1972) used the voltage induced by sapflow through a uniform magnetic field oriented at right angles to the direction of flow.

The actual flux of water (i.e. the volume of water transported in unit time) can be estimated from heat pulse methods (Marshall, 1958). If there is little use of stored water the flux will be equivalent to the amount transpired. Consequently, velocities can be converted to fluxes either from theory or by calibration against potometric determinations (e.g. Redshaw & Meidner, 1970; Cohen *et al.*, 1981).

3.2.2 Storage of water by tissues

It is evident that many plants, such as cacti and succulents, store large amounts of water in their tissues. That this water is used in the economy of the plants is scarcely questioned. However, the contribution of apoplastic water, such as that stored in the lumina of xylem vessels, is controversial. Some authors working with the water relations of cut shoots in pressure chambers (Section 4.3) claim that there is negligible exchange between the apoplast and symplast (e.g. Tyree & Karamanos, 1981) although large amounts of apoplastic water appear to characterize plants that avoid drought (P. Bannister, unpublished).

Trees store considerable amounts of water in their stems. A dense stand of Scots pine was estimated to hold the equivalent of > 20 mm of rainfall (Waring *et al.*, 1979) and, although Roberts (1976) considered that stored water did not contribute much to transpiration in cut trees, Waring *et al.* (1979) estimated that 30–50% of the transpired water was extracted from that stored in sapwood.

Changes in the water content of sapwood may be followed by extracting cores of known volume and determining their fresh (f) and dry weights (d). As resaturation of the cores is uncertain, the water content at full saturation may be obtained by estimating the available volume from the volume of the fresh sample (V_f) and the volume of structural solids (V_s). This latter can be estimated from the dry weight (d) and the density of the solids (ϱ_s), i.e.

$$V_s = d/\varrho_s \tag{13}$$

The relative water content (see equation (22)) of the sapwood (W_r) becomes

$$W_r = (f - d)/(V_f - V_s) \tag{14}$$

if the density of water is assumed to be unity (Waring & Running, 1978; Waring *et al.*, 1979).

3.3   Water loss (transpiration)

The physiological ecologist is concerned with the relationship of the water economy of the individual plant and its distribution. Much early work (e.g. Seybold, 1929) attempted to establish general relationships between transpiration and habitat, but found few consistent trends (e.g. Schratz, 1932). It was soon realized that the control of water loss and the maintenance of favourable tissue water contents was more important than the rates of loss themselves (e.g. Maximov, 1929, 1932). Consequently the physiological ecologist has directed more attention to these aspects rather than transpiration itself. However, in studies of the water economy of stands of vegetation (e.g.

Rutter, 1968) the measurement of transpiration is of considerable importance although it can increasingly be estimated from meteorological data and measurements of stomatal conductance (Section 2.2.4).

Transpiration cools leaves (Lange & Lange, 1963) and fluctuations in transpiration rate give some indication of the degree of stomatal control and the potential for the assimilation of carbon dioxide. It is, however, preferable to measure leaf temperature, stomatal conductance and assimilation rates directly rather than infer them from transpiration rates.

There are two basic methods for measuring transpiration: by measuring weight losses in intact plants or in detached plant parts, or by measuring the amount of water vapour evolved.

3.3.1   Gravimetric methods of measuring transpiration

Transpiration is readily measured by changes in weight. In potted plants this can be readily achieved by sealing the pot and soil surface so that water is lost only through the plant. Polythene bags or sheets, tied around the stem base, are often adequate for this purpose. Weighing may be automatic; Macklon & Weatherley (1965b) describe a mechanical method, but such methods have now been superceded by the use of electronic balances, many of which can be made to record periodically. Weighing lysimeters (Section 2.3) measure evapotranspiration and can be readily modified to measure transpiration alone.

Whole-plant measures of transpiration must take account of changes in leaf area that occur over a longer period of time, and this presents a problem since such measures usually involve the destruction of the plant.

The other gravimetric method of determining transpiration is by following the weight loss of cut shoots (Fig. 3.10). The method involves the excision and rapid weighing of a cut shoot, its replacement in the vegetation and reweighing after a short time interval. The shoot may be reweighed several times and the initial rate of water loss obtained by extrapolation (e.g. Willis & Jeffries, 1963). The main criticism of this method relates to the process of excision. When the shoot is cut there is a release of tension in the xylem and this may make water more readily available for transpiration; thus, transpiration rate might be expected to increase. This is the so-called Ivanov effect (Ivanov, 1928) and the above interpretation has been supported by some workers (e.g. Andersson et al., 1954). However, the effect is often greatest in turgid plants (Balasubramanian & Willis, 1969) where tensions are least. An explanation is that guard cells lose less water than the other epidermal cells and therefore open because of their relative advantage in turgor. Although increases in transpiration are often noted on excision (e.g. Weinemann & LeRoux, 1946; Andersson et al., 1954), decreases (e.g. Decker & Wien, 1960)

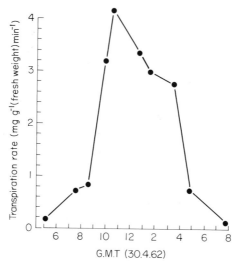

Figure 3.10 Daily course of transpiration in *Calluna vulgaris* as estimated by the quick-weighing method. (Data of Bannister, 1964b.)

or both may occur (Balasubramanian & Willis, 1969). Meidner (1965a) has described a pattern of stomatal movements in excised bean leaves that would account for both increases and decreases in transpiration. Stomatal closure is the most likely cause of decreased transpiration on excision.

As the largest changes in transpiration on excision are found in turgid plants (Balasubramanian & Willis, 1969), and plants in the field are usually suffering a water deficit, the errors on excision may not be large. All techniques have attendant disadvantages and the cut-shoot technique has been found to be as reliable as many others (Stocker, 1956; Eckardt, 1960).

3.3.2   Gasometric measurement of transpiration

Transpiration may be measured by monitoring the amount of water vapour evolved by the leaf or shoot. This usually involves enclosing the plant part in a chamber with a consequent modification of its environment. Light, temperature, humidity and air movement may all be altered and, as all these have marked effects upon both stomatal aperture and the vapour pressure gradient between leaf and air (see equation (8)) transpiration will be altered also. For this reason, short-term measurements may be preferred, and modern diffusion porometers (see Section 3.4) allow an estimate of transpiration from equation (8). Grieve & Went (1965) have described a method for the instantaneous measurement of transpiration in the field by the detection of small changes in relative humidity; this is not greatly different from the principle used in diffusion porometers.

If the problems of chamber microclimate are unimportant or can be resolved then there are a wide variety of methods for the detection of water vapour. The chief of these are psychrometric and various instruments have been described in a previous section (Section 2.1.2). Infra-red gas analysers may be used (e.g. Decker & Wien, 1960; Decker et al., 1962). As completely dry air cannot be passed over the plants without altering stomatal aperture, it is essential to monitor the air stream both before and after it passes over the plant. Otherwise the circuitry is similar to that used for the measurement of photosynthesis (see Chapter 1, this volume, Šesták et al., 1971; Slavík, 1974).

Other methods for the determination of water vapour have been largely superceded, but the evolved water may be absorbed and determined gravimetrically. Suitable absorbents are phosphorus pentoxide (Gregory et al., 1950), calcium chloride (Freeman, 1908) and magnesium perchlorate. Huber & Miller (1954) have estimated water vapour from the increase in temperature when it is absorbed by sulphuric acid.

Some older methods that are claimed to measure transpiration are more likely to measure stomatal conductance. Cobalt chloride or cobalt thiocyanate papers may be strapped to a leaf and the time taken for them to change to a standard colour is considered to be inversely proportional to water loss. Bailey et al. (1952) could not show any consistent relationship between the cobalt chloride method and gravimetric determinations of transpiration. The same criticism applies to the hydro-photographic method of Sivadjian (1952).

In potometers the uptake of water is measured, often by the movement of a bubble in a capillary (Section 3.1) and therefore will give an estimate of transpiration if uptake is considered to be equal to loss. When cut shoots are placed in water the resistance to water uptake that occurs in intact plants is lost. Consequently, stomata may show abnormal opening and rates of loss may be high. Even when rooted plants are used, the considerable resistance that exists in the soil–root interface becomes zero in water (Tinklin & Weatherley, 1966) and transpiration rates may become high, although prolonged anaerobiosis may result in a reduction in uptake. Thus, although potometers are useful in physiological studies (Gregory et al., 1950), their use in ecology is limited. They have been used to calibrate instruments for the measurement of water flux in stems (Section 3.2.1) but are unlikely to produce transpiration rates that can be related to field conditions.

Transpiration is expressed as the amount of water lost over a period of time, usually amounts are in weight but increasingly are expressed as moles of water (Section 3.4). Measures must be standardized if comparisons between different samples are to be made. Standardization per unit fresh or dry weight is simple but, as transpiration is a surface phenomenon, expression per unit area is preferred. Leaf areas may be determined by tracing onto graph paper or from silhouettes made on 'daylight' photographic paper or by photocopier. Commercial integrating leaf area meters are now commonplace and include models suitable for field use. The area of minute, folded or imbricated leaves is not easily measured. In such cases a projected area may be used or some readily measured parameter, such as weight, related to precise measurements of area made on small samples.

Leaf area is, however, not always synonymous with transpiring area (e.g. Körner *et al.*, 1979), even within leaves of the same species (Decker, 1955).

3.4   Diffusive resistance and conductance of leaves

3.4.1   Determination of diffusive resistance/conductance

An estimation of the resistance of leaves to the diffusion of water vapour and carbon dioxide is useful in the theoretical analysis and simulation of the responses of plants to the environment.

Resistance to diffusion may be divided into a boundary layer resistance (r_a), stomatal resistance (r_s) cuticular resistance (r_c) and an internal or mesophyll resistance (r_m). These resistances are combined partly in series and partly in parallel (Fig. 3.11). The usual relationship for the calculation of the various resistances to water vapour transfer is

$$E = \frac{e_L - e_A}{R} \tag{15}$$

where E = evaporation rate ($\mathrm{g\,m^{-2}\,s^{-1}}$); e_L and e_A = water concentrations at the surface of the leaf and of the air respectively ($\mathrm{g\,m^{-3}}$); and R = total resistance ($\mathrm{s\,m^{-1}}$).

r_a can be estimated by the use of model leaves made of blotting paper saturated with water (Cowan & Milthorpe, 1968; Thom, 1968), although more rigid models, particularly of awkward leaf shapes may be made from plaster. As the diffusive resistance to water vapour and heat are almost

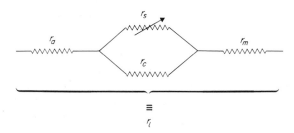

Figure 3.11 Diagrammatic representation of the various resistances to gaseous diffusion in a leaf. r_1, total leaf resistance; r_a, leaf/air resistance; r_s, stomatal resistance (variable); r_c, cuticular resistance; r_m, mesophyll resistance.

identical, r_a can also be estimated from heat loss, e.g. Grace *et al.* (1980) who used copper models of leaves of tropical trees. As the concentration of water varies with temperature it is essential that temperature of both the evaporating surface and air is known. The resistance of other components may be estimated from measurements of transpiration under defined environmental conditions (Holmgren *et al.*, 1965). Mesophyll resistance is more difficult to measure and may be taken as some function of leaf thickness, such as the ratio of internal to external surface area (Heath, 1969).

Various authors (e.g. Heath, 1969) have taken a purely anatomical approach and derived formulae for leaf resistance (R_L) such as

$$R_L = \frac{1}{D} + l_i + \frac{d_L}{8} + \frac{a_L}{na_s}\left[l_s + \frac{d_s}{4}\right] \tag{16}$$

where D = the diffusion coefficient; l_i = length of internal pathway; $d_L/8$ = depth of boundary layer of leaf diameter d_L in still air; a_L = leaf area; n = stomatal density; a_s = stomatal area; $l_s + d_s/4$ = effective length of the diffusion pathway of a single stomatal pore of depth l_s and diameter d_s; using Fick's and Stephan's laws (Heath, 1969) to describe free diffusion and diffusion away from a circular shape or pore. Parlange & Waggoner (1970) have modified the stomatal term to take account of elliptical pores and the relationships can be corrected for moving air (Penman & Schofield, 1951; Milthorpe, 1961). Without such corrections, estimates of r_a tend to be excessively high, e.g. a leaf of only 5 cm diameter would have a resistance of about 8 s cm^{-1} whereas Cowan (1977) gives 0·4 s cm^{-1} as typical for broad leaves in still air. Nevertheless, a knowledge of the components of diffusive resistance may be used to compute the transfer of water vapour, heat and carbon dioxide to and from leaves (e.g. Thom, 1968; Cowan & Milthorpe, 1968; Cowan, 1977; Monteith, 1981). These various responses may be integrated and used to predict and compare the behaviour of leaves (Gates, 1965, 1968a, b; Idle, 1970; Lewis, 1972; Parkhurst & Loucks, 1972).

The preceding definition of stomatal resistance has been recently questioned by Cowan (1977), mainly because of its assumption that the concentration or density gradient is the sole driving force. If this were the case then a system with a constant density gradient and leaf geometry would show an apparent decrease in resistance if atmospheric pressure were to decrease (as it does with increasing altitude) or if the temperature difference between leaf and air were to increase (irrespective of any corrections for the effects of temperature on density). Such anomalies are eliminated if the mole fraction gradient for water vapour is used instead of the density gradient. This gradient is the difference in vapour pressure ($p_O - p_A$) expressed as a proportion of the total atmospheric pressure P, i.e.

$$(p_O - p_A)/P \qquad\qquad (17)$$

The preferred units of water flux then become $mol\,m^{-2}s^{-1}$, rather than $g\,m^{-2}s^{-1}$ and the units of resistance become $m^2 s\,mol^{-1}$ as the mole fraction gradient is dimensionless.

For temperatures about 20°C and atmospheric pressures of about 1 bar (100 kPa) the following equivalence holds

$$R\ (\text{in } s\,cm^{-1}) \ = \ R\ (\text{in } m^2 s\,mol^{-1}) \times 2\cdot4 \qquad\qquad (18)$$

Körner *et al.* (1983) give a more general approximation

$$R\ (\text{in } s\,cm^{-1}) \ = \ R\ (\text{in } m^2 s\,mol^{-1}) \times (8\cdot31 \times T_K)/P \qquad (19)$$

where P = the atmospheric pressure (mbar); T_K = absolute temperature (K); $8\cdot31$ = a value for the gas constant ($J\,mol^{-1}K^{-1}$). The approximation assumes that leaf and air temperatures are the same which may be more or less valid for aspirated diffusion porometers (Section 3.4.2).

The reciprocal of resistance, i.e. $1/R$, is conductance and is increasingly used instead of resistance as it is directly proportional to the diffusive flow (Cowan, 1977; Jarvis, 1981). The units of conductance are usually $cm\,s^{-1}$ (or $mm\,s^{-1}$) and $mmol\,m^{-2}s^{-1}$).

3.4.2   Porometers

In porometric determinations of stomatal resistance or conductance the leaf is usually connected to the measuring apparatus by a cup attached to the leaf or a chamber or cuvette that encloses the leaf. Various types of attachment are discussed in Šesták *et al.* (1971). The microclimate within the cup or chamber must be as close to ambient as possible. Lowered carbon dioxide concentrations as a result of photosynthesis may cause stomatal opening, whereas changes in humidity and temperature may also influence stomatal aperture (Hall *et al.*, 1976). An ideal cup is one that is airtight but that can be readily removed and replaced during readings. Seals were originally effected with beeswax (Darwin & Pertz, 1911) but subsequently gelatine and, more recently, various soft plastics and synthetic rubbers have been used.

Two basic types of porometer are considered here: the mass (or viscous) flow porometers where air is drawn through the leaf and the diffusion porometers where the diffusion of gases from the enclosed surface is measured.

3.4.3   Mass flow porometers

A disadvantage of mass flow porometers is that air must be dragged through the leaf with the result that not only stomatal resistance within the cup is being measured but also the resistance of the mesophyll and of stomata outside the cup. Also, when the stomatal resistance is high, the force required to drag air through the leaf is often too great for a reading to be made; even though water is still being lost from the leaf surface.

Nevertheless, a number of instruments have been devised that are suitable for use in the field, and one of the most simple and effective is the hand porometer of Meidner (1965b; Fig. 3.12). Here the clip is attached to the leaf with the bulb deflated and the time taken for reinflation is a measure of resistance. Milburn (1979b) has described a similar porometer that uses a syringe to decrease pressure. In the field porometers of Alvim (1965, Fig. 3.12) and Bierhuizen et al. (1965), the time taken for an applied positive pressure to decrease to a predetermined level is taken as a measure of resistance. Weatherley (1966) has designed a miniaturized version of Knight's porometer (Knight, 1917).

Resistance porometers are generally more sophisticated but are usually too cumbersome for field use. However, the Gregory and Pearse porometer (1934) which has been modified by Spanner & Heath (1952) and Heath & Mansfield (1962), has been adapted for use in the field (Hsaio & Fischer, 1975). This apparatus uses two manometers to compare the leaf resistance with a fixed capillary resistance whereas in the Wheatstone Bridge porometer (Heath & Russell, 1951) the resistance of the leaf is balanced with calibrated capillary resistances by a sensitive manometer. More details of the use of mass flow porometers may be found elsewhere (Hsaio & Fischer, 1975; Jarvis, 1971; Meidner & Mansfield, 1968, etc). Mass flow porometers have also been calibrated in terms of diffusive flow (e.g. Jarvis et al., 1967; Meidner & Mansfield, 1968).

3.4.4 Diffusion porometers

Most contemporary work on stomatal resistance or conductance is carried out with diffusion porometers. These measure the diffusion of water vapour away from the leaf, although earlier porometers have measured the diffusion of exotic gases such as hydrogen (Gregory & Armstrong, 1936) or nitrous oxide (Slatyer & Jarvis, 1966).

Modern diffusion porometers are of two basic types: the 'transient' and 'steady state' porometers (Fig. 3.13). In both types the plant sample must be enclosed in a suitable chamber or cuvette. The transient type measures stomatal resistance as a function of the time taken to humidify the air by a given amount. Porometers of this type may be either unventilated (e.g. Grieve & Went, 1965; van Bavel et al., 1965; Kanemasu et al., 1969; Stiles, 1970) or ventilated (Turner & Parlange, 1970; Körner & Cernusca, 1975). Ventilation has the advantage that boundary layer resistance is reduced to a low value

The measurement of stomatal aperture

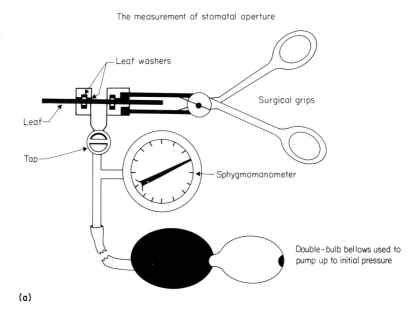

Leaf washers

Surgical grips

Leaf

Tap

Sphygmomanometer

Double-bulb bellows used to
pump up to initial pressure

(a)

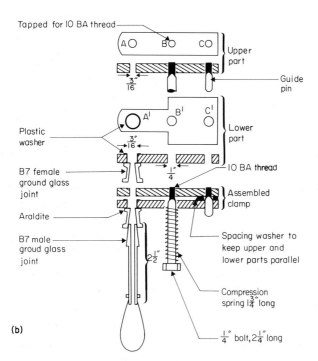

Tapped for IO BA thread

Upper
part

Guide
pin

Plastic
washer

Lower
part

IO BA thread

B7 female
ground glass
joint

Assembled
clamp

Araldite

B7 male
groud glass
joint

Spacing washer to
keep upper and
lower parts parallel

Compression
spring $1\frac{3}{4}''$ long

(b)

$\frac{1}{4}''$ bolt, $2\frac{1}{4}''$ long

Figure 3.12 (a) Diagram of an Alvim field porometer. (b) Diagram of a simple hand porometer. (From Meidner & Mansfield (1968). *Physiology of Stomata*. McGraw-Hill (UK) Ltd. Reproduced by permission.)

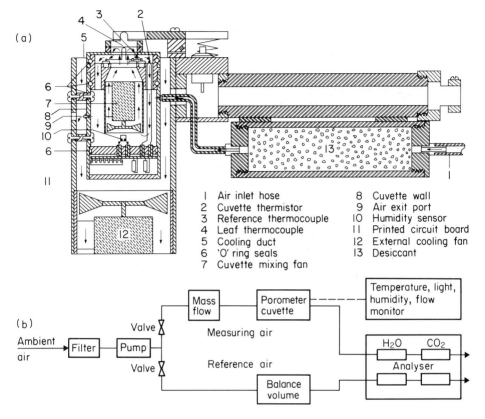

Figure 3.13 Steady-state porometer. (a) Sensor head for the LI-1600 steady state porometer. (b) Block diagram for a H_2O/CO_2 porometer based on the LI-1600 (Schulze *et al.*, 1982).

but the disadvantage that the increased evaporative loss might induce stomatal closure.

In the steady state porometer the relative humidity is maintained at or near ambient (although it may be fixed at other levels for the purpose of experiment). Stomatal resistance is derived either from the increase in humidity in an air stream passing over the sample at a constant rate (e.g. Parkinson & Legg, 1972) or from the rate of inflow of dry air needed to maintain a constant humidity (e.g. Beardsell *et al.*, 1972). The latter type is known as a null-point porometer. Steady state porometers are perforce ventilated. There is currently a considerable interest in instruments that combine the measurement of both stomatal resistance and photosynthesis (e.g. Bingham & Coyne, 1977; Schulze *et al.*, 1982; Field *et al.*, 1982) and the next generation of porometers is likely to be of such types.

Jarvis (1981) documents the rapid development of diffusion porometers during the last decade but adds a note of caution, for while the major problems (such as temperature sensitivity, hysteresis in the humidity sensor and adsorption of water vapour by the construction materials) have been largely overcome they have not been completely solved (e.g. Bloom *et al.*, 1980). Many diffusion porometers require constant attention (e.g. using completely dry desiccant) and frequent recalibration (e.g. Gay, 1983); thus, their capacity to produce large numbers of readings easily should not excuse their uncritical use.

3.4.5   Other measures of stomatal aperture

Occasionally stomata may be observed directly on the intact plant, but the observer is usually restricted to the microscopic examination of mounted material. The manipulative process, involving handling, carbon dioxide from the breath and changes in humidity and light intensity, is likely to alter stomatal aperture. Stålfelt (1932) used small strips of leaves, mounted in liquid paraffin and observed directly using an oil-immersion objective. Lloyd (1908) fixed strips of epidermis by immersion in absolute alcohol, but it is unlikely that the changes in tension caused by stripping and the trauma of immersion in absolute alcohol leave stomatal apertures unchanged from those in the intact plant. Heath (1969) considers that the method can be used only to detect broad qualitative trends whereas Slavík (1974) discourages its use altogether.

The advent of non-toxic silicone rubbers (Sampson, 1961; Groot, 1969) has revitalized techniques using impressions of the leaf surface (e.g. Buscaloni & Pollacci, 1901). Clear cellulose acetate replicas of the silicone rubber impressions are used for examination under the microscope. Cellulose acetate solutions may be applied directly to the leaf (North, 1956) which is usually damaged as a result. The main problems with the method are the degree of penetration of the stomatal pore, shrinkage of the replicas and the large number of counts needed for a representative sample. Direct observation and impressions are impossible on hairy leaves or those with sunken stomata.

Stomatal aperture may also be measured by the infiltration of liquids of low surface tension. The penetrated area appears dark or translucent, depending on the angle of view, and either the time taken for penetration or the area infiltrated after a set time, may be used to assess the relative stomatal aperture. Heath (1950) has made the penetrated area more obvious by using a solution of gentian violet in absolute alcohol (the stain persists even in dried leaves) and Michael (1969) describes a similar method for use with conifer needles. Solutions and mixtures of xylol and benzene and ethanol have also been used (Molisch, 1912) as have solutions of different viscosity (Alvim & Havis, 1954): a point is found where one solution of the series just remains on the leaf. Other workers (e.g. Fry & Walker, 1967; Lopushinsky, 1969a) have injected fluids under pressure, with the pressure just causing infiltration being used as a function of stomatal aperture. Schönherr & Bukovac (1972) have shown that penetration is not necessarily a function of stomatal aperture and thus cast serious doubt on the validity of the above procedures.

The cobalt chloride method (Stahl, 1894) and the hydrophotographic method (Sivadjian, 1952) have also been used as measures of stomatal aperture. As the loss of water from a saturated mesophyll to a sensitive paper or film is a function of diffusive conductance, such measures may provide better estimates of stomatal aperture than of transpiration (Section 3.3.3).

The various relative estimates of stomatal aperture can be made more quantitative if they are calibrated against measures of stomatal aperture or diffusive conductance made by some other means.

3.4.6   Stomatal closure and leaf water status

The relationship between stomatal aperture and the water status of the surrounding tissues may be used to differentiate between species of different ecological affinities and may be investigated by following the weight loss of excised shoots or leaves. These are brought to full saturation by placing in water and illuminating to ensure stomatal opening. Stomatal aperture can be inferred from the pattern of weight loss or measured directly by porometers.

The decline of water loss can be separated into two phases: an initial one with rapid but declining rates of loss—the stomatal phase—and a second one with a slower and more constant rate of loss—the cuticular phase. These are more apparent when expressed as rates of decline in weight, and the point of effective stomatal closure is estimated from the point of intersection of the two phases (Fig. 3.14). Other workers (e.g. Hygen, 1953; Jarvis & Jarvis, 1963) have resolved the two phases by plotting the logarithm of weight against time. The water deficit, or water potential, at stomatal closure may be used to differentiate between species (Jarvis & Jarvis, 1963; Bannister, 1964a, 1972; Lopushinsky, 1969b) or within species (Bannister, 1964a, 1971; Hutchinson, 1970). The time taken to stomatal closure in standard conditions may also be used as a basis for comparison (Bannister, 1971).

The technique depends on the stomatal closure being a function of the bulk water deficit or water potential of the tissue; thus, the more recent confirmation of a direct humidity response of stomata, independent of the water status of the surrounding tissue (e.g. Hall et al., 1976), throws some doubt on the general applicability of the technique. Recent work with pressure–volume curves (Section 4.2) has indicated that the point of stomatal closure may be coincident with the turgor loss point (Richter et al., 1981). This point has also been found to be useful in differentiating between species (Tyree & Karamanos, 1981; Jane & Green, 1983b) and, if the relationship

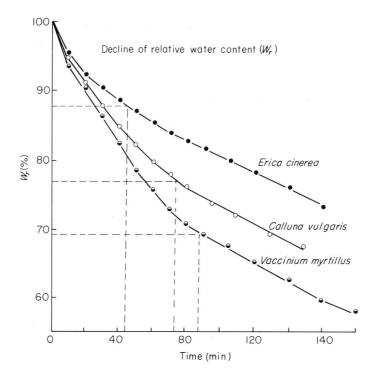

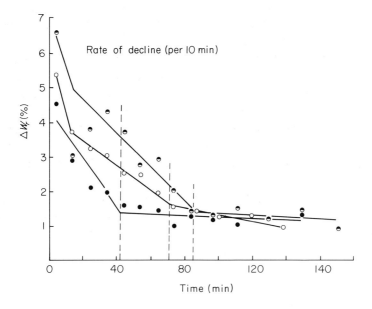

Figure 3.14 Relative water content at stomatal closure, estimated from the water loss of cut shoots. Dotted lines indicate the estimated point of closure.

between turgor loss and stomatal closure points has a general validity, could be determined simply by estimating the point of stomatal closure.

4 Status of water within the plant

Water status within plant tissues may be assessed either as water content or as water potential (Section 1.2). The methodology relevant to this section has been extensively covered by Slavík (1974) and recently been reviewed briefly by Turner (1981).

4.1 Water content

The water content is merely the difference between a fresh (f) or saturated weight (s) and dry weight (e.g. $f - d$, $s - d$). For comparative purposes it is usually expressed relative to some base, often as percentage (although given as simple ratios in equations (20)–(22) for the sake of simplicity; these may be converted to percentages by a factor of 100). Water content is often expressed per unit dry weight (W_d) as follows

$$W_d = (f - d)/d. \tag{20}$$

This expression has the disadvantage that it is strongly influenced by the fluctuations in dry weight that occur diurnally, between seasons, species, populations and even between and within similar organs on the same plant (cf. Weatherley, 1950). Consequently leaves or leafy shoots, which are often used for measures of water content show marked diurnal and seasonal changes in dry weight which would result in an apparent change in W_d even if the absolute amount of water in the tissues remained the same. Young tissues have high ratios of saturated water content to dry weight so that the use of W_d gives the impression of high water contents early in the growing season and a decline to a minimum during the dormant season (Fig. 3.15).

Relative water content (W_r) provides a measure of water content which is theoretically independent of dry weight and is the complement of the water deficit (W_s) of Stocker (1929). In both of these measures a function of field water content is related to saturated water content and is usually expressed as a percentage.

$$W_r = (f - d)/(s - d) \tag{21}$$

$$W_s = (s - f)/(s - d) = 1 - W_r. \tag{22}$$

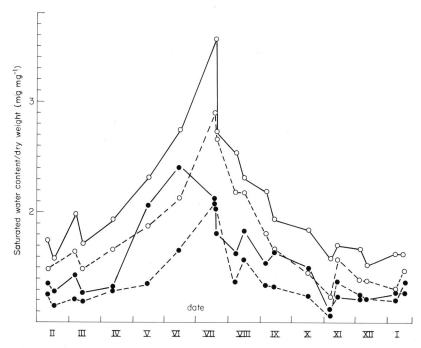

Figure 3.15 The annual course of saturated water content/dry weight in four heath species. (–O–) *Calluna vulgaris*; (--O--) *Erica cinerea*, (–●–) *Vaccinium myrtillus* (--●--) *Erica tetralix*. (Redrawn from Bannister, 1970.)

Water contents may also be expressed relative to other bases, e.g. the amount of symplastic or free water (Section 4.2.3).

4.1.1   Indirect measurement of water content

It is usually more convenient to measure water content in the laboratory than to make direct measurements in the field. However, indirect methods that are suitable for field use do exist. The foremost of these is by using a source of beta radiation (Nakayama & Ehrler, 1964; Buschbom, 1970; Jones, 1973) and relating count rates to water content (e.g. Jarvis & Slatyer, 1966). The procedure is tedious and leaves of different ages or thicknesses may need to be individually calibrated. Gamma radiation may be used to estimate the water content of more bulky organs such as tree trunks (Klemm, 1966).

The water content of leaves may also be estimated from changes in their capacitance (Slavík, 1974) or electrical conductivity (Kreeb, 1963, 1966, 1977). As these techniques usually will have to be calibrated against actual measurements of water content, it is often just as convenient to measure the water content directly.

4.1.2 Measurement of relative water content

Water content is determined directly by taking samples from the plants. The major problem of loss of moisture between excision and weighing may be minimized if samples are collected directly into tared, air-tight, containers. Thus, any water lost during transportation will be included in the initial weighing. Errors may arise also if the interval between sampling and placing in water impairs uptake. This occurs in some species after a brief period of storage (Clausen & Kozlowski, 1965) but not in others (Bannister, 1964b; Clausen & Kozlowski, 1965). The use of whole shoots or leaves is fraught: excision may cause airlocks in the xylem that prevent the rapid uptake of water, equilibration may be slow (at least 24 hours) and apparent saturation may not be achieved, especially in young leaves. It is important to examine the characteristics of the species that are used; for example, the author has found shoots of both ericaceous and epacridaceous dwarf shrubs to be amenable to the technique described in Fig. 3.16, but has experienced considerable difficulty in saturating whole fronds of small ferns.

Large samples require large containers and are destructive. For this reason discs are often preferred as they equilibrate more quickly, take up less space and cause less damage to the plant. The disdavantages of discs include rapid desiccation, sealing of cut edges during storage and the increased importance, with small samples, of errors due to blotting, losses due to respiration and gains due to growth during equilibration (Slatyer & Barrs, 1965). Weight losses can be reduced by sampling into tared containers and minimizing the time between collection and initial weighing. Weight changes due to respiration and photosynthesis can be minimized by equilibrating the discs at or near their compensation point (Slatyer, 1967) but are best controlled by shortening the period of uptake (Barrs & Weatherley, 1962; Jarvis & Jarvis, 1963).

Discs or cut sections of leaves often show greater uptake than whole leaves (Hewlett & Kramer, 1963; Clausen & Kozlowski, 1965) whereas the size of the disc may influence the uptake (Barrs & Weatherley, 1962). Extension growth of discs is a major source of error and accounts for a second phase of uptake after the initial deficit has been satisfied. It can be inhibited by floating discs at low temperature (Barrs & Weatherley, 1962; Slatyer & Barrs, 1965.)

Čatský (1960) has eliminated some of the errors that are associated with floating discs on water by placing them in contact with saturated polyurethane foam. He has also (1962, 1965) corrected for extension growth by measuring weights (s_3, s_6) after 3 and 6 hours equilibration. Using the terminology of equation (21):

$$W_r = [100(f - d)]/[s_3 - (s_6 - s_3) - d] \qquad (23)$$

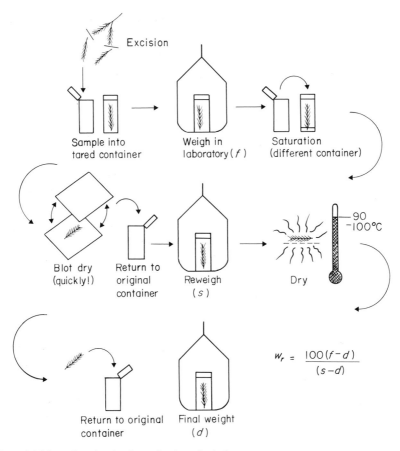

Figure 3.16 Procedure for the determination of relative water content.

This method produces similar estimates of relative water content for both mature and young tissue of the same species. Yemm & Willis (1954) have used an extrapolation procedure to determine water deficits of whole leaves.

4.2   Measurements of water potential

4.2.1   Measurements involving equilibration with water vapour

The principal problem with all techniques involving equilibration with water vapour is the maintenance of the equilibration chamber and its contained tissue, solution, vapour and measuring devices, at the same temperature. Otherwise errors introduced by convection and the distillation and condensation of water may be substantial (Slavík, 1974). An additional constraint of thermoelectric techniques is the need for highly accurate temperature measurement: for instance, differences of $<0.005°C$ must be detected in order to measure differences of < 100 kPa. Consequently, elaborate arrangements have been devised in order to achieve the required temperature stability. Slavík (1974) gives details of various designs of water bath. Such rigid control may not be necessary for techniques that do not involve thermoelectric measurements (e.g. Macklon & Weatherley, 1965a; Bannister, 1971).

While it is important to eliminate systematic errors in technique, it should be remembered that sampling error within and between replicate batches of tissue is likely to be higher than errors due to technique. For example, Slatyer (1958) estimated the error of a gravimetrically determined vapour pressure equilibration technique to be ± 20 kPa but the actual variation was in the order of ± 300 kPa (Slatyer & McIlroy, 1961). Field ecologists may not have access to sophisticated controlled systems but may wish to sacrifice a little accuracy for more adequate representation.

Major sources of error appear to be in the handling of tissue rather than in the techniques themselves. Loss of water during transfer to the chamber, the ratio of cut edge to volume (Nelson et al., 1978, Talbot et al., 1975), the time taken for equilibration, both within the tissue (Oertli et al., 1975) and between tissue and vapour (Millar, 1974; Brown, 1976), as well as heat evolved (Barrs, 1968) and dry weight lost via respiration (Kreeb & Önal, 1961) all represent potential sources of substantial error (Baughn & Tanner, 1976a, b).

Nonetheless, various temperature-compensated devices have been described and modified for field use (see below) and, with contemporary advances in microelectronics may become increasingly more practicable.

Gravimetric techniques

Tissue samples are placed in suitably designed chambers (Fig. 3.17a) so they can equilibrate with the vapour above solutions of known water potential. Sodium chloride solutions are often used and Lang (1967) gives the water potential for various concentrations of NaCl in aqueous solution over a range of temperatures. The water potential of the tissue is given by the water potential of the solution (often obtained by interpolation) that causes no change in weight. Equilibration with vapour is much slower than with tissues in direct contact with solutions (Section 4.2.2) and although 4–8 hours may suffice (Barrs & Slatyer, 1965) periods of 20–48 hours have been used (Kreeb, 1960; Kreeb & Önal, 1961; Bannister, 1971). Consequently errors due to respiratory losses may occur masking gains in weight and thus causing apparent equilibration at higher water potentials (Fig. 3.17b). A correction can be made if a duplicate set of samples is killed and dried when equilibration begins (Kreeb & Önal, 1961; Slavík, 1974). Respiratory losses and the

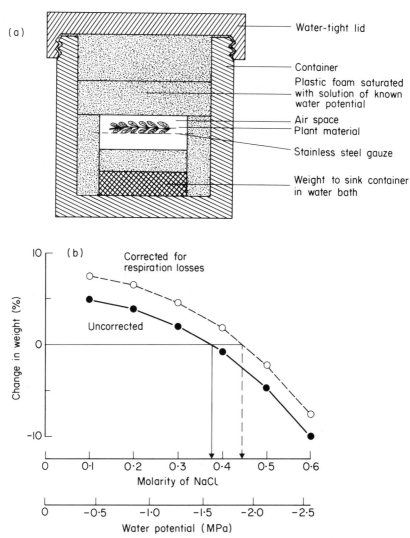

Figure 3.17 Gravimetric determination of water potential by vapour equilibration. (a) Equilibration chamber (as used by Bannister, 1971) (b) Hypothetical graph to show the effect of a 2·5% loss of weight by respiration on the estimation of tissue water potential.

probability that tissues absorb water less readily than they lose it mean that estimates of water potential by the gravimetric technique are likely to be too high rather than too low (Barrs & Slatyer, 1965; Slavík, 1974).

Microsmometer

Weatherley (1960) describes a method using the change in volume of a droplet of a test solution extruded from a micropipette and surrounded by the solution to be tested (usually absorbed on filter paper). This technique has been used to measure the water potential of both plant tissue and soil (Macklon & Weatherley, 1965a; Tinklin, 1967). The water potential of the test solution must be close (± 400 kPa) to that of the material, so that a number of trials may be necessary before the correct solution is found, and its use for tissue involves the production of rings of tissue so that the tip of the micropipette can be surrounded. It is not known how much the damage produced by this double discing and the evaporation from cut cellular surfaces interfere with the accuracy of the technique, but they would be expected to lower the value for water potential that is obtained.

Thermocouple psychrometers (Fig. 3.18)

In these the relative humidity within the sample chamber is measured thermoelectrically, usually by means of wet and dry thermocouple junctions. In one method (first employed by Spanner, 1951) a thermocouple is used to measure dry bulb temperature but then cooled (using the Peltier effect) until a drop of water condenses on the junction. The deflection produced on a galvanometer is proportional to the wet bulb depression. Other examples of these Peltier psychrometers are described by Monteith & Owen (1958), Manohar (1966) and Millar (1971a, b). Rawlins & Dalton (1967) have modified a Peltier psychrometer for use with soils *in situ* and others have designed temperature-compensated psychrometers that can, at least theoretically, be used for the direct measurement of the water potential of intact leaves in the field (e.g. Hoffman & Splinter, 1968; Millar *et al.*, 1970; Hsieh & Hungate, 1970) and also miniaturized silver-foil psychrometers (e.g. Hoffman & Rawlins, 1972; Zanstra & Hagenzieker, 1977). Despite temperature compensation, inadequate temperature control remains a problem (Hoffman & Hall, 1976).

The second major type of psychrometer, attributed to Richards & Ogata (1958), is the droplet psychrometer. The junction is permanently wet and equilibration occurs when a constant low level of output is achieved due to a steady rate of evaporation from the drop. The measured output is proportional to the water potential difference between the sample and the drop. Similar or slightly modified droplet psychrometers are described by Ehlig (1962), Lang & Trickett (1965), Barrs & Slatyer (1965).

Kreeb (1965c) describes a thermistor psychrometer which is essentially of the droplet type, while Boyer & Knipling (1965) and Boyer (1966) substitute drops of solution of known osmotic potential for the droplet of pure water and determine a point of zero deflection by interpolation. This isopiestic technique corrects for the fact that the water potential of the tissue is rarely zero, whereas that of a water droplet is zero; hence, a steady state rather than

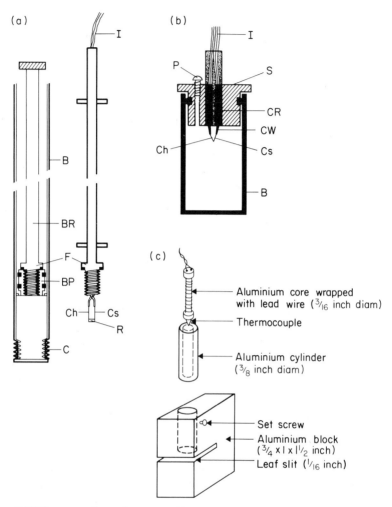

Figure 3.18 Thermocouple psychrometers. (a) Droplet type (Barrs & Slatyer, 1965, after Slavík, 1974). (b) Peltier type (after Slavík, 1974). (c) Leaf hygrometer (after Campbell & Campbell, 1974). (This may be used either as a Peltier or dewpoint hydrometer). B, Brass chamber or tube; BP, brass piston with O-rings; BR, brass rod; C, stainless steel cap with neoprene washer; Ch/Cs chromel/constantan thermocouple wires; CR, copper rod; CW copper wire; F, brass flange; I, insulated leads; P, pressure equilibration screw; R, silver ring; S, brass stopper with O-rings.

an equilibrium is attained. A related technique that uses isopiesy in conjunction with Peltier cooling is the dewpoint psychrometer of Neumann & Thurtell (1972) where dewpoint depression (rather than some function of wet bulb depression) is related to water potential. Previous psychrometers measured the electrical output of wet and dry junctions at the same temperature and are thus extremely sensitive to temperature fluctuations, whereas the dewpoint psychrometer is insensitive with an error of only 5%

for a temperature change of 10°C. This makes the technique suitable for field determinations of water potential (Campbell & Campbell, 1974; Baughn & Tanner, 1976a, b).

4.2.2 Determination of water potential by equilibration with solutions

When the water potential of a sample of plant tissue is in equilibrium with a bathing solution there is no net movement of water in and out of the tissue. This means that neither the characteristics of the tissue nor those of the solution should change. The point of equilibrium can therefore be detected by finding either of these points of no change.

The tissue (discs or sections of leaves are often used) is equilibrated in solutions of known water potential. It is essential that any loss of water between collection and equilibration is minimized, otherwise a lower estimate of water potential will be obtained. Some of the precautions mentioned in Section 4.1.1 are appropriate.

The solutions should not be taken up to any great extent by the tissue (otherwise the lowered ψ_s will result in equilibration at lower water potentials; Slatyer, 1967) and should be maintained at constant temperature (although there is no need for the very accurate temperature control needed for psychrometric and vapour equilibration techniques). Suitable, although not ideal solutions are sucrose and mannitol (Barrs, 1968). Polyethylene glycol, also known as carbowax or PEG, has been used to adjust the osmotic potential of culture solutions and could be used for equilibration if the water potential of the particular fraction is known (Lawlor, 1969). Surface deposits and exudation from cut surfaces can decrease the osmotic potential of the bathing solution leading to false estimates of water potential (Gaff & Carr, 1964); but such effects can be minimized by pre-washing the tissue in the bathing solutions (Hellmuth & Grieve, 1969). Exudation and absorption can be controlled by selecting an appropriate ratio of cut edge to surface (Hellmuth & Grieve, 1969).

A relatively short equilibration time (about 2 hours) is often sufficient. If tissue characteristics (e.g. weight, length, area, volume) are being used they must be measured both before and after equilibration and the percentage change calculated. Care must be taken, particularly when weights are used, to remove excess solution but not to use such pressure that the tissue is damaged. The water potential of the external solution that causes no change in the tissue characteristic is taken as equivalent to that of the tissue before equilibration (Fig. 3.19). Tissue characteristics cannot be used when the tissues become plasmolysed (Slatyer, 1958) as the retention of the external solution within the tissue, between the protoplast and the cell wall, prevents further changes in weight or dimensions of the tissue (Fig. 3.19). This defect

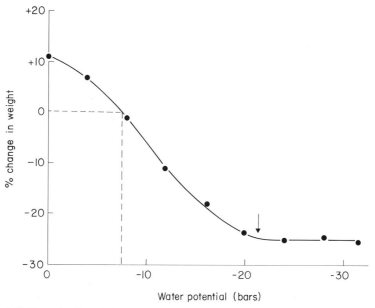

Figure 3.19 Determination of tissue water potential by solute equilibration; a hypothetical example. Dashed lines indicate the point of no change, which estimates the tissue water potential. The vertical arrow indicates the point at which the method becomes invalid because of plasmolysis (it could be used to estimate the osmotic potential of the cell sap).

can be overcome by measuring a solution characteristic or by equilibrating over vapour (Section 4.2.1).

Solution characteristics that have been used include the refractive index of sugar solutions (e.g. Gaff & Carr, 1964) and changes in the density of solutions. This latter characteristic is the basis of the Shardakov technique (Barrs (1968) traces it earliest origins), in which tissue is equilibrated with sucrose solution containing small amounts of a non-toxic, water-soluble dye (e.g. congo red, bromothymol blue). A drop of the coloured solution is released, very gently, under the surface of a colourless control solution at the original concentration; drops which have taken water up from the tissue will have become less dense and thus rise, those which fall will have given up water. Control and test solutions must be at the same temperature as their density is temperature-dependent. The technique is suitable for field use and has been critically evaluated by Hellmuth & Grieve (1969).

4.2.3 Determination of water potential by pressure chamber

The most popular instrument for the determination of water potential in the field is the pressure chamber or 'bomb' (Scholander *et al.*, 1964, 1965). Its use as an ecological tool has been critically reviewed by Ritchie & Hinkley (1975). In this method, a cut shoot or leaf is subjected to a pressure sufficient

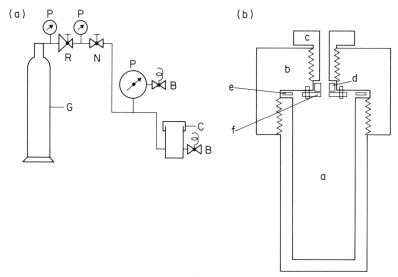

Figure 3.20 Pressure chamber or 'bomb'. (a) Diagram to show the pressure system. G, gas cylinder (N); P, pressure gauges; R, reducing valve; N, needle valve (for fine adjustment); B, bleed-off valves; C, pressure chamber. (b) Pressure chamber (not showing inlet and outlet). a, main body of chamber; b, cap screwed onto main body; c, hollow screw, various sizes; d, soft rubber gland; e, rubber washer; f, round plate with hole, various sizes.

to balance the tension in the xylem and force the retracted meniscus back to the cut surface. The shoot is inverted in a suitable pressurized chamber with the cut end protruding through a pressure-tight seal (Fig. 3.20). It is important to appreciate that the method does not provide a direct measure of leaf water potential but rather of 'xylem pressure potential', which includes frictional and gravitational components of total water potential but excludes an osmotic component in the expressed sap. A better measurement of water potential is obtained if a correction is made for the osmotic potential of the expressed sap for, while this is often negligible, in some cases it may be substantial (e.g. Kappen *et al.*, 1972).

For accurate field measurement the following precautions are advised.

(a) The time interval between excision and measurement should be minimal (e.g. Turner & Long (1980) have demonstrated losses of 0·2–0·7 MPa in 30 s). Hellkvist *et al.* (1974) have stored spruce needles in tightly rolled polythene bags for up to 24 hours with no apparent effect on balancing pressures. Karlic & Richter (1979) have prevented loss by wrapping samples in aluminium foil or immersing them in water, whereas Leach *et al.* (1982) have used clinging plastic film. Samples should not be stored without testing the possible effects of storage.

(b) If the expressed sap has a significant osmotic potential (ψ_s), it should be added to the negative of the applied pressure to determine ψ_{shoot}.

(c) End points should be determined as precisely as possible. Care should be taken to differentiate the exudation of other substances (e.g. resin and phloem sap) from water. Changes in the electrical conductivity at the surface have been used to detect the first exudation of water (Richter & Rottenburg, 1971).

(d) Pressures should be increased relatively slowly ($10–30\,\mathrm{kPa\,s^{-1}}$).

(e) Large increases in pressure cause heating and rapid pressure losses, and may cause cooling below freezing; decompression may cause mechanical damage. The chamber can be immersed in a thermostatically-controlled water bath to minimize temperature effects.

(f) Prolonged exposure to compressed nitrogen or air may cause death of the samples (Tyree *et al.*, 1973).

(g) A shoot under pressure should never be observed with unprotected eyes. At high pressures the shoot and even the seal may be ejected at high velocity.

(h) As water loss may occur to the chamber during measurement, samples have been pressurized with their leaves enclosed in a polythene bag (e.g. Tyree & Hamel, 1972; Jane & Green, 1983b).

(i) Samples should be selected with care as water potential may vary on different parts of the same plant, with height above the ground as well as with time of day. In diurnal measurements, the highest potentials are likely to be just prior to dawn and the lowest sometime between mid-morning and early evening.

4.2.4 The measurement of osmotic potential

The osmotic potential of cell sap gives an indication of water stress. Walter (1931, 1963) contends that the hydration of the cytoplasm is determined by the osmotic potential of the vacuolar sap and has produced the concept of 'hydrature'; a 'cytoplasmic relative humidity' that can be related to the vapour pressure term (p/p_0) in the equation for water potential (equation (2)). On the other hand, Slatyer (1967) and Shmueli & Cohen (1964, 1967) contend that hydration of the cytoplasm is determined by the total cell water potential which is greater (less negative) in turgid cells and equal to that of the cytoplasm only when the cell has lost turgor. As both water potential and hydrature will decrease with increasing stress, the ecological interpretation is often little affected by these theoretical considerations.

Both Slavík (1974) and Kreeb (1977) give details of suitable extraction procedures and apparatus. Kreeb (1965b) has modified hypodermic syringes for use as sap extractors—a method suitable for field use (Fig. 3.21). The usual methods of extraction involve the use of crushing, freezing, heat and toxins; but all methods are subject to error. Direct pressure may dilute the cell

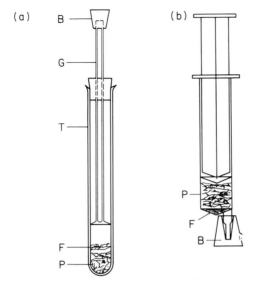

Figure 3.21 Apparatus for sampling and sap extraction of soft leaves with a high sap content (after Kreeb, 1977). (a) Made from a toughened glass test tube (b) made from a disposable plastic syringe. B, rubber bung; G, glass tube; T, test tube; F, filter paper; P, chopped plant material. Material may be killed by immersing the sampler in hot water prior to extraction.

sap by filtration or by dilution with extracellular water; freezing may extract extracellular solutes whilst boiling may hydrolyse both cellular contents and extracellular materials. In all methods solutes may be immobilized by adsorption to cellular debris.

Material must be extracted as soon as possible if deterioration and metabolic changes (e.g. changes of starch to sugar that can occur in chilled material) are to be minimized. Extracts that cannot be processed immediately may be kept deeply frozen. It is also possible to determine the fresh weight of samples, dry them rapidly, and extract the sap at a later date—adding the requisite amount of water to make up the original fresh weight. Thren (1934) claimed that osmotic values were little affected by this process and the technique may be useful for field work where other facilities are unavailable.

The most commonly used method for the determination of the osmotic potential of expressed sap is by the depression of its freezing point below that of pure water (cryoscopy). The depression of freezing point (ΔT) of an ideal molal solution with an osmotic potential of $-22\cdot4$ atm (2·21 MPa) is 1·86°C. As a direct proportionality exists

$$\frac{\text{unknown OP(atm)}}{-22\cdot4} = \frac{\text{unknown } \Delta T}{1\cdot86} \qquad (24a)$$

this simplifies to give the unknown OP as

$$-12\cdot04 \times \text{unknown } \Delta T(\text{atm}) \qquad (24b)$$

although more exact empirical formulae allow for a certain degree of non-linearity, thus the unknown OP is given by

$$0.021(\Delta T)^2 - 12.06\Delta T \text{(atm)} \tag{24c}$$

The usual method of freezing involves supercooling to about 1°C below the actual freezing point and then 'seeding' with ice crystals to induce freezing. Tables relating the freezing point depression to water potential (bars) and giving corrections for supercooling are given by Slavík (1974) and are themselves based on earlier tables (Walter, 1931, 1936; Harris & Gortner, 1914; Harris, 1915). Details of technique and of the design of microcryo-scopes are given by Slavík (1974) and Kreeb (1977). Other instruments measure melting points rather than freezing points and Ramsay & Brown (1955) have described apparatus for the determination of the melting point of very small samples ($< 10 \text{ mm}^3$).

Kreeb (1965b) has designed a microcryoscope suitable for field use and other field methods include the use of refractometry (Kreeb 1963, 1977) and electrical conductivity (Shimshe & Livne, 1967; Kreeb, 1977).

Method of 'limiting plasmolysis'

In this method, strips of tissue are mounted in suitable solutions (Section 4.2.2) of known osmotic potential and examined for plasmolysis after a short interval (20 minutes is often sufficient). The water potential (ψ_{ext}) of the solution that causes 50% plasmolysis (Fig. 3.22) is used to estimate the water potential that causes the protoplast of the 'average' cell to first lose contact with its wall; at this point turgor pressure is effectively zero but cell volume is unchanged (i.e. $\psi_p = 0$). Thus, at equilibrium,

$$\psi_{ext} = \psi_{cell} = \psi_s + \psi_p \quad \text{but} \quad \psi_p = 0$$

therefore

$$\psi_{ext} = \psi_{cell} = \psi_s. \tag{25}$$

The method is well suited to tissues containing cells with coloured contents but can be used for colourless tissues.

Once the protoplast is free of the cell wall it accommodates further decreases in water potential by shrinking in volume. Osmotic potential can be estimated from changes in volume (V_1, V_2) of the protoplast or the vacuole; as, at constant temperature, the osmotic pressure ($= -\psi_s$) is inversely proportional to volume and, at equilibrium, the osmotic pressure of a plasmolysed cell is equal to that of the external solution (π_{ext}). Thus:

$$\pi_1 \cdot V_1 = \pi_2 \cdot V_2,$$

therefore

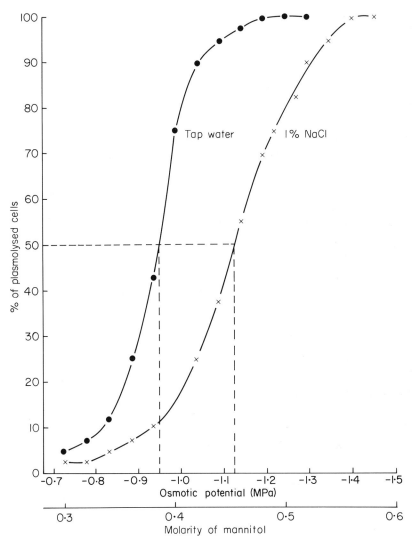

Figure 3.22 Method of limiting plasmolysis. Estimation of the osmotic potential of the cell sap of epidermal strips of *Saxifraga sarmentosa*, pretreated either with tap-water or 1% NaCl. Dashed lines indicate the mannitol concentration causing 50% plasmolysis. (Data of H. Meidner.)

$$\pi_{\text{cell}} \ = \ \pi_1 \ = \ \pi_2(V_2/V_1) \ = \ \pi_{\text{ext}}(V_2/V_1) \qquad (26)$$

This is essentially the plasmometric method of Höfler (1917).

4.3   Relationship between water content and water potential

The relationship between plant water potential and plant water content has often been given an ecological interpretation (Weatherley & Slatyer,

1957; Bannister, 1976). The advent of the pressure chamber or 'bomb' (Section 4.2.3) has simplified the production of such curves; as the volume of water expressed, or the change in weight of a sample (both functions of water content) can be plotted against applied pressure ($P = -\psi$). The volume of expressed water may be measured directly if it is collected in a capillary or micropipette (e.g. Melkonian et al., 1982) or absorbed by suitable material such as shredded paper tissue contained in a weighed detachable sleeve (e.g. Jane & Green, 1983b). Single leaves or shoots are usually subjected to a step-wise increase in applied pressure, but curves also may be constructed from spot determinations of water content made on separate samples collected either from the field (e.g. Hodges & Lorio, 1971) or artificially desiccated under controlled conditions (as in Fig. 3.23). The volume of expressed water or the weight change (V_e) may be related to the water content at full saturation (V_s), as either relative water content ($(V_s - V_e)/V_s$) or water deficit (V_e/V_s).

These pressure–volume (P–V) curves have been subjected to detailed analysis by Tyree & Hammel (1972) and Cheung et al. (1975, 1976). They invoke Boyle's and van t'Hoff's Laws and contend that, in the absence of turgor pressure, the cell volume (V) will be inversely proportional to the applied pressure (P). Hence, a plot of $1/P$ against V should yield a straight line for the portion of the plot where turgor pressure is zero. The slope of this line is a function of the osmotic potential of the cell sap and its intercept on the P-axis will give the reciprocal of osmotic pressure (π_s) at full turgor (π_0). Extrapolation to infinite pressure ($1/P = 0$) will give a point of zero cell volume and hence the maximum amount of 'free' or 'available' cell water (V_0). Any remaining water in the tissues is unavailable or 'bound' water ($= B$).

The relationship between P and V is different for turgid cells and the degree to which they resist deformation by applied pressure may be used to determine their bulk modulus of elasticity. The discrepancy between this and the linear portion of the relationship is the turgor pressure (the VAT pressure of Tyree & Hammel (1972) and Cheung et al. (1975, 1976)); whereas the point where the two phases intersect is the point of zero turgor or turgor loss point (TLP). The water deficit (V_{tlp}/V_s or V_{tlp}/V_0) and osmotic pressure (π_{tlp}) at the turgor loss point may be used to compare species (Cheung et al., 1975; Tyree & Karamanos, 1981). The cell volume at the turgor loss point (V_p) has been used in certain measures of water content (e.g. Cheung et al., 1976). The bulk modulus of elasticity is usually determined by measuring the slope of the relationship between turgor pressure (P_{vat}) and some function of V. The following relationships have been used:

$$dP_{vat}/dW_p \text{ where } W_p = (V_p - V)/V_p \quad \text{Cheung et al. (1975, 1976)}$$

$$(27a)$$

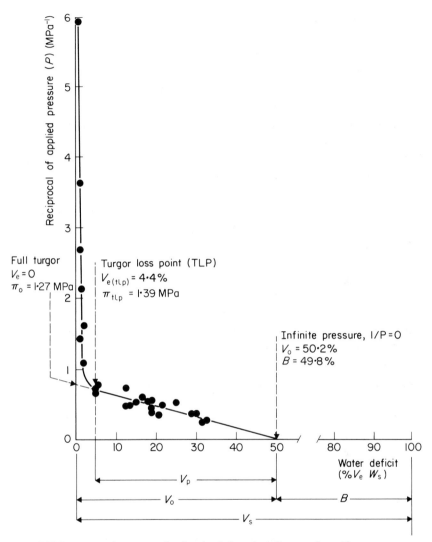

Figure 3.23 Pressure–volume curve for fronds of the polypodiaceous fern, *Phymatosorus diversifolius*. Each point represents the water deficit and the inverse of balancing pressure (P) for discrete, cut fronds, desiccated in the laboratory. For further explanation, see text. (P. Bannister, unpublished data.)

$$P_{vat}/dW_f \text{ where } W_f = V/V_0 \quad \text{Hellkvist } et \, al. \text{ (1974)} \qquad (27b)$$

$$dP_{vat}/dW_s \text{ where } W_s = V/V_s \quad \text{Jane \& Green (1983b)} \qquad (27c)$$

$$dP/dW_s \text{ where } W_s = V/V_s \quad \text{Melkonian } et \, al. \text{ (1982).} \qquad (27d)$$

The relationship between P_{vat} and V may not be linear and while linear approximations have been used (e.g. Wilson *et al.*, 1979) many authors have found increased slopes at higher turgor pressures with the bulk modulus either increasing to an asymptotic maximum (Cheung *et al.*, 1976) or showing a linear relationship with increasing turgor pressure (Hellkvist *et al.*, 1974). many of these observations and discrepancies may be explained by the fact that cells do not act in unison but lose turgor at different rates (Cheung *et al.*, 1976). The determination of bulk moduli is the least well defined aspect of the analysis of *P–V* curves (Tyree & Karamanos, 1981).

Richter (1978b) has analysed *P–V* curves by plots of *P* against $1/V$ and this and other interpretations of *P–V* curves are discussed by Richter *et al.* (1981). Plots of *P* against $1/V$ allow turgor pressures and osmotic potentials to be read directly from the graph but make other parameters (e.g. V_0) difficult to determine. Sinclair & Venables (1983) and Jane & Green (1983b) have produced algorithms that fit the whole curve and allow direct computation of the various parameters.

Graphs of water potential against water content that have been determined by other means (e.g. thermocouple psychrometer) may also be analysed by the methods used to interpret *P–V* curves (e.g. Talbot *et al.*, 1975; Richter, 1978a).

An example of a *P–V* curve is shown in Fig. 3.23; if *V* is a measure of volume so that $V > V_0$ then various relationships between the volume parameters may be summarized as follows:

$$V_e = V_0 - V, \tag{28a}$$

$$\text{relative water deficit} = W_d = V_e/V_s, \tag{28b}$$

$$\text{'free' water content} = W_f = V/V_0, \tag{28c}$$

$$W_f \text{ at turgor loss point} = V_p/V_0. \tag{28d}$$

The scale of water contents relative to V_0 (i.e. W_f) has been variously termed 'free' water content, *F*, (Hellkvist *et al.*, 1974) and osmotic or symplastic water content (e.g. Cheung *et al.*, 1975), although Cheung *et al.* (1976) use $F = (V_0 - V_p - V_e)/V_p = (V_p - V)/V_p$ as a different measure of water content. The residual water $(V_s - V_0)$ is variously known as bound or apoplastic water although its exact status is uncertain (Richter *et al.*, 1981).

4.3.1   Ecological interpretation of *P–V* curves

The following parameters have been found useful in differentiating both within and between species.

(a) Osmotic potential: particularly at full saturation $(-\pi_0)$ and at zero turgor $(-\pi_1)$. Drought-tolerant plants are generally expected to show lower (more negative) values (e.g. Hsaio *et al.*, 1976; Tyree & Karamanos, 1981). Osmotic adjustment is often an important aspect of adaptation to stress (Wilson *et al.*, 1980; Roberts *et al.*, 1980).

(b) Water content at zero turgor. Drought-resistant species might be expected to be adapted to large losses of water without loss of turgor (cf. Jane & Green, 1983b).

(c) Bulk elastic moduli. High moduli allow plants to produce large changes in water potential for only small changes in water content, hence they are able to both maintain turgor and continue water uptake. However, high bulk moduli may be associated with turgor loss at high water contents, whereas species with lower moduli may lose turgor at much lower water contents (P. Bannister, unpublished). The different strategies are in need of evaluation (Tyree & Karamanos, 1981). Some integration of turgor pressure between full and zero turgor (e.g. Roberts *et al.*, 1980) may be more useful.

(d) Overall slope of *P–V* curve. Species that show a small change in water content for a large change in water potential might be expected to be more drought-tolerant as they should be able to continue to take up water without developing large water deficits (Weatherley & Slatyer, 1957; Jarvis & Jarvis, 1963).

5   Plant resistances to drought, heat and cold

5.1   Assessment of damage

The resistance of plants to stress can be evaluated by measuring overall responses such as growth, or important physiological parameters such as photosynthesis and stomatal conductance, in stressed and unstressed plants. It is, however, not always practicable to conduct investigations with whole plants (e.g. mature trees) but valuable information can be obtained from the responses of detached plant parts. Some comparative studies—e.g. Larcher & Mair (1968) for frost and Lange (1965) for heat—have indicated no significant differences between intact and detached plant parts.

Samples can be subjected to various degrees of stress and their survival monitored. Usually the degree of stress causing 50% damage (or survival) is taken as a critical value (Fig. 3.24). Resistance may also be expressed as the degree of stress causing incipient or complete damage (Kappen, 1964; cf. Larcher, 1977b).

Where intact plants are used or the samples (e.g. grass tillers) can be readily propagated, the plants may be grown on after exposure and their survival monitored (Pearce, 1980), but generally the simplest method is to make a visual assessment of the damage produced, usually after allowing the damage to develop for some days. Detached samples may be placed in a humid atmosphere at room temperature (e.g. by wrapping in damp paper towels and enclosing in polythene bags) and left for damage to develop (usually 7–14 days but sometimes much longer periods are used; e.g. Sakai

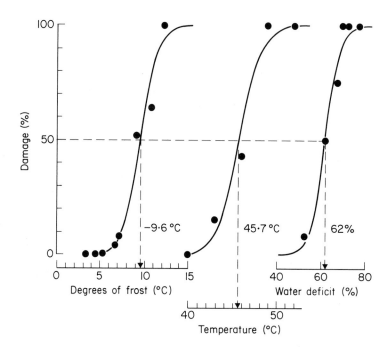

Figure 3.24 Frost, heat and drought resistance of pinnae of the fern, *Dryopteris filix-mas*, estimated by visual assessment of damage. The temperatures and the water deficit causing 50% damage are arrowed. Curves fitted by probit analysis (P. Bannister, unpublished data).

& Wardle, 1978). Damage may be assessed by visual symptoms such as browning or blackening of tissue and expressed either as a percentage of the total sample (e.g. as a percentage of leaf area or of the total number of samples exposed) or as some other scale which is usually related to % damage (e.g. Wardle, 1981; Greer & Warrington, 1982). Other methods of assessing damage include monitoring the cessation of protoplasmic streaming (Alexandrov, 1964; Alexandrov *et al.*, 1970), the colorimetric determination of abnormal respiration (e.g. Larcher, 1977b), vital staining with tetrazolium salts (e.g. Kayll, 1963; Steponkus & Lanphear, 1967) and the measurement of the conductivity of leachates from the exposed tissues (e.g. Flint *et al.*, 1967). These and other methods of assessing damage (such as the reversibility of plasmolysis and the measurement of the electrical conductivity of intact samples) are discussed by Larcher (1977b). Recently, Kyriakopoulos & Richter (1981) have investigated the use of pressure–volume curves (Section 4.3) for detecting drought and frost injury.

Measures other than direct visual estimates may be converted into indices of injury by considering the value for damaged tissue (D) in relation to that for undamaged (D_0) and completely damaged tissue (D_{100}). A simple index would be

$$\frac{D - D_0}{D_{100} - D_0} \qquad\qquad (29)$$

A more complicated version, which takes differences in sample size into account, is given by Flint *et al.* (1967). Such indices suffer if completely damaged tissue can exhibit a range of values, e.g. the conductivity of heat-killed samples (often used to determine D_{100}) is less than that for those killed by drought (Bannister, 1970). Vital staining measures the amount of living rather than dead material but similar indices to the above are readily constructed. In vital staining techniques care must be taken to avoid contamination by micro-organisms, and death by some means (e.g. toxins or heat) may not reduce the degree of staining (Mackay, 1972).

The degree of stress causing 50% damage (or any other critical level) can be assessed by plotting graphs of damage against the degree of stress. Curves may be fitted by eye, but the technique of probit analysis (Finney, 1952) allows a statistical evaluation of the curve and the fitting of standard errors to estimates.

5.2   Drought resistance

This may be determined in the field during prolonged periods of drought. Plants can be ranked in order of resistance by observing damage and survival in habitats subject to varying degrees of stress (e.g. Ogden, 1976).

Intact plants may be artificially desiccated by withholding water and monitoring the degree of stress (e.g. the water potential in the soil or in the plant) that causes damage. It is, however, often more convenient to work with excised samples (leaves, shoots, tillers, etc.) which may be desiccated by exposing them in standard conditions for differing lengths of time (e.g. Bannister, 1970, 1971) or by desiccation in atmospheres of differing humidity (e.g. Jarvis & Jarvis, 1963). Different humidities may be obtained over saturated solutions of various salts, different concentrations of the same salt or by water/sulphuric acid mixtures. Tables for all these are given by Slavík (1974).

Drought resistance is commonly divided into two components: drought avoidance, the capacity of tissues to resist the formation of damaging water deficits; and drought tolerance, the ability of tissue to tolerate high water deficits (Levitt, 1980). Avoidance may be estimated by the time taken to reach a critical water deficit or by the water deficit reached in a standard period of time: tolerance by the water deficit or water potential causing a critical amount of damage (Fig. 3.25). The two components are often inversely correlated, plants with high avoidance show low tolerance and vice versa.

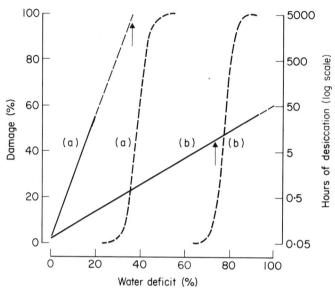

Figure 3.25 Drought tolerance and drought avoidance in two ferns; *Phymatosorus diversifolius* (a), *Blechnum penna-marina* (b). Dotted lines: relationship between water deficit and damage (= drought tolerance). Solid lines and dashed extrapolation: time taken (log scale) to develop a given water deficit (drought avoidance). Vertical arrows indicate the time taken to reach a deficit causing slight (10%) damage.

Survival may be monitored as mentioned previously (Section 5.1), but the ability of tissue to resaturate after desiccation is a further method that is often used (e.g. Bornkamm, 1958; Rychnovská-Soudkova, 1963; Bannister, 1970, 1971; Weinberger *et al.*, 1972). Undamaged material may fail to show complete resaturation although material that shows 95% resaturation is usually undamaged (Fig. 3.26). On the other hand, damaged material may show complete saturation or even supersaturation (Rychnovská, 1965); consequently it is advisable to investigate the relationship between damage and resaturation before using the latter as a measure of survival.

5.3. Heat resistance

Heat resistance is typically determined by plunging plants or plant parts into water baths maintained at a range of constant temperatures. The standard period of exposure is 30 minutes (e.g. Lange, 1961). Samples may be protected from direct contact with the hot water by exposing them wrapped in thin polythene film (e.g. Hammouda & Lange, 1962; Biebl, 1965) and the method may be used in the field if water is transported in thermos flasks (e.g. Lange, 1959). Plants may be exposed in air in thermostatically controlled chambers (e.g. Kreeb, 1965a), but unless the humidity is main-

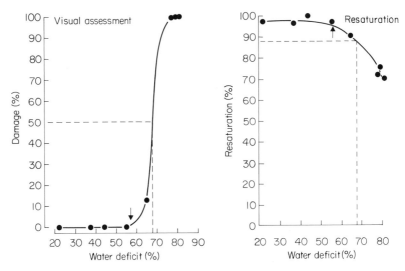

Figure 3.26 Drought resistance of *Erica cinerea*, January 1968. Dotted lines indicate the point causing 50% damage (water deficit = 68%). Arrows indicate the point of 95% resaturation (water deficit = 57%).

tained at 100% (e.g. Lange, 1953) it may not be possible to differentiate between heat and drought stress. However, if plant as well as chamber temperatures are monitored the method will allow some assessment of heat avoidance by transpirative cooling. Attached shoots in the field have been subjected to heat stress by enclosing them in darkened vessels exposed to the sun (Rouschal, 1938; Konis, 1949) but the above reservations still apply.

The physiological state of the tissue will affect the degree of heat injury, e.g. desiccated material is more resistant (Kappen, 1966; Bannister, 1970). If material is pretreated at high temperatures some heat-hardening may occur (Levitt, 1980) and exposure times other than 30 minutes may alter the degree of damage induced as a short exposure at high temperatures may be equivalent to a much longer exposure at lower temperatures (Fig. 3.27). In some instances the heat damage appears to be unaffected by shortened exposure time (Kreeb, 1970).

5.4 Frost resistance

Samples may be exposed to low temperatures by various means, such as mixtures of ice and salt (Ulmer, 1937), the forced evaporation of alcohol (Till, 1956), thermostatically controlled freezers (Irving & Lanphear, 1967; Cernusca & Larcher, 1970; Bannister, 1984) or refrigerated baths using mixtures of water and alcohol or polyethylene glycol (Polwart, 1970).

The induction and loss of hardiness is more evident for frost than for heat and drought and as a consequence various measures of resistance may be

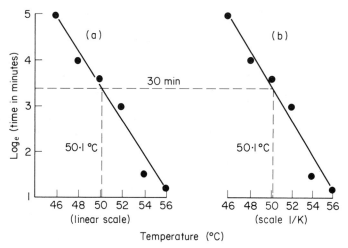

Figure 3.27 Time taken to induce 50% damage in leaves of *Griselinia littoralis* (Cornaceae) plotted (a) semi-logarithmically (after Lepeschkin, 1912) and (b) as an Arrhenius plot. (Data of R.M. Kissel & R.A. Scurr.)

made. If the actual or field resistance is to be measured then the material must be transported to the laboratory without being subjected to excessive changes in temperature and brought down to freezing temperatures relatively rapidly. Larcher (1977a) recommends rates as high as $10-20°C\,h^{-1}$, lower rates $(<5°C\,h^{-1})$ may more closely approximate natural conditions but allow some hardening to occur. Minimum frost resistance may be measured by pretreating samples for several days at temperatures of $10-20°C$ before frosting while maximum resistance may be induced by subjecting plants to stepwise cooling for several days (or longer) before exposure (e.g. Greer & Warrington, 1982; Sakai & Wardle, 1978). Larcher (1977a) gives a scheme for determining these various components of resistance.

6   References

AGGARWAL G.C. and TRIPATHI B.R. (1975) A simple and rapid method of soil water determination. *J. Soil Sci.* **26**, 437–439.

ALEXANDROV V.Y. (1964) Cytophysiological and cytoecological investigations of the resistance of plant cells towards the action of high and low temperatures. *Quart. Rev. Biol.* **39** 35–77.

ALEXANDROV V.Y., LOMAGIN A.G. and FELDMAN N.L. (1970) The responsive increase in thermostability in plant cells. *Protoplasma* **69**, 417–458.

AL-KHAFAF S. and HANKS R.J. (1974) Evaluation of the filter paper method for estimating soil water potential. *Soil Sci.* **117** 194–199.

ALLAWAY W.G., PITMAN M.G., STOREY R. and TYERMAN S. (1981) Relationships between sap flow and water potential in woody or perennial plants on islands of the Great Barrier Reef. *Pl. Cell Env.* **4**, 329–337.

ALVIM P. DE T. and HAVIS J.R. (1954) An improved infiltration series for studying stomatal opening as illustrated with coffee. *Pl. Physiol., Lancaster* **29**, 97–98.

ALVIM P. DE T. (1965) A new type of porometer for measuring stomatal opening and its use in irrigation studies. *Arid Zone Res.* **25**, 325–329.

ANDERSSON N.E., HERTZ C.H. and RUFELT H. (1954) A new fast recording hygrometer for plant transpiration measurements. *Physiol Pl.* **7**, 753–767.

BAILEY L.F., ROTHACHER J.S. and CUMMINGS W.H. (1952) A critical study of the cobalt chloride method of measuring transpiration. *Pl. Physiol., Lancaster* **27**, 563–574.

BALASUBRAMANIAN S. and WILLIS A.J. (1969) Stomatal movements and rates of gaseous exchange in excised leaves of *Vicia faba. New Phytol.* **68**, 663–674.

BANNISTER P. (1964a) Stomatal responses of heath plants to water deficits. *J. Ecol.* **52**, 423–432.

BANNISTER P. (1964b) The water relations of certain heath plants with respect to their ecological amplitude. II. Field studies. *J. Ecol.* **52**, 481–497.

BANNISTER P. (1970) The annual course of drought and heat resistance in heath plants from an oceanic environment. *Flora. Jena* **159**, 105–123.

BANNISTER P. (1971) The water relation of heath plants from open and shaded habitats. *J. Ecol.* **59**, 51–64.

BANNISTER P. (1976) *Introduction to Physiological Plant Ecology.* Blackwell Scientific Publications, Oxford.

BANNISTER P. (1984) Winter frost resistance of leaves of some plants from the Three Kings Islands, grown outdoors in Dunedin. *N.Z. J. Bot.* **22**, 303–306

BARRS H.D. (1968) Determination of water deficits in plant tissues. In *Water Deficits and Plant Growth* (ed. T.T. Kowlowski), pp. 235–368. Academic Press, New York.

BARRS H.D. and SLATYER R.O. (1965) Experience with three vapour pressure methods for measuring water potential in plants. *Arid Zone Res.* **25**, 369–384.

BARRS H.D. and WEATHERLEY P.E. (1962) A re-examination of the relative turgidity technique for estimating water deficits in leaves. *Aust. J. biol. Sci.* **15**, 413–428.

BAUGHN J.W. and TANNER C.B. (1976a) Leaf water potential: comparison of pressure chamber and *in situ* hygrometer on five herbaceous species. *Crop Sci.* **16**, 181–184.

BAUGHN J.W. and TANNER C.B. (1976b) Excision effect on the leaf water potential of five herbaceous species. *Crop Sci.* **16**, 184–190.

BAVEL C.H.M. VAN, NAKAYAMA F.S. and EHRLER W.L. (1965) Measuring transpiration resistance in leaves. *Plant Physiol.* **40**, 535–540.

BEARDSELL M.F., JARVIS P.G. and DAVIDSON, B. (1972) A null-balance diffusion porometer suitable for use with leaves of many shapes. *J. appl. Ecol.* **9**, 677–690.

BIEBL R. (1965) Temperaturresistenz tropischer Pflanzen auf Puerto Rico. *Protoplasma* **59**, 133–156.

BIERHUIZEN J.F., SLATYER, R.O. and ROSE, C.W. (1965) A porometer for laboratory and field operation. *J. exp. Bot.* **16**, 182–192.

BINGHAM F.T. and COYNE P.I. (1977) A portable, temperature-controlled, steady state porometer for field measurements of transpiration and photosynthesis. *Photosynthetica* **11**, 148–160.

BLOOM A.J., MOONEY H.A., BJORKMAN O. and BERRY J. (1980) Materials and methods for carbon dioxide and water exchange analysis. *Pl. Cell Env.* **3**, 371–376.

BORNKAMM R. (1958) Standortsbedingungen und Wasserhaushalt von Trespen-Halbtrockenrasen(Mesobromion) in oberen Leinegebiet. *Flora. Jena* **146**, 23–67.

BOUYOUCOS G.J. (1953) More durable plaster of paris moisture blocks. *Soil Sci.* **76**, 447–451.

BOUYOUCOS G.J. (1954) New type electrode for plaster of paris moisture blocks. *Soil Sci.* **78**, 339–342.

BOUYOUCOS G.J. (1972) A new electrical soil-moisture measuring unit. *Soil Sci.* **114**, 493.

BOUYOUCOS G.J. and MICK A.H. (1948) A fabric absorption unit for continuous measurement of soil moisture in the field. *Soil Sci.* **66**, 217–232.

BOYER J.S. (1966) Isopiestic technique: measurement of accurate leaf water potentials. *Science* **154**, 1459–1460.

BOYER J.S. and KNIPLING E.B. (1965) Isopiestic technique for measuring water potentials with a thermocouple psychrometer. *Proc. nat. Acad. Sci. USA* **54**, 1044–1051.

BRIGGS L.J. and SHANTZ H.L. (1912) The wilting coefficient for different plants and its indirect determination. *Bot. Gaz.* **53**, 20–37.

BROWN R.W. (1976) A new technique for measuring the water potential of detached leaf samples. *Agron. J.* **68**, 432–434.

BROWN R.W. and COLLINS S.M. (1980) A screen-caged thermocouple psychrometer and calibration chamber for measurements of plant and soil water. *Agron. J.* **72**, 851–854.

BUSCALONI L. and POLLACCI G. (1901) L'applicazione delle pellicole di colloidio allo studio di alcuni processi fisiologici delle plante ad in particolar modo pella tranpirazione. *Atti. Inst. bot. ecol., Pavia* **2**, 44–49.

BUSCHBOM U. (1970) Zur Methodik kontinuierlicher Wassergehalt-Bestimmungen an Blättern mittels β-Strahlenabsorption. *Planta* **95**, 146–166.

CAMPBELL G.S. and CAMPBELL M.D. (1974) Evaluation of a thermocouple hygrometer for measuring leaf water potential *in situ*. *Agron. J.* **66**, 24–27.

CASSEL D.K., RATCLIFF L.F. and RITCHIE J.T. (1983) Models for estimating *in situ* potential extractable water using soil physical and chemical properties. *Soil Soc. Am. J.* **47**, 764–769.

ČATSKÝ J. (1960) Determination of water deficit in disks cut out from leaf blades. *Biol. Plant.* **2**, 201–215.

ČATSKÝ J. (1962) Water saturation deficit in the wilting plant. The preference of young leaves and the translocation of water from young into old leaves. *Biol. Plant.* **4**, 306–314.

ČATSKÝ J. (1965) Leaf-disk method for determining water saturation deficit. *Arid Zone Res.* **25**, 353–360.

CERNUSCA A. and LARCHER W. (1970) A low cost program device for automatic control of laboratory refrigerators. *Cryobiology* **6**, 404–408.

CHEUNG Y.N.S., TYREE M.T. and DAINTY J. (1975) Water relations parameters on single leaves obtained in a pressure bomb and some ecological interpretations. *Can. J. Bot.* **53**, 1342–1346.

CHEUNG Y.N.S., TYREE M.T. and DAINTY J. (1976) Some possible sources of error in determining bulk elastic moduli and other parameters from pressure–volume curves of shoots and leaves. *Can. J. Bot.* **54**, 758–765.

CLAUSEN J.J. and KOZLOWSKI T.T. (1965) Use of the relative turgidity technique for the measurement of water stresses in gymnosperm leaves. *Can. J. Bot.* **43**, 305–316.

COHEN Y., FUCHS. M. and GREEN G.C. (1981) Improvement of the heat pulse method for detecting sap-flow in trees. *Pl. Cell Env.* **4**, 391–397.

COLEMAN J.D. and MARSH A.D. (1961) An investigation of the pressure membrane method for measuring the suction properties of a soil. *J. Soil Sci.* **12**, 343–362.

COLLINS B.C. (1961) A standing dew meter. *Meteorol. Mag.* **90**, 114–117.

CORNISH K.A. and GREEN G.C. (1982) An economical recording tipping-bucket rain gauge. *Agric. Meteorol.* **26**, 247–253.

COURTIN G.M. and BLISS L.C. (1971) A hydrostatic lysimeter to measure evapotranspiration under remote field conditions. *Arctic Alpine Res.* **3**, 81–89.

COWAN I.R. and MILTHORPE F.L. (1968) Plant factors influencing the water status of plant tissues. In: *Water Deficits and Plant Growth.* Vol 1. (ed. T.T. Kozlowski), pp. 127–193, Academic Press, New York.

COWAN I.R. (1977) Stomatal behaviour and environment. *Adv. Ecol. Res.* **4**, 117–228.

CRONEY D., COLEMAN J.D. and BRIDGE P.M. (1952) The structure of moisture held in soil and other porous materials. *D.S.I.R. Road Research Paper No. 24.* H.M.S.O., London.

DARWIN F. and PERTZ D.F.M. (1911) On a new method of estimating the aperture of stomata. *Proc. Roy. Soc., B* **84**, 135–154.

DAVIS W.E. and SLATER C.S. (1942) A direct weighing method for subsequent measures of soil moisture under field conditions. *J. Amer. Soc. Agron.* **34**, 285–288.

DECKER J.P. (1955) The uncommon denominator in photosynthesis as related to tolerance. *Forest Sci.* **1**, 88–89.

DECKER J.P., GAYLOR, W.G. and COLE F.D. (1962) Measuring transpiration of undisturbed tamarisk shrubs. *Plant Physiol.* **37**, 393–397.

DECKER J.P. and WIEN J.D. (1960) Transpiration surges in *Tamarix* and *Eucalyptus* as measured with an infra-red gas analyser. *Plant Physiol.* **35**, 340–343.

ECKARDT F.E. (1960) Ecophysiological measuring techniques applied to research on the water relations of plants. *Arid Zone Res.* **15**, 139–154.

EHLIG C.F. (1962) Measurement of energy status of water in plants with a thermocouple psychrometer. *Plant Physiol.* **37**, 288–290.

EVANS E.C. III and VAUGHAN B.E. (1966) New method for effecting watertight seals on corn roots. *Plant Physiol.* **41**, 1077–1078.

FALK F.O. (1966) A microwave hygrometer for measuring plant transpiration. *Z. Pflanzenphysiol.* **55**, 31–57.

FIELD C., BERRY J.A. and MOONEY H.A. (1982) A portable system for measuring carbon dioxide and water vapour exchange of leaves. *Pl. Cell Env.* **5**, 179–186.

FINNEY D.J. (1952) *Probit Analysis*, Cambridge University Press, Cambridge.

FLINT H.L., BOYCE B.R. and BEATTIE D.J. (1967) Index of injury—useful expresson of freezing injury to plant tissues as determined by the electrolytic method. *Can. J. Plant Sci.* **47**, 229–230.

FREEMAN G.F. (1908) A method for the quantitative determination of transpiration in plants. *Bot. Gaz.* **46**, 118–129.

FRY K.E. and WALKER R.B. (1967) A pressure infiltration method for measuring stomatal opening in conifers. *Ecology* **48**, 155–157.

GAFF D.F. and CARR D.J. (1964) An examination of the refractometric method for determining the water potential of plant tissues. *Ann. Bot.* **28**, 351–368.

GATES D.M. (1965) Energy, plants and ecology. *Ecology* **46**, 1–13.

GATES D.M. (1968a) Transpiration and leaf temperature. *Ann. Rev. Plant. Physiol.* **19**, 211–238.

GATES D.M. (1968b) Energy exchange in the biosphere. UNESCO, *Nat. Res. Res.* **5**, 33–43.

GAY A.P. (1983) Transit time diffusion porometer calibration: an analysis taking into account temperature differences and calibration non-linearity. *J. exp. Bot.* **34**, 461–469.

GIFFORD H.H., WHITEHEAD D., THOMAS R.S. and JACKSON D.S. (1982) Design of a new weighing lysimeter for measuring water use by forest trees. *N.Z. J. For. Sci.* **12**, 448–456.

GOLTZ S.M., BENOIT G. and SCHIMMELPFENNIG H. (1981) New circuitry for measuring soil water matric potential with moisture blocks. *Agric. Meteorol.* **24**, 75–82.

GRACE J., FASEHUN F.E. and DIXON M. (1980) Boundary layer conductance of the leaves of some tropical timber trees. *Pl. Cell Env.* **3**, 443–450.

GREER D.H. and WARRINGTON I.J. (1982) Effect of photoperiod, night temperature and frost incidence on the development of frost hardiness in *Pinus radiata*. *Aust. J. Plant. Physiol.* **9**, 333–342.

GREGORY F.G. and ARMSTRONG J.I. (1936) The diffusion porometer. *Proc. Roy. Soc. B* **114**, 477–493.

GREGORY F.G., MILTHORPE F.L., PEARSE H.L. and SPENCER H.J. (1950) Experimental studies of the factors controlling transpiration. I. Apparatus and experimental technique. *J. exp. Bot.* **1**, 1–14.

GREGORY F.G. and PEARSE H.L. (1934) The resistance porometer and its application to the study of stomatal movement. *Proc. Roy. Soc., B* **114**, 477–493.

GRIEVE B.J. and WENT F.W. (1965) An electric hygrometer apparatus for measuring water-vapour loss from plants in the field. *Arid Zone Res.* **25**, 247–257.

GROOT J. (1969) The use of silicone rubber plastic for replicating leaf surfaces. *Acta bot. Neerl.* **18**, 703–708.

GURR C.G. (1964) Calculation of soil water content from gamma ray readings. *Aust. J. Soil Res.* **2**, 29–32.

HALL A.E., SCHULZE E.-D. and LANGE O.L. (1976) Current perspectives of steady-state stomatal responses to the environment. In *Water and Plant Life* (ed. O.L. Lange, L. Kappen and E.-D. Schulze), pp. 169–188, Springer-Verlag, Berlin.

HAMMOUDA M. and LANGE O.L. (1962) Zur Hitzeresistenz der Blätter höherer Pflanzen in Abhängigkeit von ihrem Wassergehalt. *Die Naturwiss.* **49**, 500.

HANNA L.W. and SIAM N. (1980) The estimation of moisture content in the top 10 cm of soil using a neutron probe. *J. Agric. Sci.* **94**, 251–253.

HARRIS J.A. (1915) An extension to 5·99°C of tables to determine the osmotic pressure of expressed vegetable sap from the depression of the freezing point. *Amer. J. Bot.* **2**, 418–419.

HARRIS J.A. and GORTNER R.A. (1914) Notes on the calculation of osmotic pressure of expressed vegetable saps from the depression of freezing point, with a table for values of P for $\Delta = 0·001°C$ to $\Delta = 2·999°C$. *Amer. J. Bot.* **1**, 75–78.

HEATH O.V.S. (1950) Studies in stomatal behaviour. V. The role of carbon dioxide in the light response of stomata. *J. Exp. Bot.* **1**, 29–62.

HEATH O.V.S. (1969) *The Physiological Aspects of Photosynthesis*, Heinneman, London.

HEATH O.V.S. and MANSFIELD T.A. (1962) A recording porometer with detachable cups operating on four separate leaves. *Proc. Roy. Soc. B* **156**, 1–13.

HEATH O.V.S. and RUSSELL J. (1951) The Wheatstone bridge porometer. *J. exp. Bot.* **2**, 111–116.

HEATHERLEY L.G., McMICHAEL B.L. and GINN L.H. (1980) A weighing lysimeter for use in isolated field areas. *Agron. J.* **72**, 845–847.

HEINE R.W. and FARR D.J. (1973) Comparison of heat-pulse and radioisotope tracer methods for determining sap-flow velocity in stem sections of poplar. *J. exp. Bot.* **24**, 649–654.

HELLKVIST J., RICHARDS J.G.P. and JARVIS P.G. (1974) Vertical gradients of water potential and tissue water relations in sitka spruce trees measured with the pressure chamber. *J. appl. Ecol.* **11**, 637–667.

HELLMUTH E.O. and GRIEVE B.J. (1969) Measurement of water potential of leaves with particular reference to the Shardakow method *Flora, Jena* **159**, 147–167.

HEWLETT J.D. and KRAMER P.J. (1963) The measurement of water deficits in broadleaf plants. *Protoplasma* **57**, 381–391.

HINSON W.H. and KITCHING R.A. (1964) A readily constructed transistorised instrument for electrical resistance measurement in biological research. *J. appl. Ecol.* **1**, 301–305.

HIRST J.M. (1954) A method for recording the formation and persistence of water deposits on plant shoots. *Quart. J. Roy. meteorol. Soc.* **80**, 227–231.

HODGES J.D. and LORIO P.L. (1971) Comparison of field techniques for measuring moisture stress in large loblolly pines. *Forest Sci.* **17**, 220–223.

HOFFMAN G.J. and HALL A.E. (1976) Performance of the silver foil hygrometer for measuring leaf water potential *in situ*. *Agron. J.* **68**, 872–875.

HOFFMAN G.J. and RAWLINS S.L. (1972) Silver foil psychrometer for measuring leaf water potential *in situ*. *Science* **177**, 802–804.

HOFFMAN G.J. and SPLINTER W.E. (1968) Water potential measurements of an intact plant–soil system. *Agron. J.* **60**, 408–413.

HÖFLER K. (1917) Die plasmolytisch-volumetrische Methode zur Bestimmung des osmotischen Wertes lebender Pflanzenzellen. *Ber. dtsch bot. Ges.* **35**, 706–726.

HOLBO H.R. (1981) A dew point hygrometer for field use. *Agric. Meteorol.* **24**, 117–30.

HOLMGREN P., JARVIS P.G. and JARVIS M.S. (1965) Resistances to carbon dioxide and water vapour transfer in leaves of different plant species. *Physiol. Pl.* **18**, 557–573.

HSIAO T.C. and FISCHER R.A. (1975) Mass flow porometers. In *Measurement of Stomatal Aperture and Diffusive Resistance* (ed. E.T. Kanemasu), pp. 5–11, Bull. 809: Washington State Univ., Coll. of Agric. Res. Centre.

HSIEH J.J.C. and HUNGATE F.P. (1970) Temperature compensated Peltier psychrometer for measuring plant and soil water potentials. *Soil Sci.* **110**, 253–257.

HU JU G. and KRAMER P.J. (1967) Radial salt transport in corn roots. *Plant Physiol.* **42**, 985–990.

HUBER B. (1932) Beobachtung und Messung pflanzlicher Saftströme. *Ber. dtsch bot. Ges.* **50**, 89–109.

HUBER B. and MILLER R. (1954) Methoden zur Wasserdampf- und Transpirations-registrierung im laufenden Luftstrom. *Ber. dtsch bot. Ges.* **67**, 223–233.

HUBER B. and SCHMIDT E. (1937) Ein Kompensationsmethode zur thermoelektrische Messung langsamer Saftströme. *Ber. dtsch bot. Ges.* **55**, 514–529.

HUTCHINSON T.C. (1970) Lime chlorosis as a factor in seedling establishment on calcareous soils. II. The development of leaf water deficits in plants showing lime chlorosis. *New Phytol.* **69**, 143–157.

HYGEN G. (1953) On the transpiration decline of excised plant samples. *Norske Vid. Akad. Skr. I math. nat. Kl.* **1**, 1–84.

IDLE D.B. (1970) The calculation of transpiration rate and diffusion resistance of a single leaf from micro-meteorological information subject to errors of measurement. *Ann. Bot.* **34**, 159–176.

IRVING R.M. and LANPHEAR F.O. (1967) Environmental control of cold hardiness in woody plants. *Plant Physiol.* **42**, 1191–1196.

IVANOV L.A. (1928) Zur Methodik der Transpirationsbestimmung am Standort. *Ber. dtsch bot. Ges.* **46**, 306–310.

JANE G.T. and GREEN T.G.A. (1983a) Episodic forest mortality in the Kaimai Ranges, North Island, New Zealand. *N.Z. J. Bot.* **21**, 21–31.

JANE G.T. and GREEN T.G.A. (1983b) Utilisation of pressure–volume techniques and non-linear least squares analysis to investigate site induced stress in evergreen trees. *Oecologia* **57**, 380–390.

JARVIS P.G. (1971) The estimation of resistances to carbon dioxide transfer. In *Plant Photosynthetic Production. Manual of Methods*, (eds. Z. Šesták, J. Čatský and P.G. Jarvis), pp. 566–631, Junk N.V., The Hague.

JARVIS P.G. (1981) Stomatal conductance, gaseous exchange and transpiration. In *Plants and their Atmospheric Environment* (eds. J. Grace, E.D. Ford and P.G. Jarvis), pp. 175–204, Blackwell Scientific Publications, Oxford.

JARVIS P.G. and JARVIS M.S. (1963) The water relations of tree seedlings. IV. Some aspects of tissue water relations and drought resistance. *Physiol. Plant.* **16**, 501–516.

JARVIS P.G., ROSE C.W. and BEGG J.G. (1967) An experimental and theoretical comparison of viscous and diffusive flow resistances to gas flow through amphistomatous leaves. *Agric. Meteorol.* **4**, 103–117.

JARVIS P.G. and SLATYER R.O. (1966) Calibration of β-gauges for determining leaf water status. *Science* **153**, 78–79.

JENNINGS G. and MONTEITH J.L. (1954) A sensitive recording dew-balance. *Quart. J. Roy. meteorol. Soc.* **80**, 222–226.

JESSOP C.T. (1964) Development, testing and calibration of a portable evaporimeter suitable for field use. *N.Z. J. Agric. Res.* **7**, 205–218.

JONES F.E. and WEXLER A. (1960) A barium chloride film hygrometer element. *J. geophys. Res.* **65**, 2087–2095.

JONES H.G. (1973) Estimation of plant water status with the beta-gauge. *Agric. Meteorol.* **11**, 345–355.

KANEMASU E.T., THURTELL G.W. and TANNER C.B. (1969) Design, calibration and field use of a stomatal diffusion porometer. *Plant Physiol.* **44**, 881–885.

KAPPEN L. (1964) Untersuchungen über den Jahreslauf der Frost-, Hitze-, und Austrocknungsresistenz von Sporophyten einheimischer Polypodiaceen (*Filicinae*). *Flora, Jena* **155**, 123–166.

KAPPEN L. (1966) Der Einfluss des Wassergehalts auf die Widerstandsfähigkeit der Pflanzen gegenüber hohen und tiefen Temperaturen, untersucht an einiger Farne und von *Ramonda myconi*. *Flora. Jena* **B156**, 427–445.

KAPPEN L., LANGE O.L., SCHULZE E.-D., EVANARI M. and BUCHSBOM U. (1972) Extreme water stress and photosynthetic activity of the desert plant *Artemisia herba-alba* Asso. *Oecologia* **10**, 177–182.

KARLIC H. and RICHTER H. (1979) Storage of detached leaves and twigs without changes in water potential. *New Phytol.* **83**, 379–384.

KAY B.D. and LOW P.F. (1970) Measurement of the total suction of soils by a thermistor psychrometer. *Proc. Soil Sci. Soc. Amer.* **34**, 373–376.

KAYLL A.J. (1963) Heat tolerance of Scots pine seedling cambium using tetrazolium chloride to test viability. *Canad. Dep. For. Publn.* No. 1006.

KITCHING R. (1965) A precision portable electrical resistance bridge incorporating a centre zero null detector. *J. agric. eng. Res.* **10**, 264–266.

KLEMM M. and KLEMM W. (1964) Die Verwendung von Radioisotopen zur kontinuierlichen Bestimmung des Tagesverlaufes der Transpirationsstromsgeschwindigkeit bei Bäumen. *Flora. Jena* **154**, 89–93.

KLEMM W. (1966) Die Entwicklung der Durchstrahlungsmethode zur Untersuchung des Wasserhaushaltes von Bäumen. *Isotopenpraxis* **2**, 262–267.

KLUTE A. and PETERS D.B. (1962) A recording tensiometer with a short response time *Proc. Soil Sci. Soc. Amer.* **26**, 87–88.

KNIGHT R.C. (1917) The interrelations of stomatal aperture, leaf water content and transpiration rate. *Ann. Bot.* **31**, 221–240.

KNIGHT R.O. (1965) *The Plant in Relation to Water*. Dover Publications Inc., New York.

KONIS E. (1949) The resistance of maquis plants to supramaximal temperatures. *Ecology* **30**, 425–429.

KÖRNER CH., ALLISON A. and HILSCHER H. (1983) Altitudinal variation of leaf diffusive conductance and leaf anatomy in heliophytes of montane New Guinea and their interrelation with microclimate. *Flora, Jena* **174**, 91–135.

KÖRNER CH. and CERNUSCA A. (1975) A semi-automatic, recording diffusion porometer and its performance under alpine field conditions. *Photosynthetica* **10**, 172–181.

KÖRNER CH., SCHEEL J.A. and BAUER H. (1979) Maximum leaf diffusive conductance in vascular plants. *Photosynthetica* **13**, 45–82.

KREEB K. (1960) Über die gravimetrische Method zur Bestimmung der Saugspannung und das Problem des negativen Turgors. I. Mitteilung. *Planta* **55**, 274–282.

KREEB K. (1963) Untersuchungen zum Wasserhaushalt der Pflanzen unter extrem ariden Bedingungen. *Planta* **59**, 442–458.

KREEB K. (1965a) Untersuchungen über die Hitze- und Trockenresistenz an australischen Immergrünen im Keimlingstadium. *Ber. dtsch bot. Ges.* **78**, 90–98.

KREEB K. (1965b) Untersuchungen zu den osmotischen Zustandgrössen. I. Mitteilung. Ein tragbares elektronisches Microkryoscop fur ökophysiologische Arbeiten. *Planta* **65**, 269–279.

KREEB K. (1965c) Untersuchungen zu den osmotischen Zustandgrössen. II. Mitteilung. Ein elektronische Methode zur Messung der Saugspannung (NTC-Methode). *Planta* **66**, 156–164.

KREEB K. (1966) Die Registrierung des Wasserzustandes über die elektrische Leitfahigkeit der Blätter. *Ber. dtsch bot. Ges.* **79**, 150–162.

KREEB K. (1970) Methodische und öko-physiologische Untersuchungen zur Hitzeresistenz, inbesondere von *Eucalyptus*-Arten. *Angew. Bot.* **44**, 166–177.

KREEB K. (1977) *Methoden der Pflanzenökologie*. V.E.B. G. Fischer Verlag, Jena.

KREEB K. and ÖNAL M. (1961) Über die gravimetrische Methode zur Bestimmung der Saugspannung und das Problem des negativen Turgors. II. Mitteilung. Die Berucksichtigung von Atmungverlusten während der Messungen. *Planta* **56**, 409–415.

KYRIAKOPOULOS E. and RICHTER H. (1981) Pressure–volume curves and drought injury. *Physiol. Plant* **52**, 124–128.

LANG A.R.G. (1967) Osmotic coefficients and water potentials of sodium chloride solutions from 0 to 40°C. *Aust. J. Chem.* **20**, 2017–2023.

LANG A.R.G. (1968) Psychrometric measurement of soil water potential *in situ* under cotton plants. *Soil Sci.* **106**, 460–464.

LANG A.R.G. and TRICKETT E.S. (1965) Automatic scanning of Spanner and droplet psychrometers having outputs of up to 30 V. *J. sci. Instr.* **42**, 777–782.

LANGE O.L. (1953) Hitze- und Trockenresistenz der Flechten in Beziehung zu ihrer Verbreitung. *Flora, Jena* **140**, 39–97.

LANGE O.L. (1959) Untersuchungen über Wärmehaushalt und Hitzeresistenz mauretanische Wüsten- und Savannenpflanzen. *Flora, Jena* **147**, 595–651.

LANGE O.L. (1961) Die Hitzeresistenz einheimischer immer- und wintergrüner Pflanzen im Jahreslauf. *Planta. Berl.* **56**, 666–683.

LANGE O.L. (1965) The heat resistance of plants, its determination and variability. *Arid Zone Res.* **25**, 339–405.

LANGE O.L. and MATTHES U. (1981) Moisture-dependent CO_2 exchange of lichens. *Photosynthetica* **15**, 555–574.

LANGE O.L. and LANGE R. (1963) Untersuchungen über Blatttemperaturen, Transpiration und Hitzeresistenz an Pflanzen mediterraner Standorte (Costa Brava, Spanien). *Flora, Jena* **153**, 387–425.

LARCHER W. (1977a) Kälteresistenz. In *Methoden der Pflanzenökologie* (ed. K. Kreeb), pp. 51–59. VEB G. Fischer Verlag, Jena.

LARCHER W. (1977b) Vitalitätsbestimmungen. In *Methoden der Pflanzenökologie* (ed. K. Kreeb) pp. 182–195. VEB G. Fischer Verlag, Jena.

LARCHER W. and MAIR B. (1968) Das Kälteresistenzverhalten von *Quercus pubescens, Ostrya carpinifolia* und *Fraxinus ornus* auf drei thermisch verschiedenen Standorten. *Oecol. Plant.* **3**, 255–270.

LAWLOR D.W. (1969) Plant growth in polyethylene glycol solutions in relation to the osmotic potential of the root medium and the leaf water balance. *J. exp. Bot.* **20**, 895–911.

LEACH J.E., WOODHEAD T. and DAY W. (1982) Bias in pressure chamber measurements of leaf water potential. *Agric. Meteorol.* **27**, 257–263.

LEONARD R.A. and LOW P.F. (1962) A self-adjusting null-point tensiometer. *Proc. Soil Sci. Soc. Amer.* **26**, 123–125.

LEPESCHKIN W.W. (1912) Zur Kenntnis der Einwirkung supramaximaler Temperaturen auf die Pflanze. *Ber. dtsch bot. Ges.* **30**, 703–714.

LEVITT J. (1980) *Responses of Plants to Environmental Stresses. Vol 1. Chilling, Freezing and High Temperature Stresses.* Academic Press, New York.

LEWIS M.C. (1972) The physiological significance of variation in leaf structure. *Sci. Prog.* **60**, 25–51.

LEYTON L. (1970) Problems and techniques in measuring transpiration from trees. In *Physiology of Tree Crops.* (eds. L.C. Luckwill, and C.V. Cutting), pp. 101–112. Academic Press, New York.

LLOYD F.C. (1908) The physiology of stomata. *Publ. Carnegie Inst. Washington* **82**, 1–12.

LOPUSHINSKY W. (1969a) A portable apparatus for estimating stomatal aperture in conifers. *Pacific N.W. For. & Range exp. Sta.*, pp. 1–7, Portland, Oregon.

LOPUSHINSKY W. (1969b) Stomatal closure in conifer seedlings in response to leaf water stress. *Bot. Gas.* **130**, 258–263.

MACKLON A.E.S. and WEATHERLEY P.E. (1965a) A vapour pressure instrument for the measurement of leaf and soil water potential. *J. exp. Bot.* **16**, 261–270.

MACKLON A.E.S. and WEATHERLEY P.E. (1965b) Controlled environment studies of the nature and origins of water deficits in plants. *New Phytol.* **64**, 414–427.

MACKAY D.B. (1972) The measurement of viability. In *The Viability of Seeds* (ed. E.H. Roberts), pp. 172–208. Chapman and Hall, London.

MANOHAR M.S. (1966) Measurement of water potential in intact plant tissues. I. Design of a micro-thermocouple psychrometer. *J. exp. Bot.* **17**, 44–50.

MARK A.F. and HOLDSWORTH D.K. (1979) Yield and macronutrient content of water in relation to plant cover from snow tussock grassland zone of Central Otago, New Zealand. *Progress in Water Technology* **11**, 449–462.

MARK A.F. and SMITH P.M.F. (1962) A frost tolerant porous pot evaporimeter. *Proc. N.Z. Ecol. Soc.* **9**, 4–13.

MARSHALL D.C. (1958) Measurement of sap flow in conifers by heat transport. *Plant Physiol.* **33**, 385–396.

MAXIMOV N.A. (1929) *The Plant in Relation to Water.* Allen & Unwin, London.

MAXIMOV N.A. (1932) The physiological significance of the xeromorphic structure of plants. *J. Ecol.* **19**, 273–282.

MEIDNER H. (1965a) Stomatal control of transpirational water loss. In *The State and Movement of Water in Living Organisms*, Symp. Soc. exp. Biol., No. 19, pp. 185–204. Cambridge University Press, Cambridge.

MEIDNER H. (1965b) A simple porometer for measuring the resistance to air flow offered by stomata. *School Sci. Rev.* **47**, 149–151.

MEIDNER H. and MANSFIELD T.A. (1968) *Physiology of Stomata*, McGraw-Hill, London.

MELKONIAN J.J., WOLLFE J. and STEPONKUS P.L. (1982) Determination of the volumetric modulus of wheat leaves by pressure–volume relations and the effects of drought conditioning. *Crop Sci.* **22**, 116–123.

MERILL S.D., DALTON F.N., HEKELRATH W.N., HOFFMAN G.J., INGVALSON R.D., OSTER J.D. and RAWLINS S.L. (1968) Details of construction of a multipurpose thermocouple psychrometer. *US Dep. Agric., US Salinity Lab. Res. Rep.* **115**, 1–9.

METEOROLOGICAL OFFICE, LONDON (1980) *Handbook of Meteorological Instruments* (2nd edn.) Vol 5, *Measurement of Precipitation and Evaporation*, HMSO, London.

METEOROLOGICAL OFFICE, LONDON (1981) *Handbook of Meteorological Instruments* (2nd edn), Vol 3, *Measurements of Humidity*, HMSO, London.

MEYER B.S. (1945) A critical evaluation of the terminology of diffusion phenomena. *Plant Physiol.* **20**, 142–164.

MICHAEL G. (1969) Eine Methode zur Bestimmung der Spaltöffnungsweite von Koniferen. *Flora, Jena* **159A**, 559–561.

MILBURN J.A. (1979a) *Water Flow in Plants*, Longman, London.

MILBURN J.A. (1979b) An ideal viscous flow porometer. *J. exp. Bot.* **30**, 1021–1034.

MILLAR A.A., LANG A.R.G. and GARDNER W.R. (1970) Four-terminal Peltier type thermocouple psychrometer for measuring water potential in nonisothermal systems. *Agron. J.* **62**, 705–708.

MILLAR B.D. (1971a) Improved thermocouple psychrometer for the measurement of plant and soil water potential. I. Thermocouple psychrometry and an improved instrument design. *J. exp. Bot.* **22**, 875–890.

MILLAR B.D. (1971b) Improved thermocouple psychrometer for the measurement of plant and soil water potential. II. Operation and calibration. *J. exp. Bot.* **22**, 891–905.

MILLAR B.D. (1974) Improved thermocouple psychrometer for the measurement of plant and soil water potential. III. Calibration. *J. exp. Bot.* **25**, 1070–1084.

MILTHORPE F.L. (1961) Plant factors involved in transpiration. *Arid Zone Res.* **16**, 197–115.

MOLISCH H. (1912) Das Offen- und geschlossensein der Spaltöffnungen, veranschaulicht durch eine neue Methode (Infiltrationsmethode). *Z. Bot.* **4**, 106–122.

MONTEITH J.L. (1965) Evaporation and environment. In *The State and Movement of Water in Living Organisms*, Symp. Soc. exp. Biol., Vol 19, pp. 205–224, Cambridge University Press, Cambridge.

MONTEITH J.L. (1972) *Survey of Instruments for Micrometeorology*. I.B.P. Handbook No. 22, Blackwell Scientific Publications, Oxford.

MONTEITH J.L. (1981) Coupling of plants to the atmosphere. In *Plants and their Atmospheric Environment* (eds J. Grace, E.D. Ford and P.G. Jarvis), pp. 1–29. Blackwell Scientific Publications, Oxford.

MONTEITH J.L. and OWEN P.C. (1958) A thermocouple method for measuring relative humidity in the range 95–100%. *J. Sci. Instr.* **35**, 443–446.

MUNRO D.S. (1982) On determining volumetric soil moisture variations in peat by the gravimetric method. *Soil Sci.* **133**, 102–110.

NAKAYAMA F.S. and EHRLER W.L. (1964) Beta-ray gauging techniques for measuring leaf water content changes and moisture status of plants. *Plant Physiol.* **39**, 95–98.

NELMS L.R. and SPOMER L.A. (1980) Construction of serviceable, fast-response tensiometers from porous ceramic tubes. *Agron. J.* **72**, 694–695.

NELSON C.E., SAFIR G.R. and HANSON A.D. (1978) Water potential in excised leaf tissue. *Plant Physiol.* **61**, 131–133.

NEUMANN H.H. and THURTELL G.W. (1972) A Peltier cooled thermocouple dewpoint hygrometer for *in situ* measurements of water potentials. In *Psychrometry in Water Relations Research* (ed. R.W. Brown and B.P. van Haveren), pp. 193–112, Utah Agri. Exp. Stn., Utah State University.

NEWBOULD P., MERCER E.R. and LAY P.M. (1968) Estimation of changes in the water status of soils under field conditions from the attenuation of β-radiation by water held in absorbent nylon pads. *Exp. Agr.* **4**, 167–177.

NORTH C. (1956) A technique for measuring structural features of the plant surface using cellulose acetate films. *Nature* **178**, 1186–1187.

OELSEN S.E. (1973) Gamma radiation for measuring water contents in soil columns with changing bulk density. *J. Soil Sci.* **24**, 461–469.

OERTLI, J.J., ACEVES-NAVARRO E. and STOLZY L.H. (1975) Interpretation of foliar water potential measurements. *Soil Sci.* **119**, 162–166.

OGDEN J. (1976) Notes on the influence of drought on the bush remnants of the Manawatu Lowlands. *Proc. N.Z. Ecol. Soc.* **23**, 92–98.

PAINTER R.B. (1976) Climatology and environmental measurement. Ch. 7 In *Methods in Plant Ecology* (1st edn) (ed. S.B. Chapman), pp. 369–410 Blackwell Scientific Publications, Oxford.

PARKHURST D.F. and LOUCKS O.L. (1972) Optimal leaf size in relation to environment. *J. Ecol.* **60**, 505–537.

PARKINSON K.J. and LEGG B.J. (1972) A continuous flow porometer. *J. appl. Ecol.* **9**, 669–675.

PARLANGE J.Y. and WAGGONER P.E. (1970) Stomatal dimensions and resistance to diffusion. *Plant Physiol.* **46**, 337–342.

PEARCE R.S. (1980) Relative hardiness to freezing of laminae, roots and tillers of tall fescue. *New Phytol.* **84**, 449–463.

PECK A.J. and RABBIDGE R.M. (1966) Soil water potential: direct measurement by a new technique. *Science* **151**, 1385–1386.

PENMAN H.L. (1948) Natural evaporation from open water, bare soil and grass. *Proc. Roy. Soc., London* (A) **193**, 120–145.

PENMAN H.L. and SCHOFIELD R.K. (1951) Some physical aspects of assimilation and transpiration. *Symp. Soc. exp. Biol.* **4**, 115–129.

PETTERSSON S. (1966). Artificially induced water and sulfate transport through sunflower roots. *Physiol. Plant.* **19**, 581–601.

POLWART A. (1970) *Ecological aspects of the resistance of plants to environmental factors.* Unpublished Ph.D. thesis, University of Glasgow.

PREBBLE R.E. and CURRIE J.A. (1970) Soil water measurement by a low-resolution nuclear magnetic resonance technique. *J. Soil. Sci.* **21**, 273–288.

RAMSAY J.A. and BROWN R.H.J. (1955) Simplified apparatus and procedure for freezing point determination upon small volumes of fluid. *J. Sci. Instr.* **32**, 372–375.

RATCLIFF L.F., RITCHIE J.T. and CASSEL D.K. (1983) A survey of field measured limits of soil water availability and related laboratory measured properties. *Soil Sci. Soc. Am. J.* **47**, 770–775.

RAWLINS S.L. (1971) Some new methods for measuring the components of water potential. *Soil Sci.* **112**, 8–16.

RAWLINS S.L. and DALTON F.N. (1967) Psychrometric determination of soil water potential without precise temperature control. *Soil Sci.* **112**, 8–16.

REDSHAW A.J. and MEIDNER H. (1970) A thermal method for estimating continuously the sap flow through an intact plant. *Z. Pflanzenphysiol.* **62**, 405–416.

REILEY J.O., MACHIN D. and MORTON A. (1969) The measurement of microclimatic factors under a vegetation canopy—a reappraisal of Wilm's method. *J. Ecol.* **57**, 101–108.

RICE R. (1969) A fast response, field tensiometer system. *Trans. Amer. Soc. agric. Eng.* **12**, 48–50.

RICHARDS B.G. (1965) Thermistor hygrometer for determining the free energy of moisture in unstaurated soils. *Nature* **208**, 608–609.

RICHARDS L.A. (1941) A pressure-membrane extraction apparatus for soil solution. *Soil Sci.* **51**, 377–386.

RICHARDS L.A. (1948) Porous plate apparatus for measuring moisture retention and transmission by soil. *Soil Sci.* **66**, 105–110.

RICHARDS L.A. and OGATA G. (1958) Thermocouple for vapour pressure measurement in biological and soil systems at high humidity. *Proc. Soil Sci. Soc. Amer.* **2**, 55–64.

RICHTER H. (1978a) Water relations of single drying leaves: evaluation with a dew point hygrometer. *J. exp. Bot.* **29**, 277–280.

RICHTER H. (1978b) A diagram for the description of water relations in plant cells and organs. *J. exp. Bot.* **29**, 1197–1203.

RICHTER H., DUHME F., GLATZEL G., HINKLEY T.M. and KARLIC H. (1981) Some limitations and applications of the pressure–volume curve technique in ecophysiological research. In *Plants and Their Atmospheric Environment* (eds J. Grace, E.D. Ford and P.G. Jarvis), pp. 263–272, Blackwell Scientific Publications, Oxford.

RICHTER H. and ROTTENBURG W. (1971) Leitfähigkeitmessung zur Endpunktanzeige bei der Saugspannungsbestimmung nach Scholander. *Flora, Jena* **160**, 440–443.

RITCHIE G.A. and HINKLEY T.M. (1975) The pressure chamber as an instrument for ecological research. *Adv. Ecol. Res.* **9**, 165–254.

ROBERTS J. (1976) An examination of the quantity of water stored in mature *Pinus sylvestris* L. trees. *J. exp. Bot.* **27**, 473–479.

ROBERTS S.W., STRAIN B.R. and KNOERR K.R. (1980) Seasonal patterns of leaf water relations in four co-occurring forest tree species: parameters from pressure–volume curves. *Oecologia* **46**, 330–337.

ROGERS J.S. (1974) Small laboratory tensiometers for field and laboratory studies. *Proc. Soil Sci. Soc. Amer.* **38**, 690–691.

ROUSCHAL A. (1938) Zum Wärmehaushalt der Macchienpflanzen. *Oesterr. Bot. Z.* **87**, 42–50.

ROWLEY J. (1970) Lysimeter and interception studies in narrow-leaved snow tussock grassland. *N.Z. J. Bot.* **8**, 478–493.

RUTTER A.J. (1968) Water consumption by forests. In *Water Deficits and Plant Growth* (ed. T.T. Kozlowski), pp. 23–84, Academic Press, London.

RUTTER A.J. and SANDS K. (1958) The relation of leaf water deficit to soil moisture tension in *Pinus sylvestris* L. I. The effect of soil moisture on diurnal changes in water balance. *New Phytol.* **57**, 50–65.

RYCHNOVSKÁ M. (1965) Water relations of some steppe plants investigated by means of the revesibility of the water saturation deficit. In *Water Stress in Plants* (ed. B. Slavík), pp. 108–116, Czech. Acad. Sci., Prague.

RYCHNOVSKÁ-SOUDKOVÁ M. (1963) Study of the reversibility of the water saturation deficit as one of the methods of causal phytogeography. *Biol. Plant.* **5**, 175–180.

SADDLER H.D.W. and PITMAN M.G. (1970) An apparatus for the measurement of sap flow in unexcised leafy shoots. *J. exp. Bot.* **21**, 1048–1059.

SAKAI A. and WARDLE P. (1978) Freezing resistance of New Zealand trees and shrubs. *N.Z. J. Ecol.* **1**, 51–61.

SAKURATANI F. (1981) A heat balance method for measuring water flux in the stems of intact plants. *J. agric. Meteorol.* **37**, 9–17.

SALTER P.J. and HAWORTH F. (1961) The available water capacity of a sandy soil. 1. A critical comparison of methods of determining the moisture content of soil at field capacity and at the permanent wilting percentage. *J. Soil Sci.* **12**, 326–334.

SALTER P.J. and WILLIAMS J.B. (1965) The influence of texture on the moisture characteristics of soils. 1. A critical comparison of methods for determining the available-water capacity and moisture characteristic curve of a soil. *J. Soil Sci.* **16**, 1–15.

SAMPSON J. (1961) A method of replicating dry or moist surfaces for examination by light microscopy. *Nature* **191**, 932–933.

SAVAGE M.J., CASS A. and DE JAGER J.M. (1981a) Measurement of water potential using thermocouple hygrometers. *S. Afr. J. Sci.* **77**, 24–27.

SAVAGE M.J., CASS A. and DE JAGER J.M. (1981b) Calibration of thermocouple hygrometers using the dew-point technique. *S. Afr. J. Sci.* **77**, 323–329.

SCHOFIELD R.K. (1935) The pF of water in the soil. *Trans. 3rd Int. Cong. Soil Sci.* **2**, 37–48.

SCHOLANDER P.F., HAMMEL H.T., BRADSTREET E.D. and HEMMINGSEN E.A. (1965) Sap pressure in vascular plants. *Science* **148**, 339–346.

SCHOLANDER P.F., HAMMEL H.T., HEMMINGSEN E.A. and BRADSTREET E.D. (1964) Hydrostatic pressure and osmotic potentials in leaves of mangroves and some other plants. *Proc. nat. Acad. Sci. USA* **51**, 119–125.

SCHÖNHERR J. and BUKOVAC M.J. (1972) Penetration of stomata by liquids. Dependence on surface tension, wettability, and stomatal morphology. *Plant Physiol.* **49**, 813–819.

SCHRATZ E. (1932) Untersuchungen über die Beziehung zwischen Transpiration und Blattstruktur. *Planta* **16**, 17–69.

SCHULZE E.-D., HALL A.E., LANGE O.L. and WALZ H. (1982) A portable steady-state porometer for measuring the carbon dioxide and water vapour exchanges of leaves under natural conditions. *Oecologia(Berl)* **53**, 141–145.

ŠESTÁK Z., ČATSKÝ J. and JARVIS P.G. (1971) *Plant Photosynthetic Production: Manual of Methods.* Junk, The Hague.

SEYBOLD A. (1929) *Die physikalische Komponente der pflanzlicher Transpiration.* Julius Springer, Berlin.

SHERIFF D.W. (1972) A new method for the measurement of sap flux in small shoots with the magnetohydrodynamic technique. *J. exp. Bot.* **23**, 1086–1095.

SHIMSHE D. and LIVNE A. (1967) The estimation of the osmotic potential of plant sap by refractometry and conductivity. A field method. *Ann. Bot.* **31**, 506–511.

SHMUELI E. and COHEN O.P. (1964) A critique of Walter's hydrature concept and of his evaluation of water status measurement. *Israel J. Bot.* **13**, 199–207.

SHMUELI E. and COHEN O.P. (1967) In support of the critique on the hydrature concept. *Israel J. Bot.* **16**, 45–47.

SINCLAIR R. and VENABLES W.N. (1983) An alternative method for analysing pressure–volume curves produced with the pressure chamber. *Pl. Cell Env.* **6**, 211–217.

SIVADJIAN J. (1952) Recherches sur la transpiration des plantes par la methode hygrophotographique. *J. Bull. Bot. Soc. Fr.* **99**, 138–141.

SLATYER R.O. (1958) The measurement of diffusion pressure deficit in plants by a method of vapour equilibration. *Aust. J. Biol. Sci.* **11**, 349–365.

SLATYER R.O. (1967) *Plant Water Relationships.* Academic Press, New York.

SLATYER R.O. and BARRS H.D. (1965) Modifications to the relative turgidity technique with notes on its significance as an index of the internal water status of leaves. *Arid Zone Res.* **25**, 331–342.

SLATYER R.O. and JARVIS P.G. (1966) Gaseous diffusion porometer for continuous measurement of diffusive resistance of leaves. *Science N.Y.* **151**, 574–576.

SLATYER R.O. and McILROY I.C. (1961) *Practical Microclimatology, With Special Reference to the Water Factor in Soil–Plant–Atmosphere Relationships,* UNESCO, CSIRO, Australia.

SLAVÍK B. (1974) *Methods of Studying Plant Water Relations,* Academia, Prague. Springer-Verlag, Berlin.

SPANNER D.C. (1951) The Peltier effect and its use in the measurement of suction pressure. *J. exp. Bot.* **2**, 145–168.

SPANNER D.G. and HEATH O.V.S. (1952) Experimental studies of the relation between carbon assimilation and stomatal aperture. II. The use of the resistance porometer in estimating stomatal aperture and diffusive resistance. *Ann. Bot.* **5**, 319–334.

STAHL E. (1894) Einige Versuche über Transpiration und Assimilation. *Bot. Ztg.* **52**, 117–146.

STÅLFELT M.G. (1932) Die stomatäre Regulation in der pflanzlichen Transpiration. *Planta* **17**, 22–32.

STEPONKUS P.L. and LANPHEAR F.O. (1967) Refinement of the triphenyl tetrazolium chloride method of determining cold injury. *Plant Physiol.* **42**, 1423–1426.

STEWART V.I. and ADAMS W.A. (1968) The quantitative description of soil moisture states in natural habitats with special reference to moist soils. In *The Measurement of Environmental Factors in Terrestrial Ecology* (ed. R.M. Wadsworth), pp. 161–180, Blackwell Scientific Publications, Oxford.

STILES W. (1970) A diffusive porometer for field use. I. Construction. *J. appl. Ecol.* **7**, 617–622.

STOCKER O. (1929). Das Wasserdefizit von Gefässpflanzen in verschiedenen Klimazonen. *Planta* **7**, 382–387.

STOCKER O. (1956) Messmethoden der Transpiration. In *Handbuch der Pflanzenphysiologie* (1st edn.) (ed. W. Ruhland), pp. 293–311, Springer Verlag, Berlin.

STONE J.F. and SHIRAZI G.A. (1975). On the heat-pulse method for the measurement of apparent sap velocity in stems. *Planta* **122**, 169–177.

STONE J.F., KIRKHAM D. and READ A.A. (1955) Soil moisture determination by a portable neutron scatter meter. *Proc. Soil Soc. Amer.* **19**, 418–423.

SWANSON R.H. (1962) *An instrument for detecting sap movement in woody plants*. Rocky Mountain Forest and Range Exp. Sta., Fort Collins, Colorado, Paper No. 68, pp. 1–16.

TALBOT A.J.B., TYREE M.T. and DAINTY J. (1975) Some notes concerning the measurement of water potential in leaf tissue with particular reference to *Tsuga canadensis* and *Picea abies*. *Can. J. Bot.* **53**, 784–788.

TANNER C.B., ABRAMS E. and ZUBRISKI J.C. (1949) Gypsum moisture block calibration based on electrical conductivity in distilled water. *Proc. Soil. Soc. Amer.* **13**, 62–65.

TAYLOR S.A. (1955) Field determination of soil moisture. *Agric. Eng.* **36**, 657–659.

TAYLOR S.A. and SLATYER R.O. (1962) Proposals for a unified terminology in studies of plant–soil–water relationships. *Arid Zone Res.* **16**, 339–349.

THOM A.S. (1968) The exchange of momentum mass and heat between an artificial leaf and the flow of air in a wind tunnel. *Quart. J. Roy. met. Soc.* **94**, 44–55.

THREN R. (1934) Jahreszeitliche Schwankungen des osmotischen Wertes verschiedener ökologischer Typen in der Umgebung von Heidelberg. *Z. Bot.* **26**, 449–526.

TILL O. (1956) Über die Frosthärte von Pflanzen sommer-grüner Laubwälder. *Flora, Jena* **143**, 498–542.

TINKLIN R. and WEATHERLEY P.E. (1966) The role of root resistance in the control of leaf water potential. *New Phytol.* **65**, 509–517.

TINKLIN R. and WEATHERLEY P.E. (1968) The effect of transpiration rate on the leaf water potential of sand and soil rooted plants. *New Phytol.* **67**, 605–615.

TINKLIN R. (1967) Note on the determination of leaf water-potential. *New Phytol.* **66**, 85–88.

TOMAR V.S. and O'TOOLE J.C. (1980) Design and testing of a microlysimeter for wetland rice. *Agron. J.* **72**, 689–693.

TURNER N.C. (1981) Techniques and experimental apparatus for the measurement of plant water status. *Plant and Soil* **58**, 339–366.

TURNER N.C. and LONG M.J. (1980) Errors arising from rapid water loss in the measurement of leaf water potential by the pressure chamber technique. *Aust. J. Plant Physiol.* **7**, 527–537.

TURNER N.C. and PARLANGE J.Y. (1970) Analysis of operation and calibration of a ventilated diffusion porometer. *Plant. Physiol., Lancaster* **46**, 175–177.

TYREE M.T., DAINTY J. and BENIS M. (1973) The water relations of hemlock (*Tsuga canadensis*). I. Some equilibrium water relations as measured by the pressure-bomb technique. *Can. J. Bot.* **51**, 1471–1480.

TYREE M.T. and HAMEL H.T. (1972) The measurement of turgor pressure and the water relations to plants by the pressure-bomb technique. *J. exp. Bot.* **23**, 267–282.

TYREE M.T. and KARAMANOS A.J. (1981) Water stress as an ecological factor. In *Plants and Their Atmospheric Environment* (ed. J. Grace, E.D. Ford and P.G. Jarvis), pp. 237–262, Blackwell Scientific Publications, Oxford.

TYREE M.T. and ZIMMERMAN M.H. (1972) The theory and practice of measuring transport coefficients in sap flow in the xylem of red maple stems (*Acer rubrum*). *J. exp. Bot.* **22**, 1–18.

ULMER W. (1937) Über den Jahresgang der Frosthärte einiger immergrüner Arten der alpinen Stufe, sowie der Zirbe und Fichte. *Jb. wiss. Bot.* **84**, 553–592.

URSPRUNG A. and BLUM G. (1916) Zur Methode der Saugkraftmessungen. *Ber. dtsch bot. Ges.* **34**, 525–549.

VIEWEG G.H. and ZIEGLER H. (1960) Thermoelektrische Registrierung der Geschwindigkeit des Transpirationsstromes. I. *Ber. dtsch bot. Ges.* **73**, 211–226.

VISVALINGHAM M. and TANDY J.D. (1972) The neutron method for measuring soil moisture

WALTER H. (1936) Tabellen zur Berechnung des osmotischen Werter von Pflanzenpressaften, Zuckerlösungen und einigen Salzlösungen. *Ber. dtsch bot. Ges.* **34**, 328–339.

WALTER H. (1963) Zur Klärung des spezifischen Wasserzustandes im Plasma und in der Zellwand bei der höheren Pflanzen und seine Bestimmung. I. Allgemeines. II. Methodisches. *Ber. dtsch Bot. Ger.* **76**, 40–71.

WARDLE P. (1981) Winter desiccation of conifer needles simulated by artificial freezing. *Arctic and Alpine Res.* **13**, 419–423.

WARING R.H. and ROBERTS J.M. (1979) Estimating water flux through stems of Scots Pine with tritiated water and phosphorus-32. *J. exp. Bot.* **30**, 459–471.

WARING R.H. and RUNNING S.W. (1978) Sapwood water storage: its contribution to transpiration and effect upon water and conductance through the stems of old-growth Douglasfir. *Plant Cell Env.* **1**, 81–97.

WARING R.H., WHITEHEAD D. and JARVIS P.G. (1979) The contribution of stored water to transpiration in Scots Pine. *Plant Cell Env.* **2**, 309–317.

WARREN-WILSON J. (1959) Notes on wind and its effect on Arctic Alpine vegetation. *J. Ecol.* **47**, 415–427.

WATERS P. (1980). Comparison of the ceramic plate and the pressure membrane to determine the 15 bar water content of soil. *J. Soil Sci.* **31**, 443–446.

WEATHERLEY P.E. (1950) Studies in the water relations of the cotton plant. I. The field measurement of water deficits in leaves. *New Phytol.* **48**, 81–97.

WEATHERLEY P.E. (1960) A new micro-osmometer. *J. exp. Bot.* **11**, 250–260.

WEATHERLEY P.E. (1966) A porometer for use in the field. *New Phytol.* **65**, 376–387.

WEATHERLEY, P.E. and SLATYER R.O. (1957) Relationship between relative turgidity and diffusion pressure deficit in leaves. *Nature (London)* **179**, 1085–1086.

WEBSTER R. (1966) The measurement of soil water tension in the field. *New Phytol.* **65**, 249–258.

WEINBERGER P., ROMERO M. and OLIVA M. (1972) Ein kritische Beitrag zur Bestimmung des subletalen (kritischen) Wassersättigungsdefizits. *Flora, Jena* **161**, 555–561.

WEINEMANN H. and LE ROUX M. (1946) A critical study of the torsion balance of measuring transpiration. *S. Afr. J. Sci.* **42**, 147–163.

WEISS A. and HAGEN A.F. (1983) Further experiments in the measurement of leaf wetness. *Agric. Meteorol.* **29**, 207–212.

WEISS A. and LUKENS D.L. (1981) An electronic circuit for the detection of leaf wetness and a comparison of two sensors under field conditions. *Pl. Disease* **65**, 41–53.

WILLIAMS T.H. (1980) An automatic electrical resistance moisture measuring system. *J. Hydrol. Sci.* **46**, 385–390.

WILLIS A.J. and JEFFERIES E.J. (1963) Investigations on the water relations of sand dune plants. In *Water Relations of Plants* (eds. A.J. Rutter & F.H. Whitehead), pp. 168–189, Blackwell Scientific Publications, Oxford.

WILSON J.R., FISHER M.J., SCHULZE E.-D., DOLBY G.R. and LUDLOW M.M. (1979) Comparison between pressure–volume and dew-point hygrometry techniques for determining the water relations characteristics of grass and legume leaves. *Oecologia* **41**, 77–88.

WILSON J.R., LUDLOW M.M., FISHER M.J. and SCHULZE E.-D. (1980) Adaptation to water stress of the leaf water relations of four tropical forage species. *Aust. J. Plant Physiol.* **7**, 207–220.

YEMM E.W. and WILLIS A.J. (1954) Stomatal movements and changes in carbohydrate content in leaves of *Chrysanthemum maximum*. *New Phytol.* **53**, 373–396.

ZANSTRA P.E. and HAGENZIEKER F. (1977) Comments on the psychrometric determinations of leaf water potential *in situ*. *Plant and Soil* **48**, 347–368.

content—a review. *J. Soil Sci.* **23**, 499–511.

WALTER H. (1931) *Die Hydratur der Pflanzen und ihre physiologisch-ökologische Bedeutung*, Gustav Fischer Verlag, Jena.

4 Mineral nutrition

I.H. RORISON and DAVID ROBINSON

1 Introduction

Plants contain about sixteen elements that are metabolically essential: C, O, H, N, P, K, Ca, S, Mg, Mn, Mo, Fe, B, Cu, Zn, Cl. Of these, the first two originate in the atmosphere as CO_2 and O_2, respectively. The others come largely from the soil. With the exception of H, absorbed as H_2O, and N in those plants associated with diazotrophic (N_2-fixing) micro-organisms, these nutrients are absorbed from the soil solution as mineral ions. In addition, there are a number of soil elements that plants can absorb, but which in most cases are not usually essential and can be toxic at low concentrations: e.g. Pb, Cd, Al, Ni, Se, Cr, F, Na, Co. The chemistry and metabolic functions of these elements in plants have been reviewed recently (Clarkson & Hanson, 1980; Mengel & Kirkby, 1982; Marschner, 1983). The acquisition, utilization and tolerance of these elements constitute a fundamental component of a plant's physiology and can have a profound effect upon the ecology and evolution of species and populations.

Current approaches to the study of the ecological aspects of plant mineral nutrition rely upon both laboratory experiments and field investigations. The levels of biological organization considered can range from enzymes to the nutrient dynamics of ecosystems. Consequently, the techniques used often originate in disciplines that are not normally classed as 'ecology'. Some of these techniques are reviewed in this chapter, while those relating particularly to sample collection and chemical analysis are described in Chapter 6.

We have not attempted to list a series of 'recipes' in this chapter. Every investigation into the ecological aspects of plant nutrition demands a unique combination of procedures and methods, so such an approach would have been of limited use. Instead, we have provided a number of guidelines for practical research, and suggested references to original sources where specific methods are described fully. These should not be regarded necessarily as 'standard' procedures to which no modifications need to be made. Any method can be improved or adapted to suit the precise requirements of an investigation. This is particularly so in the case of methods that have been developed and tested for agricultural purposes, which may not be directly applicable for research into ecological aspects of the mineral nutrition of native species.

2   Experimental approaches

There are up to six general phenomena that could be considered as experimental variables or controlled factors, depending upon the precise objectives of the investigation.

(i) Nutrient availability.

(ii) Nutrient sources.

(iii) Uniformity of nutrient supply.

(iv) Effects of nutrients upon specific plant processes.

(v) Effects of specific environmental factors upon plants, in relation to their nutrition.

(vi) Community–nutrient relations.

2.1   Nutrient availability

2.1.1   Definition and measurement

Quantitative definitions of nutrient availability depend upon the media in which the plants are grown and, equally, the nutrient demands of the plants themselves.

In nutrient solutions, the steady-state ionic concentration in the bulk solution ($mol\,dm^{-3}$) is the usual definition of availability. Alternatively, steady-state concentrations can be combined with steady rates of solution supply ($dm^3\,d^{-1}$) to express availabilities in terms of $mol\,dm^{-3}\,d^{-1}$ or $mol\,plant^{-1}\,d^{-1}$ (Ingestad, 1982). The distinction between steady-state nutrient *concentrations* and *total* nutrient supplies during the experiment is important (Section 3.1.5). Species may differ in their requirements for minimum nutrient concentrations, and/or minimum total supplies (Rorison, 1969).

The ionic composition of the soil solution is regulated largely by processes of ion exchange and electrochemical equilibria, which have been reviewed in depth by Adams (1974) and Thomas (1974). Rather than ion concentration, it is their rates of diffusion between the bulk soil and root surface that usually limit the rate of nutrient uptake. Therefore, the ion concentration at the root surface is reduced below that in the bulk soil (Fig. 4.1). The extent of this depletion is difficult to measure *in situ*, but a small number of methods are available: including quantitative autoradiography (Sanders, 1971; Bhat & Nye, 1973), microsectioning rhizosphere soil (Bagshaw *et al.*, 1972; Kuchenbuch & Jungk, 1982), and using predictive mathematical models (Nye & Tinker, 1977, Chapters 6, 7). In circumstances where the mass flow of soil solution supplies greater quantities of an ion to the roots than can be absorbed, the root surface ion concentration will exceed that in the bulk soil,

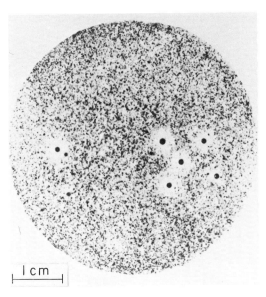

Figure 4.1 Autoradiograph showing depletions of $^{35}SO_4^{2-}$ from soil around onion roots (Sanders, 1971).

and the diffusion gradient will be away from the root. This is likely to occur only for ions present in high concentrations in soil, but that are required in small amounts by plants (e.g. Ca^{2+}, Na^+: Nye & Tinker, 1977, Chapter 6).

The soil solution ionic concentration is buffered (except in the cases of, e.g. NO_3^- and Cl^-) against depletion by the desorption of ions from clay minerals and organic matter. Thus, the potentially available soil nutrient pool can be considered to consist of the soil solution ions together with those adsorbed on the solid phase that are readily exchangeable with those in solution. The size of this pool can be determined by displacing these ions with an appropriate solution (Chapter 6, this volume) and is expressed as $mol\,cm^{-3}$ soil or $mol\,g^{-1}$ soil dry wt.

The amount of an exchangeable ion adsorbed on the solid phase relative to its concentration in the soil solution can be found by determining the ion's sorption isotherm or 'Quantity/Intensity' relationship in the soil (Beckett, 1964a, b; Bache, 1977; Mengel & Kirkby, 1982, Chapter 2). The slope of the Q/I curve is the soil's 'buffer power' with respect to the particular ion. This parameter influences the ion's diffusivity in the soil, which is an important aspect of its effective availability to roots since (Nye & Tinker, 1977, Chapter 4):

$$D = D_1\theta f_1 dC_1/dC \qquad (1)$$

where D = ion diffusion coefficient in soil ($cm^2\,s^{-1}$); D_1 = ion diffusion coefficient in free solution ($cm^2\,s^{-1}$); θ = soil water content ($cm^3\,cm^{-3}$);

f_1 = ion diffusion impedance factor ($\approx \theta$); C_1 = ion concentration in bulk soil solution (mol cm^{-3} solution); C = total exchangeable ion concentration in soil (mol cm^{-3} soil). The term dC_1/dC in equation (1) is the reciprocal of the buffer power. Thus, the more strongly an ion is adsorbed, the lower its diffusivity and the more restricted its transport to plant roots (see Baldwin, 1975).

Measurements of ion diffusivity in soil can be obtained directly, by determining the amount of ion transported over a certain distance across a plane of known area in a given time. In practice there are two general diffusion phenomena that can be measured: (a) self-diffusion and (b) counter-diffusion.

Self-diffusion

Self-diffusion involves treating a soil with an isotopically-labelled ion whose diffusivity is to be measured, placing this soil in contact with an unlabelled soil of similar ionic composition, and determining the appearance of the isotope in the originally unlabelled soil after a certain period of time (Schofield & Graham-Bryce, 1960; Rowell et al., 1967).

Counter-diffusion

When ions diffuse in soil, they do so against a counter-current of ions of similar electrical charge. Establishing a source of counter-ions at one end of a soil section will induce the diffusion of other ions through the soil towards the counter-ion source. The diffusivity of these ions can then be measured. Clarke & Barley (1968) used this method to measure the diffusivities of NH_4^+ and NO_3^-, using soils enriched with either ion placed in contact with soil containing suitable counter-ions of similar mobilities (K^+ and Cl^-, respectively). The sources and sinks for each pair (NH_4^+–K^+; NO_3^-–Cl^-) were the soils themselves. The disadvantage of this method is that it is difficult to establish uniform, intimate physical contact at two soil surfaces, and this is necessary for ion movement between the two soils to proceed without restriction at the interface. Moreover, the diffusion equations from which D is calculated (Crank, 1975; Nye & Tinker, 1977, Chapters 4, 6), assume that ions diffuse from sources of infinite capacity to sinks of infinite capacity. To apply these equations with minimal error, it is desirable that some degree of approximation to this condition should be attained in practice. This is difficult to do if soils are used as both sources and sinks. A better method is to use synthetic ion-exchange resins as sources and/or sinks. Suitable methods have been described by Vaidyanathan & Nye (1966), Tinker (1969) and Barraclough & Tinker (1981).

The best method to use for structured field soils is that of Barraclough &
Tinker (1981). In this method, D is calculated from the equation (assuming
zero ion concentration at the surface of the diffusion sink):

$$M_t = 2C_0(Dt/\pi)^{0.5} \tag{2}$$

where M_t = amount of ion per unit area ($mol\,cm^{-2}$) to diffuse from the soil
in time t; C_0 = initial ion concentration per unit soil volume ($mol\,cm^{-3}$);
t = time (s).

Recently, a method for measuring NO_3^- and NH_4^+ diffusivities and the
simultaneous mineralization of organic N was described by Darrah et al.
(1983).

There are numerous ways in which D can be expressed (e.g. relative to soil
solution or total ion concentrations; in steady- or transient-state conditions:
see Tinker, 1969). The method used to measure ion diffusivity (self- or
counter-diffusion) can also influence the calculated value of D.

A large proportion of the soil's total ion content is non-exchangeable,
being bound in organic matter or fixed in clays. Determinations of total
elemental content of soil ($mol\,g^{-1}$ soil dry wt) are required in long-term
investigations of, for example, decomposition and nutrient cycling (Swift
et al., 1979; Chapter 1, this volume) but can be of little relevance to the
immediate availability of soil nutrients to plants.

The chemical and physical heterogeneity of the soil produces local ion
concentrations that may differ significantly from the 'average' concentration.
In some circumstances, the proportion of a plant's root system that has
access to a non-uniformly distributed nutrient source is a further component
of the effective nutrient availability (Burns, 1980). This applies also to split-
root solution culture experiments in which nutrients are applied locally to
part of the root system (Section 2.3.2).

If the plant can absorb enough of a particular nutrient to realize maximum potential growth, the nutrient can be considered to be totally available
to the plant, whatever its concentration in the root medium. The plant's
innate nutrient demand (defined as the amount of a nutrient absorbed per
plant per unit time that is needed to sustain growth at its maximum potential
rate) is perhaps the ultimate determinant of nutrient availability. Consequently the true 'availability' of a nutrient has no absolute value, and can
differ between plants having different nutrient demands (arising from different maximum potential relative growth rates: Grime & Hunt, 1975), even
when growing on the same soil or in the same nutrient solution.

2.1.2 Nutrient deficiency

Nutrient deficiency occurs when the supply of nutrients is inadequate
to maintain more than a low rate of plant growth, relative to the growth of

plants of similar age under optimal conditions. Deficiency may or may not be accompanied by visible symptoms such as chlorosis, morphological deformities and increased incidence of pathogenic attack. Wallace (1961) documents characteristic deficiency symptoms for a number of crop species. A similarly comprehensive catalogue is not available for native species.

Serious nutrient deficiency leads to metabolic disorder and death. There are therefore minimum nutrient availabilities at which growth can occur. Asher (1978) and Asher & Edwards (1978) list values of steady-state nutrient concentrations in solution that have been found to be just adequate to sustain the growth of crop or pasture species. It is possible, although unconfirmed, that some slow-growing native species characteristic of infertile soils can grow at lower steady-state concentrations than potentially high-yielding crop plants.

2.1.3   Nutrient toxicity

All nutrients are potentially toxic if present in excess. Generally, trace elements and metals exert toxic effects at far lower concentrations in solution culture (c. 5×10^{-6} mol dm^{-3}: Asher & Edwards, 1978) than do major nutrients such as NO_3^--N (c. 0.14 mol dm^{-3}: Clement $et\ al.$, 1978).

In the field, effects of nutrient deficiency and excess $per\ se$ are often complicated by effects arising from nutrient imbalance. For example, NH_4^+ concentration and Al solubility increase in acidic soils, whilst the equilibrium between HPO_4^{2-} and $H_2PO_4^-$ is shifted in favour of the monovalent anion (see Rorison & Robinson, 1984).

2.1.4   Optimum nutrient availability for plant growth

Between the limits of nutrient availability at which deficiency and toxicity occur, is a range that can be considered as 'optimal' for plant growth. Plant growth at nutrient availabilities ranging from sub-optimal through to supra-optimal usually follows a classical 'response curve' (Austin & Austin, 1980; Moorby & Besford, 1983; but see also Ingestad, 1982).

The degree of plant response and the actual nutrient availabilities at which deficiency, optimality and toxicity occur, are not fixed, but vary with: plant species or population (Bradshaw $et\ al.$, 1964; Antonovics $et\ al.$, 1967; Rorison, 1968); nutrient source (Asher, 1978; Asher & Edwards, 1978); other environmental factors, such as temperature, pH, water availability, etc.; and plant age. These differences arise from the evolution of inter- and intra-specific adaptations to soil conditions such as metal contamination (Antonovics $et\ al.$, 1971), salinity (see Wright, 1977) or low nitrogen status (Antonovics $et\ al.$, 1967), for example, and from ontogenetic changes in plant

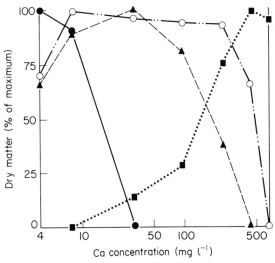

Figure 4.2 Shoot dry matter production by four plant species grown at different calcium concentrations in sand culture: (●) *Juncus squarrosus*; (▲) *Nardus stricta*; (○) *Danthonia* (= *Sieglingia*) *decumbens*; (■) *Origanum vulgare* (Jefferies & Willis, 1964).

nutrient demand (Milthorpe & Moorby, 1979, p. 88). Figure 4.2 illustrates this: the deficiency–optimum–toxicity continua of four species are displaced relative to one another along a gradient of Ca concentration. For example, the optimum Ca availability for *Nardus stricta* is toxic for the growth of *Juncus squarrosus*.

Thus, experimental investigations into plant–nutrient relationships are set against a background of ecologically-related plant responses. As Antonovics *et al.* (1967) observe, it is usually difficult to measure plant response over an entire range of nutrient availability and this can lead to misleading interpretations of experimental data. With appropriate combinations of materials and techniques, field and laboratory experiments, however, it is possible to obtain much information that is relevant ecologically (see Rorison, 1969).

2.2 Nutrient sources

AR grade inorganic salts are the most common nutrient sources in solution culture experiments (Hewitt, 1966, Chapter 8). Organic sources, such as amino acids, urea and vitamins have been used, particularly in experiments involving continuous cultures of 'simple' aquatic plants (e.g. *Lemna minor*).

In soil, there is an unavoidable combination of organic and inorganic nutrients. The relative amounts of these can be manipulated to some extent

by additions of inorganic fertilizers, leaching to remove weakly-buffered ions
(e.g. NO_3^-) or liming (additions of $CaCO_3$) to increase pH, thereby reducing
the solubility of metal ions and their potential availability to plants. It is
important to know the buffering characteristics of the soil (Section 2.1) in
order to calculate how much fertilizer or $CaCO_3$ needs to be applied to effect
a significant change in soil nutrient status. In addition, the timing of application is important, particularly in the case of N. Fertilizer applications in
early spring could lead to large losses of N from field soils through denitrification or leaching of NO_3^-.

2.3   Uniformity of nutrient supply

2.3.1   Uniform

Supplying nominally constant concentrations uniformly over the
entire root system is the most simple and widely-used means of providing
plants with nutrients in solution culture. Methods of preventing excessive
changes in solution composition during ion uptake by plants are described in
Section 3.1.

In soil, constant root-surface concentrations are impossible to achieve
(Section 2.1). Bulk soil concentrations of exchangeable ions in well buffered
soil are more likely to remain constant than in sand, for example, although
the low organic content of the latter allows for a greater degree of physical
homogeneity in the root medium. In pot experiments, ion dispersion and
spatial concentration differences can arise as a result of watering, leading to
non-uniformity of nutrient distribution. NO_3^- is particularly liable to such
dispersion. There are three ways of minimizing this effect should it be
necessary to do so:

(i) Pre-watering: supplying all water initially to at least field capacity,
ensuring that this supply is adequate for growth throughout the experiment,
and allowing equilibration to occur before planting. For long-term
(> 10 days) growth, it is likely that pots of different volumes would be needed
to accommodate the water demands of large plants (Brewster & Tinker, 1970).

(ii) Physical barriers to water movement: dividing the soil into separate
volumes using hydrophobic partitions, e.g. paraffin wax membranes. Water
movement is prevented between adjacent volumes, but unrestricted within
each volume. Water supply to each volume is achieved by pre-watering or by
supplying each volume individually from external reservoirs (Drew, 1975,
Figure 4.3). In Drew's experiment, the main root axes of barley penetrated
the wax membranes, but finer lateral roots did not.

(iii) Auto-irrigation: supplying water from within the soil in response to
transpiration losses, via buried porous filters connected to an external reser-

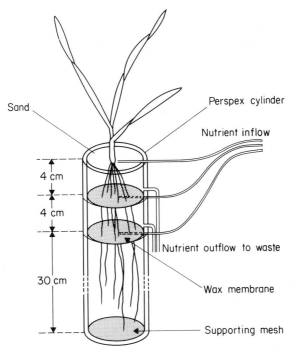

Sand

Perspex cylinder

Nutrient inflow

4 cm

4 cm

Nutrient outflow to waste

30 cm

Wax membrane

Supporting mesh

Figure 4.3 Apparatus used to supply localized nutrient treatments to barley roots in sand culture. The horizontal wax membranes divide the sand into three discrete volumes, each receiving nutrients independently from external reservoirs, with solution flow prevented between adjacent soil volumes (Drew, 1975).

voir (e.g. Brewster *et al.*, 1976). Robinson & Rorison (1985) used an auto-irrigation system that minimized gross NO_3^- dispersion between root-containing and root-free depths of a sandy soil. Further developments of this approach could be valuable, the fundamental limitations of the technique notwithstanding (Slavík, 1974). Additionally, if plants are grown in a controlled environment (Section 3.2), a high relative humidity reduces transpiration and, hence, plant water demand (e.g. Baldwin, 1975).

2.3.2 Non-uniform

In many laboratory experiments it is useful to vary the nutrient supply spatially or temporally to simulate more realistically conditions that obtain in the field, or to investigate plant competition for nutrients and water (Section 2.6).

Spatial variability in nutrient supply can be controlled by using split-root pots, in which isolated parts of the root system receive different nutrient supplies, in solution culture (e.g. Robinson & Rorison, 1983), or in solid media (Drew, 1975, Fig. 4.3). A modification of the split-root technique has been used frequently in short-term studies of ion and water absorption by discrete zones of individual roots (e.g. Russell & Sanderson, 1967; Clarkson, 1974, p. 225; McNaughton & Presland, 1983).

Controlled temporal variability in supply is more easily achieved in solution culture than in soil. Examples of this approach are the experiments of Rorison & Gupta (1974) and Clement *et al.* (1979). Variations in soil nutrient supply can be manipulated to some extent by additions of nutrients to the soil. The possibility of using an auto-irrigation system to supply nutrient solutions and flushes of distilled water alternately might be useful here.

Perhaps the most ambitious use of non-uniform nutrient supplies has been that of Crick & Grime (1983). *Agrostis stolonifera, Holcus lanatus* or *Scirpus sylvaticus* were grown in 9-chambered 'root growth arenas', in which each chamber contained a high or low concentration of nutrient solution. These spatial differences in supply were combined with temporal fluctuations in the solution concentration in each chamber. The dramatic effects of these treatments upon localized root development in *H. lanatus* can be seen in Fig. 4.4.

Figure 4.4 *Holcus lanatus* (leaves removed) after 27 days' growth in a 22 cm diameter 'root growth arena' (viewed from above). Each of the three compartments which were supplied with a full nutrient solution can be seen to contain a dense mat of roots (Crick & Grime, 1983).

2.4 Effects of nutrients upon specific plant processes

2.4.1 Growth

Nutrients have considerable effects upon the quantity, rate and form of plant growth. Changes in morphology resulting from a nutrient treatment can be the visible manifestation of internal physiological injury (e.g. abnormal root branching in Al-treated *Agrostis tenuis*: Clarkson, 1969). Alternatively, such changes may be active responses by the plant to compensate for the effects of the treatment. An example of this is an increased root : shoot ratio at low nutrient availabilities. Total nutrient fluxes into the plant are thus maintained by the production of a relatively larger root system (Davidson, 1969). Similarly, changes in root morphology (e.g. increased branching) can result in increased specific nutrient absorption rates (per unit root weight) in response to a localized nutrient supply (Robinson & Rorison, 1983). Short-term increases in specific nutrient absorption rates can also occur in such circumstances. These reflect the capacities of cellular transport systems to react metabolically to external deficiencies, rather than via morphological adjustments (e.g. Lee, 1982; Drew *et al.*, 1984). Responsive changes in root morphology, although reported frequently in solution culture experiments, are likely to be of greater functional significance in soil, where root distribution is a key factor limiting the exploitation of deficient nutrients. Interspecific differences in this respect could be important ecologically (Robinson & Rorison, 1985).

The measurement of plant growth may be destructive or non-destructive (Evans, 1972, p. 44 *et seq.*, Chapter 11). Destructive measurements involve the harvesting of a number of even-aged plants from the experiment on one or more occasions, measuring fresh and/or dry weights, lengths and areas of roots and leaves, as required. Non-destructive measurements are made on the same plants throughout the experiment. Obviously, dry weights cannot be obtained, but with careful handling, fresh weights can be measured, as can gross leaf and root dimensions. A major disadvantage of non-destructive measurements is that plants subjected to excessive physical disturbance suffer metabolic trauma, consisting of wide fluctuations in respiration rates, which can affect the dry weights of small plants to a significant extent (Evans, 1972, p. 157 *et seq.*). In experiments of long duration, this may be of no consequence, but in experiments in which short-term effects are to be examined (e.g. Hunt, 1980), this could give rise to misleading data.

Böhm (1979) describes many useful techniques for the quantitative examination of root systems. For most purposes, the following morphological characteristics should be measured: total length per plant (cm) and root diameters (cm). Root hair lengths (cm) and root hair densities (cm^{-1}) could also be measured if required. Root length is measured most conveniently by a line-intersection method (Böhm, 1979, p. 132 *et seq.*). Semi-automated procedures for root length measurement are available (Rowse & Phillips, 1974; Costigan *et al.*, 1982), but these still depend upon the manual separation and arrangement of root samples on an observation grid, and it is this procedure that is usually most laborious and time-consuming rather than the measurement itself. Root diameters and root hairs can be measured by 'sampling' the root system microscopically at a number of randomly-chosen points on the observation grid. For soil-grown plants total root length usually cannot be measured. Instead, root density (cm root cm^{-3} soil) and maximum rooting depth (cm) should be measured. In the field these parameters can be estimated from roots contained in soil cores (Böhm, 1979, Chapter 5). Large root systems can be examined by taking random subsamples of convenient weight (Hackett, 1968).

The average root radius can be estimated by assuming that root fresh weight is numerically equivalent to root volume (i.e. that the specific gravity of fresh roots is $\approx 1\cdot0$; Brewster & Tinker, 1972). Then, assuming cylindrical geometry:

$$a = (FW_R/\pi L_R)^{0\cdot5} \tag{3}$$

where a = mean root radius (cm); FW_R = root fresh wt (g); L_R = root length (cm). Alternatively, equation (3) can be used to obtain, by less tedious means, an estimate of L_R, by measuring a from a large number of samples. The separation of individual roots on the observation grid is then less critical than for root length measurement.

A crude index of root branching can be obtained from the root length : root fresh or dry wt ratio (cm g^{-1}). Low values imply that the root system contains a large proportion of thick, relatively unbranched roots; high values that the roots tend to be finer and, hence, more frequently branched (Fitter, 1976). More accurate methods of measuring branching patterns have been reported by Hackett (1968) and by Fitter (1982).

Precise measurements of the quantities and spatial distributions of roots in soil can be made by careful excavation of root systems (Böhm, 1979) or by using isotope labelling (Vose, 1980, Chapter 12). The latter method is more practicable and informative. There are two general approaches to this.

(i) Isotopic labelling of discrete zones in the soil and measuring the appearance of the isotope in plant tops. The isotope should be of a slowly-diffusing ion (e.g. ^{32}P, ^{42}K) so that the position of the label can be correlated reliably with root activity in that part of the soil, following the determination of the label in the above-ground tissues (Nye & Tinker, 1977, pp. 193–194). This is also a useful method for detecting the uptake of specific nutrients from well defined depths of a soil profile (Section 2.5.5). If the soil is labelled with radioactive isotope, autoradiographs can be used to determine root distribution (see Fig. 4.1), but for this purpose labelling of plant material rather than soil is preferred. It should be stressed, however, that the labelling of soil in the field with ^{32}P, in particular, is not to be encouraged. The incorporation of labile P into soil organic matter, which has an extremely low rate of turnover (Ottaway, 1984) means that a high residual enrichment of ^{32}P will remain in the soil indefinitely, even allowing for the 14·7 day half-life of the isotope. It is such a powerful β-emitter that soils labelled with it will retain unacceptably high levels of radioactivity for some time. Apart from the environmental hazards of using ^{32}P in the field, the application of this isotope, and others such as ^{14}C and ^{15}N, to field soil raises the background abundances of the isotopes to levels that would make subsequent work of a similar nature on the same sites unreliable.

(ii) Injecting plant tops with radio-isotopes of readily translocated nutrients such as P, K or S, allows the appearance of the label in roots, or in soil containing roots, to be detected in one of two ways. Firstly, assuming uniform translocation throughout the root system, the radioactivity in samples of soil plus roots taken from the rooting zone, should be proportional to the quantity of roots in each sample (Nye & Tinker, 1977, p. 193). Crude measurements of spatial patterns of root distribution can be obtained using this method (Bookman & Mack, 1982). Alternatively, autoradiography of soil sections containing radioactively-labelled roots allows the precise positions of roots in the soil to be recorded. Root densities are calculated from the number of roots per unit area of the autoradiograph, using equations appropriate to the degree of randomness of root orientation in the soil. The procedure has been described fully by Baldwin et al. (1971), and is suitable for both laboratory and field work. An extension of this method —injecting different plants with different isotopes—allows the distributions of inter-penetrating root systems to be determined (Baldwin & Tinker, 1972). This is important in assessing the extent of below-ground competition for water and nutrients (Section 2.6.1), and this is exceedingly difficult to do by excavation procedures.

Morphogenetic changes in growth which can occur in response, *inter alia*, to nutrients, are probably mediated by plant growth substances. Interactions between cytokinins, abscisic acid and nutrients have been reviewed by Moorby & Besford (1983). Yokota, Murofushi & Takahashi and Reeve & Crozier (in MacMillan, 1980) and Brenner (1981) describe methods to sample, analyse and quantify the action of plant growth substances.

Plant growth analysis is used frequently to compare statistically measured responses of plants to experimental treatments. Such techniques allow certain biologically-important growth functions to be derived from raw data, e.g. relative growth rate, root:shoot ratio, specific nutrient absorption rate, and specific nutrient utilization rate, amongst others (see Evans, 1972; Hunt, 1982).

2.4.2   Nutrient uptake

Ultimately, nutrient uptake is driven and regulated in the whole plant by growth (Clarkson & Hanson, 1980). The effects of nutrients that modify or impair growth, also influence nutrient uptake rate and the resulting nutrient concentration in the tissues (Jarrell & Beverly, 1981). Some examples of this were described in the previous Section. At the cellular or subcellular level, turgor maintenance (Cram, 1976) and metabolically-driven ion pumps (Clarkson, 1974, Chapter 5) appear to be primarily responsible for controlling nutrient absorption.

There are two ways of measuring nutrient uptake:
 (i) nutrient accumulation in plant tissue;
(ii) nutrient depletion from a source of known composition.
The former method is more common. Determinations of nutrient concentrations ($mol\,g^{-1}$ or $mol\,cm^{-3}$) throughout the absorbing material (organelle, excised tissue, intact root or whole plant) are required. The usual way of doing this is to analyse replicated subsamples of homogenized material, and calculate the nutrient content (mol) by multiplying concentration by the weight or volume of the entire tissue or plant. In the case of nutrients which are important osmotic constituents in plant cells (e.g. K^+), there are good physiological reasons to express their concentrations per unit of tissue water, rather than per unit weight (Leigh et al., 1983; Leigh & Wyn Jones, 1984).

The use of isotopically-labelled nutrient sources (Section 5.1) is often valuable as a means of monitoring the accumulation of a nutrient during a particular part of the experiment (e.g. NH_4^+ or NO_3^- uptake by grass species from a localized supply during the final 24 hours of the experiment, by using ^{15}N: Robinson & Rorison, 1983). In cases where no suitable isotope is available, analogues can be used: Sr^{2+} for Ca^{2+}; Rb^+ for K^+; ClO_3^- for NO_3^- (see e.g. Deane-Drummond & Glass, 1983).

The accumulation of nutrients in plants can be measured non-destructively by frequent monitoring of the nutrient concentration in solution, and recording the amounts of nutrient that must be added to the solution in order to maintain a constant nominal concentration. The total amount of nutrient absorbed by the plant is the sum of all the nutrient additions to the solution. To be a practicable method, the nutrient monitoring and addition procedures must be automated (see Section 3).

Depletion experiments allow nutrient uptake to be measured continuously and non-destructively (cf. Section 2.4.1), but the period during which such measurements can be made is usually less than 24 hours. The standard technique (Claassen & Barber, 1974) involves growing plants for a prescribed period in conventional solution culture and applying any pre-treatments that are required. Several plants are subsequently transferred to nutrient solutions, usually containing an isotopically-labelled enrichment of the element under investigation. The depletion of the element from solution is measured by sampling small volumes of the solution at intervals of, e.g. < 30 minutes for several hours. During the depletion of the element, it is important to ensure by adequate replenishments that other nutrients do not become deficient. Combined with measurements of root dimensions, short-term nutrient uptake rates in fully equilibrated tissues can be determined. The technique has been refined to a high standard of accuracy, with the use of ion-specific electrodes and micro-computer controlled recording systems (Deane-Drummond & Glass, 1983; Glass et al., 1983). Advantages of measuring nutrient depletion from solution rather than nutrient accumulation in plant tissue are that in the former, fewer plants are needed since sequential destructive harvesting is not involved, and that the same plants can be used over a wide range of external concentrations during the uptake period. An excellent account of the technique is given by Drew et al. (1984).

Until recently, the nutrient depletion technique was confined to solution-grown plants: attempting short-term depletion of soil nutrients is not practicable. However, using a modification of this technique, Bhat (1982) measured short-term depletions of P by soil-grown apple roots in a root observation laboratory. Intact roots were separated from the soil, passed through a port in the wall of the root laboratory, and into a nutrient solution. Depletions from this solution were measured as described above. This application of the technique is extremely useful in relating uptake rates in laboratory-grown material with those in field-grown plants, and also in view of the near-impossibility of extracting entire root systems from soil for use in precise physiological experiments. The extension of its applicability beyond the root laboratory and the development of a 'portable' version for use in freshly-dug soil pits or with plants grown in the laboratory in special containers, for example, has exciting possibilities for the future. The use of root observation laboratories, or 'rhizotrons', in root research has been reviewed in detail by Böhm (1979) and Huck & Taylor (1982).

Nutrient accumulation or depletion experiments can be used in solution culture experiments involving whole root systems or individual roots. Uptake by specified lengths of single roots can be measured by using special absorption cells (e.g. Russell & Sanderson, 1967; Clarkson, 1974, p. 225; McNaughton & Presland, 1983). Labelled isotopes are supplied to an isolated root segment and its accumulation in the plant or depletion from solution monitored accordingly. This approach is useful in examining spatial and temporal changes in nutrient uptake capacity that occur during ontogeny, or in relation to localized environmental conditions, such as changes in nutrient concentration. The technique can also be used to determine the minimum root length that the plant needs in order to satisfy its nutrient demands (see Burns, 1980).

The rate of nutrient uptake from dilute solutions can be related to the external ion concentration via a convenient hyperbolic equation, analogous to that used to describe Michaelis–Menten enzyme kinetics:

$$V = \frac{V_{max} \cdot C_{la}}{k_m + C_{la}} \tag{4}$$

where V = rate of nutrient uptake per unit of absorbing tissue; V_{max} = maximum potential value of V; C_{la} = ion concentration in solution at the root surface $(mol\,cm^{-3})$; k_m = value of C_{la} at which $V = V_{max}/2\,(mol\,cm^{-3})$. V_{max} and k_m can be used to describe the capacity and affinity, respectively, of an ion transport system (see Neame & Richards, 1972), but reveal nothing about the mechanism of that system (Clarkson, 1974, p. 289).

The significance in the field of interspecific differences in V_{max} and k_m values determined under laboratory conditions is dubious, in view of the primary limitation of ion diffusion to the root. Moreover, '... values for the V_{max} and k_m for the transport of various ions are often catalogued as if they were constant properties of the plants concerned. This fails to take account of the fact that the quantity of a nutrient absorbed by a plant over a given period of time is an integral of the growth which has occurred and that the rate of absorption from a given concentration is determined by the "demand" for nutrients created by growth' (Clarkson & Hanson, 1980; Section 2.1, 2.4.1). V_{max} and k_m values vary with the ion (Nye & Tinker, 1977, Table 5.3), the degree of equilibration of the absorbing tissue (Lee, 1982), and, most importantly, the method used to measure ion uptake (steady-state or depletion; equilibrated or non-equilibrated material; excised tissue or whole plants) and the length of time over which uptake is measured (Wild et al., 1979). Furthermore, interpretations of ion uptake kinetics should differentiate between ion fluxes driven by plant metabolism and those that can be accounted for by passive diffusion (see Clarkson, 1974, p. 292). Comparisons between plants based upon the kinetics of their ion uptake systems alone are therefore difficult to make with reliability. The ecological or agricultural significance of such comparisons need to be treated with circumspection.

Ion fluxes into plant cells or between subcellular compartments can be driven or prevented by differences in electrochemical potential that exist across cell membranes. It is often necessary to measure such potential differences to determine whether ion uptake involves energy consumption (active transport against the electrochemical potential) or whether it can occur passively (in the direction of the electrochemical potential). Internal and external (relative to the membrane) ion concentrations (strictly, activities) and transmembrane electrical potential differences need to be measured.

Internal ion concentrations can be determined directly, either by inserting microcapillaries into the cell and withdrawing a small volume of the sap, or by using ion-specific microelectrodes to measure concentrations *in situ* (Clarkson, 1974, p. 81). Such measurement can be made between the external medium and the cytoplasm, and between cytoplasm and vacuole, for example. Technically, these are difficult operations in higher plant cells. The giant coenocytes of certain algae are more amenable to experimentation of this type (see Hope & Walker, 1975), which is why a great deal of fundamental research has been done on genera such as *Chara, Nitella, Valonia* and *Hydrodictyon*. The usefulness of X-ray microanalysis to differentiate between ion concentrations in subcellular compartments is limited by the lack of suitable calibration procedures; only relative ion concentrations can be determined by this method (Flowers & Lauchli, 1983). The development of nuclear magnetic resonance (NMR) techniques could be useful in measuring subcellular concentrations of H^+ and P (Lee & Ratcliffe, 1983; Section 2.4.5). Crude determinations can be made in cell sap expressed under pressure from tissues frozen in liquid N_2, or in digested tissue samples (Chapter 6, this volume).

Transmembrane electrical measurements are made using microelectrodes connected to a voltmeter. Clarkson (1974, p. 83 *et seq.*) describes this technique and its advantages and disadvantages, in detail. Clarkson (1974, p. 61) also provides worked examples of its use, using data from *Hydrodictyon*.

Rates of nutrient uptake can be expressed without reference to the size of the absorbing tissue ($mol\,s^{-1}$, $mol\,plant^{-1}\,d^{-1}$, etc.). It is usual, however, to express uptake rates as a function of tissue size ($mol\,cm^{-1}\,s^{-1}$, $mol\,g^{-1}\,d^{-1}$, as appropriate). For whole plants, rates are often expressed per unit dry weight of the total root system, in order to measure the 'efficiency' of the entire system as a nutrient-absorbing organ (Hunt, 1973). This implies that all roots are equally active in nutrient uptake. In reality this is not so: old roots tend to be less effective in ion uptake than young roots. Rates derived using this assumption are therefore 'averages' of the performance of all roots as an integrated system. Uptake measured in specific parts of individual roots (see above) refer to the activity of those parts (and others of similar age and physiological history), but may not necessarily reflect the behaviour of the root system as a whole.

The use of root dry weight as the basis on which to measure nutrient uptake is of value since it allows uptake rates to be measured in terms of the net amount of 'fixed energy' (Milthorpe & Moorby, 1979, p. 5) invested in the operation of the root system, and is thus related to the growth of the whole plant. In addition, this removes effects of diurnal or ontogenetic changes in tissue water content that could confound uptake rates measured in terms of root fresh weight.

In experiments where roots are grown at low ($< 10°C$) temperatures however, it is often found that the percentage of dry matter in the root increases, probably due to the thickening of cell walls (Clarkson & Deane-Drummond, 1983). These workers argue that since this results in an increase in metabolically-inert dry weight (as far as nutrient uptake is concerned), it is more realistic in such experiments to express uptake rates per unit of root *fresh* weight, accepting that tissue water contents could change diurnally. This difficulty can be avoided if root *length* is chosen as the basis upon which to measure uptake rate.

The expression of uptake rates in terms of root weights is acceptable in solution culture experiments since, in most cases, root morphology has little functional significance in ion absorption; all that is required is some reliable measurement of the size of the root system. In soil, however, the radial depletion of a diffusible ion from the soil around a root is influenced mainly by the root length, and this is the most useful basis upon which to measure rates of nutrient uptake from soil (Brewster & Tinker, 1972).

2.4.3   Tolerance of nutrient deficiency

Tolerance of nutrient deficiency implies, by definition, a capacity to maintain some growth at low levels of nutrient supply. Experimental comparisons of tolerance are useful in interpreting the ecological distributions of plants characteristic of fertile or infertile soils.

The most direct way to make such interpretations is to compare growth, ion uptake and other processes over a range of uniform nutrient availabilities in the laboratory (Section 3.2). This has been done in nutrient solutions (see Asher & Edwards, 1983), and in fertilized sand (Bradshaw *et al.*, 1964; Austin & Austin, 1980), with plants grown in mono- or mixed culture. Similar types of experiment can be designed using non-uniform nutrient supplies, if desired (Section 3.2.2).

Difficulties may arise in maintaining the dilute nutrient concentrations constant in solutions that are needed to induce deficiencies. This problem is discussed in Section 3.1.5. If the precise nutrient concentration in the root medium is unimportant, a range of *relative* concentrations can be established using a dilution series to produce a general gradient of availability (Austin & Austin, 1980). There are, however, problems in extrapolating the results of such experiments to the field, where plant responses to both absolute and relative amounts of available nutrients may be important.

An alternative is to grow plants on a range of soils that are known to differ in fertility (e.g. McGrath, 1979, Fig. 4.5). Effects of soil nutrient availability upon growth can be measured, as before.

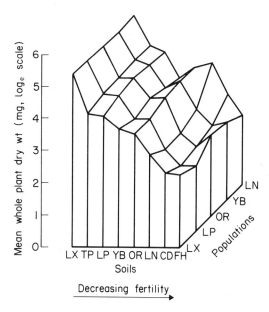

Figure 4.5 Dry matter production by individuals of five *Holcus lanatus* populations after 44 days' growth from seed on soils differing in fertility, from a number of sites in the Sheffield region. LX, Loxley; TP, Tapton Garden; LP, Lathkill Plateau; YB, Yorkshire Bridge; OR, Orgreave; LN, Lathkill north-facing slope; CD, Coombesdale; FH, Fox House. (Data taken from McGrath, 1979, Table 3.3).

Amelioration of soil nutrient deficiencies can be achieved by the application of fertilizers. This can lead to considerable floristic changes on the treated soils over a period of years. As fertility is increased, slow-growing, deficiency-tolerant, floristically-diverse vegetation is replaced by a small number of dominant species characteristic of productive habitats (Rorison, 1971). The degree of floristic change depends upon the extent to which fertilizers can modify soil nutrient status, and also the proximity of potential colonizing populations. Such experiments have been conducted in a number of different habitats: lowland meadows (Thurston, 1969); hill pastures (Bradshaw, 1969); limestone grassland (Jeffrey & Pigott, 1973); sand dunes (Willis, 1963); and salt marsh (Jefferies & Perkins, 1977).

2.4.4 Tolerance of ionic toxicity

Ionic toxicities are less common in nature than deficiencies, but where they do occur can have as great an effect upon the vegetation. Two aspects of ionic toxicity have received particular attention in recent years: metal contamination in industrial areas, and salinity in coastal or arid habitats.

Wilkins (1957, 1978) devised a method to assess the *relative* tolerances of plants to toxins: the root extension test. Seedlings are grown in a solution of

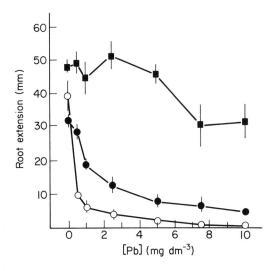

Figure 4.6 Root extension ($\pm 95\%$ confidence limits) by a lead-rake (Tideslow) population of *Festuca ovina* in relation to nominal Pb concentration in various solutions. (■) full nutrient solution (Rorison, in Hewitt, 1966, Table 30c); (○) phosphate-free nutrient solution; (●) calcium nitrate solution ($0.5\,\text{g}\,\text{dm}^{-3}$) (Shaw, Rorison & Baker, 1982).

calcium nitrate ($Ca(NO_3)_2 \cdot 4H_2O$, *c.* $1\,\text{g}\,\text{dm}^{-3}$), with the potential toxin present at a range of concentrations, starting at zero. Calcium nitrate is used to prevent precipitation of metal salts. In addition, Ca ameliorates Pb toxicity. This allows higher Pb concentrations to be used. Otherwise, toxicity would occur at Pb concentrations too low to maintain accurately, due to adsorption of Pb ions onto glassware and plant surfaces. In this test, tolerance is measured by comparing the growth of the longest roots of seedlings in calcium nitrate solution with those grown in a solution of calcium nitrate plus the toxin, over a period of several days. This method allows a large number of plants to be screened rapidly for tolerance. It has been used extensively for this purpose (e.g. Hodson *et al.*, 1981; Wong & Bradshaw, 1982).

It should be noted, however, that the technique has limitations if tolerances of *absolute* toxin concentrations are to be measured and extrapolated to potential tolerances of field conditions. Direct comparisons between toxic concentrations measured in calcium nitrate solution and the concentrations of potential toxins in the soil solution could be misleading. Figure 4.6 illustrates the effect of 'background' factors upon root extension in *Festuca ovina*. Root growth was inhibited at Pb concentrations above *c.* $4\,\text{mg}\,\text{dm}^{-3}$ when the tolerance test was conducted using calcium nitrate or phosphate-free nutrient solution. In full nutrient solution, $4\,\text{mg}\,\text{Pb}\,\text{dm}^{-3}$ had virtually no inhibitory effect upon root extension. The extraction and filtration method used to determine the ion concentration in the soil can also have a major effect upon conclusions drawn from the experiment (Shaw *et al.*, 1984).

There are many reports of plant responses to toxic concentrations of metals in solution culture: zinc (Carroll & Loneragan, 1969); aluminium (Rorison, 1985); lead (Jones, Clement & Hopper, 1973); cadmium (Jarvis *et al.*, 1976); and manganese (McGrath & Rorison, 1982).

As with nutrient deficiency, a reliable but surprisingly little-used technique is to measure the capacities of plants to grow on a range of soils that are known to contain different levels of a particular toxin (e.g. Jones, Jarvis & Cowling, 1973). A modification of this is to mix contaminated soil with potting compost, in various ratios, to produce a 'dilution' series of the potential toxin. Again, however, the problem of quantifying the *actual* concentration of the toxin in the soil remains, but at least measured responses can approximate more closely to field conditions.

Plant populations can develop genetic tolerance in response to specific toxins in the environment, e.g. metals (Antonovics *et al.*, 1971), or pre-adapted plants can colonize newly-contaminated sites, e.g. the invasion by coastal halophytes of roadside verges treated with de-icing salt (Scott & Davidson, 1982). Such vegetation changes could be manipulated in a way similar to that in which fertilizer applications are used to enhance nutrient availability (see above).

Potentially toxic environmental conditions can lead to the evolution of plants having specific metabolic specializations: metal-tolerant enzyme systems (Woolhouse, 1983); NaCl-induced accumulations of specific osmotica (Stewart *et al.*, 1979); detoxification mechanisms and capacities to regulate or exclude the uptake of potentially toxic ions (see Rorison & Robinson, 1984). Whether such mechanisms, (for which activities are often measured *in vitro*), operate similarly in plants growing in the field, needs to be confirmed experimentally.

2.4.5 Metabolism: photosynthesis, respiration, assimilation and translocation

The influence of nutrients upon these processes has been reviewed recently by Moorby & Besford (1983). As with nutrient uptake (Section 2.4.2), apparent effects of nutrients upon the overall rates of these metabolic processes can be complicated by simultaneous effects upon growth.

Photosynthesis

Generally, low nutrient availabilities limit photosynthetic rates in C_3 and C_4 plants (Wong, in Osmond *et al.* 1982). Methods for the measurement of photosynthetic rate are described in Chapter 1.

It should be emphasized that effects of nutrients upon photosynthetic rates occur usually through effects upon specific partial processes (e.g. N deficiency upon ribulose bisphosphate carboxylase-oxygenase (EC 4.1.1.39) activity; Mg and Fe deficiencies upon chlorophyll synthesis; K deficiency upon stomatal activity and, hence, upon rates of gas exchange), rather than upon 'photosynthesis' as a whole. Which partial processes to investigate in this respect depends upon which nutrients are of interest, and whether the plant exhibits C_3, C_4 or CAM photosynthesis.

Respiration

Increased rates of respiration (brought about by, e.g. increased O_2 partial pressures) can result in higher rates of nutrient uptake (see Mengel & Kirkby, 1982, p. 116). Respiration rates usually are not critically dependent upon nutrient availabilities except in the case of those metals that are essential to the electron transport system (Ca, Fe, Zn). Moorby & Besford (1983) report instances where dark respiration rates in crop species are influenced by nutrient supply. It is unlikely that ecologically significant differences in such effects will exist between species, since respiration usually appears to be 'protected' against adverse environmental conditions in all plants (G.A.F. Hendry, pers. comm.).

Assimilation

This involves the incorporation of certain nutrients (chiefly, N, P and S) into organic molecules via enzyme-mediated reactions. Studies of nutrient assimilation can include:
 (i) identification of assimilatory pathways;
(ii) determination of enzyme activities.
The usual way of identifying the component steps in nutrient assimilation pathways is to use 'pulse-chase' experiments. This method involves supplying the isotopically-labelled nutrient to the absorbing material in a brief 'pulse', followed by the nutrient supplied in the unlabelled form, and determining the appearance of ('chase') the label in specific compounds over a period of time, usually by fractionating tissue extracts chromatographically.

Of greater importance ecologically, however, is the measurement of *how much* of a certain compound is synthesized from a given nutrient input, under various environmental conditions. Such studies are still in their infancy, but all to date have employed pulse-chase methods. Attempts to integrate C and N assimilatory pathways on a quantitative basis in several crop species have been made using simultaneous $^{15}N/^{14}CO_2$ labelling (Pate, 1980, 1983). Fentem *et al.* (1983) followed the amounts of ^{15}N to be incorporated in various amino acid pools in barley. Bieleski & Ferguson (1983) reported measurements of P transformations and turnover in *Spirodela*.

The methodological difficulties of such experiments are considerable. Only by employing such methods, however, can the potential importance of, for example, the mobilization of storage compounds under different environmental conditions be evaluated. Recent technological developments could provide powerful analytical methods to investigate such problems. For example, computer-controlled gas chromatography–mass spectrometry systems (GC–MS: Brenner, 1981; Rhodes *et al.*, 1981) allow the assimilation of ^{15}N into various molecules to be followed with great accuracy, but demand that samples can be volatilized prior to analysis. The conversion of sample compounds into volatilizable derivatives is often the most difficult part of this procedure (J.H. Peterkin, pers. comm.). GC–MS has been used to investigate differences in amino-N turnover between plants capable of absorbing N and growing during winter, and plants in which N uptake and growth are restricted at low temperatures (Peterkin *et al.*, 1982). Future applications and developments in high-performance liquid chromatography systems should provide more detailed information of plant nutrition at the biochemical level. Another potentially useful technique in this respect is ^{31}P-NMR (Lee & Ratcliffe, 1983), which offers the possibility of determining *in vivo* concentrations and spatial localizations of ^{31}P-labelled molecules. Preliminary work to apply this to an experiment on the effects of Al upon P assimilation in Al-tolerant and non-tolerant plants is in progress (Rorison & Spencer, 1984). Often, however, the full potential of such analytical techniques is limited in ecological investigations because of the low rate of sample throughput that can be achieved, and the large numbers of samples that even 'small' ecologically-based experiments generate. For example, thirty ^{15}N analyses per day by semi-automated mass spectrometry, could be considered to be an optimistic figure. Careful evaluation of the likely numbers of samples requiring analysis should be made at the planning stage of any experiment.

Enzymological studies (Section 3.3.5) are far more common, mainly because of the less complex analytical procedures that are involved. Ecological interest has concentrated in recent years upon the enzymes of inorganic N assimilation (Lee & Stewart, 1978; Pate, 1983). *In vivo* activities of nitrate reductase (NR; EC 1.6.6.1) in plants in the field can provide a general indication of the NO_3^- availability in the soil, since NR is substrate-inducible, and its activity is related directly to NO_3^- concentrations at its site(s) of action. Plants characteristic of calcareous, NO_3^--rich soils tend to have higher *in vivo* NR activities than species from acidic, NO_3^--poor soils. Applications of NO_3^--fertilizer to soils can result in large increases in NR activity in both calcareous and acid-soil plants. Only in very few species (mainly in the Ericaceae) are inducible NR activities innately low, indicating a small capacity to assimilate NO_3^- and, therefore, a physiological dependence upon NH_4^+ as a N source. There are no comparable effects on NH_4^+ availability upon NH_4^+-assimilating enzymes (Taylor & Havill, 1981); these, and most other plant enzymes, are not substrate-inducible and so cannot respond readily to environmental fluctuations in substrate availability. Differences in plant distribution in relation to soil P availability have been correlated with the capacities of plants to synthesize root-surface phosphatase enzymes and (it has been assumed) thereby solubilize P in the rhizosphere (Woolhouse, 1969; Atkinson, 1983; McCain & Davies, 1984). Assessments of enzymatic functioning in plants should take into account effects of environmental acclimatization upon, and diurnal fluctuations of, enzyme activities in various parts of the plant (see e.g. Clarkson & Deane-Drummond, 1983).

Translocation

The internal transport of absorbed nutrients or assimilates can be of ecological significance. For example, whether NO_3^- and NH_4^+-N are assimilated in roots or shoots can affect internal pH regulation and the extent of rhizosphere pH changes (Raven & Smith, 1976). Ecological differences between plants can often be related to their pattern of N metabolism (Pate, 1983) and to tolerances of soil reaction factors that can result (Rorison & Robinson, 1984). Examinations of translocated substances can reveal differences in sites of nutrient assimilation. The dynamics of assimilate fluxes between roots and shoots can also be investigated with accuracy.

The contents of phloem sap is sampled most effectively via aphid stylets (Dixon, 1975; Peel, 1975; Raven, 1983), since these penetrate individual phloem elements, cause minimal disruption to translocation, and result in less contamination of the extracted fluid. A practical guide to the use of the exuding-stylet method has been published recently (Fisher & Frame, 1984). Alternatively, phloem sap can be extracted from external incisions into the stem, but a problem with this method is that sample contaminations are likely (see Pate, 1980).

To sample xylem sap, Pate (1980) recommends the insertion of microcapillaries into exposed ends of individual vessels. More common, is to collect 'bleeding sap' from the cut ends of roots and shoots, but again, this can result in sample contamination.

2.4.6   Seed germination

External nutrient supplies are not essential for seed germination. NO_3^- is an exception, however, in that it can stimulate the germination of seeds of many species, although the reason for this is obscure. Certain species have seeds that show tolerance of potentially injurious nutrient concentrations. For example, some succulent halophytes have seeds that can tolerate long periods of immersion in high NaCl concentrations and germinate after being transferred to pure water. Interactions between nutrients and other factors have been studied infrequently.

Germination experiments are usually performed in the laboratory where light, temperature and water supply can be controlled. Seeds can be subjected to various nutrient treatments by adding solutions of the required compositions to Petri dishes containing filter paper and the seeds. Effects of nutrients can be quantified by measuring the time taken for 50% of the seeds to germinate (see Grime *et al.*, 1981). Rorison (1967) measured the germination (expressed as radicle emergence) of seeds of different species on a range of soils differing in pH and associated chemical characteristics. It was found that on acidic soils, the germination rate of calcareous-soil seeds was inhibited, relative to germination on soils upon which the seeds originated.

2.4.7 Flowering/seed production

Effects of nutrients upon long-term events such as flowering and seed production are most easily conducted on soil-grown plants. In many annual species (e.g. *Poa annua*), flowering can occur within 4 weeks after germination, depending upon the prevailing environmental conditions (Peterkin, 1981). In such cases, solution cultures can be used without difficulty. Flowering responses to, *inter alia*, nutrient availabilities can be an important aspect of a plant's capacity to survive unfavourable conditions.

2.5 Effects of specific environmental factors upon plants, in relation to their nutrition

2.5.1 Water supply

Effects of water supply upon plants and the methods used to investigate these are discussed fully in Chapter 3. Here, specific effects upon nutrition will be considered briefly.

A low soil water content reduces ion diffusivity in soil (see equation (1)). This means that an ion is effectively less accessible to roots in dry soil than in moist soil. Low water supplies can also reduce transpiration, which would reduce the rate of mass flow of nutrient supplies.

At the other extreme, in waterlogged soils, anoxia alters the redox potential of the soil, producing reducing conditions. This causes denitrification, which can lead to low inorganic N availability, the production of potentially-toxic Mn^{2+}, Fe^{2+} ions and sulphides, and low O_2 availabilities, which can impair root growth. This is further complicated by the potential ameliorating effect of Ca upon Mn toxicity (see Rorison & Robinson, 1984). Difficulties in interpreting unequivocally the effects of waterlogging upon plants could arise. Measurements of soil redox potential, metal (Chapter 6) and O_2 availabilities (Greenwood & Goodman, 1967) should be made.

In the case of aquatic plants, the rate of water flow over plant surfaces is important in their nutrition, particularly in haptophytes (see Raven, 1981). The formation of hairs on the outer surfaces of haptophytes can penetrate boundary layer depletion zones, thereby contributing to nutrient acquisition. Hair formation is sensitive to nutrient supply and, hence, water flow rate (see Section 3.1.5).

2.5.2 Temperature

Extremes of temperatures can affect nutrient supplies, nutrient uptake and plant growth differentially in ecologically-distinct species (Rorison et al., 1983). It is important to define *exactly* the temperature treatment to be used: constant or fluctuating temperatures; the extent of diurnal temperature changes; root and shoot differential temperatures; and the absolute temperature range.

Seasonal temperatures can affect the availabilities of nutrients whose mineralization depends upon microbial activity (e.g. NH_4^+, NO_3^-, P). Patterns of mineralization in any soil vary with altitude, aspect, topography and depth in the soil profile, all of which influence soil temperature, aeration and moisture content. Sampling soil and measuring nutrient concentrations on a number of occasions throughout the year can reveal such effects (e.g. Davy & Taylor, 1974; Gupta & Rorison, 1975; Taylor et al., 1982). The effects of specific temperatures upon nutrient availability can be measured by incubating soil samples at various temperatures in the laboratory, and determining the production of available nutrients at the end of the incubation period (Kowalenko & Cameron, 1976; see also Chapter 6, this volume).

In controlled-environment plant growth facilities, shoot temperatures can be maintained precisely using programmable air-conditioning systems (e.g. Sutton & Rorison, 1980). Using culture solutions, roots can be maintained at precise temperatures by flowing the solutions through a refrigeration or heating unit before entry into the plant-growth containers (Bhat, 1980) or by immersing the plant-growth containers in thermostatically-controlled water-baths (Peterkin, 1981).

In soil, control over the root temperature is more difficult to achieve, due to the insulating capacity of the soil itself. It is perhaps better to allow the soil to respond naturally to ambient temperature changes and to record corresponding changes in soil temperature at various depths using a continuous data-logging system (Rafarel & Brundson, 1976). Measured plant responses can then be related to the root temperature at their time of occurrence.

Seasonal temperature and nutrient availability changes can be simulated in the laboratory by increasing or reducing temperature during the experiment (to simulate, e.g. 'spring' or 'autumn' conditions). A sudden temperature change of more than a few degrees could induce thermal shock in the plants. A period of acclimatization is necessary, so temperature changes should be made gradually over a period of hours or days.

Changes in air temperature affect the saturated vapour pressure deficit. To compensate for this in controlled-environments, relative humidity should be adjusted according to the figures given by Hoffman (in Tibbitts & Kozlowski, 1979).

2.5.3 pH

Soils can be inherently basic, acidic or neutral, depending upon the chemical nature of the parent material. Plants themselves can influence the pH of their root environment by excreting H^+ or OH^-/HCO_3^- ions during uptake of NO_3^- or NH_4^+ respectively (Raven & Smith, 1976; Marschner & Römheld, 1983), or as a means of solubilizing P (Grinsted et al., 1982).

Conventional methods of pH measurement are given in Chapter 6. Localized differences in rhizosphere pH can be measured in situ by impregnating soil with agar containing a pH-indicator. Marschner & Römheld (1983) describe this technique in detail and provide remarkable evidence of its precision.

Using solution cultures, it is possible to vary pH almost independently of other factors. True independence cannot be achieved, due to the solubilities of metal ions in solution being pH-sensitive (Hewitt, 1966, Chapter 8). The maintenance of constant solution pH is therefore of prime importance. Methods by which solution pH can be controlled are described in Section 3.1.5.

2.5.4 Nutrient source—pH interactions

When investigating the effects of soil chemical factors upon plants, it is necessary to deal with a multiplicity of interactions between nutrient source, nutrient availability, metal toxicity, ion balance and pH. The greatest difficulty here is the need to find a compromise between field conditions, in which all these factors interact freely, and a simplified experimental system, in which individual factors can be varied or held constant as required. In the former, there is virtually no control over the intensities of each factor, so any observed plant responses will be difficult to explain unequivocally. In the latter, if the effects of only one variable are examined, there is a danger of drawing erroneous conclusions from plant responses to unrealistic conditions.

For example, in temperate, acidic soils, Al^{3+} and NH_4^+ occur in relatively high concentrations and are considered to be major determinants of differences in plant distribution between acidic and calcareous soils (see Kinzel, 1983). In soils of higher pH, Al^{3+} solubility is virtually zero, and NH_4^+ is mineralized to NO_3^-. An experimental design to investigate the effects of Al^{3+} availability, N source and pH upon plants in solution culture (not soil, because of the difficulty in manipulating its composition) might consist of the following:

2 N sources (NH_4^+ or NO_3^-) $\times$ 2 Al^{3+} treatments (present or absent)
$\times$ constant pH $\times$ 3 species (restricted to acidic or calcareous
soils or of widespread distribution) $\times$ 5 replicates.

Of particular importance in such experiments is the prevention of pH shifts in the bulk solution (cf. Section 3.4.3). In the field, large, rapid changes in pH in the bulk soil are usually prevented by the soil's buffering capacity. Shifts in nutrient solution pH are therefore unrealistic. Thus, experiments (e.g. Foy & Fleming, 1978) in which plants are supplied with NH_4NO_3 and Al^{3+} at a low pH, and allowed to raise the solution pH by means of NO_3^- uptake/OH^- excretion, reveal an interesting physiological phenomenon, but provide little *realistic* evidence of how the plants would respond to conditions of low pH and high Al^{3+} concentration in the field where the likely N-source would be predominantly NH_4^+.

Similar caution needs to be exercised when dealing with other nutrient–pH interactions: e.g. P $\times$ Ca $\times$ pH (Nassery & Harley, 1969); NO_3^-/NH_4^+ $\times$ pH (Gigon & Rorison, 1972).

2.5.5   Plant–micro-organism interactions

Most laboratory experiments and all field studies of plant nutrition involve interactions with micro-organisms. Unless seeds are surface-sterilized and seedlings maintained subsequently in axenic culture, bacteria, and possibly fungi, will be associated with the roots. Normally, it is convenient and acceptable to ignore this, and to assume that 'roots' behave as if sterile conditions prevailed. In certain circumstances, however, unwanted colonization of roots by micro-organisms can seriously influence experimental results. Barber (1968) reported that non-sterile barley roots absorbed P

100–300% faster than sterile roots, in the P concentration range 1–10 × 10^{-9} mol cm^{-3}. It was found that this difference was due to the absorption of P by bacteria on the surfaces of non-sterile roots, rather than by the roots themselves (see also Clarkson, 1974, Chapter 10).

In studies of plant–micro-organism interactions, four stages of investigation can be identified.

(i) Identification and isolation of the micro-organisms concerned. This is done using conventional microbiological techniques (see e.g. Aaronson, 1970, Chapters 4, 7; Parkinson et al., 1971; Page, 1982) and light or electron microscopy (Rovira et al., 1983).

(ii) Establishing the extent of the plant–micro-organism association in the field: e.g. Read et al. (1976); Newman et al. (1981); Mosse et al. (1982); Moser & Haselwandter (1983).

(iii) Establishing the nature of the plant–micro-organism association. This might consist of the direct transfer of nutrients, water or assimilates between the plant and a mycorrhizal endophyte (Harley & Smith, 1983, Chapters 16, 17), or between the plant and symbiotic or free-living diazotrophic bacteria for example. Indirect transfers of nutrients could be important, such as the release of nutrients from bacteria in the rhizosphere and the subsequent uptake by the plant. The use of micro-organisms or plants grown on, or in media containing isotopically-labelled nutrients (Section 3.5) is valuable in detecting nutrient transport between plants and associated micro-organisms. The primary nutritional relationship between plants and mycorrhizal fungi may be one of enhancing root geometry in soil, thereby increasing the quantity of nutrients that are potentially available to the plant (in which case it is necessary to compare nutrient uptake rates (per unit root length) of mycorrhizal and non-mycorrhizal plants: Sanders & Tinker, 1973), one in which the fungal hyphae are capable of utilizing nutrient sources that are not immediately available to the plant (e.g. organic N: Stribley & Read, 1980), or one in which the endophyte might confer tolerance of potentially toxic metals upon the plant (e.g. Cu, Zn: Bradley et al., 1982).

(iv) Comparing the growth of axenically-grown plants with plants grown in association with the micro-organism, in order to demonstrate whether the association has positive, negative or neutral effects upon either organism. This might include measurements of plant growth responses, nutrient uptake and morphological changes (particularly in the roots, e.g. Bowen & Rovira, 1961) that could be attributed to the micro-organism. Studies of this type are numerous, particularly with crop plants and mycorrhizal fungi, e.g. Stribley et al. (1980). A normal pre-requisite for such experiments is the establishment of specific plant–micro-organism associations. For free-living bacteria, this involves the straightforward inoculation of the sterile rooting medium of the plant with an isolate of the appropriate bacterium. Establishment of associations between plants and symbiotic or free-living diazotrophic micro-organisms in field and laboratory experiments are described in detail by Vincent (1970) and Bergersen (1980). Sprent (1979, pp. 172–182) gives a useful review of appropriate techniques in this field. The establishment of legume-root nodule bacteria symbioses in solution cultures can be achieved by adding suspensions of rhizobia to the solution bathing the roots of uninfected plants. Specific inocula can be obtained from appropriate culture collections (e.g. see Bergersen, 1980, Table 10). Non-specific inocula can be obtained by crushing nodules removed from soil-grown legumes, and adding these to the solution (J.H. Macduff, pers. comm.).

For mycorrhizas, the method of infection depends upon the type of endophyte: vesicular-arbuscular (Mosse & Phillips, 1971; Hepper, 1981; St John et al., 1981); ericaceous (Pearson & Read, 1973); orchidaceous (Beardmore & Pegg, 1981); or ectomycorrhizas (Fortin et al., 1983). General experimental methods for mycorrhizal research can be found in Schenck (1982).

Experimental work on the nutritional relationship between plants and mycorrhizal fungi has increased considerably in the past 10 years, although ecological aspects of such relationships have been considered relatively infrequently (see Harley & Smith, 1983, Chapter 19). Experimental confirmation of the possibility that most plants in the field are interconnected via their mycorrhizas, and that nutrient, water and assimilate transport can occur between them (see Newman, 1983), would radically alter present perceptions of plant–nutrient relationships.

An important aspect of experimental work on mycorrhizas is the quantification of the extent of infection of roots by the endophyte. This is usually measured as the percentage of the root length in which fungal structures can be observed. Various staining and counting methods have been described: Phillips & Hayman (1970); Giovanetti & Mosse (1980); Stribley et al. (1980) and Biermann & Lindermann (1981). The recovery of the external mycelium of endomycorrhizas is extremely difficult. Sanders & Tinker (1973) reported that this could be achieved with limited success by careful wet-sieving, manual removal of foreign matter, and rapid measurement of fresh weight. Mycelial length was calculated using similar assumptions as those for equation (3) (Section 2.4.1).

Although mycorrhizas can be synthesized in solution cultures (Asher & Edwards, 1983), the maintenance of the necessary sterile conditions is difficult in systems in which large volumes of solutions are re-circulated. Smaller self-contained culture units in which mycorrhizal plants can be grown in solid media and supplied intermittently with nutrient solutions are more reliable (e.g. Bigg & Alexander, 1981, Fig. 4.7). The synthesis of mycorrhizas in solution cultures is dependent upon nutrient concentrations. High concen-

(a)

(b)

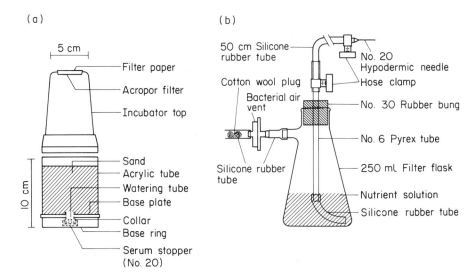

Figure 4.7 (a) Culture unit for aseptic culture of mycorrhizal seedlings. (b) Watering device for aseptic renewal of nutrient solution in culture unit (Bigg & Alexander, 1981).

trations inhibit the establishment of the symbiosis. Asher & Edwards (1983) suggested that the inhibition of root hair production in plants grown at high nutrient availabilities can reduce the incidence of root nodulation in legumes, and this could also explain why mycorrhizal establishment is suppressed under similar conditions, root hairs being primary sites of infection by mycorrhizal fungi.

2.4.6 Plant–animal interactions

Direct nutritional relationships between plants and animals occur when animals feed upon plants or when plants feed upon animals.

Herbivory can take the form of the total or partial destruction of above-ground plant material by grazing insects and mammals, the selective consumption of phloem and xylem contents by hemipteran insects (Raven, 1983), or the grazing on roots and associated micro-organisms by insect larvae, mammals, nematodes (McNaughton, 1983) and collembolans (Warnock, *et al.*, 1982).

Phyto-carnivory is a relatively rare phenomenon, being restricted to less than twenty genera, and such plants are often found in acidic, nutrient (especially N)-deficient habitats (Lüttge, 1983).

Grazing can affect plant nutrition through reductions in root growth caused by the disruption of carbohydrate supplies from the shoot (Alcock, 1962). This effect can be simulated in experiments where foliage is clipped to a certain height at regular intervals over a period of months. Experiments such as these will not simulate the effects of selective grazers that show preferences for vegetation of different nutritional values (Hughes *et al.*, 1962). In this respect, differences in the capacities of plants subjected to grazing to synthesize secondary metabolites as anti-herbivore compounds (Swain, 1977) could be important. Analytical methods for the determination of such compounds are described by Harborne (1973). Sampling methods for animal (insect) populations on vegetation in terrestrial or aquatic habitats have been described by Southwood (1978, Chapters 4, 5, 6).

2.6   Community nutrient relationships

2.6.1   Competition

Nutrient competition (*sensu* Harper, 1961) experiments are usually conducted in the laboratory and involve soil-grown plants. This is because in the field, it is difficult to isolate plants from their neighbours in order to obtain non-competing controls, and because in soil the spatial distribution of roots determines whether nutrient competition can occur or not, an effect lost in solution cultures.

The classical experimental design for competition experiments is that of Donald (1958) in which the effects of competition for nutrients, water and/or light upon the growth of plants in specially-designed pots (Fig. 4.8) is measured. An important aspect of any such experimental design is the duration over which competition is to occur. Differences between plants in terms of their relative growth rates or seed sizes, for example, can lead to differences in their absolute growth during the course of competition. Experiments of too short a duration would not necessarily detect such differences and could lead to erroneous conclusions (see Newman, 1983).

A further aspect that should be considered carefully is nutrient availability. The outcome of competition between plants growing in soil containing abundant nutrient supplies could be different from that resulting from competition for a low nutrient supply, particularly if the plants involved had

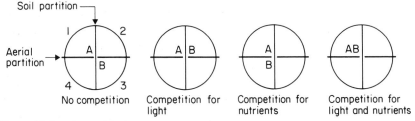

Figure 4.8 Experimental design used by Donald (1958) to investigate effects of root and shoot competition for nutrients, water or light by two species, A and B, arranged in the pots as shown.

widely different ecological characteristics. Competition between species can sometimes result in the replacement of potentially fast-growing plants by slower-growing plants (the 'Montgomery effect': Van den Bergh, 1969). The morphology, and, hence, spatial distribution, of the roots of ecologically different plants grown in isolation, can respond differently to low or high availabilities of N, for example (Robinson & Rorison, 1985). Such differences in growth response could enhance or obviate interspecific competition for nutrients depending upon the species involved. Although this is of considerable ecological interest, little work has been done on the problem.

Competition for nutrients occurs when the ion concentration depletion zones around adjacent roots overlap (cf. Section 2.1; Nye & Tinker, 1977, p. 218 et seq.). Interpretations of the results of nutrient competition experiments should determine, firstly, whether this condition occurred during the experiment. To do this, measurements of root densities and root radii are required (see Section 2.4.1), and measurements of the diffusive and convective mobilities of the ions in soil should be made (see Section 2.1.1). For slowly-diffusing ions (e.g. PO_4^{3-}), the radial extent of concentration depletion around a root will, on average, be small, and after an initially rapid depletion, will spread only slowly from the root into the bulk soil. For NO_3^-, however, its depletion zone will extend some distance from the root surface and will continue to spread rapidly through the soil. Thus, for inter-root competition to occur for PO_4^{3-}, roots need to lie in very close proximity in the soil, whereas even widely-spaced roots are likely to compete for NO_3^- after a very short time. The minimum root density needed for inter-root competition to occur for a particular ion can be calculated from the equation

$$L_V = 1/\pi[2(Dt)^{0.5} + a]^2 \tag{5}$$

where L_V = root density $(cm\,cm^{-3})$; D = ion diffusion coefficient in soil $(cm^2\,s^{-1})$; t = time after start of depletion (s); a = average root radius (cm). If root densities measured at time t are greater than the value of L_V thus calculated, then inter-root competition for the ion in question would have occurred.

Secondly, the effects of nutrient competition upon the growth of the plants involved need to be assessed. In most published accounts of such experiments, the growth of only above-ground plant material is measured, because of the difficulty in separating completely the intermingled root systems of different plants grown in the same soil.

Competition experiments usually reach a conclusion of the type: 'species x is a better competitor than species y', with no attempt to identify with certainty, which characteristics of species x and y were responsible for the observed outcome. Admittedly, it is difficult to obtain definite answers to this question experimentally. The theoretical approach of Baldwin (1976), however, suggests which parameters could be measured during competition experiments. In addition, this approach offers a means to simulate how varying one characteristic independently of the others could contribute to the growth of the plant and, hence, influence the outcome of competition with another plant possessing a similar set of defined characteristics (see Section 3.8).

The problems of running nutrient competition experiments in the field are considerable (see Newman, 1983). The major difficulty is the elimination of above-ground interference (*sensu* Harper, 1961) between plants. Such experiments are too laborious and time-consuming to compensate for the small amount of useful information that might be produced.

2.6.2   Nutrient cycling

In contrast to many of the experimental methods outlined above, nutrient cycling experiments are conducted largely in the field. Nutrient cycling through ecosystems is linked fundamentally with decomposition processes. These are considered in Chapter 1 of this volume, and by Swift *et al.* (1979), who also provide a succinct review of appropriate methods. Here, attention will be restricted to nutrient 'balance sheets' of a simple soil–plant system.

The objects of such an investigation are to determine how much nutrient is potentially available to plants, how much of this is taken up by plants, and what happens to the nutrients that are not taken up, usually over a minimum time period of a growing season (e.g. spring–autumn) or a complete annual cycle. The dynamics of a simple system can be visualized as shown in Fig. 4.9. At any time, the total size of the soil nutrient pool (which includes organically-bound, microbial, adsorbed and available fractions) is the difference between input processes (wet and dry deposition, deposition of animal remains, fertilizer applications, litter production, etc.) and output processes (leaching and volatilization in the case of N, plant uptake, and grazing). In theory, this value, calculated from the measurements of the partial processes in the cycle, should equal the value for total soil nutrient content as determined by chemical analysis. If this is the case, all the nutrient fluxes through the systems should have been accounted for (hence 'balance sheet'). As is more often the case, the two values are not equal, which would suggest that the system consists of fluxes other than those measured, or inaccuracies of measurement.

The measurement of nutrient fluxes in such systems is made most conveniently using a lysimeter system. This consists of volumes of soil that can be sampled periodically for chemical analyses, and from which throughputs of water and nutrients can be recorded (see Knowles, 1980).

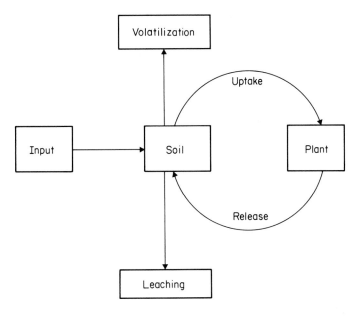

Figure 4.9 Schematic representation of a simple nutrient cycle through the soil–plant–atmosphere system.

The lysimeter could consist of the field soil *in situ* (Chapter 1, pp. 44–46), monoliths of field soil encased within special containers, or field soil which has been sieved and dried before packing into containers (e.g. Rorison *et al.*, 1983). The size of the lysimeter depends upon the duration of the experiment and the expected sizes of the plants involved at the end of the experiment. For a 2-year experiment with perennial grasses grown from seed, Rorison *et al.* (1983) used soil volumes of $1 \, dm^3$, each containing a single plant.

The essential elements of a lysimeter experiment are shown in Fig. 4.10. For comparative purposes it is necessary to make continuous measurements of solar radiation using a solarimeter (Painter, 1976) and of air, soil surface, and subsurface temperatures. Thermistors connected to a data-logging system (Rafarel & Brundson, 1976) are suitable for this purpose. Periodic measurements of plant growth and nutrient content and soil water and nutrient content should be made throughout the experiment. More frequently, volumes and nutrient contents of water throughput from lysimeters containing soil with plants, soil without plants, and without soil or plants should be measured. Volatilization of N is a vital but frequently ignored component of the N cycle. Gaseous N emissions can be measured directly from small areas ($0.5 \, m^2$) of soil contained within enclosed chambers, or from larger areas using micrometeorological models. These techniques have been

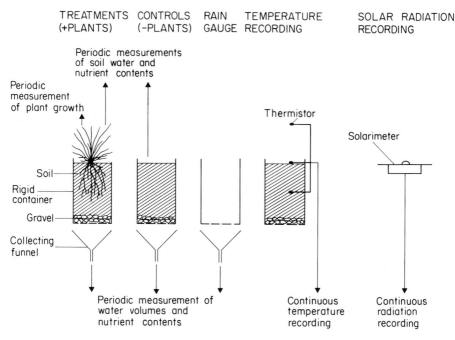

Figure 4.10 The components of a lysimeter experiment, designed to investigate nutrient 'balance sheets' in the soil–plant–system.

described fully by Ryden & Rolston and by Denmead (in Freney & Simpson, 1983).

2.6.3 Field surveys

Whilst laboratory conditions can provide the means to conduct fully controlled experiments, field surveys provide the necessary perspective with which the results of such experiments should be viewed. This has been mentioned previously in this chapter, in connection with the extent of plant–micro-organism associations (Section 2.5.5). The difference between field surveys and field experiments is that in the former, measurements are made of plant and environmental characteristics as they are in the undisturbed state; in the latter, plants and environment within a defined area are manipulated in certain ways before or during such measurements.

Methods to describe the sample floristic and soil characteristics are described in Chapters 5 and 9 of this volume. Appropriate methods of material collection, storage and chemical analysis are given in Chapter 6.

For nutrient surveys in particular, the following points should be noted:

(i) It is practicable to measure nutrient concentrations only in above-ground plant material. Chemical analyses of roots extracted from soil are not reliable, due to contamination (see Section 3.1.1).

(ii) Samples of above-ground plant material should differentiate between green leaf, flower, bark, wood and dead material, since nutrient concentrations in each fraction differ with the age of the material.

(iii) Soil samples for chemical analysis should be taken from the entire depth of the rooting zone, if possible, by core sampling (Section 2.4.1).

(iv) If the survey area covers a number of widespread sites, it is often difficult to make sample collections on the same day. Foliar, soil and water nutrient concentrations vary over time. To make meaningful comparisons, therefore, whenever possible, samples should be collected over a short a time period as possible on each sampling occasion.

3 Experimental methods

3.1 Root growth media

As indicated in the previous section, plants can be grown for experimental purposes in solid or liquid rooting media. There is no universally acceptable material for this purpose. The type of rooting medium used will depend upon (and, to some extent, limit) the scope of the experiment (see Asher, 1978).

3.1.1 Soil

Soil is most suitable for long-term experiments (Section 2.4). Its main advantage is that nutrients are unlikely to be depleted (except in deficient, unbuffered soils). Consequently, the amount of routine management a soil-based experiment requires is minimal, consisting mainly of maintaining an adequate water supply. For terrestrial plants, soil is obviously the most realistic rooting medium. Effects of ion diffusion, root morphology and root distribution can be examined. However, the extraction of entire root systems from soil is exceedingly difficult. The complete removal of soil particles from recovered roots is impossible. These difficulties are exacerbated with increasing clay and organic matter content of the soil. Soil cannot be manipulated to produce exact nutrient compositions, so responses of plants to precisely-defined availabilities cannot be measured.

Changes in temperature, water content, compaction, drying, re-wetting, exposure to air, storage and general handling all affect nutrient mineralization processes and, hence, soil nutrient concentrations.

Sterile soils are required for experiments on plant–micro-organism interactions (Section 2.5.5), and for those in which it is necessary to minimize chemical transformations caused by soil micro-organisms (e.g. $NH_4^+ \rightarrow NO_3^-$). Soil can be sterilized by using γ-radiation, autoclaving, microwave radiation (Wainwright et al., 1980), or specific microbial inhibitors (Slangen &

Kerkhoff, 1984). The degree of sterilization depends upon the duration of exposure to, or concentration of, the sterilizing agent. Microbial activity will recover its original level faster after the soil has been treated with low doses of radiation or low concentrations of inhibitors. Ideally, the radiation dose or inhibitor concentration should be limited to that needed to minimize microbial activity for the duration of the experiment, but no longer. The use of high concentrations of 'N-serve' to inhibit nitrification (Ashworth *et al.*, 1980) is to be avoided because of its extreme toxicity.

An important point is that sterilization procedures cause the release of nutrients from killed micro-organisms. Determinations of soil nutrient concentrations should be made before and after sterilization, to check the magnitude of the nutrient release.

3.1.2   Sand

Whilst sand provides much of the physical realism of an inorganic soil, its innately low buffering capacity and available nutrient content make it more amenable to experimental manipulation. Although sand can be regarded as being chemically inert for most purposes, for micronutrient studies it should first be purified by acid washing (Hewitt, 1966, p. 56 *et seq.*). Nutrient treatments are usually imposed by regular applications of nutrient solutions (see Asher, 1978). Coarse sand is freely-draining and well aerated, but this makes it susceptible to rapid nutrient depletion. Using finer sand improves water holding capacity, but this reduces aeration. The removal of roots from sand is relatively straightforward and it can be regarded as a useful compromise between the inaccessible reality of organic soil and the wholly artificial environment (for terrestrial plants, at least) of solution cultures. It is a medium that would appear to be particularly well suited to growing morphologically-diverse plants for periods of weeks at a range of nominal nutrient availabilities (e.g. Bradshaw *et al.*, 1964; Grime & Hunt, 1975; Austin & Austin, 1980).

3.1.3   Vermiculite and perlite

These materials are suitable as uniform, inert substrates upon which seeds can be germinated and seedlings raised. They are unsuitable for long-term growth experiments, since their water-holding capacities are low (see Evans, 1972, p. 71).

3.1.4   Agar

Agar is the ubiquitous medium for the culture of micro-organisms and certain algae (Aaronson, 1970; Stein, 1973). Its use for higher plants is limited, however, by the small volumes (e.g. Petri dishes) in which liquified agar will solidify uniformly, and by the difficulties of aeration. Agar has been used with higher plants in short-term experiments involving the transport of nutrients via mycorrhizal hyphae (Pearson & Tinker, 1975; Section 2.5.6). The maintenance of slow-growing plants (e.g. tree seedlings) or tissue cultures on agar slopes in test tubes is common practice.

3.1.5 Solution

Solution cultures overcome many of the problems involved in growing plants in solid or semi-solid media. The ionic composition of the root environment can be defined, manipulated and measured with precision. However, this bring with it major *dis*advantages in that nutrient solutions are liable to rapid changes in ionic composition due to ion uptake/efflux by plant roots, and that solution culture experiments demand more routine maintenance than those using soil-grown plants.

Principles, techniques, details of materials, equipment and practical procedures are discussed in the exhaustive review of the subject by Hewitt (1966). Chemical composition of nutrient solutions are also given by Hewitt (1966, Chapter 8), and Asher (1978). Distilled/de-ionized water should always be used for culture solutions.

In all solution culture systems, there are four general technical difficulties that must be overcome:

(i) *pH maintenace* (cf. Sections 2.5.3, 2.5.4). This is essential in the majority of solution culture experiments. pH can change rapidly in an unbuffered medium due to differential anion–cation uptake by roots. Glass *et al.* (1983) reported a decrease in pH from 5·2 to 4·9 in 40 minutes during NH_4^+ and NO_3^- uptake by barley roots. The direction and magnitude of such changes depend not only upon N source and charge balance required,

Table 4.1 pH drift in 48 hours caused by one plant in $0·5 dm^3$ nutrient solution (from Gigon & Rorison, 1972)

N source	*Rumex acetosa*				*Deschampsia flexuosa*		
	Initial pH	Final pH	Mean RGR (d^{-1})	Final dry wt $(mg\ plant^{-1})$	Final pH	Mean RGR (d^{-1})	Final dry wt $(mg\ plant^{-1})$
NH_4	4·2	3·4	0·12	240	3·8	0·13	94
	5·8	3·6	0·14	389	4·3	0·10	78
	7·2	3·3	0·15	656	6·6	0·15	117
NO_3	4·2	6·5	0·17	922	4·2	0·10	80
	5·8	6·5	0·16	887	6·0	0·12	73
	7·2	7·2	0·18	996	7·1	0·05	13

but also on plant size, species and nutritional status (Table 4.1). Daily manual monitoring of pH and appropriate acid/base additions to the nutrient solutions are usually adequate to minimize serious pH shifts. In some experiments, however, this is unacceptable, and automated sampling, monitoring and acid/base addition systems are required. These can be part of a large solution culture installation (e.g. Clement *et al.*, 1974; Breeze *et al.*, 1982), or as additions to small, specific-ion monitoring systems, such as that described by Glass *et al.* (1983).

(ii) *Prevention of algal contamination.* If nutrient solutions are exposed to air and light they, and surfaces in contact with them, will become contaminated by green algae. If sterile conditions are not observed this is virtually impossible to prevent, but it can be minimized by using black, opaque materials wherever possible. Covering pots, tubes, reservoirs, etc., with aluminium foil also helps to exclude light from the solutions. The inclusion of algicidal chemicals in the solutions could be inhibitory to plant growth.

(iii) *Oxygenation of solutions.* Roots require adequate O_2 concentrations to grow properly. The solubility of O_2 in water is very low ($2 \cdot 76 \times 10^{-4}$ mol dm^{-3} at $20°C$), so some provision for solution aeration should be made in order to prevent O_2 depletion. Bubbling compressed, filtered air through the nutrient solution is the most effective way of doing this. Alternatively, rapid circulation of solution around the roots, using magnetic stirrers or continuous flow systems (see below) encourages O_2 dissolution at the air–solution interface, and hence, to some extent maintains solution O_2 concentrations.

(iv) *Prevention of localized concentration depletions.* In static solutions, rapid nutrient uptake can cause depletion zones to form between the solution at the root surface and the bulk solution. Ion diffusion from the bulk solution to the root will then limit uptake rate as in the soil (Section 2.1). This effect can also occur when solutions are well stirred, within a dense mat of roots where bulk solution movement is restricted, although this effect is likely to be significant only at concentrations below *c.* $2 \cdot 5 \times 10^{-5}$ mol dm^{-3} (Nye & Tinker, 1977, p. 112). The effect can be minimized by rapid stirring and aeration of solutions (see above).

There are two basic methods of supplying nutrients to plants growing in solution cultures, i.e. intermittent renewal and continuous flow.

Intermittent renewal

As the name implies, this involves the partial or total replacement of solutions from which nutrients have been removed by plants, with freshly-prepared, non-depleted solutions, at specific times during the experiment. It is inevitable that some nutrient depletion will occur between solution changes, and this must be taken into account when planning the experiment. At initially high concentrations, such depletions are unlikely to be serious if renewals are frequent. When dilute solutions are used, depletions could be significant even with daily or twice-daily solution changes.

The minimum time between solution changes can be estimated from the following equation (Asher & Edwards, 1983):

$$t_{min} = \frac{d_{max} \cdot V_1 \cdot C_{li}}{100 \, W_R \cdot A} \tag{6}$$

where t_{min} = minimum time between solution changes (d); d_{max} = maximum acceptable concentration depletion (%); V_1 = volume of solution per pot or per plant (dm^{-3}); C_{li} = initial ion concentration in solution ($mol \, dm^{-3}$); W_R = root dry (or fresh) wt per pot or per plant (g); A = specific absorption rate of the ion at concentration C_{li} ($mol \, g^{-1}$ root dry wt (or fresh wt) d^{-1}). If values of t_{min} are calculated for all the ions in the nutrient solution, the smallest of these values should be used as the interval between solution changes. Nutrient concentrations in intermittently renewed systems tend to be much higher than the minimum concentrations needed to maintain maximum growth (see Asher & Edwards, 1978), because of the need to prevent the occurrence of growth-limiting depletions.

Supplying the same amounts of nutrients at each solution change (as is common practice) is unlikely to have the same effect upon the plants throughout the experiment, since the requirements of plants for nutrients increase as each plant gets bigger. Nutrient supplies can be aligned more closely with plant requirements by using 'programmed nutrient additions' (see Asher & Edwards, 1983).

This technique involves frequent, step-wise nutrient additions, the rates of which increase incrementally with time, and are determined previously from growth curves and patterns of nutrient uptake by plants growing under near-optimum conditions. Deficiency or toxicity treatments can be established by adding fixed percentages of the optimum nutrient additions. A more flexible version of this approach aims to maintain constant concentrations in the plant (after Ingestad, 1982):

$$M_t = \frac{[X] \cdot 10^4}{a_x \cdot W_i (\exp (R_M \cdot t))(\exp (R_M) - 1)}$$

where M_t = addition rate of nutrient X at time t ($mol \, d^{-1}$); $[X]$ = concentration of X in the plant (% dry (or fresh) wt); a_x = atomic wt of X; W_i = initial whole-plant dry (or fresh) wt (g); R_M = relative nutrient addition rate (d^{-1}), and numerically equivalent to relative growth rate; t = time (d).

Continuous flow

By flowing nutrient solution continuously past the roots, concentration depletions can be minimized and nominal concentrations maintained approximately constant. Thus, the *raison d'être* of continuous flow systems is the elimination of serious concentration differences between root surfaces and the bulk solution. Low nutrient concentrations can be used in continuous flow systems compared with intermittently renewed systems (Asher & Edwards, 1978).

Edwards & Asher (1974) identified two basic types of continuous flow system. Firstly, the Van den Honert-type (in Hewitt, 1966, p. 288 *et seq.*; Fig. 4.11a), in which the uptake of nutrients by the roots is balanced by compensatory increases in the flow rate of a nutrient solution of the same composition. Precise regulation of flow rate is critical to the success of this system. It must be adjusted frequently and individually to each pot to account for differences in plant size, etc. between pots. This means that a solution flow to only a few pots can be regulated effectively at any one time, which limits the applicability of this system to ecological work. It has certain advantages, however, in that it requires only low rates of solution flow and small volumes of solution per plant.

The second, more widely-used, system is the Asher-type (Asher *et al.*, 1965; Fig. 4.11b). In this system, large numbers of pots can be used, but this requires the rapid flow in a recirculation system of large volumes of solution. Depletions are supplemented by controlled additions from stock solutions to the flowing solution before reaching the roots. Without recirculation, the volumes of solutions needed would be impossibly great, but recirculation could result in the build-up of potentially allelopathic exudates in the nutrient solution. For this reason, and to minimize microbial contaminations, the entire system is drained completely at intervals of a week or more.

Such systems can be installed as a permanent facility in a glasshouse (Asher *et al.*, 1965; Clement *et al.*, 1974), and interfaced with automated monitoring equipment and a microcomputer control system (Breeze *et al.*, 1982). The cost of this is considerable, however. A portable version has been designed (Bhat, 1980) for use in controlled-environment growth rooms, but this suffers from being unable to monitor and supplement solutions automatically, and being restricted to circulating solution of only one composition at a time. Again, the applicability of this system to ecological problems is limited. The incorporation of both rapid computer-controlled monitoring (e.g. Glass *et al.*, 1983) and nutrient addition equipment (Rorison *et al.*, 1981, Fig. 4.11c) into such a continuous flow system could allow relatively small solution volumes and/or low flow rates to maintain a wide range of nutrient concentrations, but such developments have yet to be made.

Specialized types of continuous flow systems have been devised. Ingestad & Lund (1979) grew birch seedlings in units in which the roots were suspended continuously in an aerosol of nutrient solution: 'mist culture'. Moorby & Nye (1983) describe a useful nutrient-film technique in which plants are supplied with nutrients in shallow sloping troughs. This technique has also been used for commercial crop production (Asher & Edwards, 1983).

Edwards & Asher (1974) define the minimum solution flow rate needed to maintain a certain concentration as:

$$f = \frac{100 W_R \cdot A}{d_{max} \cdot C_1} \tag{8}$$

where f = solution flow rate ($dm^3 \, pot^{-1} \, d^{-1}$); C_1 = nominal nutrient concentration ($mol \, dm^{-3}$); other symbols as in equation (6).

Using equation (8), the practical limitations of a continuous flow system can be evaluated for a particular experiment. For example, if $W_R = 1 \, g \, pot^{-1}$, $A = 5 \times 10^{-3} \, mol \, g^{-1} \, d^{-1}$, $d_{max} = 5\%$, and C_1 ranges from 0·001 to $1 \times 10^{-3} \, mol \, dm^{-3}$, the minimum flow rates needed to maintain these concentrations are given in Table 4.2. The daily *total* amounts of the nutrient that the calculated flow rates would supply are the same at each concentration. However, the maximum flow rate that has been achieved in practice is $c. \, 5000 \, dm^3 \, pot^{-1} \, d^{-1}$ (Asher & Edwards, 1983, Table 4). If this flow rate was used for all the concentrations in this example, the daily total nutrient supplies are as shown in the fourth column of Table 4.2. It can be seen that at the most dilute concentrations, some depletions would occur.

It follows that the same daily total nutrient supplies could be delivered *en masse* to the plants in single, daily solution renewals, rather than as a continuous flow (see Rorison, 1969). The difference would be that in the former case, no attempt is made to maintain constant concentrations, and depletions between solution changes are expected. Since plants growing in continuous flow systems will deplete dilute solutions in any case, the use of intermittent solution renewal might be equally useful, provided that the differences between the two systems are fully appreciated. Differences in short-term nutrient uptake rates could be found between plants grown in constant flow and those grown in intermittently-renewed systems (Wild *et al.*, 1979). Comparisons between plants grown in different systems are unlikely to be meaningful in this respect.

Solution cultures are totally artificial growth media for terrestrial plants, so the relevance of such experiments to their ecological characteristics should be considered with care (see Rorison, 1969). For aquatic plants, however, solution cultures can provide a more realistic simulation of natural conditions. This is particularly so for those plants that do not penetrate the substrate (planktophytes, pleusophytes and haptophytes: Raven, 1981).

Mineral nutrients are as potentially limiting to plant growth and distribution in aquatic ecosystems (Haslam, 1978; Raven, 1981) as they are in terrestrial habitats. Many aquatic plants can be maintained in the laboratory (Stein, 1973) and used in nutrition experiments in much the same way as terrestrial plants, provided that a number of special precautions are observed.

Although obvious in the case of terrestrial plants, the parts of the plants that are submerged in the solution and which are exposed, will depend on the particular species (cf. Haslam, 1978; pp. 3–4). Certain intertidal marine algae require alternations between submersion and emersion in order to maintain maximum growth (Schonbeck & Norton, 1979). The containers in which the plants are grown should be transparent, or have transparent lids, so that the solutions can be illuminated adequately. This is obviously of particular importance for completely submerged plants. Since in illuminated solutions the growth of algal contaminants is encouraged (see below), it is advisable to maintain axenic conditions for experiments with aquatic plants. This is, in any case, essential in experiments on micronutrient or vitamin requirements (J.A. Raven, pers. comm.). Suitable procedures for the maintenance of axenic culture conditions are described in Stein (1973). The importance of solution flow rate over plant surfaces has been mentioned in Section 2.5.1. Sea water,

---

Figure 4.11 Schematic representations of continuous-flow solution culture systems: solid arrows = solution flow; broken arrows = information flow and feedback controls.

(a) Van den Honert-type

Solution composition	: constant
Solution flow rate	: low, adjusted according to analysis
Solution overflow	: run to waste
No. pots per system	: few
Cost	: low

(b) Asher-type

Solution composition	: modified by stock solution additions according to analysis
Solution flow rate	: high
Solution overflow	: recirculated, system drained completely at regular intervals
Solution volume	: large
No. pots per system	: many
Cost	: high

(c) Rorison proto-type

Solution composition	: modified by stock solution additions according to analysis; nutrient supplements injected into individual pots ensures rapid response time
Solution flow rate	: low
Solution overflow	: run to waste
Solution volume	: small
No. pots per system	: few, but with developments, more would be included
Cost	: moderate

(a)

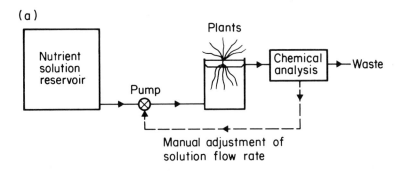

(b)

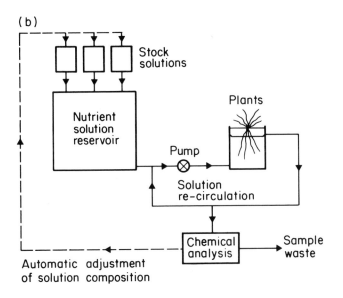

(c)

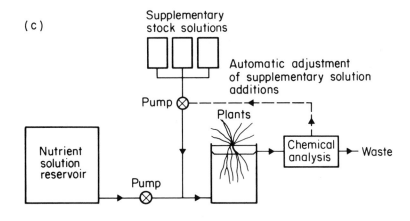

Table 4.2 Relationships between nutrient concentrations, solution flow rates and total daily nutrient supplies in a hypothetical continuous flow system

Nutrient concentration (mol × 10^{-3} dm^{-3})	Minimum solution flow rate (dm^3 pot^{-1} d^{-1})	Total amount of nutrient supplied with minimum flow rate (mol × 10^3 pot^{-1} d^{-1})	Total amount of nutrient supplied with flow rate = 5000 dm^3 pot^{-1} d^{-1} (mol × 10^{-3} pot^{-1} d^{-1})	Percentage depletion of concentration with flow rate = 5000 dm^3 pot^{-1} d^{-1}
0·001	100 000	100	5	100
0·01	10 000	100	50	10
0·1	1 000	100	500	1·0
1·0	100	100	5000	0·1

fresh water, and appropriate mixtures of these, can be used as 'natural' growth media for marine, freshwater and estuarine plants, respectively. Davenport (1982) describes techniques, developed for experiments on marine and estuarine animals, by which the magnitude of temporal fluctuations in salinity, temperature and O_2 concentration of sea/fresh water mixtures can be controlled in the laboratory.

3.2   Controlled environments

Laboratory experiments lose much of their potential ecological value if they are conducted in an environment that cannot be defined, measured and reproduced. Glasshouses are widely used but the control of diurnal fluctuations of temperature and irradiance is not possible without supplementary air-conditioning and lighting installations.

The control of environmental temperature and humidity becomes more difficult with increasing air volume. For this reason, the most precise control over environmental conditions is achieved in small growth chambers, suitable types of which are now available commercially. Principles and applications of controlled-environment facilities have been reviewed by Rees *et al.* (1972) and Tibbitts & Kozlowski (1979).

In order to compare experiments conducted in different controlled environments, some means of standardizing the measurement and recording of environmental variables is required. At present, no single set of standard procedures (e.g. Table 4.3) has been adopted universally.

3.3   Experimental materials

3.3.1   Communities or multi-species mixtures

Obviously, only gross effects of nutrients upon productivity and growth can be measured at the community level. Floristic changes occurring in communities following fertilizer applications have been discussed in Section 2.4.3. The uptake of isotopically-labelled nutrients (Section 2.4.1) placed at various depths in the soil beneath a plant community can be detected in the tops of the component species, indicating the approximate spatial distributions of roots in the soil. Responses of individual plants grown in monoculture are often different quantitatively from those of plants grown in mixed cultures (Austin & Austin, 1980).

3.3.2   Whole-plants

In most cases, a 'whole-plant' is the entity resulting from the germination of a single seed. Post-germination vegetative growth can give rise to

Table 4.3 Proposed guidelines for measuring and reporting the environment for plant studies (after Tibbitts & Kozlowski, 1979)

Parameter	Units	Measurements		
		Where to take	When to take	What to report
Radiation PAR (photosynthetically active radiation) (a) Quantum flux density 400–700 nm with cosine correction	$\mu E\,s^{-1}\,m^{-2}$	At top of plant canopy. Obtain average over plant growing area.	At start and finish of each study and biweekly if studies extend beyond 14 days.	Average over containers at start of study. Decrease or fluctuation from average over course of the study.
or				
(b) Irradiance 400–700 nm with cosine correction	$W\,m^{-2}$	(Same as quantum flux density)	(Same as quantum flux density)	(Same as quantum flux density)
Total irradiance 280–2800 nm with cosine correction	$W\,m^{-2}$	(Same as quantum flux density)	At start of each study.	Average over containers at start of study.
Spectral irradiance 250–850 nm in <20 nm bandwidths with cosine correction	$\mu E\,s^{-1}\,m^{-2}\,nm^{-1}$ or $W\,m^{-2}\,nm^{-1}$	At top of plant canopy in centre of growing area.	At start of each study.	Graph of irradiance for separate wavebands at start of study.
Photometric 380–780 nm with cosine correction	klx	(Same as quantum flux density)	At start of each study.	(Same as total irradiance)
Daylength	hours			Length of photoperiod in 24 h cycle.

Temperature

Air				
Shielded and aspirated ($> 3\,\mathrm{m\,s^{-1}}$) device	$^\circ\mathrm{C}$	At top of plant canopy. Obtain average over plant growing area.	Hourly over the period of the study. (Continuous measurement advisable)	Average of hourly average values for the light and dark periods of the study with range of variation over the growing area.
Soil and liquid	$^\circ\mathrm{C}$	In centre of representative container.	Hourly during the first 24 h or the study (hourly measurements over the period of the study advisable).	Average of hourly average values for the light and dark periods for the first day or over entire period of the study if taken.
Atmospheric moisture				
Shielded and aspirated ($> 5\,\mathrm{m\,s^{-1}}$) psychrometer, dewpoint sensor or infra-red analyser	$\tau\ \mathrm{RH}$ or dewpoint or $\mathrm{g\,m^{-3}}$	At top of plant canopy in centre of plant growing area.	Once during each light and dark period, taken at least 1 h after light changes (hourly measurements over the course of the study advisable).	Average of once daily readings for both light and dark periods with range of diurnal variation over the period of the study (or average of hourly values if taken).
Air velocity	$m\ \mathrm{s^{-1}}$	At top of plant canopy. Obtain maximum and minimum readings over plant growing area.	At start and end of studies. Take 10 successive readings at each location and average.	Average and range of readings over containers at start and end of the study.
Carbon dioxide	$m\ \mathrm{moles\ m^{-3}}$	At top of plant canopy.	Hourly over the period of the study.	Average of hourly average readings and range of daily average readings over the period of the study.

Continued.

Table 4.3 Continued.

Parameter	Units	Measurements		What to report
		Where to take	When to take	
Watering	ml		At times of additions.	Frequency of watering. Amount of water added per day and/or range in soil moisture content between waterings.
Substrate				Type of soil and amendments. Components of soilless substrate.
Nutrition	Solid media: kg m^{-3} Liquid culture: micro nutrients = μ moles l^{-1} macro nutrients = m moles l^{-1}		At times of nutrient additions.	Nutrients added to solid media. Concentration of nutrients in liquid additions and solution culture. Amount and frequency of solution addition and renewal.
pH	pH units	In liquid slurry for soil and in solution of liquid culture.	Start and end of studies in solid media. Daily in liquid culture and before each pH adjustment.	Mode and range during study.
Conductivity	dS m^{-1} (decisiemens per meter)	In liquid slurry for soil and in solution of liquid culture.	Start and end of studies in solid media. Daily in liquid culture.	Average and range during study.

genetically-identical units (e.g. grass tillers, *Lemna* clones, etc.), each of which is capable of growth independently of the parent plant. The aforementioned definition of 'whole-plant' then becomes less viable. Physiologically, however, a 'whole-plant' can be considered to consist of: (a) a root system capable of absorbing nutrients and water; and (b) a shoot–leaf system, capable of fixing CO_2 and mediating gas exchange. If the transport of materials can occur between roots and shoots, those roots and shoots can be considered to be part of the same organism. If, as recent evidence (Section 2.5.5) suggests, plants in the field are interconnected via mycorrhizal hyphae, even this definition is meaningless, but this is a complication that for most laboratory experiments can be conveniently overlooked.

It is usual for seeds collected from the field to be germinated in the laboratory, and seedlings thus produced to be used as experimental material. This is always the case for solution culture experiments, since the seedlings can be maintained under clean (sterile, if necessary) conditions, which is not possible if rooted material collected from the field is used. In addition, the environmental history of seedlings raised in the laboratory is known. This is not necessarily the case with tiller material that may have been maintained in an environment very different from that of the experiment (see Evans, 1972, p. 24 *et seq.*).

It is necessary to select a number of seedlings from those that germinate for use in the experiment. Natural plant populations tend to be more variable in characteristics such as time for 50% germination (Grime *et al.*, 1981) and seedling vigour (Hunt, 1984) than agricultural varieties. At the time of seedling selection, therefore, it is likely that there will be a range of seedling size. For some solution culture systems, seedlings must be of a minimum size in order for them to be supported effectively in the pots (Fig. 4.12a, b). In these cases, seeds are germinated on floats of plastic beads (Robinson & Rorison, 1983) or in Petri dishes, and seedlings of a suitable size are transferred to the solution culture pots. In others (Fig. 4.12c), seeds are germinated *in situ* and are thinned to the required number when enough seedlings have been produced. This is also the usual procedure for soil or sand-culture experiments.

Seedling selection procedures such as these obviously bias the population sample used in the experiment towards those phenotypes showing greatest vigour. This is unavoidable if enough material is to be available for adequate statistical replication and for chemical analyses. It can be minimized, however, by discarding the *most* vigorous seedlings, as well as the least. This removes extreme phenotypes from the experimental sample, which, for screening purposes (Section 3.7) might not always be desirable.

Usually, seedlings aged less than 2 months are used, since during this period maximum relative growth rate is achieved (Grime & Hunt, 1975).

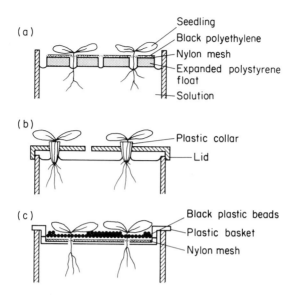

Figure 4.12 Methods of providing mechanical support for seedlings growing in solution cultures.

Obviously, if flowering or senescent plant material is required, seedlings must be maintained throughout their vegetative phase before the experiment. A similarly long period of pre-experimental maintenance is necessary if root systems are to be large enough to be divided between different compartments of multi-chambered pots (J.C. Crick, pers. comm.; cf. Fig. 4.4).

Natural populations usually exhibit a greater degree of genotypic variation than agricultural populations. Genetically-uniform plant material can be obtained by using cuttings of, for example, soil-rooted tillers (Hunt, 1984), or batch-cultured *Lemna* clones (Jefferies *et al.*, 1969).

3.3.3   Excised roots, leaf discs and cultured tissues

These materials have been used extensively in the study of short-term ion uptake. They have the advantage of reaching electro-chemical equilibrium with the external medium rapidly, since absorbed ions cannot be removed from the absorbing tissue by long-distance transport processes. In addition, only a few plants may be needed to provide enough material for an entire short-term experiment.

The classical studies of ion uptake kinetics (Section 2.4.2) were made using excised roots and leaf discs. Excised mycorrhizas were used to elucidate patterns of ion transport by roots and associated fungi (Harley & Smith, 1983, p. 240). Cultured tissues have been used less frequently in experiments of this type, probably because they require specialized growth facilities (see

Street, 1977; Jakoby & Pastan, 1979). A useful, concise guide to tissue culture techniques is given by Reinert & Yeoman (1981). The advantage of using cultured callus tissue in nutrition experiments is that the processes of ion and water uptake, turgor maintenance, cell division and cell expansion are not complicated by simultaneous tissue differentiation (Mott & Steward, 1972).

3.3.4 Isolated organelles

Organelles isolated from cells provide the means to study the permeability of specific membranes to specific ions, and analyse the subcellular compartmentation of absorbed ions.

Methods for the isolation of mitochondria, chloroplasts, vacuoles, cell membranes and nucleii are given in Hall & Moore (1983).

3.3.5 Enzymes

Assay procedures for particular enzymes are given in appropriate handbooks (e.g. Colowick & Kaplan, 1955–63; San Pietro, 1980). General enzymological techniques are described in Smith (1977).

Enzyme activities are expressed in terms of the rate of substrate disappearance or the rate of product synthesis (mol g^{-1} fresh wt hour^{-1}). Specific enzyme activities are expressed per mg total protein. Differences in interpretation of *in vivo* and *in vitro* activities are important. The measurement of *in vivo* activities involves the homogenization of the plant material and the determination of enzyme activity in the crude, filtered preparation. This could be complicated by the exposure of enzymes to potential inhibitors (e.g. phenolics: Rhodes in Smith, 1977) that in the intact cell, are localized in separate compartments. *In vivo* activities can be considered to reflect more closely (but not necessarily identically) enzyme characteristics *in situ*. *In vitro* assays require that the enzyme be separated from other cellular constituents (by, e.g. chromatography or electrophoresis; Colowick & Kaplan, 1955–63; Smith, 1977) and assayed in an artificial medium containing the necessary substrate, co-factors and buffer. Thus, *in vitro* activities do not necessarily reflect enzyme characteristics *in situ*, but allow the assay conditions to be defined precisely, and the maximum potential rate of catalysis to be expressed under ideal conditions.

Attempting biochemically-detailed work on organelles or enzymes isolated from widely diverse species is, however, fraught with difficulties. To quote Leech (in Smith, 1977): 'Many attempts to relate biological findings to biochemical changes have foundered because the plants used have yielded either totally inactive cellular fractions or cell homogenates with such high phenolic content, acidity, or viscosity as to be unsuitable for further investigation.' For this reason, most investigations of this type have involved a limited number of crop species which have been found to have peculiar morphological or physiological features, making them particularly suitable as experimental material.

3.4   Chemical analysis

Methods for the chemical analysis of soils, solutions and plants are given in Chapter 6. Here, the relevance of chemical analysis to mineral nutrition studies will be considered.

In all mineral nutrition studies it is necessary at some stage, and in varying degrees of detail, to measure the amounts of nutrient in the plants' environment, and/or the amounts of nutrient in the plants themselves.

For solution cultures, measurements of ion concentration should be made before any seedlings are grown in them, as a check on solution composition. During plant growth, it is important to adjust pH drifts and to replenish depleted solutions (Section 3.1.5).

When nutrient uptake is being measured on the basis of depletion from solution (Section 2.4.2), the concentration of the solution should be measured at the start and end of the depletion period. As a precaution, analyses of plant nutrient content at the start and end of the depletion period should also be made, by destructively harvesting replicate plants (Section 2.4.1).

Since most chemical analyses of soils involve the transportation of samples from the field to the laboratory, the chemical changes that such handling can cause should be noted (Section 3.1.1). Collection of soil solutions can be made without disturbing the soil, using pressure-vacuum lysimeters (Chapter 1, pp. 44–46). Measurements of the nutrient status of soils at a single time of the year are valid only for that time. Correlations between long-term vegetation changes, annual productivity and soil nutrient status, should be made on the basis of analyses carried out throughout the year, to take into account seasonal fluctuations in soil nutrient concentration (Davy & Taylor, 1975; Gupta & Rorison, 1975; Taylor *et al.*, 1982). Foliar nutrient analyses are useful in 'screening' the nutrient status of plants in the field, but differences in age, species and the supply of other nutrients, should be taken into account (Mengel & Kirkby, 1982; Chapter 2). In general, chemical analyses of both soil and plants are preferred.

3.5   Isotope labelling

The use of labelled isotopes in plant nutrition studies is now common practice. Reference has been made in previous Sections of this Chapter to the use of isotopes in investigations of root growth and distribution in soil (2.4.1), nutrient uptake (2.4.2), assimilation (2.4.5), plant–micro-organism interactions (2.5.5), and nutrient cycling (2.5.6). In an excellent book, Vose (1980) covers the principles and practical approaches of isotope-labelling techniques.

Usually, isotope labelling involves the enrichment of the plant's environment, or the plant itself, with an isotope of different atomic weight from that present naturally, and measuring the subsequent abundance of the isotope in the material of interest, in excess of that occurring naturally. Radio-isotopes (e.g. ^{32}P, ^{42}K, ^{45}Ca) can be detected qualitatively (e.g. by autoradiography, see Fig. 4.1), or quantitatively (e.g. by scintillation or Geiger-Muller counting). Stable isotopes (e.g. ^{15}N) can only be detected on a quantitative basis (e.g. by mass or optical emission spectroscopy).

The majority of isotopes that are used in plant nutrition studies are available from commercial manufacturers. In some cases, however, the availability of an isotope of very short half-life (e.g. ^{13}N: $t_{\frac{1}{2}} = 10$ minutes) depends upon the experimental material being in close proximity to a cyclotron, in which the isotope is produced, in order for radioactive decay to be minimized before the isotope is used (see McNaughton & Presland, 1983). The use of such isotopes imposes strict limitations upon the type and duration of an experiment. For this reason, they have been used very infrequently in plant nutrition research.

Labelling nutrient solutions with an isotope is straightforward. The enriched element is substituted in the nutrient solution for an equivalent weight of the unlabelled element. The isotopically-enriched salt containing the labelled element is dissolved in water and made up into a standard solution. This is added to the nutrient solution in place of a volume of unlabelled stock solution containing an equivalent weight of the element of interest. This gives a labelled nutrient solution of the same chemical composition as an unlabelled solution, but with a certain, known enrichment of the isotope label.

When labelling soils (see p. 157), it is essential, in most cases, to ensure uniform distribution of the isotope throughout the soil. For pot experiments this can be done by mixing the soil vigorously during labelling, the isotope being applied as a fine spray of solution before the pots are filled. In the field, isotopes can be applied to soil as solutions or granular fertilizers. It is usual to apply a low concentration of a radioisotope of high specific activity (counts per minute: see Vose, 1980, p. 4), in order to affect as little as possible the labile nutrient concentration in the soil. If localized enrichments are required, the isotope can be injected as a solution at the required location, or placed as solid granules of the enriched salt at the required depth.

For some purposes, it is necessary to use isotopically-enriched gases (e.g.

$^{15}N_2$ in studies of N_2-fixation). Supplying labelled gases to plants allows the assimilation of gaseous elements to be determined directly but a system enclosing plants, soil and atmosphere is required (Stefanson, 1970; Knowles, 1980). Where this is not practicable, an alternative method for measuring N_2-fixation is to supply ^{15}N-labelled salts to soil containing N_2-fixing plants and non-fixing controls. The relative 'dilution' of ^{15}N in the N_2-fixing plant compared with its abundance in non-fixing control plants is taken to represent the acquisition of N derived from atmospheric N_2 rather than soil N. The assumptions underlying the validity of this technique, and possible difficulties in the interpretation of the results of such experiments, have been discussed by Witty (1983) and Domenach & Corman (1984). A useful guide to the analysis of data from ^{15}N-labelling experiments has been produced by the IAEA (1983).

3.6   Bioassays

In the present context, bioassays involve measuring the effects of a range of soils differing in one or more physico-chemical characteristics, or of specific nutrients in solution culture, upon the growth of certain 'test' species. The emphasis is upon maximizing the number of treatments (soil types, nutrient sources, etc.), rather than the number of species or populations (cf. Section 3.7). For bioassays involving only one test species, the species chosen should be tolerant of a wide range of edaphic conditions, rather than one restricted in its distribution to particular soil types (cf. Fig. 4.2).

Bioassay techniques can be used in the laboratory or in the field. Rorison (1967) assessed the effects of a number of soils, differing in pH, nutrient availability, and texture, upon the germination, growth and seedling mortality under laboratory conditions, of a number of herbaceous species differing in distribution with respect to soil type. Principal component analyses were used to identify which plant–soil relationships were correlated most significantly with observed responses. Gupta & Rorison (1975) used the growth and P uptake of *Rumex acetosa*, a species of wide edaphic distribution in the British Isles, as a bioassay to assess P availability in a range of soils collected from the field.

Characteristic floras can be used as 'natural bioassays' to estimate the distribution of particular chemical features of soils in the field. For example, characteristic assemblages of species can be identified for calcareous, acidic, saline, serpentine and metal-contaminated soils (Brooks, 1983; Chapter 3). 'Indicator plants', individual species that are characteristic of soils overlying a certain geological substrate, can be used in a similar way (Brooks, 1983; Chapter 4). Although useful, this approach should ideally be combined with chemical analysis of soil.

3.7   Screening programmes

Screening programmes are almost the converse of bioassays (cf. Section 3.6), in that their aim is to examine certain features of a large number of plants, under standardized growth conditions. This allows broad, intra- and inter-specific comparisons to be made between plants on the basis of large samples. Thus, general groupings of similar physiological attributes can be identified in plants characteristic of particular ecological conditions. Examinations of a small number of plants would not be sufficiently reliable for this purpose.

Plants have been screened in the laboratory for such attributes as seedling relative growth rate (Grime & Hunt, 1975), potassium uptake rate (Glass, Siddiqi & Giles, 1981), seed germination characteristics (Grime et al., 1981), type of CO_2-assimilation system (see Osmond et al., 1982), salt or metal tolerance, and efficiency of nutrient utilization (see Wright, 1977), for example. In the field, plants have been screened for nutrient concentrations in foliage (UCPE, unpublished), in vivo nitrate reductase activity (Lee & Stewart, 1978; see Section 2.4.5), and storage amino acids in aestivating structures (see Page, 1983).

To maximize the scope of such programmes, the screening procedures used should be as rapid as is compatible with the attribute being measured. 'Short-cut' procedures have been used widely in screening programmes for this reason. Examples of these are: root extension tests to screen for metal tolerance (Section 2.4.4); short-term H^+ efflux rate to screen for K^+ uptake rate (Glass et al., 1981); $^{13}C/^{12}C$ isotope discrimination to screen for C_3 or C_4 photosynthetic systems (see Osmond et al., 1982).

Screening for plant growth characteristics (e.g. relative growth rate) usually have involved fairly long-term measurements (e.g. Grime & Hunt, 1975). As Hunt (1980) commented, however, the reasons for this were practical and statistical, rather than biological, and suggested that measurements of seedling growth from hourly, destructive harvests over a period of 2–3 days, can yield information comparable with that obtained from weekly harvests. In the experiment described by Hunt (1980), plants were harvested manually using shifts of workers. To be practicable on a wide scale, however, this procedure requires automated harvesting equipment which, to date, has yet to be developed. Nevertheless, this technique offers the possibility of running screening programmes with greater efficiency of labour and equipment deployment than hitherto.

3.8   Mathematical models

Mathematical modelling is an increasingly-used approach to plant nutrition research, as reference to Milthorpe & Moorby (1979) and Vol. 12

of the *Encyclopedia of Plant Physiology*. New Series (1983) indicates. There are two reasons for this. Firstly, experimental data can explain the operation of observed phenomena to only a limited extent. More complete interpretations require the development of theoretical, quantitatively-based analyses of the processes involved. The accumulation of experimental data has, however, proceeded more rapidly than the development of theoretical concepts, but the importance of the latter has become more widely appreciated in recent years. Secondly, recent rapid technological advances have made inexpensive, versatile and powerful microcomputers almost standard as laboratory equipment. Thus, the quantitative simulation of theoretical models can be accomplished with less practical difficulty than had been the case previously.

The principles of mathematical modelling as a tool in plant nutrition research have been outlined by Thornley (1976) and Penning de Vries (1983). Mathematical models have found widest use in simulating crop production in relation to environmental factors (Milthorpe & Moorby, 1979; Rose & Charles-Edwards, 1981; Penning de Vries, 1983), plant–soil nutrient interactions (Nye & Tinker, 1977), and nutrient cycling processes (e.g. McGill *et al.*, 1981). Mathematical models are indispensible in the investigation of such processes, since each consists of a multiplicity of interacting and counteracting components. It is difficult to control these components individually in an experimental system, as cause-and-effect relationships cannot be identified. This is often an intractable barrier to progress. Such control *is* possible, however, using appropriate theoretical models.

Mathematical models are usually formulated in terms of algebraic equations transposed into computer programs written in general high-level languages (Fortran, Pascal, Basic, etc.) or specialized simulation languages (e.g. CSMP). Models can be based upon empirical functions that are appropriate to describe observed phenomena or upon functions derived from the mechanistic relationships between observed phenomena (e.g. Baldwin, 1976). Parameter values taken from experimental data can be introduced directly into the equations and model results calculated from these, or can be generated internally by the program at each simulated time-step.

The work of Baldwin (1975, 1976) provides a particularly fine illustration of how theoretical concepts, couched in mathematical terms, can be used in the analysis of plant–nutrient interactions.

4   Acknowledgments

We are indebted to the following for their help during the preparation of the manuscript: A.J.M. Baker; G.A.F. Hendry; J.G. Hodgson; M.J.

Hopper; J.H. Macduff; J.H. Peterkin; J.A. Raven; N. Ruttle; and to the following for permission to use material: J.C. Crick; S.P. McGrath; S.C. Shaw.

5   References

AARONSON S. (1970) *Experimental Microbial Ecology*. Academic Press, New York.

ADAMS F. (1974) Soil solution. *The Plant Root and its Environment* (ed. E.W. Carson), pp. 441–481. University Press of Virginia, Charlottesville.

ALCOCK M.B. (1962) The physiological significance of defoliation on the subsequent regrowth of grass–clover mixtures and cereals. In *Grazing in the Terrestrial and Marine Environments* (ed. D.J. Crisp), pp. 25–41. Symposia of the British Ecological Society, No. 4. Blackwell Scientific Publications, Oxford.

ANTONOVICS J., BRADSHAW A.D. and TURNER R.G. (1971) Heavy metal tolerance in plants. *Adv. Ecol. Res.* **7**, 1–85.

ANTONOVICS J., LOVETT J. and BRADSHAW A.D. (1967) The evolution of adaptation to nutritional factors in populations of herbage plants. In *Isotopes in Plant Nutrition and Physiology*, pp. 549–567. International Atomic Energy Agency, Vienna.

ASHER C.J. (1978) Natural and synthetic culture media for spermatophytes. In *C.R.C. Handbook Series in Nutrition and Food. Section G: Diets, culture, media, food supplements* (ed. M. Recheigl, Volume III, pp. 575–609. C.R.C. Press, Cleveland.

ASHER C.J. and EDWARDS D.G. (1978) Critical external concentrations for nutrient deficiency and excess. In *Proceedings of the 8th International Colloquium on Plant Analysis and Fertilizer Problems* (ed. A.R. Ferguson, R.L. Bieleski and I.B. Ferguson), pp. 13–28. DSIR Information Series No. 134, Wellington, New Zealand.

ASHER C.J. and EDWARDS D.G. (1983) Modern solution culture techniques. In *Inorganic Plant Nutrition. Encyclopedia of Plant Physiology*. New Series, Vol. 15A, (ed. A. Läuchli and R.L. Bieleski), pp. 94–119. Springer-Verlag, Berlin.

ASHER C.J., OZANNE P.G. and LONERAGAN J.F. (1965) A method for controlling the ionic environment of plant roots. *Soil Sci.* **100**, 149–156.

ASHWORTH J., PENNY A., WIDDOWSON F.V. and BRIGGS G.G. (1980) The effects of injecting nitrapyrin ('N-Serve'), carbon disulphide or trithiocarbonates, with aqueous ammonia, on yield and % N of grass. *J. Sc. Fd. Agric.* **31**, 229–237.

ATKINSON C.J. (1983) Phosphorus acquisition in four co-existing species from montane grassland. *New Phytol.* **95**, 427–437.

AUSTIN M.P. and AUSTIN B.O. (1980) Behaviour of experimental plant communities along a nutrient gradient. *J. Ecol.* **68**, 891–918.

BACHE B.W. (1977) Practical implications of quantity–intensity relationships. In *Proceedings of the International Seminar on Soil Environment and Fertility Management in Intensive Agriculture*, pp. 777–787. Society of the Science of Soil and Manure, Tokyo.

BAGSHAW R., VAIDYANATHAN L.V. and NYE P.H. (1972) The supply of nutrient ions by diffusion to plant roots in soil. V. Direct determination of labile phosphate concentration gradients in a sandy soil induced by plant uptake. *Pl. Soil* **37**, 617–626.

BALDWIN J.P. (1975) A quantitative analysis of the factors affecting plant nutrient uptake from soils. *J. Soil. Sci.* **26**, 195–206.

BALDWIN J.P. (1976) Competition for plant nutrients in the soil; a theoretical approach. *J. agric. Sci., Camb.* **87**, 341–356.

BALDWIN J.P. and TINKER P.B. (1972) A method for estimating the lengths and spatial pattern of two interpenetrating root systems. *Pl. Soil* **37**, 209–213.

BALDWIN J.P., TINKER P.B. and MARRIOTT F.H.C. (1971) The measurement of length and distribution of onion roots in the field and the laboratory. *J. appl. Ecol.* **8**, 543–554.

BARBER D.A. (1968) Micro-organisms and the inorganic nutrition of higher plants. *A. Rev. Pl. Physiol.* **19**, 71–88.

BARRACLOUGH P.B. and TINKER P.B. (1981) The determination of ionic diffusion coefficients in field soils. I. Diffusion coefficients in sieved soils in relation to water content and bulk density. *J. Soil Sci.* **32**, 225–236.

BEARDMORE J. and PEGG G.F. (1981) A technique for the establishment of mycorrhizal infection in orchid tissue grown in aseptic culture. *New Phytol.* **87**, 527–535.

BECKETT P.H.T. (1964a) Studies on soil potassium. I. Confirmation of the Ratio Law. *J. Soil Sci.* **15**, 1–8.

BECKETT P.H.T. (1964b) Studies on soil potassium. II. The immediate Q/I relation of labile potassium in the soil. *J. Soil Sci.* **15**, 9–23.

BERGERSEN F.J. (ed.) (1980) *Methods for Evaluating Biological Nitrogen Fixation.* John Wiley, Chichester.

BHAT K.K.S. (1980) A low-cost, easy-to-install flow culture system suitable for use in a constant environment cabinet. *J. exp. Bot.* **31**, 1435–1440.

BHAT K.K.S. (1982) Determination of the relationship between nutrient uptake rate and solution composition at the root surface under field conditions: ^{32}P-orthophosphate uptake by apple roots. *J. exp. Bot.* **33**, 190–194.

BHAT K.K.S. and NYE P.H. (1973) Diffusion of phosphate to plant roots in soil. I. Quantitative autoradiography of the depletion zone. *Pl. Soil* **38**, 161–175.

BIELESKI R.L. and FERGUSON I.B. (1983) Physiology and metabolism of phosphate and P compounds. In *Inorganic Plant Nutrition. Encyclopedia of Plant Physiology.* New Series, Volume 15A, (ed. A. Läuchli and R.L. Bieleski), pp. 422–449. Springer-Verlag, Berlin.

BIERMANN B. and LINDERMAN R.G. (1981) Quantifying vesicular-arbuscular mycorrhizae: a proposed method towards standardization. *New Phytol.* **87**, 63–67.

BIGG W.L. and ALEXANDER I.J. (1981) A culture unit for the study of nutrient uptake by intact mycorrhizal plants under aseptic conditions. *Soil Biol. Biochem.* **13**, 77–78.

BÖHM W. (1979) *Methods of Studying Root Systems. Ecological Studies* 33. Springer-Verlag, Berlin.

BOOKMAN P.A. and MACK R.N. (1982) Root interaction between *Bromus tectorum* and *Poa pratensis*: a three-dimensional analysis. *Ecology* **63**, 640–646.

BOWEN G.D. and ROVIRA A.D. (1961). Effects of micro-organisms on plant growth. I. Development of roots and root hairs in sand and agar. *Pl. Soil* **15**, 166–188.

BRADLEY R., BURT A.J. and READ D.J. (1982) The biology of mycorrhiza in the Ericaceae. VIII. The role of mycorrhizal infection in heavy metal resistance. *New Phytol.* **91**, 197–209.

BRADSHAW A.D. (1969) An ecologist's viewpoint. In *Ecological Aspects of the Mineral Nutrition of Plants* (ed. I.H. Rorison), pp. 415–427. Symposia of the British Ecological Society, No. 9. Blackwell Scientific Publications, Oxford.

BRADSHAW A.D., CHADWICK M.J., JOWETT D. and SNAYDON R.W. (1964) Experimental investigations into the mineral nutrition of several grass species. IV. Nitrogen level. *J. Ecol.* **52**, 665–676.

BREEZE V.G., CANAWAY R.J., WILD A., HOPPER M.J. and JONES L.H.P. (1982) The uptake of phosphorus by plants from flowing nutrient solution. I. Control of phosphate concentration in solution. *J. exp. Bot.* **33**, 183–189.

BRENNER M.L. (1981) Modern methods for plant growth substance analysis. *A. Rev. Pl. Physiol.* **32**, 511–538.

BREWSTER J.L., BHAT K.K.S. and NYE P.H. (1976) The possibility of predicting solute uptake and plant growth response from independently measured soil and plant characteristics. V. The growth and phosphorus uptake of rape in soil at a range of phosphorus concentrations and a comparison of the results with the predictions of a simulation model. *Pl. Soil* **44**, 295–328.

BREWSTER J.L. and TINKER P.B.H. (1970) Nutrient cation flows in soil around plant roots. *Soil Sci. Soc. Am. Proc.* **34**, 421–426.

BREWSTER J.L. and TINKER P.B.H. (1972) Nutrient flow rates into roots. *Soils Ferts.* **35**, 355–359.

BROOKS R.R. (1983) *Biological Methods of Prospecting for Minerals.* John Wiley, New York.

BURNS I.G. (1980) Influence of the spatial distribution of nitrate on the uptake of N by plants: a review and a model for rooting depth. *J. Soil Sci.* **31**, 155–173.

CARROLL M.D. and LONERAGAN J.F. (1969) Response of plant species to concentrations of zinc in solution. I. Growth and zinc content of plants. *Aust. J. agric. Res.* **19**, 859–868.

CLAASSEN N. and BARBER S.A. (1974) A method for characterizing the relation between nutrient concentration and flux into roots of intact plants. *Pl. Physiol., Lancaster* **54**, 564–568.

CLARKE A.L. and BARLEY K.P. (1968) The uptake of nitrogen from soils in relation to solute diffusion. *Aust. J. Soil Res.* **6**, 75–92.

CLARKSON D.T. (1969) Metabolic aspects of aluminium toxicity and some possible mechanisms for resistance. In *Ecological Aspects of the Mineral Nutrition of Plants* (ed. I.H. Rorison), pp. 381–397. Symposium of the British Ecological Society, No. 9. Blackwell Scientific Publications, Oxford.

CLARKSON D.T. (1974) *Ion Transport and Cell Structure in Plants.* McGraw-Hill, London.

CLARKSON D.T. and DEANE-DRUMMOND C.E. (1983) Thermal adaptation of nitrate transport and assimilation in roots? In *Nitrogen as an Ecological Factor* (ed. J.A. Lee, S. McNeill and I.H. Rorison), pp. 211–224. Symposia of the British Ecological Society, No. 22. Blackwell Scientific Publications, Oxford.

CLARKSON D.T. and HANSON J.B. (1980) The mineral nutrition of higher plants. *A. Rev. Pl. Physiol.* **31**, 239–298.

CLEMENT C.R., HOPPER M.J., CANAWAY R.J. and JONES L.H.P. (1974) A system for measuring the uptake of ions by plants from flowing solution of controlled composition. *J. exp. Bot.* **25**, 81–99.

CLEMENT C.R., HOPPER M.J. and JONES L.H.P. (1978) The uptake of nitrate by *Lolium perenne* from flowing nutrient solution. I. Effect of NO_3^- concentration. *J. exp. Bot.* **29**, 453–464.

CLEMENT C.R., JONES L.H.P. and HOPPER M.J. (1979) Uptake of nitrogen from flowing nutrient solution: effect of terminated and intermittent nitrate supplies. In *Nitrogen Assimilation of Plants* (ed. E.J. Hewitt and C.V. Cutting), pp. 123–133. Academic Press, London.

COLOWICK S.P. and KAPLAN N.O. (Eds) (1955–63) *Methods in Enzymology* (6 vols). Academic Press, New York.

COSTIGAN P.A., ROSE J.A. and McBURNEY T. (1982) A microcomputer based method for the rapid and detailed measurement of seedling root systems. *Pl. Soil* **69**, 305–309.

CRAM W.J. (1976) Negative feedback regulation of transport in cells. The maintenance of turgor, volume and nutrient supply. In *Transport in Plants II. Part B. Tissues and Organs. Encyclopedia of Plant Physiology.* New Series, Vol. 2 (ed. U. Lüttge and M.G. Pitman), pp. 284–316. Springer-Verlag, Berlin.

CRANK J. (1975) *The Mathematics of Diffusion*, 2nd edn. Clarendon Press, London.

CRICK J.C. and GRIME J.P. (1983) Plant 'foraging'. *Annual Report of the Unit of Comparative Plant Ecology* (*NERC*), *University of Sheffield 1983*, pp. 5–13.

DARRAH P.R., NYE P.H. and WHITE R.E. (1983) Diffusion of NH_4^+ and NO_3^- mineralized from organic N in soil. *J. Soil Sci.* **34**, 693–707.

DAVENPORT J. (1982) Environmental simulation experiments on marine and estuarine animals. *Adv. Mar. Biol.* **19**, 133–256.

DAVIDSON R.L. (1969) Effects of soil nutrients and moisture on root/shoot ratios in *Lolium perenne* L. and *Trifolium repens* L. *Ann. Bot.* **33**, 571–577.

DAVY A.J. and TAYLOR K. (1974) Seasonal patterns of nitrogen availability in contrasting soils in the Chiltern Hills. *J. Ecol.* **62**, 793–807.

DEANE-DRUMMOND C.E. and GLASS A.D.M. (1983) Short-term studies of nitrate uptake into barley plants using ion-specific electrodes and $^{36}ClO_3^-$. I. Control of net uptake by NO_3^- efflux. *Pl. Physiol., Lancaster* **73**, 100–104.

DIXON A.F.G. (1975) Aphids and translocation. In *Transport in Plants. I. Phloem Transport. Encyclopedia of Plant Physiology.* New Series, Vol. 1 (ed. M.H. Zimmerman and J.H. Milburn), pp. 154–170. Springer-Verlag, Berlin.

DOMENACH A.M. and CORMAN A. (1984) Dinitrogen fixation by field grown soybeans: statistical analysis of variations in $\delta^{15}N$ and proposed sampling procedure. *Pl. Soil* **78**, 301–313.

DONALD C.M. (1958) The interaction of competition for light and for nutrients. *Aust. J. agric. Res.* **9**, 421–435.

DREW M.C. (1975) Comparison of the effects of a localized supply of phosphate, nitrate, ammonium and potassium on the growth of the seminal root system, and the shoot, in barley. *New Phytol.* **75**, 479–490.

DREW M.C., SAKER L., BARBER S.A. and JENKINS W. (1984) Changes in the kinetics of phosphate and potassium absorption in nutrient deficient barley roots measured by a solution depletion technique. *Planta* **160**, 490–499.

EDWARDS D.G. and ASHER C.J. (1974) The significance of solution flow rate in flowing culture experiments. *Pl. Soil* **41**, 161–175.

EVANS G.C. (1972) *The Quantitative Analysis of Plant Growth.* Blackwell Scientific Publications, Oxford.

FENTEM P.A., LEA P.J. and STEWART G.R. (1983) Ammonia assimilation in the roots of nitrate- and ammonia-grown *Hordeum vulgare* (cv. Golden Promise). *Pl. Physiol., Lancaster* **71**, 496–501.

FISHER D.B. and FRAME J. (1984) A guide to the use of the exuding-stylet technique in phloem physiology. *Planta* **161**, 385–393.

FITTER A.H. (1976) Effects of nutrient supply and competition from other species on root growth of *Lolium perenne* in soil. *Pl. Soil* **45**, 177–189.

FITTER A.H. (1982) Morphometric analysis of root systems: application of the technique and influence of soil fertility on root system development in two herbaceous species. *Plant, Cell, Environ.* **5**, 313–322.

FLOWERS T.J. and LÄUCHLI A. (1983) Sodium versus potassium: substitution and compartmentation. In *Inorganic Plant Nutrition. Encyclopedia of Plant Physiology.* New Series, Vol. 15B (ed. A. Läuchli and R.L. Bieleski), pp. 651–681. Springer-Verlag, Berlin.

FORTIN J.A., PICHÉ Y. and GODBOUT C. (1983) Methods for synthesising ectomycorrhizas and their effect on mycorrhizal development. *Pl. Soil* **71**, 275–284.

FOY C. and FLEMING A.L. (1978) The physiology of plant tolerance to excess available aluminium and manganese in acid soils. In *Crop Tolerance to Suboptional Land Conditions* (ed. C.A. Jung), pp. 301–328. American Society of Agronomy, Madison.

FRENEY J.R. and SIMPSON J.R. (Eds) (1983) *Gaseous loss of Nitrogen from Plant–Soil Systems.* Martinus Nijhoff/Dr W. Junk Publishers, The Hague.

GIGON A. and RORISON I.H. (1972) The response of some ecologically distinct plant species to nitrate- and to ammonium-nitrogen. *J. Ecol.* **60**, 93–102.

GIOVANETTI M.S. and MOSSE B. (1980) An evaluation of the techniques for measuring vesicular-arbuscular mycorrhizal infection in roots. *New Phytol.* **84**, 489–500.

GLASS A.D.M., SIDDIQI M.Y. and DEANE-DRUMMOND C.E. (1983) A multichannel microcomputer-based system for continuously measuring and recording ion activities of uptake solutions during ion absorption by roots of intact plants. *Plant, Cell, Environ.* **6**, 247–253.

GLASS A.D.M., SIDDIQI M.Y. and GILES K.I. (1981) Correlations between potassium uptake and hydrogen efflux in barley varieties. A potential screening method for the isolation of nutrient efficient lines. *Pl. Physiol., Lancaster* **68**, 457–459.

GREENWOOD D.J. and GOODMAN D. (1967) Direct measurements of the distribution of oxygen in soil aggregates and in columns of fine soil crumbs. *J. Soil Sci.* **18**, 182–196.

GRIME J.P. and HUNT R. (1975) Relative growth-rate: its range and adaptive significance in a local flora. *J. Ecol.* **63**, 393–422.

GRIME J.P., MASON G., CURTIS A.V., RODMAN J., BAND S.R., MOWFORTH M.A.G., NEAL A.M. and SHAW S. (1981) A comparative study of germination characteristics in a local flora. *J. Ecol.* **69**, 1017–1059.

GRINSTED M.J., HEDLEY M.J., WHITE R.E. and NYE P.H. (1982) Plant-induced changes in the rhizosphere of rape (*Brassica napus* var. Emerald) seedlings. I. pH change and the increase in P concentration in the soil solution. *New Phytol.* **91**, 19–29.

GUPTA P.L. and RORISON I.H. (1975) Seasonal differences in the availability of nutrients down a podzolic profile. *J. Ecol.* **63**, 521–534.

HACKETT C. (1968) A study of the root system of barley. I. Effects of nutrition on two varieties. *New Phytol.* **67**, 287–299.

HALL J.L. and MOORE A.L. (Eds) (1983) *Isolation of Membranes and Organelles from Plant Cells.* Academic Press, London.

HARBORNE J.B. (1973) *Phytochemical Methods.* Chapman & Hall, London.

HARLEY J.L. and SMITH S.E. (1983) *Mycorrhizal Symbiosis.* Academic Press, London.

HARPER J.L. (1961) Approaches to the study of plant competition. In *Mechanisms in Biological Competition* (ed. F.L. Milthorpe), pp. 1–39. Symposia of the Society of Experimental Biology, No. 15. Cambridge University Press, Cambridge.

HASLAM S.M. (1978) *River Plants. The Macrophytic Vegetation of Water Courses.* Cambridge University Press, Cambridge.

HEPPER C. (1981) Techniques for studying the infection of plants by vesicular-arbuscular mycorrhizal fungi under axenic conditions. *New Phytol.* **88**, 641–647.

HEWITT E.J. (1966) *Sand and Water Culture Methods Used in the Study of Plant Nutrition,* 2nd edn. Technical Communication No. 22, Commonwealth Agricultural Bureaux, Farnham Royal.

HODSON M.J., SMITH M.M., WAINWRIGHT S.J. and ÖPIK H. (1981) Cation co-tolerance in a salt-tolerant clone of *Agrostis stolonifera* L. *New Phytol.* **90**, 253–261.

HOPE A.B. and WALKER N.A. (1975) *The Physiology of Giant Algal Cells.* Cambridge University Press, Cambridge.

HUCK M.G. and TAYLOR H.M. (1982) The rhizotron as a tool in root research. *Adv. Agron.* **35**, 1–35.

HUGHES R.E., MILNER C. and DALE J. (1962) Selectivity in grazing. In *Grazing in the Terrestrial and Marine Environments* (ed. D.J. Crisp), pp. 189–202. Symposia of the British Ecological Society, No. 4. Blackwell Scientific Publications, Oxford.

HUNT R. (1973) A method of estimating root efficiency. *J. appl. Ecol.* **10**, 157–164.

HUNT R. (1980) Diurnal progressions in dry weight and short-term plant growth studies. *Plant, Cell, Environ.* **3**, 475–478.

HUNT R. (1982) *Plant Growth Curves: the Functional Approach to Plant Growth Analysis.* Edward Arnold, London.

HUNT R. (1984) Relative growth rates of cohorts of ramets cloned from a single genet. *J. Ecol.* **72**, 299–305.

IAEA (1983) A guide to the use of nitrogen-15 and radioisotopes in studies of plant nutrition: calculations and interpretation of data. I.A.E.A.—TECDOC—288. International Atomic Energy Agency, Vienna.

INGESTAD T. (1982) Relative addition rate and external concentration; driving variables used in plant nutrition research. *Plant, Cell, Environ.* **5**, 443–453.

INGESTAD T. and LUND A.B. (1979) Nitrogen stress in birch seedlings. I. Growth technique and growth. *Physiologia Pl.* **45**, 137–148.

JAKOBY W.B. and PASTAN I.H. (Eds) (1979) *Cell Culture. Methods in Enzymology,* Vol. LVIII. Academic Press, New York.

JARRELL W.M. and BEVERLY R.B. (1981). The dilution effect in plant nutrition studies. *Adv. Agron.* **34**, 197–224.

JARVIS S.C., JONES L.H.P. and HOPPER M.J. (1976) Cadmium uptake from solution by plants and its transport from roots to shoots. *Pl. Soil* **44**, 179–191.

JEFFERIES R.L., LAYCOCK D., STEWART G.R. and SIMS A.P. (1969) The properties of mechanisms involved in the uptake and utilization of calcium and potassium by plants in relation to an understanding of plant distribution. In *Ecological Aspects of the Mineral Nutrition of Plants* (ed. I.H. Rorison), pp. 281–307. Symposia of the British Ecological Society, No. 9. Blackwell Scientific Publications, Oxford.

JEFFERIES R.L. and PERKINS N. (1977) The effects on the vegetation of the additions of inorganic nutrients to salt marsh soils at Stiffkey, Norfolk. *J. Ecol.* **65**, 867–882.

JEFFERIES R.L. and WILLIS A.J. (1964) Studies on the calcicole–calcifuge habit. II. The influence of calcium on the growth and establishment of four species in soil and sand culture. *J. Ecol.* **52**, 691–707.

JEFFREY D.W. and PIGOTT C.D. (1973) The response of grasslands on sugar-limestone in Teesdale to application of phosphorus and nitrogen. *J. Ecol.* **61**, 85–92.

JONES L.H.P., CLEMENT C.R. and HOPPER M.J. (1973) Lead uptake from solution by perennial ryegrass and its transport from roots to shoots. *Pl. Soil* **38**, 403–414.

JONES L.H.P., JARVIS S.C. and COWLING D.W. (1973) Lead uptake from soils by perennial ryegrass and its relation to the supply of an essential element (sulphur). *Pl. Soil* **38**, 605–619.

KINZEL H. (1983) Influence of limestone, silicates, and soil pH on vegetation. In *Physiological Plant Ecology III. Responses to the Chemical and Biological Environment. Encyclopedia of Plant Physiology*. New Series, Vol. 12C (ed. O.L. Lange, P.S. Nobel, C.B. Osmond and H. Ziegler), pp. 201–244. Springer-Verlag, Berlin.

KNOWLES R. (1980) Nitrogen fixation in natural plant communities and soils. In *Methods for Evaluating Biological Nitrogen Fixation* (ed. F.J. Bergersen), pp. 557–582. John Wiley, Chichester.

KOWALENKO C.G. and CAMERON D.R. (1976) Nitrogen transformations in an incubated soil as affected by combinations of moisture content and temperature and adsorption-fixation of ammonium. *Can. J. Soil Sci.* **56**, 63–70.

KUCHENBUCH R. and JUNGK A. (1982) A method for determining concentration profiles at the soil–root interface by thin slicing rhizosphere soil. *Pl. Soil* **68**, 391–394.

LEE J.A. and STEWART G.R. (1978) Ecological aspects of nitrogen assimilation. *Adv. Bot. Res.* **6**, 1–43.

LEE R.B. (1982) Selectivity and kinetics of ion uptake by barley plants following nutrient deficiency. *Ann. Bot.* **50**, 429–449.

LEE R.B. and RATCLIFFE R.G. (1983) Phosphorus nutrition and the intracellular distribution of inorganic phosphate in pea root tips: a quantitative study using ^{31}P-NMR. *J. exp. Bot.* **34**, 1222–1244.

LEIGH R.A., STRIBLEY D.P. and JOHNSTON A.E. (1983) How should tissue nutrient concentrations be expressed? In *Proceedings of the 9th International Mineral Nutrition Colloquium* (ed. A. Scaife), pp. 317–322. Commonwealth Agricultural Bureaux, Farnham Royal.

LEIGH R.A. and WYN JONES R.G. (1984) A hypothesis relating critical potassium concentrations for growth to the distributions and functions of this element in the plant cell. *New Phytol.* **97**, 1–13.

LÜTTGE U. (1983) Ecophysiology of carnivorous plants. In *Physiological Plant Ecology III. Responses to the Chemical and Biological Environment. Encyclopedia of Plant Physiology*. New Series, Vol. 12C (ed. O.L. Lange, P.S. Nobel, C.B. Osmond and H. Ziegler), pp. 489–517. Springer-Verlag, Berlin.

MCCAIN S. and DAVIES M.S. (1984) Effects of pretreatment of phosphate in natural populations of *Agrostis capillaris* L. II. Interactions with aluminium on the acid phosphatase activity and potassium leakage of intact roots. *New Phytol.* **96**, 589–599.

MCGILL W.B., HUNT H.W., WOODMANSEE R.G. and REUSS J.O. (1981) Phoenix, a model of the dynamics of carbon and nitrogen in grassland soils. In *Terrestrial Nitrogen Cycles. Processes, Ecosystem Strategies and Management Impacts* (ed. F.E. Clark and T. Rosswall). *Ecological Bulletins*, **33**, pp. 49–115. Swedish Natural Science Research Council (NFR).

MCGRATH S.P. (1979) *Growth and distribution of* Holcus lanatus *populations with reference to nitrogen source and aluminium*. PhD Thesis, University of Sheffield.

MCGRATH S.P. and RORISON I.H. (1982) The influence of nitrogen source on the tolerance of *Holcus lanatus* L. and *Bromus erectus* Huds. to manganese. *New Physol.* **91**, 443–452.

MACMILLAN J. (Ed.) (1980) *Hormonal Regulation of Development. I. Molecular Aspects of Plant Hormones. Encyclopedia of Plant Physiology*. New Series, Vol. 9. Springer-Verlag, Berlin.

MCNAUGHTON G.S. and PRESLAND M.R. (1983) Whole plant studies using radioactive 13-nitrogen. I. Techniques for measuring the uptake and transport of nitrate and ammonium ions in hydroponically grown *Zea mays*. *J. exp. Bot.* **34**, 880–892.

MCNAUGHTON S.J. (1983) Physiological and ecological implications of herbivory. In *Physiological Plant Ecology. III. Responses to the Chemical and Biological Environment. Encyclopedia of Plant Physiology*. New Series, Vol. 12C (ed. O.L. Lange, P.S. Nobel, C.B. Osmond and H. Ziegler), pp. 657–677. Springer-Verlag, Berlin.

MARSCHNER H. (1983) General introduction to the mineral nutrition of plants. In *Inorganic*

Plant Nutrition. Encyclopedia of Plant Physiology. New Series, Vol. 15A (ed. A. Läuchli and R.L. Bieleski), pp. 7–60. Springer-Verlag, Berlin.

MARSCHNER H. and RÖMHELD V. (1983) *In vivo* measurement of root-induced pH changes in the soil root interface: effect of plant species and nitrogen source. *Z. Pflanzenphysiol.* **111**, 241–251.

MENGEL K. and KIRKBY E.A. (1982) *Principles of Plant Nutrition*, 3rd edn. International Potash Institute, Berne.

MILTHORPE F.L. and MOORBY J. (1979) *An Introduction to Crop Physiology*, 2nd edn. Cambridge University Press, Cambridge.

MOORBY H. and NYE P.H. (1983) A nutrient film technique for the simultaneous measurement of root growth and nutrient uptake. *Pl. Soil* **70**, 151–154.

MOORBY J. and BESFORD R.T. (1983) Mineral nutrition and growth. In *Inorganic Plant Nutrition. Encyclopedia of Plant Physiology*. New Series, Vol. 15B (ed. A. Läuchli and R.L. Bieleski), pp. 481–527. Springer-Verlag, Berlin.

MOSER M. and HASELWANDTER K. (1983) Ecophysiology of mycorrhizal symbioses. In *Physiological Plant Ecology III. Responses to the Chemical and Biological Environment. Encyclopedia of Plant Physiology*. New Series, Vol. 12C (ed. O.L. Lange, P.S. Nobel, C.B. Osmond and H. Ziegler), pp. 391–421. Springer-Verlag, Berlin.

MOSSE B. and PHILLIPS J.M. (1971) The influence of phosphate on the development of vesicular-arbuscular mycorrhiza in culture. *J. Gen. Microbiol.* **69**, 157–166.

MOSSE B., STRIBLEY D.P. and LeTACON F. (1982) Ecology of mycorrhizae and mycorrhizal fungi. *Adv. Microbial Ecol.* **5**, 137–210.

MOTT R.L. and STEWARD F.C. (1972) Solute accumulation in plant cells. V. An aspect of nutrition and development. *Ann. Bot.* **36**, 915–937.

NASSERY H. and HARLEY J.L. (1969) Phosphate absorption by plants from habitats of different phosphate status. I. Absorption and incorporation of phosphate by excised roots. *New Phytol.* **68**, 13–20.

NEAME K.D. and RICHARDS T.G. (1972) *Elementary Kinetics of Membrane Carrier Transport*. Blackwell Scientific Publications, Oxford.

NEWMAN E.I. (1983) Interactions between plants. In *Physiological Plant Ecology III. Responses to the Chemical and Biological Environment. Encyclopedia of Plant Physiology*. New Series, Vol. 12C (ed. O.L. Lange, P.S. Nobel, C.B. Osmond and H. Ziegler), pp. 679–710. Springer-Verlag, Berlin.

NEWMAN E.I., HEAP A.J. and LAWLEY R.A. (1981) Abundance of mycorrhiza and root-surface micro-organisms of *Plantago lanceolata* L. in relation to soil and vegetation: a multi-variate approach. *New Phytol.* **89**, 95–108.

NYE P.H. and TINKER P.B. (1977) *Solute Movement in the Soil–Root System*. Blackwell Scientific Publications, Oxford.

OSMOND C.B., WINTER K. and ZIEGLER H. (1982) Functional significance of different pathways of CO_2 fixation in photosynthesis. In *Physiological Plant Ecology II. Water Relations and Carbon Assimilation. Encyclopedia of Plant Physiology*. New Series, Vol. 12B (ed. O.L. Lange, P.S. Nobel, C.B. Osmond and H. Ziegler), pp. 479–547. Springer-Verlag, Berlin.

OTTAWAY J.H. (1984) Persistance of organic phosphates in buried soil. *Nature* **307**, 257–259.

PAGE A.L. (Ed.) (1982). *Methods in Soil Analysis (Part 2). Chemical and Microbiological Properties*, 2nd edn. *Agronomy* No. 9. American Society of Agronomy, Madison.

PAINTER R.B. (1976) Climatology and environmental measurement. In *Methods in Plant Ecology*, 1st edn. (ed. S.B. Chapman), pp. 369–410. Blackwell Scientific Publications, Oxford.

PARKINSON D., GRAY T.R.G. and WILLIAMS S.T. (Eds.) (1971) *Methods for Studying the Ecology of Soil Micro-organisms*. IBP Handbook, No. 19. Blackwell Scientific Publications, Oxford.

PATE J.S. (1980). Transport and partitioning of nitrogenous solutes. *A. Rev. Pl. Physiol.* **31**, 313–340.

PATE J.S. (1983) Patterns of nitrogen metabolism in higher plants and their ecological significance. In *Nitrogen as an Ecological Factor* (ed. J.A. Lee, S. McNeill and I.H. Rorison), pp. 225–255. Symposia of the British Ecological Society, No. 22. Blackwell Scientific Publications, Oxford.

PEARSON V. and READ D.J. (1973) The biology of mycorrhiza in the Ericaceae. I. The isolation of the endophyte and synthesis of mycorrhiza in aseptic culture. *New Phytol.* **72**, 371–379.

PEARSON V. and TINKER P.B. (1975) Measurement of phosphorus fluxes in the external hyphae of endomycorrhizas. In *Endomycorrhizas* (ed. F.E. Sanders, B. Mosse and P.B. Tinker), pp. 277–287. Academic Press, London.

PEEL A.J. (1975) Investigations with aphid stylets into the physiology of the sieve tube. In *Transport in Plants. I. Phloem transport. Encyclopedia of Plant Physiology.* New Series, Vol. 1 (ed. M.H. Zimmerman and J.H. Milburn), pp. 171–195. Springer-Verlag, Berlin.

PENNING DE VRIES F.W.T. (1983) Modelling of growth and production. In *Physiological Plant Ecology IV. Ecosystem Processes: Mineral Cycling, Productivity and Man's influence. Encyclopedia of Plant Physiology.* New Series, Vol. 12D (ed. O.L. Lange, P.S. Nobel, C.B. Osmond and H. Ziegler), pp. 117–150. Springer-Verlag, Berlin.

PETERKIN J.H. (1981) *Plant growth and nitrogen nutrition in relation to temperature.* PhD Thesis, University of Sheffield.

PETERKIN J.H., RORISON I.H. and CLARKSON D.T. (1982) Plant growth and nitrogen nutrition —relative to temperature. *Annual Report of the Unit of Comparative Plant Ecology (NERC),* University of Sheffield *1982,* p. 16.

PHILLIPS J.M. and HAYMAN D.S. (1970) Improved procedures for cleaning roots and staining parasitic and vesicular-arbuscular mycorrhizal fungi for rapid assessment of infection. *Trans. Br. mycol. Soc.* **55**, 158–161.

RAFAREL C.R. and BRUNDSON G.P. (1976) Data collection systems. In *Methods in Plant Ecology,* 1st edn. (ed. S.B. Chapman), pp. 467–506. Blackwell Scientific Publications, Oxford.

RAVEN J.A. (1981) Nutritional strategies of submerged benthic plants: the acquisition of C, N and P by rhizophytes and haptophytes. *New Phytol.* **88**, 1–30.

RAVEN J.A. (1983) Phytophages of xylem and phloem: a comparison of animal and plant sap-feeders. *Adv. Ecol. Res.* **13**, 137–234.

RAVEN J.A. and SMITH F.A. (1976) Nitrogen assimilation and transport in vascular land plants in relation to intracellular pH regulation. *New Phytol.* **76**, 415–431.

READ D.J., KOUCHEKI H.K. and HODGSON J. (1976) Vesicular-arbuscular mycorrhiza in natural vegetation. I. The occurrence of infection. *New Phytol.* **77**, 641–653.

REES A.R., COCKSHULL K.E., HAND D.W. and HURD R.G. (Eds.) (1972) *Crop Processes in Controlled Environments.* Academic Press, London.

REINERT J. and YEOMAN M.M. (1982) *Plant Cell and Tissue Culture. A Laboratory Manual.* Springer-Verlag, Berlin.

RHODES D., MYERS A.C. and JAMIESON G. (1981) Gas chromatography—mass spectrometry of N-heptafluorabutyryl isobutyl esters of amino acids in the analysis of the kinetics of $[^{15}N]H_4^+$ assimilation in *Lemna minor* L. *Pl. Physiol., Lancaster* **68**, 1197–1205.

ROBINSON D. and RORISON I.H. (1983) A comparison of the responses of *Lolium perenne* L., *Holcus lanatus* L., and *Deschampsia flexuosa* (L.) Trin. to a localized supply of nitrogen. *New Phytol.* **94**, 263–273.

ROBINSON D. and RORISON I.H. (1985) A quantitative analysis of the relationship between root distribution and nitrogen uptake from soil by two grass species. *J. Soil. Sci.* **36**, 71–85.

RORISON I.H. (1967) A seedling bioassay on some soils in the Sheffield area. *J. Ecol.* **55**, 725–741.

RORISON I.H. (1968) The response to phosphorus of some ecologically distinct plant series. I. Growth rates and phosphorus absorption. *New Phytol.* **67**, 913–923.

RORISON I.H. (1969) Ecological inferences from laboratory experiments on mineral nutrition. In *Ecological Aspects of the Mineral Nutrition of Plants* (ed. I.H. Rorison), pp. 155–175. Symposia of the British Ecological Society, No. 9. Blackwell Scientific Publications, Oxford.

RORISON I.H. (1971) The use of nutrients in the control of the floristic composition of grassland.

In *The Scientific Management of Animal and Plant Communities for Conservation* (ed. E. Duffey and A.S. Watt), pp. 65–77. Symposia of the British Ecological Society, No. 11. Blackwell Scientific Publications, Oxford.

RORISON I.H. (1985) Nitrogen source and tolerance of *Deschampsia flexuosa, Holcus lanatus* and *Bromus erectus* to aluminium during seedling growth. *J. Ecol.* **73**, 83–90.

RORISON I.H. and GUPTA P.L. (1974) The growth of seedlings in response to variable phosphorous supply. In *Proceedings of the 7th International Colloquium on Plant Analysis and Fertilizer Problems* (ed. J. Wehrmann), pp. 373–382. German Society of Plant Nutrition, Hanover.

RORISON I.H., GUPTA P.L. and SPENCER R.E. (1983) Nutrient balance and soil exhaustion in relation to plant growth and uptake by contrasting species. *Annual Report of the Unit of Comparative Plant Ecology (NERC)*, University of Sheffield *1983*, p. 26.

RORISON I.H., PETERKIN J.H. and CLARKSON D.T. (1983) Nitrogen source, temperature and plant growth. In *Nitrogen as an Ecological Factor* (ed. J.A. Lee, S. McNeill and I.H. Rorison), pp. 189–209. Symposia of the British Ecological Society, No. 22. Blackwell Scientific Publications, Oxford.

RORISON I.H. and ROBINSON D. (1984) Calcium as an environmental variable. *Plant, Cell, Environ.* **7**, 381–390.

RORISON I.H. and SPENCER R.E. (1984) Nuclear magnetic resonance (NMR) studies. *Annual Report of the Unit of Comparative Plant Ecology (NERC)*, University of Sheffield 1984, p. 26.

RORISON I.H., SPENCER R.E., SUTTON F. and GUPTA P.L.(1981) Auto-control of the composition of nutrient flow culture. *Annual Report of the Unit of Comparative Plant Ecology (NERC)*, University of Sheffield, pp. 19–20.

ROSE D.A. and CHARLES-EDWARDS D.A. (Eds.) (1982) *Mathematics and Plant Physiology*. Academic Press, London.

ROVIRA A.D., BOWEN G.D. and FOSTER R.C. (1983) The significance of rhizosphere microflora and mycorrhizas in plant nutrition. In *Inorganic Plant Nutrition. Encyclopedia of Plant Physiology*. New Series, Vol. 15A (ed. A. Läuchli and R.L. Bieleski), pp. 61–93. Springer-Verlag, Berlin.

ROWELL D.L., MARTIN M.W. and NYE P.H. (1967) The measurement and mechanism of ion diffusion in soils. III. The effect of moisture content and soil solution concentration on the self-diffusion of ions in soils. *J. Soil Sci.* **18**, 204–222.

ROWSE H.R. and PHILLIPS D.A. (1974) An instrument for measuring the total length of root in a sample. *J. appl. Ecol.* **11**, 309–314.

RUSSELL R.S. and SANDERSON J. (1967) Nutrient uptake by different parts of the intact roots of plants. *J. exp. Bot.* **18**, 491–508.

SANDERS F.E. (1971) *Effect of root and soil properties on the uptake of nutrients by competing roots*. DPhil. Thesis, University of Oxford.

SANDERS F.E. and TINKER P.B. (1973) Phosphate flow into mycorrhizal roots. *Pest. Sci.* **4**, 385–395.

SAN PIETRO A. (Ed.) (1980) *Photosynthesis and Nitrogen Fixation. Methods in Enzymology*, Vol. 69. Academic Press, New York.

SCHENCK N.C. (Ed.) (1982) *Methods and Principles of Mycorrhizal Research*. American Phytopathological Society, St. Paul.

SCHOFIELD R.K. and GRAHAM-BRYCE I.J. (1960) Diffusion of ions in soils. *Nature* **188**, 1048–1049.

SCHONBECK M. and NORTON T.A. (1979) The effects of brief periodic submergence on intertidal fucoid algae. *Est. & Coast. Mar. Sci.* **8**, 205–211.

SCOTT N.E. and DAVISON A.W. (1982) De-icing salt and the invasion of road verges by maritime plants. *Watsonia* **14**, 41–52.

SHAW S.C., RORISON I.H. and BAKER A.J.M. (1982) Physiological mechanisms of heavy metal tolerance in plants. *Annual Report of the Unit of Comparative Plant Ecology (NERC)*, University of Sheffield 1982, pp. 13–15.

SHAW S.C., RORISON I.H. and BAKER A.J.M. (1984) Solubility of heavy metals in lead mine spoil extracts. *Environmental Pollution (Series B)* **8**, 23–33.

SLANGEN J.H.G. and KERKHOFF P. (1984) Nitrification inhibitors in agriculture and horticulture: a review. *Fert. Res.* **5**, 1–76.

SLAVÍK B. (Ed.) (1974) *Methods of Studying Plant–Water Relations. Ecological Studies,* **9**. Springer-Verlag, Berlin.

SMITH H. (Ed.) (1977) *Regulation of Enzyme Synthesis and Activity in Higher Plants.* Academic Press, London.

SOUTHWOOD T.R.E. (1978) *Ecological Methods: with Particular Reference to the Study of Insect Populations.* 2nd edn. Chapman & Hall, London.

SPRENT J.I. (1979) *The Biology of Nitrogen-fixing Organisms.* McGraw-Hill, Maidenhead.

STEFANSON R.C. (1970) Sealed growth chambers for studies on the effects of plants on the soil atmosphere. *J. agric. Engng. Res.* **15**, 295–301.

STEIN J.R. (Ed.) (1973) *Handbook of Phycological Methods: Culture, Methods and Growth Measurements.* Cambridge University Press, Cambridge.

STEWART G.R., LARHER F., AHMAD I. and LEE J.A. (1979) Nitrogen metabolism and salt tolerance in higher plant halophytes. In *Ecological Processes in Coastal Environments* (ed. R.L. Jefferies and A.J. Davy), pp. 211–227. Symposium of the British Ecological Society, No. 19. Blackwell Scientific Publications, Oxford.

ST JOHN T.V., HAYS R.I. and REID C.P.P. (1981) A new method for producing pure vesicular-arbuscular mycorrhiza–host cultures without specialized media. *New Phytol.* **89**, 81–86.

STREET H. (Ed.) (1977) *Plant Tissue and Cell Culture*, 2nd edn. Botanical Monographs, Vol. 11. Blackwell Scientific Publications, Oxford.

STRIBLEY D.P. and READ D.J. (1980) The biology of mycorrhiza in the Ericaceae. VII. The relationship between mycorrhizal infection and the capacity to utilize simple and complex organic nitrogen sources. *New Phytol.* **86**, 261–266.

STRIBLEY D.P., SNELLGROVE R.C. and TINKER P.B. (1980) Effect of vesicular-arbuscular mycorrhizal fungi on the relations of plant growth, internal phosphorus concentration and soil phosphate analysis. *J. Soil Sci.* **31**, 655–672.

SUTTON F. and RORISON I.H. (1980) An interface, allowing a commercial paper-tape reader to be used in the programming of growth-room temperature. *J. exp. Bot.* **31**, 691–696.

SWAIN T. (1977) Secondary compounds as protective agents. *A. Rev. Pl. Physiol.* **28**, 479–501.

SWIFT M.J., HEAL O.W. and ANDERSON J.M. (1979) *Decomposition in Terrestrial Ecosystems.* Blackwell Scientific Publications, Oxford.

TAYLOR A.A., DE FELICE J. and HAVILL D.C. (1982) Seasonal variation in nitrogen availability in an acidic and calcareous soil. *New Phytol.* **92**, 141–152.

TAYLOR A.A. and HAVILL D.C. (1981) The effect of inorganic nitrogen on the major enzymes of ammonium assimilation in grassland plants. *New Phytol.* **87**, 53–62.

THOMAS G.W. (1974) Chemical reactions controlling soil solution electrolyte concentration. In *The Plant Root and its Environment* (ed. E.W. Carson), pp. 483–506. University Press of Virginia, Charlottesville.

THORNLEY J.H.M. (1976) *Mathematical Models in Plant Physiology.* Academic Press, London.

THURSTON J.M. (1969) The effects of liming and fertilizers on the botanical composition of permanent grassland, and on the yield of hay. In *Ecological Aspects of the Mineral Nutrition of Plants* (ed. I.H. Rorison), pp. 3–10. Symposia of the British Ecological Society, No. 9. Blackwell Scientific Publications, Oxford.

TIBBITTS T.W. and KOZLOWSKI T.T. (Eds) (1979) *Controlled Environment Guidelines for Plant Research.* Academic Press, New York.

TINKER P.B. (1969) A steady state method for determining diffusion coefficients in soil. *J. Soil Sci.* **20**, 336–345.

VAIDYANATHAN L.V. and NYE P.H. (1966) The measurement and mechanism of ion diffusion in soils. II. An exchange resin paper method for measurement of the diffusive flux and diffusion coefficient of nutrient ions in soils. *J. Soil Sci.* **17**, 175–183.

VAN DEN BERGH J.P. (1969) Distribution of pasture plants in relation to chemical properties of the soil. In *Ecological Aspects of the Mineral Nutrition of Plants* (ed. I.H. Rorison), pp. 11–23. Symposia of the British Ecological Society, No. 9. Blackwell Scientific Publications, Oxford.

VINCENT J.M. (1970) *A Manual for the Study of Root Nodule Bacteria.* IBP Handbook, No. 15. Blackwell Scientific Publications, Oxford.

VOSE P.B. (1980) *Introduction to Nuclear Techniques in Agronomy and Plant Biology.* Pergamon Press, Oxford.

WAINWRIGHT M., KILLHAM K. and DIPROSE M.F. (1980) Effect of 2450 MHz microwave radiation on nitrification, respiration and S-oxidation in soil. *Soil Biol. Biochem.* **12**, 489–493.

WALLACE T. (1961) *The Diagnosis of Mineral Deficiencies in Plants.* H.M.S.O., London.

WARNOCK A.J., FITTER, A.H. and USHER M.B. (1982) The influence of a springtail *Folsomia candida* (Insecta, Collembola) on the mycorrhizal association of leek *Allium porrum* and the vesicular-arbuscular endophyte *Glomus fasiculatus. New Phytol.* **90**, 285–292.

WILD A., WOODHOUSE P.J. and HOPPER M.J. (1979) A comparison between the uptake of potassium by plants from solutions of constant potassium concentration and during depletion. *J. exp. Bot.* **30**, 697–704.

WILKINS D.A. (1957) A technique for the measurement of lead tolerance in plants. *Nature* **180**, 37–38.

WILKINS D.A. (1978) The measurement of tolerance to edaphic factors by means of root growth. *New Phytol.* **80**, 623–633.

WILLIS A.J. (1963) Braunton Burrows: the effects on the vegetation of the addition of mineral nutrients to the dune soil. *J. Ecol.* **51**, 353–374.

WITTY J.F. (1983) Estimating N$_2$-fixation in the field using ^{15}N-labelled fertilizer: some problems and solutions. *Soil Biol. Biochem.* **15**, 631–639.

WONG M.H. and BRADSHAW A.D. (1982) A comparison of the toxicity of heavy metals, using root elongation of ryegrass, *Lolium perenne. New Phytol.* **91**, 255–261.

WOOLHOUSE H.W. (1969) Differences in the properties of the acid phosphatase of plant roots and their significance in the evolution of edaphic ecotypes. In *Ecological Aspects of the Mineral Nutrition of Plants* (ed. I.H. Rorison), pp. 357–380. Symposia of the British Ecological Society, No. 9. Blackwell Scientific Publications, Oxford.

WOOLHOUSE H.W. (1983) Toxicity and tolerance in the responses of plants to metals. In *Physiological Plant Ecology III. Responses to the Chemical and Biological Environment. Encylopedia of Plant Physiology.* New Series, Vol. 12C (ed. O.L. Lange, P.S. Nobel, C.B. Osmond and H. Ziegler), pp. 245–300. Springer-Verlag, Berlin.

WRIGHT M.J. (Ed.) (1977) *Plant Adaptation to Mineral Stress and Problem Soils.* Cornell University, Ithaca.

5 Site and soils

D.F. BALL

1 Introduction

For most plant ecological field studies an outline understanding of the geomorphological and geological characteristics of a site is adequate. Soils, which are one of the direct controls on plant distribution and performance, demand more detailed consideration.

It is impossible to summarize all aspects of soil science here. A useful general account is that of White (1979) and an encyclopedic summary of soil composition is provided by Fairbridge & Finkl (1979). An essential basis for most plant ecological investigations is considered to be an appreciation of the pedological approach, which itself is an ecological view of soils as bodies responding to environmental factors. The pedological view treats soils as a complex natural body, a thin skin covering the rocks, which has evolved and continues to develop in different ways in response to physical and biotic environmental influences. Within this approach, the morphological criteria considered in field descriptions of soils, and their use in grouping soils in classification schemes, are emphasized. Soil chemical analyses are treated in Chapter 6, so receive little attention here, but physical and mineralogical methods of likely interest to the ecologist are covered.

The pedological view of soil formation and distribution can be studied in works by Bunting (1967), Ganssen (1972), Cruikshank (1972), Duchaufour (1976, 1977, 1982), Bridges (1979), Fitzpatrick (1980), Buol et al. (1980) and Bridges & Davidson (1982). Though the virtues of the pedological approach in understanding the relationship of plants to environment are accepted and stressed here, a caution should be noted about over-reliance on it. Morphologically-based soil classes rely on features which can persist over climatic or nutrient ranges sufficient to produce substantial differences in performance and survival of plant species. The morphological recognition of distinct soil classes—though one key to plant ecological associations, and representing a more persistent aspect of soil character than its nutrient status—includes only part of the influence exerted by soils on plant distribution and growth, as discussed, for example, by Dumanski et al. (1982). Nutrient supply by soils to plants, dealt with in Chapter 4, can only be broadly related to soil class differences, since, within limits, chemistry can be temporarily modified within a soil class, while classes which differ in many ways from each other may have some chemical characteristics in common. Among texts which concentrate on nutrient transfers between soils and plants are Cooke (1967), Rorison (1969), Russell (1974), Nye & Tinker (1977), Lindsay (1979) and Bohn *et al.* (1979). Trudgill (1977) links the atmosphere, soils, and plants, by emphasizing a modelling approach to nutrient transfer.

Soil physics, especially considerations of water movement, are treated by Childs (1969), Marshall & Holmes (1979), Hillel (1980, 1982) and Hanks & Ashcroft (1980). Soil organic matter is covered at a specialist level by Schnitzer & Khan (1978).

Soil-dwelling organisms and micro-organisms, and their influence on soil processes and on nutrient and energy cycling, are discussed in works by Kühnelt (1961), Garret (1963), McLaren & Peterson (1967), Burges & Raw (1967), Wallwork (1970), Russell (1971), Phillipson (1971), Parkinson *et al.* (1971), Griffin (1972), Richards (1974), Walker (1975), Alexander (1977) and Brown (1978).

The plant ecologist may become involved in the rather different civil engineering–soil mechanics approach to soils through issues such as the restoration of disturbed land. It is as well to be aware that the terminology and analytical procedures used for soils viewed as engineering materials differ in some respects from those of the soil-genesis oriented pedologist or the plant response and production-oriented soil chemist and agronomist. For a treatment of this geotechnical, engineering consideration of soils the ecologist can consult Bowles (1979) and study the engineering classification of soils described by Dumbleton (1981).

2   Site description

2.1   Location

Before considering anything else in the field, the location of a sampling or observation point should be clearly identified with a precision that is appropriate to the scale of the particular study. Sites for which a relatively exact location is needed should be accurately marked on an appropriate map and/or air photo, where these are available, and this marked point supported where possible by recording bearings and distances measured from permanent landscape features. If precise site re-location is essential, secure permanent field markers must be used, of course only with the agreement of the site owner (Chapter 9 considers permanent quadrats in vegetation surveys). When a general location is acceptable, grid references are adequate for areas with appropriate maps. If suitable maps are unobtainable, sketches identifying the site (e.g. Fig. 5.1) are essential, and they are also useful for supplementary information in all cases.

Appreciation of the virtues of identifying a retrievable location must be

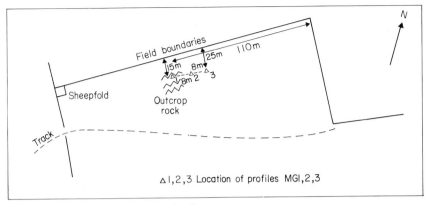

Figure 5.1 An example site location sketch.

balanced by recognition of the typical point-to-point variations in vegetation, and the known scale of significant variability in soils. Major changes in soil type can be found within a distance of a few metres, and substantial soil chemical differences can occur in samples of a morphologically uniform soil taken at as little as 10 cm apart. Problems caused by variability, sampling and site identification are general for the ecologist, but they are increased in soil work since any observation and/or sampling of soil modifies or destroys it, preventing further observations on hitherto undisturbed material at the precise site.

2.2 Physiography

Physiography can be defined as the natural form of the land surface. Topography is sometimes used as if synonymous with physiography, but is better confined to man-made features in the landscape (e.g. buildings, roads, field boundaries). Clarke (1971) included *altitude, slope, aspect, land form* and *microrelief* as key physiographic factors that should be recorded at all sites from which soils are described. These site characteristics can also usefully be recorded by the plant ecologist because, as well as their influence on soil processes and types, they have direct and indirect effects on vegetation.

Where maps are available, the *altitude* of a site can often be obtained with sufficient accuracy from interpolation between mapped contours. Otherwise altitude can be determined using an altimeter which records air pressure change, provided this is calibrated frequently against nearby known altitudes.

Slope angle should be given quantitatively in degrees (from 0° horizontal to 90° vertical), rather than qualitatively, since a slope which is 'moderate' to a worker in the plains might be described as 'gentle' or 'almost level' in mountain regions. A simple instrument for measuring slope is the 'Dr Dollar' pocket clinometer. Formerly commercially available, a version of this could be made in most workshops. A perspex plate has an inset semi-circular groove in which a steel ball-bearing can run. The dimensions of plate and groove match those of a good quality 180° perspex protractor, or a specially inscribed perspex plate, which is correctly positioned above the groove in the base plate and screwed or otherwise fixed permanently to the grooved plate. If a 15 cm diameter protractor is used, the base plate can be $c.$ 17 × 9 × 0·8 cm. Slope angle can be quickly measured with reasonable accuracy from the position taken up by the steel ball when the edge of the clinometer is laid parallel to the ground surface, either directly on it, or better, on a longer object laid on the ground, such as an auger or map-board.

The pantometer of Pitty (1968) is a larger instrument, measuring slope angle and length, that can be used to draw slope profiles, as in an ecological and land-use study in North Wales by Armitage (1973). An Abney level, used with sighting rods and measuring tapes, can give comprehensive slope profiles (e.g. King, 1966; Pitty, 1969; Young, 1972).

The *aspect* of a sloping site can be recorded as either a compass bearing in degrees, or as a sector (e.g. N, NE, E), by taking a compass reading parallel to the direction of, and facing outwards from, maximum slope.

Specialist geomorphological description and mapping of landforms is a complex procedure. In some studies the ecologist will require such a specialist view and will then turn to expert advice or the works suggested below. In most ecological work a more simple approach is adequate. Landform, defined as the general surface form of the country around the sample site, can be recorded using descriptive phrases such as: mountain ridge crest; alluvial floor of river valley; or gently rolling, low hills.

Although *microrelief* (definable as the land form immediately adjacent to a study location) was among the physiographic factors given by Clarke (1971) as important to soil development, what is more generally useful is a somewhat wider identification of local relief (say covering a radius of some 10 m). Descriptive terms such as: uniform slope; steep face of rocky outcrop; small depression in hummocky terrain; can be used to supplement the major landform description. Although such descriptive terminology may appear fundamentally unsatisfactory, it is generally adequate when coupled with quantitative values for altitude, slope and aspect. It can usefully be supported on data record cards by a sketch profile like that of Fig. 5.2, indicating the general relationship of description or sampling site to land form.

The description of relief in soil studies was outlined by Curtis *et al.* (1965). Methods in geomorphology are treated by King (1966), Goudie (1981) and Gardiner & Dackombe (1981). An interesting integration of soils and landforms has been provided recently by Gerrard (1981). Broader geomorphology texts are those of Holmes (1965), Sparks (1972) and Rice (1977).

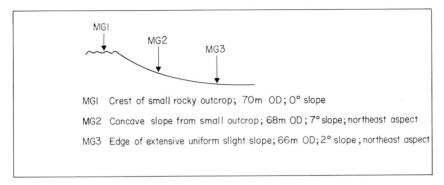

MGI   Crest of small rocky outcrop; 70m OD; 0° slope

MG2   Concave slope from small outcrop; 68m OD; 7° slope; northeast aspect

MG3   Edge of extensive uniform slight slope; 66m OD; 2° slope; northeast aspect

Figure 5.2 An example landform sketch.

2.3   Geology

Read & Watson (1962), provide an introduction to both geomorphology and geology. Because the mineralogical and chemical composition of rocks directly affects soil type and soil chemistry, the ecologist needs always to consider the origin of the solid or unconsolidated rock material that forms the parent material of the soil. Often the plant ecologist can hope to find a geological map that identifies the rock system at sites of interest, but the availability of a map is no guarantee of sufficient geological information.

Most geological maps use stratigraphic categories (giving geological age) for mapping units, rather than directly showing the distribution of rock properties more immediately relevant to soil scientists and ecologists, such as their mineralogy, chemistry, particle size and weathering regime. Two sandstones of widely differing ages have more in common as soil parent materials than a sandstone and a limestone within the same stratigraphic unit. As age classes are the obvious feature to be drawn from most geological maps, and there are occasions when age differences correlate with significant chemical or 'weatherability' properties in rocks of similar lithology, stratigraphic age should, however, be included in any site description.

Turning to rocks as soil parent materials, soils can form directly from weathering of massive or shattered rock *in situ*, or from unconsolidated deposits, such as glacial drifts of different types, alluvium, or wind-blown deposits. It is advisable, when transported material immediately underlies the soil at a site, to record, where possible, identifiable characteristics of both this and the solid rock beneath it. Though the types of stones in a soil of course suggest at least something of its parent material, it can happen that those which remain are only one part of a soil parent material, i.e. the part that has not weathered to smaller particles, or has done so only very slowly. In such cases the bulk of a soil may have formed from other more rapidly-weathering rocks of different geological origin. Mineralogical (Section 6.3) and chemical analyses must be used to identify such situations.

Recognition of rock types from hand-specimens alone is not easy without geological training and, even with this, is often not possible with certainty for fine-grained rocks. In general the ecologist will need to have specialist knowledge or seek it elsewhere if geological information beyond that provided by available maps is needed. For accounts of petrology (the composition, origin and relationships of rocks) Harker (1962) and Hatch (1971, 1972) provide simpler and more comprehensive texts, while Fitzpatrick (1980) gives a good account of geological deposits as soil parent materials.

2.4   Climate

Chapter 5 deals with the measurement of microclimatic conditions around the growing plant. In giving general climatic data for study sites, since there are relatively few climatic recording stations even in developed countries, studies often have to be made in areas where local long-term measurements of a wide range of climatic variables are not available.

Typically, an extrapolated value for mean annual rainfall at the site, obtainable from small-scale regional or national maps, is all that is recorded on a soil description sheet. When sites need to be accurately contrasted in terms of their climatic variables—among which annual and seasonal rainfall and temperatures, wind-speeds and directions, radiation input, and the duration and intensity of frost-action are important—only costly and lengthy instrumental measurements at the site for several years can provide a solution. Even with such data, although short-term influences of climate can be of immediate significance to plants, and factors such as incidence of frost or the distribution of rainfall throughout the year are of direct importance to soil properties, the general nature of a soil profile may have been influenced by climate over hundreds or thousands of years. Accurate data for present-day climates at sites are thus not necessarily completely adequate to explain soil type differences, since the present climates may not be the same as those of the sites over the whole period of soil development.

An important ecological aspect of the total climate at a site is *exposure*, the result of a combination of physiographic factors such as altitude, slope and aspect, together with local climatic factors of wind-speed supplemented by rainfall and temperature. Qualitative measures of relative exposure between sites can be obtained by several methods: using 'tatter-flags' and noting the damage caused to them by wind over measured time intervals (Lines & Howell, 1963); observing the degree of damage visible in tree growth forms (Thomas, 1973); or by calculating a site topographic exposure ('topex') index from summing measurements of the angle of the skyline at the eight major compass points (Pyatt *et al.*, 1969).

3   Soil description and sampling

3.1   Profile preparation

There is no uniform international terminology for soil description. Most versions however are variants, with more or less detail, of the approach covered in USDA (1951). In Britain the latest recommendations for soil description are given in Hodgson (1974) (see also Hodgson, 1978). The requirements set out there are more detailed than the recommendations given below.

The basis for soil description and classification is the *soil profile* (Section 3.2), a vertical section through the soil from ground surface to underlying pedologically unaltered parent material. A definition of 'pedologically unaltered' is impossible to formulate without many qualifications but a commonsense judgement of where essentially 'unaltered' parent material or underlying rock begins is usually possible. Soil profiles can be studied in established natural exposures but although these give useful background information, the untypical condition of a long-established free soil face makes them unsuitable for representative description and sampling. Recent excavations (field drains, pipeline courses, road works) may provide temporary displays of the local three-dimensional soil variability. A rapid idea of the range of soil profile character of a study area can be got by examination from point to point using a hand auger (Section 3.3). Occasionally in some soft stoneless soils a manual coring auger can obtain an uncompressed profile. For a wider range of soils, a relatively undisturbed core can be extracted using a power-operated auger, for example mounted on a Land-rover (Wells, 1959), but such a system is not usually available. In general, soil profile pits (Fig. 5.3) are required to permit comprehensive soil description and sampling.

The maximum depth to which a profile pit is dug in Britain is normally about 1 m, less than this if soil parent material or underlying hard rock has previously been reached. For some purposes involving deep-rooting species, and in other regions, this depth will not be enough. To expose a face to 1 m, a pit of surface area about 1 m × 60 cm is needed, but often a pit 60 m^2 is adequate in Britain. In order to limit site damage, the surface soil and vegetation from the pit should so far as possible be carefully removed in a way enabling their later correct replacement in the restored site. Excavated soil from subsequent depths should be separately piled on a stout polythene sheet laid up to one face of the pit. Select the side of the pit receiving the maximum light as the face to be examined, particularly if photography is intended, and avoid trampling over this side. When exposed fully, the study face should be carefully picked back a few centimetres further than its initial position with a trowel or sheath knife, starting from the top, to give a natural

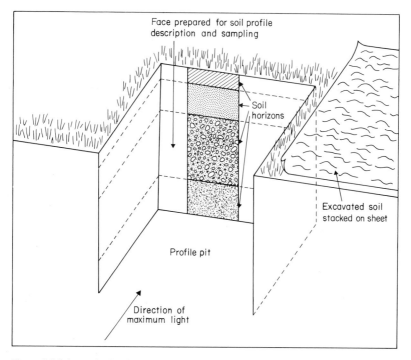

Figure 5.3 Schematic sketch of soil profile pit.

clean face on which observations and measurements can be made and from which samples can be taken. The cleaned face can be used immediately, or after allowing some drying in order to make structural units more apparent. Photographs of the soil profile should include a suitable scale object (such as a tape, trowel or sheath knife).

3.2   Profile description

3.2.1   General considerations

Section 4 discusses the difficulties that prevent recommendation of a single universally applicable soil classification. The classifications favoured here rely on the concept of soil as a natural body which has developed into morphologically distinct types of *soil profiles* by the interaction of physical, chemical and biological processes on *parent materials*, usually mineral but occasionally organic. Soil profiles consist of one or more *horizons* (Fig. 5.3); these are more-or-less well-defined layers, distinguished from each other by properties such as colour, texture and structure.

Before discussing the information conventionally used to define soil horizons it should be considered how such data can best be recorded. A form of the type illustrated in Fig. 5.4 employing word entries, a simple version of a type of form widely used in soil mapping, is convenient. The top of the form covers information on site and general profile characteristics, and the lower part contains horizon data.

An alternative form is completed by ticking one of a number of boxes, or entering codes, for grades or measures of each observed property. Figure 5.5 shows a version used in the Institute of Terrestrial Ecology, Great Britain. Numerical entries are used for features such as grid reference and altitude, while letter and number codes are used for other characteristics like geology, profile drainage, and soil class. When the example was completed, the horizon symbols used on it differed from those recommended later here. A more detailed coded soil description card used by the Soil Survey of England and Wales is reproduced in Hodgson (1974) and discussed, in relation to data handling, by Thompson (1979). Other approaches to numerical coding of profile data have been made (see e.g. Muir & Hardie, 1962; Barkham & Norris, 1970; John *et al.*, 1972). The use of punched feature-cards for soil profile data storage and retrieval on a local scale has been described by Rudeforth & Webster (1973).

A written form is most suitable for small numbers of sites and directly gives a conventional description for publication. A 'ticked box' coded format enables direct data transfer for computer storage and analysis, and is particularly useful for extensive data collection. The latter method enforces firm decisions because of the either/or nature of coded entries, but alternatively there are advantages in being able to show nuances of character by appropriate wording on the written form. No amount of apparent increase in precision resulting from the selected method of data recording should be allowed to obscure the fact that at present many facets of soil description cannot be easily defined objectively and quantitatively.

3.2.2 Features recorded

In the following outline of information typically recorded in soil profile descriptions, the features are dealt with in the general order in which they are entered on the form illustrated in Fig. 5.4.

The profile is initially identified by an *index number*, its *location* given in words and by *grid reference* and/or sketch map, and the *sampling date* noted. When the soil described is related to an area covered by a soil survey, the *mapping symbol* used on relevant soil maps can be given. *Altitude*, *slope* and *aspect* are quantitatively entered, and *relief* (general and local landform) is described in words supported by a sketch.

The heading *site drainage* covers a subjective estimate of the balance likely between surface water received at the site, and lost from it by run-off. This

Horizon depth, classi-fication,and sharpness of horizon boundary	Colour	Field texture	Stone quantity and type	Soil structure	Porosity	Handling consis-tency	Organic matter estimate	Root distri-bution	Field moisture	Earth-worms	Horizon number and notes

Profile number

Site location

Altitude
Slope
Aspect

Relief

Major soil group
(or sub-group)

Soil series
Phase
Type

Map symbol

Site drainage

Profile drainage

Date of sampling

Map reference

Solid geology

Soil parent material

Site rainfall

Vegetation

Figure 5.4 Typical soil profile description form.

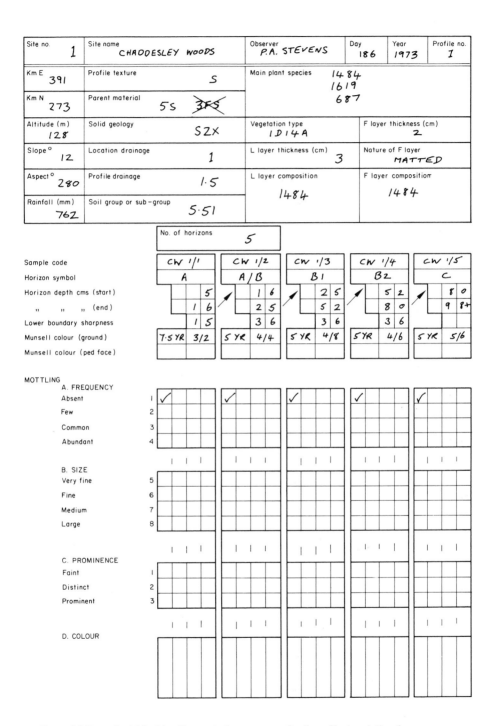

Figure 5.5 Part of a 'ticked box'/numerical entry type of soil profile description form.

is an interpretation of site relief and an observation of ground surface condition, rather than a definitive hydrological measurement, using the following terms: normal (inflow and run-off waters in approximate balance); shedding (run-off significantly greater than inflow); receiving (inflow surface water likely to exceed run-off, (a) run-off present but low, (b) no run-off).

Solid geology is entered from field observation where possible and/or from available maps. *Parent material*, which may be identical to, or differ from, the solid geology recorded for the underlying rock, is again recorded from available maps, or from field identification of the geological character of the weathering solid rock or unconsolidated deposit on which the soil has formed.

The inadequate attention paid in conventional soil descriptions to *climate* and *vegetation* is clear from the spaces provided on this typical form. Only a figure for annual rainfall when a map source is available, and broad vegetation categories such as 'chalk grassland', 'scrub birch woodland', or '*Sphagnum* bog' are entered when such forms are completed by the average soil scientist.

Major soil group, *soil series*, *phase* and *type* may be entered in the field in situations where the local soil pattern is familiar to the person completing the form, but this is often done later in accord with a chosen classification system (Section 4.2).

Profile drainage is again a subjective classification based on observed soil wetness and on profile colours, textures and structures. Current British Soil Survey procedure is not to prejudge conclusions, that should be drawn from adequate experimental data, by allocating drainage classes from visual field characteristics. Some value remains in attempting such assessment, though its limitations must be understood. Morphology of the soil profile may have been inherited from earlier drainage conditions rather than result from the current situation. Drainage classes typically used are:

Excessive. Soils with very low moisture retention such as occur in very permeable materials and in shallow soils overlying fissured or permeable rocks. Shallow soils with low moisture capacity overlying indurated subsoils are excessively drained in dry periods but waterlogged in wet periods because of low subsoil permeability, so that they should be treated as a special class of complex drainage.

Free. In these soils there is moderate to good water retention but no persistent waterlogging of soil pore space. There is thus no mottling of the matrix colour with grey or rusty colours brought about by reducing conditions within the depth to the parent material, or to an arbitrary depth, often taken as about 75 cm. Of course, uniform grey colours can occur in particular horizons of some freely-drained soils, or be inherited from parent materials, so colours must be interpreted in relation to the soil parent material and profile type.

Imperfect. These soils have surface horizons dominated by the matrix colour found in freely drained soils on similar parent material, but show fine rusty mottles, often along root channels, in upper horizons, and this mottle can pass below to stronger grey mottled horizons where water saturation of pore space is more persistent. About 40 cm has been conventionally taken as the depth below which the stronger mottling resulting from prolonged water-logging may occur in 'imperfectly drained' soils.

Poor. Soils of poor drainage have dominantly grey colours, due to the transformation of generally brownish ferric iron oxides, characteristic of aerobic situations, to ferrous iron minerals as a result of reducing conditions caused by long seasonal periods of waterlogging up to or near the soil surface. In mineral soils which have not developed organic surface horizons, domi-nantly grey horizons with some yellow mottle are typically found below a thin dark grey horizon which has strong rusty mottling. Again parent material must be considered, as in some strongly red-coloured parent materials the grey colours develop only moderately, even in very poorly drained soils.

Very poor. This drainage class is used for soils which are virtually per-manently saturated with water to the surface. Such soils are often peaty throughout, or have surface peaty horizons overlying pale grey permanently waterlogged mineral soil.

The lower part of the soil description form covers individual horizons. It is based on Clarke (editions from 1936 to 1971) and USDA (1951). In *Guidelines for Soil Description* (FAO, 1968) it is considered that soils can be best described using terms 'which have received wide acceptance among soil surveyors'. The view that it is preferable at present for the non-specialist to use relatively simple conventional terminology rather than to seek greater precision through new terms or measures is followed here. Because of the wide range of alternative descriptive terms available, it is always desirable to indicate the terminology that is being followed, when qualitative soil descrip-tions are used in any publication.

Horizon depth, classification and boundaries

Horizon depths should be measured from the surface downwards (e.g. 0–5, 5–18, 18–24 cm). In the field when the soil is familiar, or following laboratory analyses when it is not, a genetic classification of horizons can be shown by symbols (Section 3.2.3). Sharpness of the boundary between horizons can be recorded as: sharp (change occurs over less than 2 cm); clear (change occurs between 2 and 5 cm); gradual (change occurs over more than 5 cm). Whether the boundary is level or undulating is also noted. In the case of soils in which no distinct horizons are recognizable within a thickness of > 50 cm, it can be desirable to sample convenient depth zones (e.g. 0–15, 15–50 cm) separately when soil chemical analyses are to be carried out.

Colour

Soil colour should be identified (either in the field, or later but using field-moist samples) by comparing fresh unrubbed soil with the small colour squares of standard colours given in 'the Munsell system' (Munsell Soil Color Charts, Munsell Color Co., Baltimore, Maryland, USA). Colour chips on these charts, bound in a book covering virtually all the possible soil colour range, are identified by 'hue', 'value' and 'chroma', so that '2·5 YR 5/4, reddish-brown', for example, is a complete 'Munsell colour' of hue 2·5 YR, value 5 and chroma 4, with the standard Munsell name for this combination. Without a colour book, descriptive subjective names must be used, but only the use of standard charts can convey consistent information, so far as individual differences in matching colours allow. Mottling of the matrix colour with other colours is recorded as: faint or strong; occasional (< 5% of surface area occupied); frequent (5–25% of surface area) or abundant; and as fine (< 5 mm diameter), medium (5–15 mm) and large (e.g. 'faint, occasional, fine mottle of 10 YR 5/6, yellowish-brown').

Field texture

This is an assessment, made by sight and feel, of the relative proportions in which primary particles of different size ranges occur in the soil. Laboratory analysis (Section 6.2.1) determines these proportions quantitatively, but field texture determinations by feel can be quite well related to textural classes based on laboratory data. Standard laboratory particle size analyses are carried out on the 'fine-earth' fraction of soils (i.e. on the fraction of < 2 mm 'equivalent spherical diameter' (see Section 6.2.1), after removal of 'stones' (> 2 mm)). Particle size classes are as follows.

Sand (2·0–0·05 mm): divisible into coarse sand (2·0–0·2 mm), which has individual particles that are very obvious to the eye and give a strongly gritty feel when rubbed between the fingers, and fine sand (0·2–0·05 mm), in which grains are still obvious, but which has a less gritty feel when rubbed.

Silt (0·05–0·002 mm): individual grains are not clearly felt but give a smooth and soapy feel with only slight stickiness when rubbed.

Clay (< 0·002 mm): when moist feels sticky and is easily rolled between the fingers into coherent mouldable threads (if dried, considerable moistening and working may be needed to restore plasticity).

The 0·05 mm limit is that of the widely used 'US scale' for the boundary between silt and 'fine sand'. The older 'International' limit, still sometimes used, is 0·02 mm, while currently the British Soil Surveys (Avery, 1973) use 0·06 mm, which is in accord with soil mechanics practice in civil engineering.

Boundaries between named textural classes are derived from a triangular diagram (e.g. USDA, 1951; Fig. 5.8) of which the corners represent 100% sand, silt, clay. Sectors with limits chosen to relate to observed field properties are given texture class names so that any soil can be allotted to its textural class by plotting laboratory particle size analyses on the diagram. In the field, however, apparent texture is certainly influenced by 'stone' content, especially of fine gravel slightly larger than the laboratory 'sand' size, and by organic matter, so that qualifying terms such as 'gravelly' or 'organic' are used with field texture class names.

The texture classes of the triangular diagram of Fig. 5.8, based on laboratory analysis, can be quite closely identified in the field, with experience, from the following general characteristics.

Sand: loose when dry, and not at all sticky when wet.

Loamy sand: has a small degree of cohesion and plasticity when wet.

Sandy loam: sand fraction feel is obvious but the soil moulds readily when sufficiently moist, without stickiness.

Loam: the soil moulds readily when moist but sticks slightly to fingers, though some grittiness from a sand fraction is still obvious.

Silty loam: has moderate plasticity but little stickiness, with the smooth soapy feel of silt being conspicuous.

Sandy clay loam: has sufficient clay to be quite sticky when moist, but the presence of a gritty sand fraction is still an obvious feature.

Clay loam: again, sticky when moist, with just a slight sand fraction still detectable by its gritty contribution.

Silty clay loam: less sticky than silty clay or clay loam, with elements of a slight sandy feel and a soapy feel due to the silt fraction.

Sandy clay: plastic and sticky when moistened, and the sand fraction is obvious, but there is no modifying soft silty influence.

Clay: very sticky when moist, and very hard when dry.

Silty clay: very low content of sand but the smooth silt contribution reduces the stickiness of the clay.

Silt: dominated entirely by smooth soapy feel of silt fraction.

Organic: high organic matter content, not fitting into any of the above classes.

Only experience in comparing field feel with known laboratory determinations can give confidence to quick assessments of more than broad sandy, loamy, silty or clayey categories. The terms light, medium, and heavy soils are often used agriculturally for soils of sandy, loam, and clay dominated texture categories. Differences between the USDA class limits on a triangular diagram of soil texture and those currently used in Soil Survey in Britain (Hodgson, 1974) are mentioned later (Section 6.2.1).

Stone quantity and type

The *quantity* of 'stones' (> 2·0 mm) can be approximately assessed in the field as none, occasional (< 5% by volume), frequent (5–25%) or abundant. It is quite straightforward to determine quantitatively the stone content of a soil by weight or volume, by passing the total soil through sieves in the field and the laboratory (6.2.1), but this is not generally considered necessary. Stone size can be given as small, medium or large (these classes alternatively can be called 'gravel', 'stones' and 'boulders' respectively) using size limits of approximately 1 and 10 cm mean diameter. Stone shape can be subjectively estimated as rounded, subangular or angular. Where the rock types of conspicuous stones are obvious, they should be noted in the description.

Soil structure

Soil structure refers to the natural aggregates ('peds') into which primary soil particles are combined. Many fine subdivisions have been proposed. Quantitatively the shape, size and strength of aggregates can be determined in the laboratory (Section 6.2.4) but the following field classification terms are normally adequate.

Single grain, structureless: no aggregation as, e.g. in a loose sand or gravel.

Crumb: roughly rounded, soft, porous aggregates, often subdivided by size into small (< 2 mm), medium (2–5 mm) and large (> 5 mm). This is the dominant structure of medium-textured freely-drained soils.

Granular: units of crumb shape but relatively hard, difficult to crush between the fingers. These are infrequent in most temperate zone soils but are the typical structure of calcareous soils of Rendzina type.

Blocky (or alternatively *Cloddy*): roughly equidimensional, rather flat-faced and angular aggregates (which may themselves be broken down into crumb structure units), graded as small (< 10 mm), medium (10–20 mm), or large (> 20 mm). This structure occurs in medium- and heavy-textured soils of imperfect drainage, and is also found in freely-drained soils with low levels of organic matter that have been compacted by stock, machinery or other surface impact.

Prismatic: angular, flat-faced units which have their vertical dimension much larger than their horizontal dimensions. This structure occurs in seasonally waterlogged horizons of heavy-textured soils, as a result of periods of wetting and expansion alternating with times of drying and contraction.

Platy: a structure of relative thin, horizontally-disposed laminar units.

This occurs in some alluvial soils, particularly silts, and is also present in soils which have inherited this structure from the effect of ice crystal formation, as in former areas of perennially frozen ground.

Massive, structureless: no individual structural units are obvious in a homogeneous coherent horizon.

Structure strength may be given as: (i) weak, describing a situation when structure units are just visible in the profile face, and easily destroyed by handling; (ii) moderate, when the units are not markedly distinct in the undisturbed soil but remain intact when lightly handled; (iii) strong, when the structural units are obvious in undisturbed soil, and remain intact when the soil is handled, unless appreciable pressure is used.

Porosity

Porosity again can and should be determined experimentally, but as a field observation it is an assessment of the size, quantity and distribution of cavities within the soil mass. It is a quality related to texture and structure and provides an estimate of ease of water movement through the soil. Classes used in field assessments are: poor (very dense fabric with few medium or larger interconnecting pores or fissures); moderate (some interconnecting medium and large pores but a relatively dense fabric); good (fabric of the soil has a conspicuous interconnecting series of medium and large pores or occasionally an extensive permanent fissure system); excessive (very open fabric with abundant large interconnecting pores).

Handling consistency

This is another secondary quality, identified subjectively by feel, that is closely allied to texture but also modified by organic matter content, stoniness and structure. Classes used are:

Loose: single-grain structure with no impression of any binding or sticky character.

Fluffy: soft and approaching 'loose', but the material is aggregated, often being of small crumb structure in soils of high organic matter.

Friable: breaks easily into crumbs or other small aggregates but does not stick to the hand.

Sticky: adheres to hands and implements.

Compact: massive and firm without stickiness or ready fracture into small aggregates.

Indurated: very hard to break with hands and to dig initially, although after digging the material may break apart more rapidly.

Organic matter estimate

A broad indication of the amount and type of organic matter in each horizon can be gained in the field as very high, high, moderate or low, largely on a basis of soil colour, since darker-coloured soils have a higher humus content. In relation to laboratory analyses, divisions between these ranges can be placed at about 30, 15 and 5 weight per cent of organic matter.

The type of organic matter in surface horizons can be identified as:

Mor: organic matter with little or no mineral admixture. Subdivisions of mor that have frequently been used (a variant of these as supplementary horizon symbols is given in Table 5.1) are the terms L layer (freshly fallen litter), F layer (partially decomposed litter, but with its origin still recognizable), H (completely decomposed, humified organic matter).

Moder: an intergrade between mor and mull, dominated by relatively high quantities of decomposed organic matter but containing a substantial mineral admixture, either in very small crumb units which often show a gradation from darker to lighter zones passing from surface to core of the unit ('mull-like moder'), or as bleached sand grains.

Mull: dominantly mineral material in which the organic matter is completely altered from its original fabric and is intimately and fully incorporated with the mineral soil, only a darker colour distinguishing it from horizons of lower organic matter.

For peaty materials there is a qualitative subdivision based on handling consistency and degree of humification.

Fibrous: composed of recognizable plant residues, which retain their individual character when handled.

Pseudofibrous: plant remains are recognizable initially, but they are partially decomposed and do not retain their character when handled.

Amorphous: recognizable plant remains are absent.

Root distribution

Detailed root counts by number or weight are not often made by soil scientists. Usually only an estimate of abundance is given, without any exact quantitative limits in mind, as: rare, few, frequent, or abundant. Root type, whether fibrous, fleshy or woody, and any relationship of distribution to profile features, is also noted. Chapter 1 shows up the weakness of this limited approach and makes it clear that the soil scientist should aim for improvements in this aspect of soil data recording.

Field moisture

The wetness state of the soil when it is being described, which may influence colour determination or structural assessment, is usually noted simply as dry, moist, or wet.

Earthworms

The lack of ecological expertise of the average pedologically-oriented soil scientist concerned with field studies is again shown up by conventional soil descriptions, in which soil faunal observations are generally limited to the presence or absence of earthworms as a group. Usually again only an approximate estimate of the numbers of earthworms is given as few, frequent, or abundant. In horizons where earthworm channels are conspicuous features, although the animals themselves are not seen, the presence of the channels is recorded.

Horizon number and notes

In this column soil horizons are conveniently numbered sequentially (e.g. ME 78/1, 2, 3) for subsequent easy labelling of samples. Other characteristics not covered on the standard form can also be recorded in this space.

3.2.3 Horizon samples

Soils, in the pedological approach favoured here, are described as profiles consisting of sequences of horizons. To assist the interpretation of soil descriptions through profile comparisons and to support soil classification, individual horizons can be indexed by symbols. These symbols carry genetic connotations which may be recognized in the field or may be determined or confirmed by subsequent laboratory analysis. Soils with similar horizon sequences can be grouped together (Section 4.2), regardless of differences in horizon thickness, morphological detail and parent material. A recommended set of horizon symbols is given in Table 5.1, based on Kubiena (1953), the International Society of Soil Science (1967), and past British Soil Survey practice. An initial division is into six groups identified by capital letters, which are then further defined by the use of prefix numerals and suffix letters.

Current practice in the Soil Survey of England and Wales (Hodgson, 1974) uses a more detailed set of symbols, differing in part from the system given here. A more complex terminology is the basis for the United States classification system, *Soil Taxonomy* (USDA, 1975).

Table 5.1 A system of soil horizon symbols

Master horizon symbols

O   Horizon which, although it may contain some mineral admixture, is dominated by the organic fraction (loss-on-ignition values > 30%).

A   Horizon at or near the soil surface, consisting of an intimate mixture of organic and mineral material, that fails to fulfil the definition of the O master horizon (loss-on-ignition values < 30%).

E   Horizon from which sesquioxides (Fe and Al) and/or clay particles have been reduced or removed, (eluvial horizon), occurring below O or A horizons. (In earlier conventions, these were often termed A_2 horizons.)

B   Horizon of mineral material, modified by physical, chemical and biological alteration or deposition so that it is differentiated by, e.g. structure, colour or texture from horizons above or below.

C   Mineral material which has been little altered by pedological processes.

R   Unaltered rock, which, even when moist, is too hard to be dug with a spade.

Intergrades between master horizons may be indicated as A/B, B/C, etc. Only suffixes given below as applicable to all horizons would be applied to such intergrade horizons.

Prefixes and suffixes

Arabic number prefixes (2-, 3-, etc.) are used to indicate buried soil profiles or horizons, the superimposition of horizons formed in two or more episodes of soil formation. By convention, 1 is omitted so that, e.g. a sand-dune section might have horizons A, C, 2A, 2C, 3A, 3C. This would indicate a superimposed sequence of three soils of AC type, developed at different periods of time.

Roman number prefixes (II-, III-, etc.) are used to indicate original geological (rather than pedological) discontinuities, within a soil profile which has developed during a single formative stage. For example, wind-blown sand overlying boulder-clay would give a soil with horizon sequence of A, E, B, IIC.

Supplementary horizon symbols

Suffixes applicable to any master horizon

g   Horizon showing evidence in structure (e.g. cloddy) and colour (e.g. grey-brown with rusty-brown fine mottle) of the influence of seasonal periods of waterlogging.

G   Horizon dominated by structural (e.g. prismatic structure) and colour (e.g. grey with large yellow-brown or orange-brown mottle) effects resulting from long-term waterlogging.

c   Horizon containing residual calcium carbonate.

k   Horizon containing deposits of secondary calcium carbonate.

n   Horizon containing excess of sodium in the exchangeable cations, or free sodium chloride.

x   Horizon having a massive consistency due to induration.

Suffixes applicable to O horizons

o   Horizon having a loss-on-ignition value > 60%.

l   Horizon of little altered plant remains ('fibrous').

f   Horizon of partially broken down and decomposed plant remains which are still recognizable to the naked eye ('pseudofibrous').

h   Horizon of decomposed humified plant remains with no recognizable original structure ('amorphous').

p   Ploughed or otherwise cultivated horizon.

Table 5.1 (*Continued*)

Suffixes applicable to A horizons

h   Horizon visibly darkened by having a high content of organic matter, but not fulfilling the requirement of the O master horizon.
he   As for h, but also including bleached sand grains or rock fragments.
p   Defined as applicable to O horizons.
an   Surface horizon artificially deepened or modified by the addition of material by man.

Suffixes applicable to E horizons

a   Horizon from which sesquioxides have been translocated down the profile.
b   Horizon from which clay has been translocated down the profile.

Suffixes applicable to illuvial B horizons resulting from deposition of translocated material from above

h   Horizon with level of humic organic matter which is high compared to horizons above and below.
s   Horizon with levels of sesquioxides (of Fe and/or Al) which are high compared to horizons above and below.
t   Horizon with level of clay which is high compared to horizons above and below.
fe   Iron pan (hard, thin Fe-oxide rich band).

Suffixes applicable to C horizons

r   Horizon predominantly composed of shattered or weathered material derived from underlying solid rock (R horizon).

3.3   Soil sampling

For reconnaisance observations a screw auger, the standard tool of the soil surveyor concerned with making a soil map, can provide an initial picture of the soil at and around an area of interest, without the degree of disturbance and potential unpopularity in many situations that can be caused by digging a rash of small pits. Such screw augers are typically locally constructed using a wood-boring large diameter (c. 2·5 cm) bit with a screw thread c. 20 cm in length, welded to a steel rod of c. 1·2 cm diameter fitted with a T handle to make a tool of total length c. 1 m. In use the auger is screwed into the soil while applying only moderate pressure, and is then withdrawn by means of a steady pull at 15–20 cm depth intervals to allow material retained on the screw thread to be examined. Its use in very loose or stony soils is difficult and such horizons are likely to be obscured as material falls from the screw when the auger is withdrawn, but it is generally rapid, usually effective, and is relatively non-destructive compared to a soil pit. In organic soils the thickness of peat or of a peaty surface horizon can be determined quickly by the greater ease with which an auger can be pushed directly down, rather than screwed, through peat compared to its passage through mineral soil.

Having gained an idea of the range of soil character in an area of interest by auger survey, sites for detailed examination using profile pits can be chosen subjectively, or in accord with objective sampling constraints. The excavation of a soil profile pit was described in Section 3.1. Horizon sampling from such a pit should be carried out by cutting, with a straight-sided planting trowel or sheath-knife, a slice of constant cross-sectional area from the full depth of the horizon, rather than by sampling centrally in each horizon as is sometimes recommended. Working upwards from the lowest horizon is generally quickest, as otherwise it is necessary to re-clean the face of the pit after each horizon has been sampled. The collection of soil samples in more-or-less undisturbed field condition, as required for some analyses, is considered later.

Block-bottom bags made of bitumenized brown paper were widely used to carry and store samples, being strong, easier to label and more resistant to cutting by stones than polythene bags. However, where such paper bags are still used, care must be taken (see Chapter 6, Section 2.1) if a lengthy period is likely before the soils reach the laboratory, because of the interaction between soil fauna and flora and the paper in the bag. Polythene bags are now the usual method of carrying samples. The typical range of routine analyses can be carried out on quite small quantities of soil; on < 50 g air-dry 'fine-earth' (< 2 mm) for chemical analyses alone, or on 150–200 g if particle-size analysis is also required. However, unless transport is difficult, considerably larger samples are normally taken as standard, often giving an effective sample volume up to c. 1500 ml and a weight of 1–2 kg before drying and sieving treatment. The larger sample has some 'smoothing' effect in reducing the influence of soil variability, and gives a reserve of material for additional studies. Large and medium size stones can be hand-picked in the field to reduce weight. If quantitative data on stone content are required, then large samples are certainly needed, and therefore these determinations are preferably made in the field (Section 6.3.1).

A straight-sided graduated planting trowel is useful also for obtaining near-surface samples if these are all that are wanted, being well adapted to extracting samples of uniform cross-section and constant depth. However, sampling of surface zones, if the soil is not particularly stony, can be most easily carried out using tube corners. A simple version of these can be made from short lengths of heavy gauge brass tube (c. 2·5–5 cm diameter), with one end chamfered for easier penetration, and the other having holes drilled opposite each other to enable a short removeable rod to be used to give leverage to the corer as it is pressed into, and then withdrawn from, the soil.

The outer surfaces of the corer tubes are permanently marked at lengths required to control the depth or volume of sampling (e.g. 5, 10, or 15 cm depth and/or 100, 200 ml soil volume). The rod 'handle' can conveniently have a short plunger fixed at one end, of diameter slightly less than the tube bore, for use in pushing the cores from the tube plunger into sample bags. Tube corers are particularly useful for collecting subsamples of surface soil for compositing into large samples for analysis; for constant-volume sampling; and, to a limited extent, for obtaining undisturbed samples for certain physical determinations, though most soils cannot be extracted from simple one-piece tube corers without compression. Where constant-volume samples of individual horizons are required, then a tube corer can be used horizontally to sample from horizons exposed in a profile pit.

Complex tube corers with an inner split container can be used to provide samples in relatively undisturbed field conditions (e.g. Blake, in Black, 1965). A hand-operated corer capable of collecting soil cores to 2 m has been described by Foale & Upchurch (1982). As an alternative, samples which have had minimum disturbance, such as are required for determination of bulk density (Section 6.3.2), or for preparation of thin sections (Section 6.4.1), are best collected from the face of a profile pit by inserting a sampling tin into either a vertical horizon face or a horizontally cut-back horizon surface. Sampling tins which have removable lids on two opposing sides can be made from galvanized steel or aluminium sheet. A standard size is about 8×5 cm cross-section $\times$ 5 cm deep. With both lids removed, the four-sided frame of the tin is, in the simplest case, pushed (or in less amenable soils, progressively gently hammered in and cut around) into a required location in the soil face, until the outer edge is flush with the soil surface. One lid is placed over this exposed soil in the tin, and the tin with its contained soil is carefully removed from the profile face by cutting around its sides and behind it with a knife, then levering and easing it away, with its contained soil, from the sampling face. The orientation of the sample (upper and lower sides), as well as its location, is marked on the tin (on the frame rather than on the lids). After trimming the exposed soil in the open side of the tin flush with the edge of the frame, the second lid can be pushed on to enclose the sampled soil, and both lids secured with elastic bands. Such sampling is not always straightforward in practice, especially in stony soils, in which there is no easy solution to the problem of inserting the frame of the tin. For sampling indurated soils, a large soil fragment removed with hammer and chisel may be packed securely in a container, again making sure that its orientation has been marked. Larger tins of similar type can be used to collect profile sections for display, although practical sampling problems are correspondingly greater. From such sections, more permanent display monoliths can be produced by impregnation of the soil in the tin with a suitable resin (e.g. Wright, 1971).

An aspect of soil sampling related to analysis presentation and nutrient budget calculations (see Chapter 1, Section 4) is the collection of material so that analyses can be given on a volume basis, either directly as analysed quantity/ml soil or, for example, converted to kg/ha of soil to a standard depth (e.g. Mehlich, 1972, 1973). Most soil analyses are still made on a weight basis on air-dry soil which has passed a 2 mm sieve but the weight (field and air dry) of measured soil volumes can give weight:volume conversions for each sample.

Vegetation sampling for analysis is treated in Chapters 1 and 6, and many general considerations mentioned there also apply to sampling soils. Although soil data should ideally be obtained by sampling sufficient profiles on a horizon basis to ensure comprehensive site coverage, this is not in general economically or practically feasible. Soil chemical variability has been found to be substantial even if samples are taken as close as 10 cm apart within one plant community on a soil which is uniform at series level (e.g. Ball & Williams, 1968). This small-scale spatial variability includes the greater part of the total variability present over tens or hundreds of metres within a single soil-vegetation association. Significant variability has also been shown in physical properties, such as bulk density and moisture parameters, within one area of a single soil type (Nielsen *et al.*, 1972). Because of this, intensive sampling may be required to obtain results with a desired precision. General considerations of soil chemical variability have been discussed by Peterson & Calvin (in Black, 1965); Ball & Williams (1971) and Beckett & Webster (1971). The importance of appreciating spatial variability in determining and interpreting soil chemical data, although accepted in principle, has been widely ignored in practice by soil scientists and ecologists alike.

Correlation of soil chemistry with plant species or community distribution can often be sought through analyses of standard depth samples from surface zones of soils. These can be collected by short coring tools, and it is suggested that, in many field plant–ecological studies, soil profile data should be used as a foundation for an interpretation of surface soil chemistry determined from a more intensive sampling programme. The standard depths employed may be 0–5, 0–10 or 0–15 cm, the last being generally most useful.

Type profile analyses give the general chemical composition of a soil and allow internal comparisons to be made between horizons within the profile. Analysis of a composite sample, on a depth or horizon basis, increases the precision with which the mean of a soil chemical quantity for an area or horizon has been determined. However, only if a number of samples are analysed independently, rather than as a single bulked sample, can the range and confidence limits of the quantity about this mean be given for a site. Clearly, an intensive ecological investigation of one site might justify a large-scale soil sampling programme to determine chemical factors precisely but, if many sites are involved, a compromise between effort and precision is essential.

Ball & Williams (1971) compared the mean chemical quantities obtained from several sampling programmes for an uncultivated, unfertilized brown earth against the 'best' values for the site as determined from 275 samples covering a single block of contiguous plots, each sample being a composite of ten subsamples of the 0–15 cm soil depth. The least laborious sampling programme involved determining the mean and range of soil chemical properties from composited samples on six plots only. The difference in effort involved was in a ratio of 45:1 between the time required to carry out the most comprehensive sampling and analysis programme and the six-plot sampling, though the latter scheme covered a range about the mean values of most analysed quantities of the order of 50% of the total site variation. Arising from this study, a sampling programme in which ten subsamples are taken and combined in each of six plots of $6 \, \text{m}^2$, set out about centres 20 m apart on a transect 100 m in length, has subsequently been employed on sites including sand-dunes (Ball & Williams, 1974) and chalk grassland (Wells et al., 1976). There is nothing sacrosanct about the dimensions or arrangement of this sampling design but something similar, adapted to individual sites and problems, can achieve a balance between sampling and analysis effort and data precision (see e.g. Falck, 1973). Reynolds (1975) concluded that analyses of ten individual samples gives a reasonable assessment of the average chemistry of a soil.

4   Soil classification

4.1   General considerations

Theoretical and practical considerations of soil classification have been widely discussed (Kubiena, 1953; USDA, 1938, 1975; Muir, 1962; Smith, 1965; Webster, 1968; Fitzpatrick, 1971, 1980; Buol et al., 1980). No attempt is made here to cover the historical development of soil classifications or review comprehensively the range of available options. Many of the problems facing the soil scientist choosing a classification scheme are shared by ecologists concerned with vegetation classification (see Chapter 9) since an attempt is being made to impose relatively simple, definable classes on a natural body with very high three-dimensional variabilities.

It is assumed here, however, that a soil classification is of practical value to ecologists and that the most generally applicable schemes are based on profile morphology, which results from the interaction of a wide range of chemical, physical and biological parameters. As noted in the introduction to this chapter, non-pedological soil classifications may well be in use in situations in which an ecologist is involved, for example when collaborating with civil engineers. An example of a soil mechanics approach to classification is the British system for engineering purposes introduced in a British Standard code of practice for site investigations (Dumbleton, 1981). For some problems, prior knowledge may suggest that it is acceptable to classify soils simply in terms of a single parameter, such as surface horizon texture or organic matter content, but this is generally too limiting to use other than as a supplement to a wider grouping.

Arguments have been put cogently by Fitzpatrick (1971, 1980) against hierarchical systems which accept explicit or implicit correlations between soil classes and pedogenetic trends related to environment. These arguments emphasize the inability of such systems to cover effectively a medium which is a continuum, rather than being evolutionary; the necessity they impose of forcing intergrade soils into a rigid class structure; and the inadequacy of a class name to communicate complete information about the soil. In spite of the validity, to some extent, of these objections, there is sufficient justification for, and value in, local or regional 'natural' genetically-based, hierarchical systems, to recommend their use.

An alternative or supplementary approach to soil classification is to use computer analysis of recorded soil properties to seek the statistically soundest division of a population of soils (e.g. Norris, 1971; Norris & Loveday, 1971; Moore *et al.*, 1972; Webster, 1975). Although many of the raw data employed may remain subjective in their selection or assessment, these methods have a greater apparent objectivity in data handling. However, they have not yet generally achieved soil classifications which are sufficiently clearer or more readily applicable and interpretable to justify replacing those given by morphological and genetic approaches.

4.2   Soil classification systems

4.2.1   International soil classifications

The USDA system, under the title *Soil Taxonomy* (USDA, 1975), which is an extended hierarchical system with its own quantitatively-based horizon terminology, is the most detailed scheme applicable to world soils, from a basis of the very wide range of soils in the USA. It has not yet shown sufficient superiority over existing national and regional classifications to supplant them and it is complex to use, but soil types referred to by national class names are often also identified in published work by their *Soil Taxonomy* class. Practical applications in Tanzania and Sudan have been discussed by Mitchell (1973). A clear presentation of the ten 'soil orders' of this system and their subdivisions is given by Buol *et al.* (1980). Other summaries of the scheme can be found in Cruikshank (1972), Brady (1974) and Bridges (1979). Application of *Soil Taxonomy* classes to British soils has been considered by Ragg & Clayden (1973).

An alternative world soil classification is that of the FAO (1974) described clearly in Fitzpatrick (1980). This is in use for the completed 1 : 5000000 series of soil maps of the world and is particularly widely used in resource surveys. An increasing convention in publications of potential international interest is to identify soil types by their FAO as well as *Soil Taxonomy* and local names. Duchaufour (1976) illustrates a range of world soil types and correlates their classification in the French system with these two international alternatives. Gerasimov (1980) relates the FAO system to the soil classification in use in the USSR, while Clayden, in Bridges & Davidson (1982) compares FAO, USDA and national schemes. Fitzpatrick (1971) devised a classification based on horizon characteristics and thicknesses which did not place reliance on genetic relationships. His original terminology introduced to carry the scheme has not achieved sufficient acceptance for its possible advantages to be explored adequately.

Broad studies can use simple systems, such as that described by Ball in Peterken (1967) (after Aubert & Duchaufour, 1956; Aubert, 1965; Duchaufour, 1956; see also Duchaufour, 1970). This system has been applied by ecologists to classify soils in a world assessment of areas of conservation importance. Comment by users has led to the version given in Table 5.2a being slightly amended from that previously published. Six soil 'categories' and a total of fourteen 'subcategories' based on horizon sequences are included, and a key (Table 5.2b) enables selection of the appropriate subcategory from limited chemical and morphological data. For many purposes such systems are inevitably over-simplified. They cannot substitute for effective national or regional classifications but, used with discretion, they provide a simple broad framework in which to set a more detailed classification.

4.2.2 National soil classifications

A regional or national soil classification system is essential in ecological studies that need to identify in any detail the soil classes present at study sites. Among regional systems, Kubiena (1953) in a classic work, gave a system widely applicable throughout Europe, although it is not now in standard use in this form in any European country. Classifications of African soils were discussed by D'Hoore in Moss (1968) and by Ahn (1970), while tropical soils have been more widely considered by Young (1976).

In most countries, national systems are periodically modified, current or recent examples being for Britain (Avery, 1973, 1980); Germany

Table 5.2 (a) Outline world soil classification

Soil category	Subcategory	Soil categories Main profile type(s) (horizon sequences)	Examples of soil subgroups in each category	Subcategory symbol
Saline soils (S)	—	Ang, C An, Bn, C	Solonchak Solonetz	S_1 S_2
High sesquioxide (ferritic) soils (Fe)	—	A, B, C	Terra rossa, rotlehm, laterite, ferritic brown earth	Fe
Organic soils (O)	—	O	Organic soil, peat, fen, bog	O
Well-drained non-saline, non-ferritic soils with good profile development (F)	With calcareous surface or subsurface horizon	Ac, C	Rendzina, chernozem, chestnut soil (kastanozem)	F_1
		A, Bc, C	Brown calcareous soil, terra fusca	F_2
	Non-calcareous throughout profile	A, Cr	Brown ranker, skeletal brown earth	F_3
		A, B, C; A, Bg, C	Brown earth (braunerde), sol brun acide, brown earth with gleying	F_4
		A, Bs, C; O, Ea, Bh and/or Bs, C; Eb, Bt, C	Podzol, peaty podzol, brown podzolic soil (sol brun podzolique), sol lessivé	F_5
Poorly-drained non-saline, non-ferritic soils with good profile development (P)	With calcareous surface or subsurface horizon	Ag, BGe, C; O, BG, C	Calcareous gley, fen marl, peaty calcareous gley	P_1
	Non-calcareous throughout profile	Ag, BG, C; O, BG, C	Non-calcareous gley, peaty gley	P_2
Soils with weak profile development (I)	Lack of profile development controlled by climatic factors	A/C, C	Serosem, burosem, desert soil	I_1
	Lack of profile development controlled by limitations of time available for profile development	A/C, C	Unconsolidated parent materials; e.g. recent alluvium, raw warp soil, grey warp soil, regosol	I_2
		A/C, R; O/C, R	Massive rock parent materials; e.g. lithosols, raw mineral soils on rock surfaces and coarse boulder accumulations	I_3

Table 5.2 Continued.

(b) Key to soil categories. This should assist identification of the category to which a soil could be allocated. (Where a soil cannot readily be placed in a single category, two or more must be given; e.g., a shallow AC profile on rock in a mountain area might be doubtfully placed between F_1 or F_3 and I_3.)

1	Soil containing a high concentration of alkaline salts.	2
	Soil without high concentration of alkaline salts.	3
2	Saline gley soil with water-soluble salts in upper horizons.	S_1
	Saline soil with water-soluble salts in lower horizon and high exchangeable sodium in the surface.	S_2
3	Soil with high concentration of iron oxides.	Fe
	Soil with normal concentration of iron oxides.	4
4	Soil with dominantly organic surface horizon at least 60 cm deep. If total soil depth less than 60 cm, then surface organic horizon directly succeeded by unaltered rock.	O
	Soil without dominantly organic surface horizon or with organic surface horizon succeeded by mineral soil horizon at less than 60 cm depth.	5
5	Well drained (i.e. no evidence of strongly impeded drainage above 40 cm depth).	6
	Poorly drained (i.e. evidence in mottled colours or other features, of strongly impeded drainage nearer the surface than 40 cm).	11
6	Immature profile, that is with weakly developed and shallow soil formation, possibly with little biological activity.	7
	Well developed horizon sequence with moderate to strong biological activity.	8
7	Immaturity resulting from climatic factors, e.g. very low rainfall and/or temperature.	I_1
	Immaturity resulting from lack of time for soil formation to proceed:	
	(a) on fine-textured material such as recent alluvium, dune-sands and eroded surfaces of fine-textured drift.	I_2
	(b) on very coarse-textured materials such as boulder accumulations, or on massive rock.	I_3
8	Calcareous in one or more soil horizons.	9
	Non-calcareous throughout profile.	10
9	Shallow or simple profiles of A horizons overlying parent material.	F_1
	A, B, C profiles.	F_2
10	Shallow or simple profiles of A horizons overlying parent material.	F_3
	A, B, C profiles.	F_4
	Profiles including Bh, Bs or Bt horizons, showing accumulation of translocated humus, iron and aluminium sequioxides, or clay.	F_5
11	Calcareous in one or more soil horizons.	P_1
	Non-calcareous throughout profile.	P_2

(Mückenhausen, 1962, 1965); France (Duchaufour, 1970, 1977, 1982); Australia (CSIRO, 1983)); Canada (Canada Soil Survey Committee, 1978; Strong & Limbird, 1981); New Zealand (Taylor & Pohlen, 1968); and for the USA prior to *Soil Taxonomy* of the USDA (1975) (Baldwin *et al.* (in USDA, 1938, pp. 979–1001); Thorpe & Smith, 1949). Butler (1980) helpfully summarizes and comments on a number of older and newer national systems. A wide review of many national soil classification systems then in use is contained in the section on 'Soil Classification and Soil Fertility', (pp. 278–551) in International Society of Soil Science (1962). Where the ecologist may be concerned with information from older literature, the fluidity, in concept and detail, of some national systems of soil classification must be borne in mind.

Classification systems and terminology have been even more dynamic in their changes of direction and emphasis than soil development is itself. Added difficulties in interpreting existing soil information have followed. In Britain, for example, soil class terminology has varied in detail in local soil memoirs according to their date of publication, while soil class and series names differ in Scottish soil memoirs from those of England and Wales. The variety of available options makes it impossible to set out the range of alternative classifications here. As an example of a level of detail thought generally useful to ecologists, the classification of British soils set out in Table 5.3 is based on the approach of Kubiena and on past conventional British practice. Seven major soil groups are divided into thirty-two subgroups named from what has become fairly widely used terminology. The definitions of the classification categories employed here and subsequently are:

Major soil group: a class distinguished by the presence, or absence, of specified master horizons.

Soil subgroup: a class within a major Soil Group, defined by a particular horizon sequence.

Soil series: a class within a subgroup, distinguished by being formed from a specific soil parent material.

Soil type: a class within a Soil series, defined as a textural variant, e.g. as sandy loam or loamy sand types of a single series.

Soil phase: a class within a Soil series defined by a correlation with a site or soil factor insufficient to cause a change at a higher classification level, but significant to soil performance, e.g. a 'steep phase' or 'stony phase' of a soil series.

Figure 5.6 illustrates schematically the main horizon sequences of soils in these seven major soil groups.

The latest classification of soils in England and Wales arising from Soil Survey experience (Avery, 1973, 1980) has 10 'major soil groups', 43 soil 'groups' and 126 soil 'subgroups'. The definition of any subgroup must

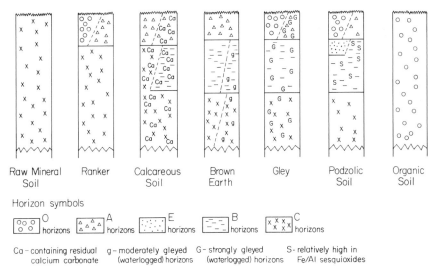

Horizon symbols

Ca – containing residual   g – moderately gleyed   G – strongly gleyed   S – relatively high in
        calcium carbonate   (waterlogged) horizons   (waterlogged) horizons   Fe/Al sesquioxides

Figure 5.6 Schematic representation of main horizon sequences in major soil groups in Britain.

clearly be more precise when 126 classes are involved compared to the 32 of Table 5.3, but more classes create greater problems of decision at class boundaries. A balance must be assessed by the ecologist, in relation to each study, between the effort to be expended in soil data collection and interpretation, and the degree of soil definition required for the communication of adequate information. The highest categories of the system of Table 5.3 and that of Avery are equated in Table 5.4.

To summarize, the ecologist should identify soils by a national or regional system of reasonable but not excessive subdivision using, if possible, generally understood terminology. Equivalence of class names in different languages (for example: brown earth ≡ sol brun acide ≡ braunerde; brown podzolic soil ≡ sol brun podzolique) can be arguable in detail but such names generally have sufficiently uniform meanings for the national systems to be used with advantage.

4.2.3   Local soil classifications

Soil surveys at scales of 1 : 50 000 or larger are mostly made at soil series level, the soil series conventionally being divisions of a soil subgroup formed from a specific parent material, though current thinking is to move away from parent material geology as a primary defining criterion for soil classes. Series are usually named from geographic locations where they were first described and defined. At smaller scales, complexes of associations of different series are often used as mapping units. When official surveys are

Table 5.3 A classification of British soils

Major soil groups

(1)	Raw mineral soil	Soils lacking continuous lateral development of any master horizon other than C.
(2)	Ranker	Non-calcareous soils with O or A and C horizons but lacking B horizons. Some variants have incipient E horizons.
(3)	Calcareous soil	Soils with Ac and Cc horizons or with A, Bc and Cc horizons.
(4)	Brown earth	Mineral soils, generally of A, B (other than Bh or Bs) and C horizons, non-calcareous in the B horizon, but including variants with an Eb horizon.
(5)	Gley	Soils which have horizons of G type within 40 cm of the surface.
(6)	Podzolic soil	Soils with Bs, and/or Bh horizons, which generally, but not always, underlie an E horizon.
(7)	Organic soil	Soils with no master horizons other than O present at a depth of less than 60 cm.

Subgroups

The horizon sequences shown are intended as examples, not to cover all variations possible within a subgroup.

Soil subgroups		*Horizons in profiles*
1	RAW MINERAL SOIL	
1.1	Non-gleyed raw mineral soil	C
1.2	Gleyed raw mineral soil	CG
2	RANKER	
2.1	Peaty ranker	O, C or Cr
2.2	Humic ranker	Ah, C or Cr
2.3	Brown ranker	A, C or Cr
2.4	Podzol ranker	A, Ea, C or Cr; O, Ea, C or Cr
2.5	Gley ranker	A or Ag, Cg
3	CALCAREOUS SOIL	
3.1	Humic rendzina	Ahc, Cc, or Ccr
3.2	Rendzina	Ac, Cc or Ccr
3.3	Humic brown calcareous soil	Ah, Bc, Cc or Ccr
3.4	Brown calcareous soil	A, Bc, Cc or Ccr
3.5	Gleyed brown calcareous soil	A, Bcg, Ccg or Ccr
4	BROWN EARTH	
4.1	Eutrophic brown earth	A, B, C or Cr
4.2	Acid brown earth	A, B, C or Cr
4.3	Humic brown earth	Ah, B, C or Cr
4.4	Gleyed brown earth	A, Bg, Cg, or Cr
4.5	Leached brown earth (sol lessivé)	A, Eb, Bt, C or Cr
4.6	Gleyed leached brown earth	A, Ebg, Btg, Cg or Cr

The distinction between 'eutrophic' and 'acid' brown earth is an arbitrary one based on pH levels of $>$ or $< 6\cdot0$ in the A horizon. It could also be applied to separate gleyed brown earths into two subclasses.

5	GLEY	
5.1	Non-calcareous gley	Ag, BG, CG or Cr
5.2	Calcareous gley	As 5·1 but with BGc and/or CGc

Table 5.3 (*Continued*)

5.3	Humic gley	Variants of 5·1 and 5·2 with Ah horizons
5.4	Peaty gley	Variants of 5·1 and 5·2 with O or Oo surface horizons
5.5	Podzolic gley	Ah, Eag, Bsg, CG

A distinction can be drawn between gleys resulting from a high water-table (ground-water gleys) and those due to impedance of surface water drainage through the profile (surface-water gleys). Many gleys have characteristics partly attributable to both causes so that general use of this distinction is difficult.

6	PODZOLIC SOIL	
6.1	Brown podzolic soil	A or Ah, Bs, C or Cr
6.2	Iron podzol	O or A, Ea, Bs, C or Cr
6.3	Humus podzol	O or A, Ea, Bh, C
6.4	Humus-iron podzol	O or A, Ea, Bh, Bs, C
6.5	Peaty gley podzol	O, Eag, Bfe, Bs, C or Cg
7	ORGANIC SOIL	
7.1	Eutrophic peaty soil	O
7.2	Acid peaty soil	O
7.3	Eutropic peat	Oo
7.4	Acid peat	Oo

Within the organic soil group, the presence of IIC, or IIR horizons between 60 and 100 cm should be noted, as giving divisions important to some forms of land-use or potential land-use. The distinction between 'eutrophic' and 'acid' soils is again an arbitrary one based, as above, on pH values of > or < 6·0.

Table 5.4 Correlation of major soil groups in classifications of British soils

Major soil group (Avery, 1980)	Major soil group (Table 5.3)
Terrestrial raw soils	Raw mineral soil
Raw gley soils	Raw mineral soil
Lithomorphic soils (non-calcareous)	Ranker
Lithomorphic soils (calcareous)	Calcareous soil
Pelosols (non-calcareous)	Brown earth
Pelosols (calcareous)	Calcareous soil
Brown soils (non-calcareous)	Brown earth
Brown soils (calcareous)	Calcareous soil
Podzolic soils	Podzolic soil
Surface-water gley soils	Gley
Ground-water gley soils	Gley
Man-made soils	—
Peat soils	Organic soil

available for a study area or for a climatically and geologically comparable district, the series names provide a compact means of conveying information. For example, the Denbigh Series, defined from North Wales and since mapped more widely, is, in the classification given here, an acid brown earth on drift derived from hard argillaceous sedimentary rock (shale) of Silurian age. Initial use of both a full description and, when available, the series name, is recommended in ecological reports. The series name allows more concise repetition of soil identification, and also extrapolation of any interpretation to other surveyed areas; the former caters for those without ready access to the original definitive publication specifying the series name. Within a soil group, application-oriented classifications may sometimes be preferred, as Toleman (1973) considered for a subdivision of organic soils in respect of forestry use in Britain.

5   Soil and terrain maps

5.1   Soil maps

The ecologist is unlikely to wish to carry out soil surveys aimed at producing regional or local soil maps, but will frequently turn for information to published maps and their accompanying reports. Most published maps employ mapping units equivalent to the soil series. The distinction between the soil series as a classification unit and as a mapping unit must be appreciated. Classification units are conventionally described as a permitted range of variation about selected 'modal' profiles which have been chosen to emphasize the distinguishing horizon criteria of the series in their most clear-cut form (in many cases however, only the modal profiles, without quantitative data for permitted variability, are given in legends or accompanying memoirs). In the real world of the field, mapping units named from the central concept of a series include this range of variation, defined or unspecified, so that the area they cover contains intergrades to other series, in addition to inliers of different soil series of an extent and variety dependent on the local soil pattern and the scale of the soil map. Thus, a map unit may typically contain at a series of sampled points only some 65% or so of soils reasonably characteristic of the central concept of the series (Courtney, 1973). Even at mapping scales of 1:10 000, the degree of profile variation from the ideal classification unit of the soil series, and the rapidity of change, even at soil subgroup level, over distances of metres, can make soil mapping units into complexes of several series rather than being map units dominated by a single series (e.g. Ball *et al.*, 1969). Because of the difficulties of transforming even single-series units into quantitative assessments of particular soil properties it is frequently claimed (e.g. Trudgill & Briggs, 1981) that what is needed in place of the conventional series-based maps are data stores containing descriptions and analyses at a set of data points, from which maps of single attributes or chosen combinations of attributes can be drawn for specific needs. Certainly such an option is desirable, but how such a quantity of data could be economically gathered other than for widely spaced sampling patterns, is not clear. The recent Soil Survey 1 : 250 000 mapping programme in England and Wales, for example, had an observation density of some two points per km², with described and sampled profiles at a density of one per 25 km², on a 5 × 5 km grid, to give a systematic National Soil Inventory.

The existence even of a reasonable standard of soil map is thus no substitute for the ecologist's own observations at a particular site, when accurate local detail is essential to a study, though it clearly reduces the range of soils likely to be found there. However, to turn from unthinking confidence to total pessimism about the use of published soil maps would be to go too far. For Britain and Ireland a valuable series of example interpretations of existing soil maps in relation to contemporary and historic land use and to ecology and landscape (Curtis *et al.*, 1976) shows what is possible.

For those who wish to consider the methods used in soil survey, these are treated in the classic works of Clarke (editions from 1936 to 1971) and USDA (1951). Recent views on different survey approaches can be found in Webster (1975), Dent & Young (1981), USDA (1981), and Olson (1982). The application of air photography to soil survey is covered by Carroll *et al.* (1977) and the relationship of soil classifications to the objectives of a soil survey is considered by Butler (1980).

5.2   Terrain maps

It can be useful for an ecologist who wishes to relate vegetation distribution to land character to consider, instead of a single-factor approach linking vegetation to soils, a wider relationship of vegetation with what are called 'land systems' and 'land units', or 'terrain units', on 'terrain maps'. Mitchell (1973b) and Gerrard (1981) summarize different variants of terrain mapping. One variant is still a single-factor approach, but using physiography, with units derived from geomorphological mapping (e.g. Cooke & Doornkamp, 1974; Chartres, 1982). What are essentially physiographic units are derived from terrain mapping based on air photograph interpretation (Webster & Beckett, 1970) and recently on satellite imagery, the application of which to terrain mapping is considered by Townsend (1981). Other forms of terrain mapping use more complex groupings, based on a combination of some or all of physiography, soils, geology, climate and vegetation (Beckett & Webster, 1965; Beckett *et al.*, 1972; Rowe & Sheard, 1981). When vegetation is an element in a land system and unit classification (Van den Broek *et al.*, 1981), the ecologist will have been prominent in developing the scheme. Alternatively, from a land classification based on physical factors, the plant ecologist will be concerned with testing the value of the terrain classification in explaining and interpreting vegetation distribution.

Terrain classifications are often devised to assist single objectives; for example, the assessment of potential forest land in upland Britain (Rowan, 1977). Multi-factor schemes have been widely used in forest site evaluation. A long-established German scheme, involving physiography, soils and vegetation as defining criteria for the terrain units, has been applied recently to a study area in the United States (Barnes *et al.*, 1982).

As an alternative to air photo or ground mapping of terrain units, quantitative physiographic, topographic, climate, geology and soil data in different combinations, as recorded for cell units, such as, in Britain, those provided by Ordnance Survey national grid squares of appropriate scale, can be used in computer analyses in order to classify the cells into land classes (Inst. Terr. Ecol., 1978; Bunce *et al.*, 1983; Moss, 1985). Such land classes can be related to vegetation distributions as recorded at sampling points in the grid squares (Ball *et al.*, 1981, 1982).

6   Physical and mineralogical analysis of soil

6.1   General considerations

Figure 5.7 illustrates the steps that are required, after soils have been sampled, to bring field samples to the condition in which different analyses are carried out. As Chapter 4 deals with the mineral nutrition of plants, Chapter 1 includes consideration of nutrient budgets, and Chapter 6 is concerned with chemical analyses of soils and plants, only a couple of points about soil chemistry are made here. A comprehensive approach to the chemistry of soil processes and constituents is given in the volumes edited by Greenland & Hayes (1978, 1981) and a valuable guide to French work is provided by the English translation of Bonneau & Souchier (1982). Major texts which cover soil analysis are the two-volume compendium published by the American Society of Agronomy (Black, 1965), the FAO recipe-book of Dewis & Freitas (1970), and the volume by Hesse (1971). The use of physical methods to study soil and plant chemistry is included in Tinker (1980). The methods in use in agricultural advisory work in Britain are covered by ADAS (1981), and those used by the Soil Survey of England and Wales are given in Avery & Bascomb (1974). Because techniques develop rapidly as people aim for more precision or speed, or for less costly methods, the monthly publication *Soils and Fertilizers* (S & F), one of the bibliographies produced by the

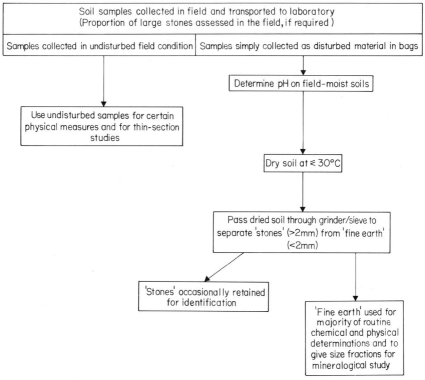

Figure 5.7 Outline flow chart for treatment of soil samples collected for analysis.

Commonwealth Bureaux of Soils, Harpenden, is an essential source for locating recent work. Periodic compilations of abstracts covering particular topics, as abstracted in S & F, are also produced on a wide range of subjects by the Commonwealth Bureaux of Soils, under the series title *Annotated Bibliographies*. A scan through available titles in this series can often reduce much individual searching of journals.

The preceding sections have emphasized field examination and consequent classification of soils but the ecologist wants quantitative associations between plant distribution and performance, and measurable characteristics of soil chemistry and physics, as well as being interested in the more qualitative correlations of vegetation with soil class. However, while conventional chemical analyses give reproducible results when carefully applied to subsampling 'homogeneous' samples, as illustrated for example by the relatively low variability of 'bulk' sample analyses given in Ball & Williams (1968), great attention is needed to subsampling procedures if analyses are to be representative of the bulk sample (Mullins & Hutchinson, 1982). Ecologists and soil scientists also need to appreciate the limitations of extrapolating such data, however reproducible, to field interpretations for large bodies of soil. These limitations include inherent chemical heterogeneity, even in apparently uniform soils, and the broader issues of how far the standard soil analyses give data which as a general rule correlate with growth responses of the plants being studied. Oertli (1973), considering the use of chemical potentials as an assessment of nutrient availability to plants, has concluded that difficulties in choosing a relevant nutrient potential make 'it appear unlikely that a practical and scientifically-rigorous criterion for nutrient availability to crops will soon be introduced in soil management'. Hesse (1971) combines consideration of the theoretical background and a practical method for each chemical analysis with a discussion of why, in a soil context, it is useful. Hesse comments in his preface that 'the myth of soil analyses being the answer to the farmer's problem is not kept up' (in his book) and that 'the reader will repeatedly be reminded that field experiments are essential'. With these reservations applying to the study of crop plants, for which growth responses to nutrient conditions have been intensively studied, the interpretation of quantitative chemical data at any detailed level in relation to native plant species, about which so much less is generally known, becomes even more a subject for caution. Epstein (1972) emphasized that 'nutritional ecological studies' or 'edaphic (soil-related) plant ecology', based on contrasting soil types, offers 'situations ideally suited to ecology studies'; and that this aspect of ecology, then felt to be 'entering a period of development', has not received 'the attention it merits'.

6.2   Physical analyses

Chemical analyses, closely followed by soil moisture measurements, are most often considered by plant ecologists, but it has become increasingly recognized that there are other ecologically-significant soil physical parameters. Marshall & Holmes (1979), Hillel (1980, 1982), and Koorevaar et al. (1983) deal with the principles of soil physics, while an older but compact text on agricultural aspects is that of Rose (1966). Chapters in Vol. 1 of Black (1965) give practical methods for the majority of soil physical determinations.

6.2.1   Particle size distribution

Soil texture has previously been defined as a field assessment of the proportion in which particles of different size ranges are present in the soil. Particle size determination, less accurately but widely termed 'mechanical analysis' (see Day, in Black, 1965) gives a more reproducible basis for textural classification.

Normally, particle size analyses, as with other standard soil analyses, are made on the fine-earth fraction of the total soil, that is on material of less than 2 mm 'equivalent spherical diameter' (see below). Field soil properties are however affected by the size and quantity of larger particles (i.e. 'stones' by definition). When quantitative assessments are required, the larger stones, in soils that are relatively dry, stony and friable, can most conveniently be weighed in the field. Soil from a large field sample can be passed through a 10 mm mesh nylon sieve to separate medium and large stones from fine earth plus small stones. Weighing of separated fractions is simple using a spring balance to weigh material in polythene sacks. If a series of measurements is to be made, a collapsible light metal tubular tripod with a hook to support the spring balance and its added weight, is worth the effort of construction. Small 'stones' must normally be further separated from fine-earth in the laboratory, as sieving of undried soils through smaller mesh sieves is not usually practical in the field.

In the laboratory, soil samples are initially dried in air or in an oven at $c.$ 30°C, then passed through a 2 mm sieve manually or through a mechanical grinder which combines a mild aggregate-crushing action with a sieve. This dried fine-earth fraction will generally include structural aggregates as well as primary particles. Dispersal of these, normally in aqueous suspension, is essential before particle size analysis. Effective dispersion of aggregates in suspension may require chemical pre-treatment to remove organic matter and/or iron or other oxides and hydroxides acting as cementing media (Kunze, in Black, 1965). Where possible however, dispersion should be carried out without such pre-treatment, by taking 50 or 100 g of air-dry fine-earth in a bottle, and adding to it 400 ml of water and 25 ml of 5% 'Calgon' (sodium hexametaphosphate) at pH 9 as a dispersing agent. Mechanical shaking of the suspension for several hours is then required, or an ultrasonic vibrator can be used (Watson, 1971; Pritchard, 1974; Mikhail & Briner, 1978). The objective is to ensure that all primary particles are dispersed in the suspension, but without too vigorous handling which could break these down from their original size to smaller sizes during the treatment. Because of this, extra care must be taken when dealing with soils containing fragile rock and mineral fragments. A mixer mill technique has been proposed (Lim, 1982) as an alternative, more effective means than ultra-sonic dispersion for reaching a plateau of recovered clay-size particles.

For highly organic soils, particle size analysis of the mineral fraction requires initial destruction of the organic matter. In most soils this is done by carefully adding 30% hydrogen peroxide to the wetted soil, followed by a time interval for the reaction of peroxide with organic matter, carrying out this operation a sufficient number of times to ensure total destruction of organic matter, and completing the final digestion on a hotplate at 90°C. Hydrogen peroxide is effective only in an acid medium so that carbonates, if present, must also be removed by acid treatment prior to organic matter destruction.

In general, when organic matter and/or carbonate levels are high, particle size analysis of the non-carbonate mineral fraction becomes an even more arbitrary procedure than usual, largely divorced from field properties of the soil, and of value mainly in studies relating to the origin of the mineral fraction rather than of ecological interest. It is therefore suggested that textural analysis—which can be more quantitative than field assessment—is most likely to be useful to ecologists in cases where minimal pre-treatments are needed.

After dispersal, the soil suspension is transferred to a litre cylinder, made up to the litre mark, and thoroughly mixed. The simplest method for determining the percentage of particles in different size classes uses a specifically graduated 'Bouyoucos' hydrometer. This is carefully inserted into the suspension to be read at settling time intervals of 46 s for 'American silt' and clay; 4 min 48 s for 'International silt' and clay, and 5 h for clay, the values for material in suspension after these settling times being read directly from the graduated stem of the hydrometer as grams per litre. If 50 g of soil were taken, the values are multiplied by 2 to give weight percentages. The settling times are calculated as for a pure water suspension at a given temperature, so that the measured values require small corrections for the amount of Calgon in the suspension and for a temperature difference from 18°C. Hydrometer measurements are reasonably reproducible and of adequate precision for most purposes. It must be appreciated that the whole method accepts various assumptions and uncertainties so that very careful efforts to be precise about the necessary corrections are not justified. Nomographs from which settling times for different particle sizes and suspension temperatures can be obtained directly are given in several sources, including Tanner & Jackson (1948) and British Standards Institution (1967). Kaddah (1974) has compared the hydrometer method of mechanical analysis with the more laborious sampling, drying and weighing needed to determine material in suspension by a pipette method which samples the suspension at specific depths after specific time intervals.

Table 5.5 gives size classes used in soil particle size analyses. Conventionally, only three fractions are distinguished: total sand, silt, and clay. For more detailed work, a larger number of size classes is needed within the sand fraction, obtained by sieve separations (Ingram, 1971; ASTM, 1972). Cumulative percentage curves, in which particle sizes are plotted against the percentage of the total soil material having a smaller estimated diameter, enable visual comparison of particle size distributions. Actual size values are plotted on logarithmic paper or, more conveniently, a logarithmic size scale, such as the 'phi' scale given in Table 5.5, is used with ordinary graph paper.

Table 5.5 Soil particle size classification

Size fraction	British Standard sieve mesh numbers*	Mesh aperture of sieve (microns = 0·001 mm)	Assumed diameter of particles just passing through mesh (mm)	log₁₀ of particle diameter (microns)	'Phi' value† (−log₂ of particle diameter (mm))
Large stones >100 mm	—	100 000 (10 cm)	100	5·0	−6·6
Medium stones 100–10 mm	—	10 000 (1 cm)	10	4·0	−3·3
Small stones 10–2 mm	8	2,057	2	3·3	−1·0
Coarse sand 2–0·2 mm	16	1,003	1	3·0	0
	30	500	0·5	2·7	1·0
	60	251	0·25	2·4	2·0
	Alternatives 72	211	0·21	2·3	2·3
	85	178	0·18‡	2·3	2·5
Fine sand 0·2–0·05 mm	120	124	0·12	2·1	3·1
	150	106	0·1	2·0	3·3
	Fine sieves become increasingly delicate, expensive and unsuitable 240	63	0·07	1·8	3·8
	300	53	0·05	1·7	4·3
Silt (American) 0·05–0·002 mm	—	(50)	0·05	1·7	4·3
Silt (International) 0·02–0·002 mm	—	(20)	0·02	1·3	5·6
Clay <0·002 mm	—	(2)	0·002	0·3	9·0

*The same mesh numbers in other systems (e.g. ASTM, 1972) have different aperture sizes to those quoted for the British Standard sieves (as manufactured by Endecotts).
†'Phi' values are conveniently used for plotting cumulative particle size curves on ordinary (non-logarithmic) paper (see Krumbein & Pettijohn, 1938).
‡Generally used as the preferred sieve for '0·2' mm.

The nomenclature of the particle size ranges used is not internationally consistent (Section 3.2.2). Table 5.5 emphasizes the USDA (1951) limit of 0·05 mm for the silt/sand boundary, but the older 'International' scale had an upper size limit for silt of 0·02 mm, and Avery (1973) has recommended a division at 0·06 mm. With routine methods, differences between measurements at nominal separations of 0·06 and 0·05 mm would not be significant for the great majority of soils. It must be borne in mind that, in particle size analysis, settling speeds in suspension are calculated on untrue assumptions, such as that soil particles are uniform spherical particles of uniform specific gravity. Similarly, in sieve analysis, long lath-like fragments of larger volume can just pass the same sieve aperture as spherical particles of smaller volume. For these reasons, all soil particle size limits are correctly termed 'equivalent spherical diameters' rather than actual diameters.

Sand, silt and clay, as percentages of fine-earth, can be plotted on a triangular diagram, and textural class names (as discussed for field estimates in Section 3.2) obtained from the location of points representing soil horizons. The USDA system (1951) is illustrated in Fig. 5.8. An English variant (Hodgson, 1974; Avery, 1980) differs by the loss of 'loam' and 'silt' as texture class names, by repositioning the boundaries for silt loam, clay loam, silty clay and silty clay loam classes, and by the introduction of sandy silt loam as a texture class. Where particle size data are given it should be stated which size limit is being used for 'silt' and which triangular diagram convention is being followed.

Particle size distribution gives a standardized means of comparing a basic characteristic of soils that is related to key properties such as permeability. It has been found that reasonable estimates of available water capacity (the total quantity of water that a soil can store in a form accessible to plants, see Section 6.2.5) can be made by calculation from analyses of particle size distribution and organic matter content (Salter *et al.*, 1966) and that this can also be done from field textural determinations (Salter & Williams, 1967).

6.2.2 Bulk density

Bulk density (or volume weight) (e.g. Blake, in Black, 1965) is the ratio of the dry weight of a soil sample to the total volume it occupies in field condition (i.e. the moist volume of a soil). It can be used indirectly to assess differences in soil structure and porosity caused by natural processes or by management. Bulk density and porosity determinations can often be related to penetrability by roots (e.g. Mirreh & Ketcheson, 1972) or to the effect of a surface soil crust on germination. Bulk density also can be correlated with available water, retained water and air capacity (Reeve *et al.*, 1973).

Bulk density is determined from the air-dry mass of a known field volume of soil. This volume can be obtained directly with the use of a volumetric

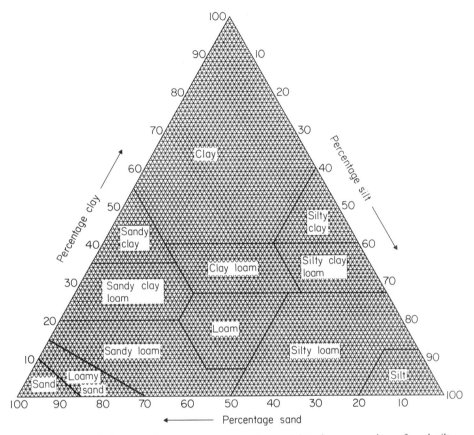

Figure 5.8 Soil texture class names (from USDA, 1951) based on proportions of sand, silt and clay size fractions.

corer (Section 3.3 and see Lelievre, 1981). An alternative method is to remove and retain for weighing a quantity of soil as a disturbed sample, then determine the volume of the randomly-shaped excavated hole by filling it with a measured volume of sand (British Standards Institution, 1967) or by closely lining it with an impermeable rubber or plastic film (Shipp & Matelski, 1964) and pouring water or sand of measured volume in until the cavity is filled. These infill methods are particularly useful in gravelly soils for which core samples are impractical, but require transport or availability on the site of water or sand, and volumetric measuring containers. Some problems in their application are discussed by Rimmer (1982). Instrumental *in situ* field measurements of bulk density by radiation methods are possible (Blake, in Black, 1965).

A direct laboratory method of determining bulk density—applicable to small core samples, to samples of random shape cut from a transported block of soil, and to large, strong, structural aggregates—is to seal the outer surface of the sample with an impermeable resin, then immerse it consecutively into a series of liquids made up to known specific gravities until one is found in which it just floats (Campbell, 1973). The bulk density of structural units can be determined by measuring the volume change when a known weight of soil is mixed with a known volume of fine powder (Bisal & Hinman, 1972).

Bulk density measurements are used to convert analyses made on a weight basis to a volume basis. The volume occupied by a known weight of total soil and the proportions of total stones to fine earth in this soil are both needed to convert the standard weight presentation (analysed quantity/100 g fine-earth) to a field volume basis. Without measurement on individual soils it is possible to give correction factors of a generally applicable approximate order of magnitude. The close relationship between bulk density and organic matter content is relevant. Using weight/volume ratios of air-dry loosely-packed fine-earth, rather than true field bulk densities of total soil, Ball et al. (1969) suggested conversion factors ranging from × 1·0 (0–5% organic matter), through × 0·7 (21–30% organic matter) to × 0·3 (91–100% organic matter) to give a relative volumetric basis from weight-based figures. Jeffrey (1970) and Adams (1973) give more precise considerations of the relationship between organic matter, true and bulk densities while Rawls (1983) deals with the estimation of bulk density from particle size and organic matter content. In estimating bulk density of soils from organic matter content determined by loss-on-ignition, Harrison & Bocock (1981) emphasize the desirability of local equations for particular ranges of soils, rather than reliance on a general formula.

6.2.3   Pore size distribution

The appropriate chapter in Black (1965), by Vomocil, again gives a comprehensive review of this topic. Total porosity, which is the percentage of the bulk soil volume not occupied by solid material, can be calculated from the difference between the volume of the undisturbed sample and the volume of soil solids, or as: $100 (a - b)/a$, where a = mean soil particle density and b = soil bulk density. Mean soil particle density and soil solid volume can be determined from the volume of liquid displaced by a weighed, dis-aggregated dry soil sample.

The proportion of total porosity that is in different size pores is vital to consideration of soil water regimes. A particular total porosity gives a very different pattern of water retention and release depending on whether it is due to a few large pores or many small pores. The proportion of pores in different size grades is normally determined by measuring the quantities of water drained from a soil at given suction pressures (Section 6.3.5), assuming a capillary model to represent soil pore space. The volume of water extracted from saturated soil at a given suction is controlled by the proportion of soil pore space that is occupied by pores of larger radius than that of the smallest capillary from which this suction is able to extract water.

Cary & Hayden (1973) suggest an index for pore size distribution based on the change in soil water content as suction pressure increases. The method requires laboratory measurement of soil moisture on samples subjected to three levels of suction, and gives a single value to represent pore size distribution and therefore, indirectly, soil moisture characteristics. Thomasson (1978) provides a classification of the structural condition of soils based on the volumes of two ranges of pore size.

6.2.4 Structural determinations

The primary particles of some soils either form a loose agglomeration of single grains or may occur as a compact mass. Both of these categories are conventionally considered 'structureless' being respectively 'single grain' and 'massive'. The majority of soils have their primary particles arranged into aggregate units. The size and shape of these provide a basis for subjective morphological classifications of soil structure (Section 3.1.2). For greater precision in structural assessment, the size distribution (Kemper & Chepil, in Black, 1965), and stability of aggregates (Kemper, in Black, 1965) require quantitative determination.

For the determination of the size range of structural aggregates, a soil sample, preferably in a fairly dry state, is collected and transported carefully to the laboratory, dried thoroughly, and then passed through a nest of sieves on a shaker. Method V-41 of de Boodt (1967) uses sieves with apertures of 80, 40, 20, 10, 5 and 2 mm. Structural unit weights retained on the different sieves can then be plotted in cumulative curves as for primary particle size distributions (Section 6.2.1), or calculated as single-value parameters (Kemper & Chepil, in Black, 1965). The size of aggregates is important to porosity and aeration of soils. When reasonably large size is combined with stability, structural aggregates have a strong influence on resistance to erosion by wind or water.

Stability of aggregates is normally measured by determining the proportion of selected size fractions retained on a sieve after a standardized wetting procedure. Kemper has noted that limitation of stability tests to a single size group of aggregates (say 1–2 mm) is usually as effective as determination on several size classes. The method used to wet the aggregates is adjusted according to the purpose for which stability is required to be measured. Immersion wetting simulates surface irrigation, so the stability of aggregates wetted in this way relates to their performance in a surface soil. Subsurface wetting is more closely imitated when aggregates adsorb water gently under tension or vacuum. Methods of this type are discussed by Williams *et al.* (1966). Ones in which the stability of individual crumbs is assessed by a simulated rainfall 'falling drop' technique are given by Low (1954, repeated in de Boodt, 1967). The method of Emerson (1967), which by a standard technique classifies the stability of structural aggregates into eight classes, has been shown to be useful in identifying soils which may have structural problems in the field (Greenland *et al.*, 1975). In clay-rich soils a method of determining aggregate stability has been proposed that uses as an index the ratios of clay obtained after dispersal in water with and without chemical pre-treatment and dispersing media (Harris, 1971). Resistance of dry aggregates to wind erosion can be assessed (Chepil, 1958, 1962) by determining the rate at which a chosen aggregate size-class breaks down with time on prolonged dry rotary sieving. Bryan (1971) made a comprehensive study of the accuracy with which various indices of aggregate stability reflect differences between soils.

In general, the standardization of results between different laboratories in the area of soil structure measurements, even when the same techniques are employed, has proved elusive. Where such assessments are required, one of a wide number of possible variations of the general principles outlined can be adopted to provide, for well replicated samples, an internally consistent system. In comparing results, only major contrasts in aggregate size ranges or stabilities are likely to be significant factors in soil or treatment differences.

6.2.5   Soil water

General discussions of soil moisture are to be found in Marshall (1959), Black (1968), Childs (1969), Hillel (1971) and Marshall & Holmes (1979). Four main aspects of water relationships in soils are involved: the rate at which water moves into and infiltrates through soils; the quantity of water in a soil (soil wetness); the force with which the water is held (soil moisture suction); and for the botanist, related to all these aspects, the availability of soil moisture to plants. Chapter 3 discusses this last aspect, the water relationships of plants from the physiological viewpoint. An outline of some methods used to study the first three of these aspects is given here. A wide range of techniques is covered in Black (1965), Curtis & Trudgill (1974) and by Kramer (1983). A recent summary of methods of soil moisture assessment, including remote sensing approaches, is that of Schmugge *et al.* (1980).

Infiltration and movement of water through the soil

Bertrand (in Black, 1965) emphasizes that many of the methods proposed for the study of infiltration have been developed for special purposes and may not be widely applicable. Childs (1969) points out that, although laboratory methods measuring percolation rates through soil cores are capable of refinement and control and are suitable for the study of principles, field determinations are to be preferred when the results need to be applied to field situations. Hall *et al.* (1977) describe methods in use by the Soil Survey of England and Wales.

Methods of determining water flow rates through soils in the field depend on whether the measurements are to be made above or below the water table (the level up to which the soil horizons are saturated with water). Klute (in Black, 1965) describes the measurement of hydraulic conductivity below a water table by creating an unlined cavity and determining the rate of inflow from surrounding soil by pumping water out at a rate necessary to balance inflow. Rates of water flow in soil above a water table (Boersma, in Black, 1965) are measured most accurately by the 'double-tube' method. In this, two concentric tubes are inserted into the soil to a particular horizon depth. Changes in water level in one tube are related to imposed water levels in the second tube, and a formula for hydraulic conductivity can be applied using the observations made. More simply but less accurately, the infiltration rate from water applied to a known surface area within a constraining surround can be observed.

A metal cylinder is driven into the soil and allowed in part to project above the soil surface. After initial saturation with water applied via the cylinder, the rate of infiltration of further volumetric additions of water into the soil is observed as a determination of the time required for a known volume of water added to the tube to pass through a known surface area (Hills, 1970). This procedure can be repeated with the cylinder inserted to different depths or into different horizons of the soil profile. Knapp (1973) has described the complexities that are involved in running a fully instrumented installation for determining water flow in soil on a horizon basis.

Soil leachate studies using lysimeters (see Chapter 1) to measure rates of flow of water through soils in the field and to collect leachates for analysis have been described by Bourgeois & Lavkulich (1972a, b). Glass fibre casings can be used to transport soil monoliths for lysimetry (Belford, 1979). In a simpler way, Reynolds (1966) applied fluorescent dyes to determine patterns of percolation of rainwater through the soil.

In a different approach to assessing hydraulic conductivity of soil horizons, Anderson & Bouma (1973) and Bouma & Anderson (1973) correlated conductivity measurements in sample soils with point-count measurements of pore space observed in soil thin sections (Section 6.3.1) and found the latter method reasonable for assessing hydraulic conductivity indirectly.

Where fluctuations of the water table in a soil are of interest, then sets of access tubes can be inserted to different depths or horizons and left open for observation. Such installations and examples of their use in the study of water table location are described by Rutter (1955), Fourt (1961), Thomasson & Robson (1967), Thomasson (1971), and Hall *et al.* (1977). A float device which allows the determination of water table level in a tube at the time of inspection, and its maximum and minimum heights since the previous recording, has been described by Boggie (1977). In using access tubes it is important not to rely solely on one deeply-placed tube. This may tap a moisture source of higher hydrostatic potential than that of the soil surrounding the upper part of the tube, in which case water will rise higher in the tube than the water table actually reaches in the soil. Ideally, the water table should be identified by means of tubes inserted to different depths, as the level obtained in the shallowest tube that contains free water. This will not always be possible when a number of sites are to be monitored. The possible errors in determining water tables in some soils by access tubes should be recognized (as discussed by Visvalingam, 1974).

Determination of water content of soil

Soil water content ('wetness') is generally defined as the water content of the soil that can be driven off by heating a sample at 105°C. It can be expressed as volume or weight per cent of soil. Although less careful sampling is required to determine soil water on a dry weight basis, it is often considered more meaningful to determine it on a volume basis (Section 6.2.2).

Gravimetric analysis (Gardiner, in Black, 1965; Reynolds, 1970) is the simplest method of measuring soil water, and provides the standard reference technique for calibrating other methods. A soil sample, preferably of known field volume, is taken, weighed at the earliest opportunity, and then re-weighed after it has been dried in an oven at 105°C. The weight difference, the dry weight and the known volume allow the percentage of water in the soil to be given as weight and volume per cent. The advantages of the method are that it is cheap, simple and requires no calibration; the disadvantages are that each replicate sample actually differs because of soil heterogeneity, and, where regular measurements are required, sampling can change size conditions and ultimately destroy a sampling site.

The instrumental, neutron scatter technique of measuring soil moisture (Visvalingam & Tandy, 1972; Bell, 1969, 1973; Greacen, 1981) requires specialized equipment, experienced careful use and established sites. It depends upon the fact that when fast high energy neutrons are emitted into the soil they collide elastically with the various atomic nuclei with a loss of kinetic energy. This energy loss is greatest when a neutron collides with a particle of similar mass. The hydrogen nuclei are the particles nearest in mass to the neutron, and the rate of conversion of fast neutrons to slow (thermal) neutrons is proportional to the hydrogen content of the soil. Although hydrogen is present in organic matter and in the structural water of clay and hydroxide materials, for soils other than peats most of the hydrogen is contained in soil water, so that an observed count rate of slow neutrons is proportional to the soil moisture content. Calibration of the method must be carried out, for each soil and soil horizon studied, against gravimetric determinations of soil moisture. Once calibrated, a fast neutron source and slow neutron detector can be lowered into permanently installed access tubes to provide rapid and repeatable measures of moisture content of soil at a site. Modifications must be employed for near-surface measurements of soil water with the neutron probe when its sphere of measurement, approximately 120 mm in wet soil and 200 mm in very dry soil, breaks the soil surface (Hanna & Siam, 1980). Soil water content regimes determined by neutron probe for a range of soil types have been reported by Gardner (1981).

A wide range of other methods have been used to measure soil wetness (see e.g. reviews by Shaw & Arble, 1959; Cope & Trickett, 1965). These include measures of electrical capacitance (Thomas, 1966) and its thermal conductivity (Cornish *et al.*, 1973). Rutter & Sands (1958) modified a method of Davis & Slater (1942) to determine soil moisture. Wedges of gypsum were replaced in cavities lined with filter-paper at the bottom of tubes sunk into the soil. Moisture content of the wedges was considered to become equilibrated with that of the soil, and was determined by weighing the gypsum blocks at required intervals; but their water content is a function of soil moisture suction (see below) as well as water content. A variant directly using filter paper in contact with soil has been examined (Al-Khafaf & Hanks, 1974).

The force with which soil water is held

Water availability to plants depends not only on the absolute quantity of water in a soil but also on the force (energy) with which this water is held in the soil matrix. Water in soil pores is retained over a range of soil moisture tension (or 'soil suction') so that only applied pressures (or suctions) in excess of a particular tension are able to remove a particular part of the total soil water. Looked at simply, the energy required to remove water from a soil pore is greater the smaller the pore, so that two soils with different pore-size distributions have different water availability at any given water content; or alternatively, as a soil dries, successive equal increments of water require more energy for their withdrawal because the remaining water is held with greater tension in smaller pores and as films on particles. Thus, a consideration of water relationships between soils and plants requires measures of the absolute water content and of the force with which this water is held.

In classic work, Schofield (1935) introduced, by analogy with pH, the term pF as $\log_{10}$ (suction in cm of water) where 'suction' is the force required to extract water from soil. A pF curve then was a plot of water content against suction to give a picture of the moisture release characteristics of a soil and its range of 'available water'. Suction in this terminology ranged from 0 when a soil was saturated (pF $= 0$) to $c.$ 10^7 cm water suction (pF 7·0) when a soil was dried at 105°C. Soil moisture suctions are generally given now either as pressures (more correctly as negative pressures) in millibars (mb) and bars (b), or as their equivalent in SI units (1 mb $= 100$ Pa (Pascals), e.g. 50 mb $= 5$ kPa). Alternatively, soil moisture suctions can be treated as 'water potentials' based on energy units (Slatyer & Taylor, 1960; Woodward & Sheehy, 1983).

The range of soil water available to plants is conventionally considered to extend between two arbitrarily defined states. As water percolates from a soil surface, it progressively displaces air from pore space. If the surface supply ceases, water drains rapidly away (assuming subsoil or sideways drainage permits this) from larger pores to a condition at which gravitational flow falls off sharply and tends to zero. At this stage a soil is described as at 'field capacity' ('the maximum equilibrium moisture content under free drainage' (Winter, in ADAS, 1975) or 'the water retained 48 hours after saturation' (Smith & Thomasson, in Avery & Bascomb, 1974). Field capacity corresponds to a suction pressure of about 50 mb. As a generalization, 'plants' are considered able to abstract water from soils over a range of moisture suction between this state and a 'permanent wilting point' at about 15 bars suction. The quantity of water between these imprecisely defined 'points' is termed the 'available water capacity' of a soil. Wilting point actually varies, in a given soil, with plant species and atmospheric conditions. The relationship between soil moisture tension and soil water content depends on soil particle size, structure and porosity as noted above, and also upon whether it is determined in a wetting or drying cycle. Comprehensive calibration curves used to relate suctions to water content must be constructed for both situations.

The soil moisture suction as typically measured in routine studies, to be considered below, is the matric suction attributable to the capillary forces acting on water in soil pores as outlined above. Total soil water suction, so far as the plant is concerned, is a combination of this matric suction and osmotic suction pressure (Krahn & Fredlund, 1972). This latter is the pressure differential required to balance diffusion of water from a zone of lower concentration of dissolved salts to a zone of higher concentration through a semi-permeable membrane that permits water, but not dissolved ions, to pass it. Problems of independent measurement of osmotic and matric potential (suction) are discussed in an assessment of osmotic potential measurement methods by Sands & Reid (1980). In the laboratory, the measurement of matric soil moisture suction requires two methods to cover the full range (see e.g. Smith & Thomasson, in Avery & Bascomb, 1974). Water content of suctions up to 0·5–1 bar can be measured on suction tables in which the soil sample is placed in close contact with a sand or kaolin surface. Moisture equilibrium is established in the soil sample under an applied suction, and the moisture content of the soil under this suction is then determined gravimetrically. Samples should be replicated for each horizon or depth. At higher tensions, pressure membrane apparatus must be used. The range covered depends on the specific apparatus but observations can be made at pressures from 2 to 15 bar (the 'wilting point'). Soil samples, typically cut from a soil core, rest on a cellulose membrane supported by a sintered bronze disc in a stainless steel pressure chamber. Pressure is applied to the chamber from a compressor or a nitrogen cylinder to give a pressure difference across the membrane. After equilibrium has been attained, the water content of the samples is again determined gravimetrically.

The most common field measurement method for soil moisture tension uses tensiometers. The basic tensiometer (Richards, in Black, 1965; Berryman et al., 1976) is a water-filled porous ceramic pot placed in contact with the soil at a desired depth or in a desired horizon. Soil water in pores and films in the soil is in hydraulic contact with water in the porous pot through the pot walls. The tension of the soil water causes flow into or out of the pot but the system is only effective to a pressure of some 800 mb, above which air enters the porous pot. The pot is connected by an air-tight system to a pressure measuring device. The range that tensiometers cover includes some 50–75% of the available water capacity of soils. A number of commercial designs are available, mainly in use in crop response studies. A system that could be assembled from readily available components was reported by Webster (1966) and a small variant suitable for assessment of rapid changes over short distances has been described by Rogers (1974). Manually read systems are cheap and simple but tensiometers do require regular inspection and servicing. Bearing this reservation in mind, an automatic recording tensiometer system has been described by Williams (1978).

An alternative assessment of moisture status, which is related to soil moisture tension, though calibrated by water content, can be obtained from electrical resistance measurements on resistance blocks buried in the soil (Bouyoucos, 1953, 1954). These porous blocks contain a pair of electrodes, resistance between which varies according to the moisture content of the block, which in turn depends principally on soil moisture tension as a control on moisture inflow. Gypsum (Ca $SO_4 \cdot 2H_2O$) blocks are used in drier soils, and nylon or fibreglass blocks (Bouyoucos & Mick, 1948; Bouyoucos, 1972) are better at the lower tensions of wetter soils. Accurate calibration of resistance blocks is affected by a number of factors, and blocks require individual calibration. In use they have been considered as providing general rather than quantitative indicators of soil moisture conditions. Sources of error include poor contact between block and soil, slow response, hysteresis between wetting and drying cycles, and the influence of soil-water chemistry. However, interest continues, and Williams (1980) has described an apparently reliable electrical resistance system for measuring soil moisture, while Strangeways (1983) has combined the use of gypsum blocks with a data logging system.

More exact measures of total soil moisture tension (matric + osmotic) can be carried out over a wide range of tensions using measurements of vapour pressure, in the field or the laboratory, by thermocouple psychrometers (Spanner, 1951; Richards, in Black, 1965; Williams, 1968; Globus, 1972; Woodward & Sheehy, 1983). Modifications and applications of the technique have been described in the volume edited by Brown & van Haveren (1972), and by Easter & Sosebee (1974).

Finally, available water capacity, defined above as the quantity of water that can be held in a soil between field capacity and wilting point, can be estimated indirectly, with reasonable confidence for some purposes, from data for the particle size distribution and organic matter content of a soil (Salter & Williams, 1965a, b; Salter et al., 1966; Salter & Williams, 1967; Gupta & Larson, 1979).

6.2.6   Soil temperature

Thermal behaviour of soils has been reviewed by Gilman (1977). Soil temperatures directly affect microbiological activity and thus organic matter cycling and nutrient availability, as well, of course, as affecting soil water content through evaporation. Temperatures depend on radiation input but also on the effects of colour, texture, drainage and other soil characteristics. Because comprehensive soil temperature records need to consider daily and seasonal variations, semi-permanent installations, continuously or intermittently recorded, or read frequently, are likely to be needed. Recording methods are treated by Taylor & Jackson, in Black (1965). Frequently used instruments include standard mercury thermometers, electric resistance instruments using wires or thermistors, and thermocouples. In all cases the quality of contact between instrument and soil is critical. Mercury thermometers are more useful as relative standards for routine meteorological measurements than for precise determinations of a soil temperature, because of this contact problem in the access tubes in which they are generally used.

The alternatives are typically small, insertable in precise locations, and readable without further site disturbance. A particular wire resistance thermometer has been recently devised by Savage (1980), while the operation of a large assembly of thermistor resistance thermometers was reported by

Bocock (1973). Using data from his study site, the estimation of soil temperatures from air temperatures and other climatic variables has been described (Bocock *et al.*, 1977). The use of thermocouple thermometers for surface and soil temperature measurements has been considered by Rodskjer (1975).

In a quite different approach, variants of a method for determining mean temperature over a period of exposure, which depend on laboratory measurement of the inversion of sucrose in solution that has taken place in sealed capsules placed in field sites, have been applied to soil temperature measurements as an alternative to conventional thermometry (Jones & Court, 1980).

6.2.7 Mechanical strength of soils

Measurements of the mechanical strength of soils are more particularly the concern of civil engineers (e.g. British Standards Institution, 1967; ASTM, 1970), in relation to the capacity of soil materials to support engineering loads. They are, however, also of interest and importance to the agronomist concerned with the impact of cultivations and grazing pressures on such soil properties as structure and permeability, and to the ecologist studying, for example, the effect of trampling by man and animals on soil conditions and hence on vegetation, and the factors controlling natural and induced erodibility.

Chapters in Black (1975) provide their usual good review of these aspects of the analysis of soils. Brief reference is given here to some methods used to measure soil strength, particularly as manifested in soil compaction, and to means of assessing the relative erodibility of different soils.

Freitag (1971) and Saini (1980) cover methods of measuring soil compaction. *In-situ* soil strength tests rely on devices of the 'shear-vane' type in which a measurement is made of the torque required to turn a bladed device inserted in the soil. Pressure-sensitive cells placed in the soil can be used to determine directly stresses brought about by natural or enhanced surface pressures and other causes of soil movement (Blackwell & Soane, 1981). The simplest method of comparison of compaction effects on soils is the use of versions of the 'cone penetrometer'. This in essence is a rod with a pointed end, equipped with a means of measuring the force needed to push the rod into the soil to a standard depth. It can be mechanically driven (Soane, 1973) or hand-held (Gerrard, 1982; Armstrong & Davies, 1983); if the latter, its speed of use allows a number of readings to be taken quickly from an area of interest. The theoretical basis for what is being measured is not certain, though resistance to penetration results from a combination of structural cohesive characteristics and the frictional properties of a soil. Thus, an increase in penetration resistance accompanies an increase in soil density and a decrease in pore space as a result of increased compaction. Because the cone penetrometer provides a useful empirical measure of the mechanical effect of compaction on the soil, attention has turned to improving the reproducibility of readings by modifying the simple hand-held penetrometer. Anderson *et al.* (1980) describe the satisfactory use of a solid-state recording hand-held penetrometer which gives measurements of the force required to penetrate soil at standard depths.

Turning to erodibility measures, Kirkby & Morgan (1980) cover processes and methods of measuring wind and water erosion of soils. Erosion, and its frequent corollary of localized deposition, can result from mass-movement, from water transport, or from particulate movement caused by wind and, in some cases, frost. The application and value of different indices of soil erodibility to water, derived from normal soil analytical data, have been reviewed by Bryan (1968, 1971). Direct observation of local gains or losses of soil material by wind erosion can be made by trapping transported sediment (e.g. Buckley, 1982); by following the movement of tracer particles; and by observing the burial or exhumation of markers fixed securely at known levels (e.g. Kinako & Gimingham, 1980).

6.3   Mineralogical and fabric analyses

Laboratory investigations of the mineralogy of soils, and of the interrelationships of component particles, aggregates and voids investigated on a micro-scale by thin-section techniques, are of specialist interest. Although thin-section studies can provide information that is directly applicable by soil ecologists, in general plant ecologists need only be aware of the range of mineralogical and fabric analysis techniques that are available.

6.3.1   Thin-section techniques

Mineralogical and fabric studies of rocks using the polarizing microscope are the classic laboratory technique of the petrologist. Mineral identification of individual grains, or of constituent minerals in rock thin sections, depends on a knowledge of the optical properties of rock-forming minerals (e.g. Kerr, 1959). A standard mineralogical text is that of Deer *et al.* (1966), while minerals in soils are comprehensively covered by Gieseking (1975b). Thin sections can be prepared directly from massive rocks, but friable rocks require an initial impregnation with a resin compound to give adequate mechanical strength during grinding and polishing of a section. Since the classic work of Kubiena (1938), this thin-section approach applied to soils has developed into a vigorous branch of soil science, under the name of soil micromorphology. Using a carefully collected soil sample in undisturbed field condition in a sampling tin (Section 3.3), the impregnating resin is introduced in a way that replaces air and water in pore spaces, with minimal modification of the original soil fabric. When the resin has been hardened, slices can be cut from the impregnated soil block using a diamond saw, and a face of the cut slice polished. The slice is then mounted on a microscope slide and ground to a standard thickness, using the conventional slide preparation procedures of the petrologist concerned with microscopic examination of rock thin-sections.

Kubiena (1953, 1970) used the products of such methods of thin-section preparation to supply data for the interpretation of soil-forming processes and to assist understanding of soil correlation and classification. The comprehensive study of soil microfabrics in thin section by Brewer (1964) has provided a terminology and definitions for this field of soil science that are now widely applied. Osmond & Bullock (1970) give practical details on thin-section preparation and methods for their study, together with applications to pedological investigations. A series of 'position papers' in Rutherford (1974) outline the status of soil thin-section studies then. Comprehensive cover of a co-operative research study in this field, and of techniques and applications with particular attention to porosity determinations, is given in a special issue of the journal *Geoderma* (Bisdom & Ducloux, 1983).

Fitzpatrick (1971, 1980) has made extensive use of micromorphological information in his discussions of soil formation and relationships. Thin sections provide a direct means of study of the relationships and composition of different soil constituents in a relatively undisturbed state. In addition to observations and identifications using the optical microscope, finer detail can be studied by electron microscopy, while the chemical composition of particular features can be determined by X-ray spectrography and the electron microprobe. The plant ecologist is likely to be particularly interested in results of the application of thin-section techniques to the study of relationships between biological and mineralogical components of soils. An early example of such work was the illustration by Hepple & Burges (1956) of a fungal hypha bridging a soil pore as it crossed between humus coatings on two quartz sand grains. Barratt (1964, 1968) approached the classification of soil humus through micromorphological studies. Babel, in Gieseking (1975a) has provided a review of micromorphological studies of soil organic matter. An effective means of overcoming the problems of impregnating peat and highly organic soil horizons with resin without destroying their natural fabric has been developed by Fitzpatrick & Gudmundsson (1978). Thin sections have been studied as a means of identifying the contribution of soil organisms to the structural elements of soil fabrics, for example the role of soil microarthropods in producing characteristic structural elements in rendzina and ranker soils (Rusek, 1975).

6.3.2   Sand-grain mineralogy

The main purposes of qualitative and quantitative identification of mineral species in the sand fraction of soils are (i) to determine the nature of the soil parent material and (ii) to follow the effects of weathering within a soil profile or between contrasting profiles (Cady, in Black, 1965). Sand size-fractions are obtained by washing dispersed soil in suspension through appropriate sieves (Table 5.5). A separated sand fraction is conventionally subdivided into 'light' and 'heavy' minerals by flotation in a heavy liquid, usually bromoform. The light fraction, dominated by quartz, is much larger in quantity, but less diagnostic of parent material origin. The subordinate heavy fraction generally contains a wider range of minerals which can be used to identify the parent material. Study of sand grain material is covered comprehensively by Milner (1962), while Marshall (1964) includes a chapter reviewing the mineralogy and chemistry of the major minerals of the sand and silt fractions of soils, and McCrone (1982) summarizes aspects of mineral identification for determining the origins of soil materials.

6.3.3   Silt and clay mineralogy

The microscope techniques used to investigate sand-grain mineralogy can also be applied to examination of the silt fraction, though less easily because the particles are smaller. Alternatively, silt mineralogy can be studied using the non-optical methods needed for the very small particles of the clay fraction of soils. Silt and clay size fractions are separated from dispersed suspensions of a soil (i) by sedimentation through specified depths over specified times; (ii) by sampling a suspension by pipette, again at depths and times related to settling rates of particular particle size ranges; or (iii) by centrifuging a suspension at given speeds for given times (Chapters by Day, Kunze, and Whittig, in Black, 1965; and see also Jackson, 1956). Clay and silt fractions thus separated are dried, often after chemical pre-treatments, to give material for analysis. The silt size fraction, as well as being intermediate in size between sand and clay, is often intermediate in mineralogical character, containing small particles of minerals that occur in the sand fraction as well as 'large' crystals of some clay minerals.

Minerals of the clay fraction are not simply smaller particles of the common minerals of the silt and sand size-fractions of the soil but are dominantly a suite of alumino-silicate minerals—collectively termed 'the clay minerals' and mineralogically classified as phyllosilicates (Grim, 1968; Gieseking, 1975b). These are a suite of minerals characterized by sheet structures, a principle component of which is a sheet composed of SiO_4 tetrahedra. In different clay mineral structures these silica sheets combine in alternative ways with octahedral sheets of alumina or magnesia and, in some cases, also with potassium ions as a structural constituent. The extent of isomorphous substitution that is possible within these structures can lead to charges on the sheet surface that are satisfied by cations held external to the sheet layer. Different structural combinations thus give clay mineral species with characteristic cation-exchange and/or swelling properties. These properties and differences are responsible for the active role soil clays play in nutrient cycling and the physical properties of soils, and for the contrasting behaviour of the clay fraction of different soils.

The clay fraction can also contain subordinate quantities of minerals not of the 'clay mineral' sheet-structure type, including crystalline oxides and hydroxides and non-crystalline amorphous alumina-silicates. Identification of clay minerals in the complex assemblage of mineral species and intergrades characteristic of most soils is principally by X-ray diffraction methods (Brindley & Brown, 1980). While investigation of the clay minerals present in soils, and of their physical chemistry, is not directly applicable to most plant ecological studies, it is an essential consideration if nutrient cycling and supply contrasts between different soils are to be explored in detail.

7   Soils, site history, and plant distribution

It is often the case that, when soil data are included in ecological publications, or vegetation data referred to by soil scientists in their papers, the impression can be drawn that such cross-references have been made from a sense of duty rather than from a conviction that both aspects are essential contributions to adequate understanding of either field of study. This book is concerned essentially with methods rather than applications. However, in order to direct readers to a few of the approaches which have aimed more positively at linking soil data with the history or present distribution of vegetation types and individual plant species, some references to published work are given here. More guidance in this respect is included in other chapters, particularly those concerned with site history and mineral nutrition.

Eyre (1975) provides a world review of the broad associations between present vegetation and soil classes, while Cox et al. (1976) summarize the relationships between some major soil groups and plant communities. For Great Britain, among works which deal with particular regions or vegetation categories, those by Good (1948) and Dony (1967) (discussed in Seddon, 1971) illustrate the correlation between vegetation distribution and soil and geological site characteristics in parts of lowland England (Dorset and Hertfordshire respectively); McVean & Ratcliffe (1962) in their major monograph, relate soil groups to plant communities of the Scottish Highlands; Birse & Robertson (1976) compare vegetation classes, climatic zones and soils in the Scottish lowlands and Southern Uplands; and Smith (1980) describes soil characteristics in relation to plant communities and species occurring on soils over chalk. Soils and soil parent materials of the surviving 'ancient woodlands' in England have been discussed by Rackham (1980). In this field, Ball & Stevens (1981) suggested a programme through which variants of selected soils under 'ancient woodland' and other land uses might be studied, on the one hand to assist an understanding of woodland history, and on the other to investigate the contrasting impacts of woodland and other uses on soils of the same series.

Much activity has of course been applied to investigating the historic development of soils in relation to vegetation and to man's impact on both vegetation and soils. This is particularly the case in archaeological work, the study of soils in this context being comprehensively covered by Limbrey (1975). Papers in Evans et al. (1975), which deals with man's impact on the upland landscape of Britain, consider the general relationships between man, vegetation, and soil change, with particular reference to the effects of early man on the woodland–moorland balance. Palaeo-ecological investigations linking soils, vegetation and their history (e.g. Chambers, 1982) must rely on pollen analysis (see Chapter 1). Illustrations from Britain and Ireland of the part played by soil conditions in influencing natural and man-made eco-systems in the past and at the present day, and of the effects of vegetation on soil development, are included in Curtis et al. (1976). As examples of local studies of the interactions between soil profile development, soil history, vegetation, and change over relatively short periods of recent time in Britain, the formation of stabilized soil profiles from mobile sand under dune grass-land and woodland over some 100 years on an expanding dune system has been outlined (Ball & Williams, 1974); an effect of old buried surface horizons in other dune soils on contemporary tree growth has been shown (Gauld, 1981); soil composition changes in relation to recent cultivation history and the vegetation pattern form part of a study on the chalk land of Porton Down, Wiltshire by Wells et al. (1976); changes in vegetation and soils after cultivation had been abandoned on what had formerly been heath soils on the Lizard, Cornwall, have been studied by Marrs & Proctor (1979); and the survival of a historic Dartmoor wood has been shown to be in part related to particular soil conditions, though not controlled by these alone, since historic and contemporary stock-grazing regimes are also a key part of the story (Barkham, 1978).

In considering present vegetation distribution it is often the case that morphologically classified soil types can be correlated more clearly with vegetation than soil chemical parameters can, although the latter of course cannot be ignored. As an example, a study in western Montana, USA (Munn et al., 1978) found that soil classes of the USDA *Soil Taxonomy* (Section 4.2.1) derived from soil morphology, were more effective components in modelling studies of vegetation productivity than either climatic data or soil nutrient levels. Similarly, soil profile classes in small areas of sharp contrast in parent material and landform, such as those described for the mountain locality of Snowdon, North Wales (Ball *et al.*, 1969), correlate strongly with equally sharp contrasts in grassland, moorland and bog vegetation, though standard chemical analyses on their own give less clear-cut explanations for vegetation differences. It is, however, necessary to be aware that other natural environmental factors, such as local climate, can cause apparently identical soils to support quite contrasting vegetation, as with podzols under lichen-heath in Alaska, which, although they have no evidence of a woodland past, have been found (Ugolini *et al.*, 1981) to be indistinguishable—morphologically and by standard pedologic chemical criteria—from neighbouring podzols under spruce; a conclusion throwing doubt on assumptions that the occurrence of such podzols in the region indicated a previous woodland phase in their history.

Turning to soil chemical properties, and returning to British examples, McVean & Ratcliffe (1962) gave the ranges of surface soil chemistry found under vegetation types and individual species in the Scottish Highlands, with particular reference to soil acidity (pH) and calcium content. In general, soil pH and organic matter content are the readily-determined chemical quantities most likely to be clearly relatable to plant distribution. Grime & Lloyd (1973) show the pH ranges found in surface soils supporting a large number of English grassland plant species, distinguishing those species which have limited ranges of soil pH from those which have wide tolerances. In considering upland grassland and moorland vegetation classes, rather than individual species, Ball *et al.* (1981) found a strong correlation between the pH of surface (0–10 cm) soils, and the frequency of vegetation classes within improved pasture, rough pasture, grassy heath and shrubby heath groups. Vegetation stability or change in a range of chalk grassland and heath types at Porton Down, Wiltshire (Wells *et al.*, 1976) could be related to different levels of organic matter within a single soil class, these different levels bringing about alterations in faunal activity, nutrient cycling and moisture status of the rendzina soils. This situation was one in which the influence of soil characteristics on vegetation, and of vegetation on soil development, were of comparable importance.

Among studies of plant influences on soils, one frequently considered effect is the role of tree species in increasing or counteracting an inherent trend to podzolization in relatively low temperature, high rainfall locations in relatively nutrient-poor soils. In the Belgian Ardennes, spruce (*Picea abies*) has been said to produce, over some 80 years, a distinguishable degree of podzolization on acid brown earths (Herbouts & de Buyl, 1981). Conversely, the reverse pedogenic trend from a podzolic soil with mor humus, towards a brown earth with mull humus and consequently different nutrient and organic matter regimes, has been claimed to develop on some sites when birch (*Betula pendula* and *B. pubescens*) becomes established on podzolic moorland soils (Miles & Young, 1980; Miles, 1981). However, in the case of the British site from which this effect was first proposed (Dimbleby, 1952) this result has not begun to show morphologically or in soil faunal changes under 30-year old birch (Satchell, 1980; Miles, 1981). Within one plant community, individuals of some species can have a local effect on soil chemistry, as shown in symmetrical patterns of soil pH change that have been demonstrated across individual plants of gorse (*Ulex europaeus*), heather (*Calluna vulgaris*) and bell heather (*Erica cinerea*) in a grass-heath community on soils over chalk (Grubb *et al.*, 1969).

Finally, a point that may perhaps be useful as a reminder to soil scientists. Plant species can develop ecotypes adapted to particular soil conditions, and this can confuse simplistic associations between a species and soil chemistry. This is not only the case with rather short-lived plants such as white clover (*Trifolium repens*) (Snaydon, 1962) but can occur also with longer-lived woody species, as has been shown for heather (*Calluna vulgaris*) sampled from chemically sharply contrasting soils (Marrs & Bannister, 1978).

8 References

ADAMS W.A. (1973) The effect of organic matter on the bulk and true densities of some uncultivated podzolic soils. *J. Soil Sci.* **24**, 10–17.

ADAS (1975) *Soil Physical Conditions and Crop Production.* Tech. Bull. 29. Ministry of Agriculture, Fisheries, and Food. HMSO, London.

ADAS (1981) *The Analysis of Agricultural Material* (*2nd Edn*). Agricultural Development and Advisory Service, Ministry of Agriculture, Fisheries, and Food. HMSO, London.

AHN P.M. (1970) *West African Soils* (West African Agriculture Vol. 1) Oxford University Press, Oxford.

ALEXANDER M. (1977) *Introduction to Soil Microbiology.* Wiley, New York.

AL-KHAFAF S. and HANKS R.J. (1974) Evaluation of the filter paper method for estimating soil water potential. *Soil Sci.* **117**, 194–199.

ANDERSON G., PIDGEON J.D., SPENCER H.B. and PARKS R. (1980) A new hand-held recording penetrometer for soil studies. *J. Soil Sci.* **31**, 279–296.

ANDERSON J.L. and BOUMA J. (1973) Relationships between saturated hydraulic conductivity and morphometric data of an argillic horizon. *Soil Sci. Soc. Amer. Proc.* **37**, 408–413.

ARMITAGE P.L. (1973) *Aber Mountain: A land and land-use study.* M.Sc. thesis, University College, London.

ARMSTRONG A.C. and DAVIES P.A. (1983) Use of the cone penetrometer. *Area* **15**, 132–135.

ASTM (1970) *Special Procedures for Testing Soil and Rock for Engineering Purposes.* American Society for Testing and Materials. Philadelphia.

ASTM (1972) *Manual on Test Sieving Methods.* American Society for Testing and Materials. Philadelphia.

AUBERT G. (1965) Classification des sols. *Cahiers ORSTOM, sér. Pédologie* III (3), 269–288.

AUBERT G. and DUCHAUFOUR PH (1956) Projet de classification des sols. *Trans. 6th Int. Congr. Soil Sci. Paris.* Vol. E, 597–604.

AVERY B.W. (1973) Soil classification in the Soil Survey of England and Wales. *J. Soil Sci.* **24**, 324–338.

AVERY B.W. (1980) *Soil Classification for England and Wales (Higher Categories)*. Tech. Mon. 14, Soil Survey of England and Wales, Harpenden.

AVERY B.W. and BASCOMBE C.L. (Eds) (1974) *Soil Survey Laboratory Methods*. Tech. Mon. 6, Soil Survey of England and Wales, Harpenden.

BALL D.F., DALE J., SHEAIL J., DICKSON K.E. and WILLIAMS W.M. (1981) *Ecology of Vegetation Change in Upland Landscapes, Part I: General Synthesis*. Bangor Occasional Paper 2, ITE, Bangor.

BALL D.F., DALE, J., SHEAIL J. and HEAL O.W.H. (1982) *Vegetation Change in Upland Landscapes*. ITE, Cambridge.

BALL D.F., MEW G. and MACPHEE W.S.G. (1969) Soils of Snowdon. *Fld. Stud.* **3**, 69–107.

BALL D.F. and STEVENS P.A. (1981) The role of 'ancient' woodlands in conserving 'undisturbed' soils in Britain. *Biol. Cons.* **19**, 163–176.

BALL D.F. and WILLIAMS W.M. (1968) Variability of soil chemical properties in two uncultivated Brown Earths. *J. Soil Sci.* **19**, 379–391.

BALL D.F. and WILLIAMS W.M. (1971) Further studies on variability of soil chemical properties: efficiency of sampling programmes on an uncultivated Brown Earth. *J. Soil Sci.* **22**, 60–68.

BALL D.F. and WILLIAMS W.M. (1974) Soil development on coastal dunes at Holkham, Norfolk, England. *Trans. 10th Int. Congr. Soil Sci., Moscow*, **VI(II)**, 380–386.

BARKHAM J.P. (1978) Pedunculate oak woodland in a severe environment: Black Tor Copse, Dartmoor. *J. Ecol.* **66**, 707–740.

BARKHAM J.P. and NORRIS J.M. (1970) Multivariate procedures in an investigation of vegetation and soil relations of two beech woodlands. Cotswold Hills, England. *Ecology* **51**, 630–639.

BARNES B.V., PREGITZER K.S., SPIES T.A. and SPOONER V.H. (1982) Ecological forest site classification. *J. For.* **80**, 493–498.

BARRATT B.C. (1964) A classification of humus forms and microfabrics of temperate grasslands. *J. Soil Sci.* **15**, 342–356.

BARRATT B.C. (1968) A revised classification and nomenclature of microscopic soil materials with particular reference to organic components. *Geoderma* **2**, 257–271.

BECKETT P.H.T. and WEBSTER R. (1965) *Field Trials of a Terrain Classification System: Units and Principles*. MEXE Report **872**, M. Eng. Exp. Est., Christchurch, G.B.

BECKETT P.H.T. and WEBSTER R. (1971) Soil variability: A review. *Soils Fertil., Harpenden* **34**, 1–15.

BECKETT P.H.T., WEBSTER R., MCNEIL G.M. and MITCHELL C.W. (1972) Terrain evaluation by means of a data bank. *Geogrl. J.* **138**, 430–456.

BELFORD R.K. (1979) Collection and evaluation of large soil monoliths for soil and crop studies. *J. Soil Sci.* **30**, 363–373.

BELL J.P. (1969) A new design principle for neutron soil moisture gauges: The 'Wallingford' neutron probe. *Soil Sci.* **108**, 160–164.

BELL J.P. (1973) *Neutron Probe Practice*. Rept. 19. Institute of Hydrology, Wallingford.

BERRYMAN C., THORBURN A. and TRAFFORD B.D. (1976) *Soil Water Tensiometers*. Tech. Bull. 76–7. ADAS Field Drainage Exp. Unit, Ministry of Agriculture, Fisheries, and Food, HMSO, London.

BIRSE E.L. and ROBERTSON J.S. (1976) *Plant Communities and Soils of the Lowland and Southern Upland Regions of Scotland*. Soil Survey of Scotland, Aberdeen.

BISAL F. and HINMAN W.C. (1972) A method for estimating the apparent density of soil aggregates. *Can. J. Soil Sci.* **52**, 513–514.

BISDOM E.R.A. and DUCLOUX J. (1983) *Submicroscopic Studies of Soils. Geoderma* **30**, (1–4) special issue.

BLACK C.A. (Ed.-in-chief) (1965) *Methods of Soil Analysis. Part I, Physical and Mineralogical Properties: Part 2, Chemical and Microbiological Properties*. American Society of Agronomy.

BLACK C.A. (1968) *Soil–Plant Relationships*. Wiley, Chichester.

BLACKWELL P.S. and SOANE B.D. (1981) A method of predicting bulk density changes in soils resulting from compaction by agricultural traffic. *J. Soil Sci.* **32**, 51.

BOCOCK K.L. (1973) *The Collection of Soil Temperature Data in Intensive Study of Ecosystems.* Merlewood Research and Development Paper 47, ITE, Grange-over-Sands.

BOCOCK K.L., JEFFERS J.N.R., LINDLEY D.K., ADAMSON J.K. and GILL C.A. (1977) Estimating woodland soil temperature from air temperature and other climatic variables. *Agric. Met.* **18**, 351–372.

BOGGIE R. (1977) A simple device for recording maximum and minimum water-table levels in soils. *J. Appl. Ecol.* **14**, 283–285.

BOHN H.L., McNEAL B.L. and O'CONNOR E.A. (1979) *Soil Chemistry.* Wiley, Chichester.

BONNEAU M. and SOUCHIER B. (Eds) (1982) *Constituents and Properties of Soils* (Translated by V.C. Farmer. Originally published as *Pédologie; 2, Constituents et Propriétés du Sol.*) Academic Press, London.

BOUMA J. and ANDERSON J.L. (1973) Relations between soil structure characteristics and hydraulic conductivity. In *Field Soil Water Regime*, SSSA Special Publication 5, Soil Science Society of America.

BOURGEOIS W.W. and LAVKULICH L.M. (1972a) A study of forest soils and leachates on sloping topography using a tension lysimeter . *Can. J. Soil Sci.* **52**, 375–391.

BOURGEOIS W.W. and LAVKULICH L.M. (1972b) Application of acrylic plastic tension lysimeters to sloping land. *Can. J. Soil Sci.* **52**, 288–290.

BOUYOUCOS G.J. (1953) More durable plaster of Paris moisture blocks. *J. Soil Sci.* **76**, 447–451.

BOUYOUCOS G.J. (1954) New type electrode for plaster of Paris moisture blocks. *J. Soil Sci.* **78**, 339–342.

BOUYOUCOS G. (1972) A new electrical soil-moisture measuring unit. *J. Soil Sci.* **114**, 493.

BOUYOUCOS G.J. and MICK A.H. (1948) Fabric absorption unit for continuous measurement of soil moisture in the field. *J. Soil Sci.* **66**, 217–232.

BOWLES J.E. (1979) *Physical and Geochemical Properties of Soil.* McGraw-Hill, New York.

BRADY N.C. (1974) *The Nature and Properties of Soils.* Macmillan, London.

BREWER R. (1964) *Fabric and Mineral Analysis of Soils.* Wiley, New York.

BRIDGES E.M. (1979) *World Soils.* Cambridge University Press, Cambridge.

BRIDGES E.M. and DAVIDSON D.A. (Eds) (1982) *Principles and Applications of Soil Geography.* Longman, Harlow.

BRINDLEY G.W. and BROWN G. (1980) *Crystal structures of clay minerals and their X-ray identification.* Mineralogical Society, London.

BRITISH STANDARDS INSTITUTION (1967) *Methods of Testing Soils for Engineering Purposes.* British Standard 1377, London.

BROWN A.L. (1978) *Ecology of Soil Organisms.* Heinemann Educational Books, London.

BROWN R.W. and VAN HAVEREN B.P. (Eds) (1972) *Psychrometry in Water Relations Research.* Utah State University.

BRYAN R.B. (1968) The development, use and efficiency of indices of soil erodibility. *Geoderma* **2**, 5–26.

BRYAN R.B. (1971) The efficiency of aggregation indices in the comparison of some English and Canadian soils. *J. Soil Sci.* **22**, 166–178.

BUCKLEY R.C. (1982) Soils and vegetation of central Australian sandridges. IV. Soils. *Austr. J. Ecol.* **7**, 187–200.

BUNCE R.G.H., BARR C.J. and WHITTAKER H. (1983) A stratification system for ecological sampling. In *Ecological Mapping for Ground, Air and Space* (ed. R. Fuller). ITE, Cambridge.

BUNTING B.T. (1967) *The Geography of Soil.* Hutchinson, London.

BURGES A. and RAW F. (1967) *Soil Biology.* Academic Press, London.

BUOL S.W., HOLE F.D. and McCRACKEN R.J. (1980) *Soil Genesis and Classification.* Iowa State University Press.

BUTLER B.E. (1980) *Soil Classification for Soil Survey.* Clarendon Press, Oxford.

CAMPBELL D.J. (1973) A flotation method for the rapid measurement of the wet bulk density of soil clods. *J. Soil Sci.* **24**, 239–243.

CANADA SOIL SURVEY COMMITTEE (1978) *The Canadian System of Soil Classification*. Canada Development of Agriculture Publication 1646. Ottawa.

CARROLL D.M., EVANS R. and BENDELOW V.C. (1977) *Air Photo-interpretation for Soil Mapping*. Tech. Mon. 8, Soil Survey of England and Wales, Harpenden.

CARY J.W. and HAYDEN C.W. (1973) An index for soil pore size distribution. *Geoderma* **9**, 249–251.

CHAMBERS F.M. (1982) Two radiocarbon-dated pollen diagrams from high-altitude blanket peats in South Wales. *J. Ecol.* **70**, 445–459.

CHARTRES C.J. (1982) The use of landform–soil associations in irrigation soil surveys in northern Nigeria. *J. Soil Sci.* **33**, 317–328.

CHEPIL W.S. (1958) *Soil Conditions that Influence Wind Erosion*. Tech. Bull. 1185. United States Dept. of Agriculture, Washington.

CHEPILL W.S. (1962) A compact rotary sieve and the importance of dry sieving in physical soil analysis. *Soil Sci. Soc. Amer. Proc.* **26**, 4–6.

CHILDS E.C. (1969) *An Introduction to the Physical Basis of Soil Water Phenomena*. Wiley, New York.

CLARKE G.R. (1971) *The Study of Soil in the Field* (5th edn, assisted by P. Beckett). Clarendon Press, Oxford.

COOKE G.W. (1967) *The Control of Soil Fertility*. Crosby Lockwood, London.

COOKE R.U. and DOORNKAMP J.C. (1974) *Geomorphology in Environmental Management*. Oxford University Press, Oxford.

COPE F. and TRICKETT E.S. (1965) Measuring soil moisture. *Soils Fertil., Harpenden* **28**, 201–208.

CORNISH P.M., LARYEA K.B. and BRIDGE B.J. (1973) A non-destructive method of following moisture content and temperature changes in soils using thermistors. *Soil Sci.* **115**, 309–314.

COURTNEY F.M. (1973) A taxonometric study of the Sherborne soil mapping unit. *Trans. Inst. Brit. Geog.* **50**, 113–124.

COX C.B., HEALEY I.N. and MOORE P.D. (1976) *Biogeography: An Ecological and Evolutionary Approach*. Blackwell Scientific Publications, Oxford.

CRUIKSHANK J.G. (1972) *Soil Geography*. David and Charles, Newton Abbot.

CSIRO (1983) *Soils, An Australian Viewpoint*. Academic Press, London.

CURTIS L.F., DOORNKAMP J.C. and GREGORY K.J. (1965) The description of relief in field studies of soils. *J. Soil Sci.* **16**, 16–30.

CURTIS L.F. and TRUDGILL S. (1974) *The Measurement of Soil Moisture*. Tech. Bull. 13, British Geomorphological Research Group.

CURTIS L.F., COURTNEY F.M. and TRUDGILL S. (1976) *Soils in the British Isles*. Longman, Harlow.

DAVIS W.E. and SLATER C.S. (1942) A direct weighing method for subsequent measurements of soil moisture under field conditions. *J. Amer. Soc. Agron.* **34**, 285.

DE BOODT M. (Secretary-general, editing committee) (1967) *West-European Methods for Soil Structure Determination*. Ghent, Belgium.

DEER W.A., HOWIE R.A. and ZUSSMAN J. (1966) *An Introduction to the Rock-forming Minerals*. Longman, Harlow.

DENT D. and YOUNG A. (1981) *Soil Survey and Land Evaluation*. Allen and Unwin, London.

DEWIS J. and FREITAS F. (1970) *Physical and Chemical Methods for Soil and Water Analysis*. FAO, Rome.

DIMBLEBY G.W. (1952) Soil regeneration on the northeast Yorkshire moors. *J. Ecol.* **40**, 331–341.

DONY J.G. (1967) *Flora of Hertfordshire*. Hitchen Museum.

DUCHAUFOUR PH. (1956) *Pédologie: Applications Forestières et Agricoles*. Nancy, Ecole Nat. des Eaux et Forets.

DUCHAUFOUR PH. (1970) *Précis de Pédologie* (3rd edn). Nancy, Ecole Nat. des Eaux et Forets.

DUCHAUFOUR PH. (1976) *Atlas Ecologique des Sols du Monde*. Masson, Paris.

DUCHAUFOUR PH. (1977) *Pédologie: I, Pédogenèse et Classification*. Masson, Paris.

DUCHAUFOUR P.L. (1982) *Pedology* (translated by T. Paton, originally published as Duchaufour, 1977). Allen and Unwin, London.

DUMANSKI J., BRITTAIN L. and GIRT J.L. (1982) Spatial association between agriculture and the soil resource in Canada. *Can. J. Soil Sci.* **62**, 375–385.

DUMBLETON M.J. (1981) *The British Soil Classification System for Engineering Purposes: its Development and Relation to other Comparable Systems.* Transport and Road Research Laboratory Report, 1030, Dept of the Environment, London.

EASTER S.J. and SOSEBEE R.E. (1974) Use of thermocouple psychrometry in field studies of soil–plant–water relationships. *Plant and Soil* **40**, 707–712.

EMERSON W.W. (1967) A classification of soil aggregates based on their coherence in water. *Aust. J. Soil Res.* **5**, 47–57.

EPSTEIN E. (1972) *Mineral Nutrition of Plants: Principles and Perspectives.* Wiley, New York.

EVANS J.G., LIMBREY S. and CLEERE H. (1975) *The Effect of Man on the Landscape: The Highland Zone.* Research Report II, Council for British Archaeology, London.

EYRE S.R. (1975) *Vegetation and Soils—A World Picture.* Edward Arnold, London.

FAIRBRIDGE R.W. and FINKL C.W. JR (Eds) (1979) *The Encyclopedia of Soil Science Part I, Physics, Chemistry, Biology, Fertility and Technology.* (Encyclopedia of Earth Sciences Series, Vol XII.) Dowden, Hutchinson and Ross, Stroudsburg, Pennsylvania.

FALCK J. (1973) *A Sampling Method for Quantitative Determination of Plant Nutrient Content of the Forest Floor.* (Swedish with English summary.) Research Notes No 1, Royal College of Forestry, Stockholm.

FAO (1968) *Guidelines for Soil Description.* FAO, Rome.

FAO-UNESCO (1974) World Soil Classification. In *Legend to Soil Map of the World, Vol. 1.* UNESCO, Paris.

FITZPATRICK E.A. (1971) *Pedology. A Systematic Approach to Soil Science.* Oliver and Boyd, Edinburgh.

FITZPATRICK E.A. (1980) *Soils, their Formation, Classification and Distribution.* Longmans, Harlow.

FITZPATRICK E.A. and GUDMUNDSSON T. (1978) The impregnation of wet peat for the production of thin-sections. *J. Soil Sci.* **29**, 585–587.

FOALE M.A. and UPCHURCH D.R. (1982) A soil coring method for sites with restricted access. *Agron. J.* **74**, 761–763.

FOURT D.F. (1961) The drainage of a heavy clay site. *Forestry Commission Report on Forest Research for the year ended March, 1960*, pp. 137–150. HMSO, London.

FREITAG D.R. (1971) Methods of measuring soil compaction. In *Compaction of Agricultural Soils* (eds K.K. Barnes *et al.*), pp. 47–105. Am. Soc. Agric. Engs., USA.

GANSSEN R. (1972) *Bodengeographie.* Koehler, Stuttgart.

GARDINER V. and DACKOMBE R.V. (1981) *Geomorphological Field Manual.* Allen and Unwin, London.

GARDNER C.M.K. (1981) *The Soil Moisture Data Bank.* Rept. 76, Institute of Hydrology, Wallingford.

GARRET S.D. (1963) *Soil Fungi and Soil Fertility.* Pergamon, Oxford.

GAULD J.H. (1981) The Soils of Culbin Forest, Morayshire: their evolution and morphology, with reference to their forestry potential. *Appl. Geog.* **1**, 199–212.

GERASIMOV I.P. (1981) FAO–UNESCO world soils units in the light of the concept of elementary soil processes. *Soviet Soil Science* **12**, 148–157.

GERRARD A.J. (1981) *Soils and Landforms.* Allen and Unwin, London.

GERRARD J. (1982) The use of hand-operated soil penetrometers. *Area* **14**, 227–234.

GIESEKING J.E. (1975a) *Soil Components.* Vol. 1, *Organic Components.* Springer, Berlin.

GIESEKING J.E. (1975b) *Soil Components.* Vol. 2, *Inorganic Components.* Springer, Berlin.

GILMAN K. (1977) *Movement of Heat in Soils.* Rept. 44, Institute of Hydrology, Wallingford.

GLOBUS A.M. (1972) Design, operation and temperature sensitivity of a thermocouple psychrometric moisture potentiometer based on the Peltier effect. *Soviet Soil Sci.* **4**, 745–752.

GOOD R. (1948) *A Geographical Handbook of the Dorset Flora.* Dorset Natural History Society, Dorchester.

GOUDIE A. (editor) (1981) *Geomorphological Techniques*. Allen and Unwin, London.

GREACEN E.L. (Ed.) (1981) *Soil Water Assessment by the Neutron Method*. CSIRO, Australia.

GREENLAND D.J. and HAYES M.H.B. (Ed.) (1978) *The Chemistry of Soil Constituents*. Wiley, New York.

GREENLAND D.J. and HAYES M.H.B. (Ed.) (1981) *The Chemistry of Soil Processes*. Wiley, New York.

GREENLAND D.J., RIMMER D. and PAYNE D. (1975) Determination of the structural stability class of English and Welsh soils, using a water coherence test. *J. Soil Sci.* **26**, 294–303.

GRIFFIN D.M. (1972) *Ecology of Soil Fungi*. Chapman and Hall, London.

GRIM R.E. (1968) *Clay Mineralogy*. McGraw-Hill, New York.

GRIME J.P. and LLOYD P.S. (1973) *An Ecological Atlas of Grassland Plants*. Arnold, London.

GRUBB P.I., GREEN H.E. and MERRIFIELD R.C.J. (1969) The ecology of chalk heath: its relevance to the calcicole–calcifuge and soil acidification problems. *J. Ecol.* **57**, 175–212.

GUPTA S.C. and LARSON W.E. (1979) Estimating soil water retention characteristics from particle size distribution, organic matter percent and bulk density. *Water Resources Research* **15**, 1633–1635.

HALL D.G.M., REEVE M.J., THOMASSON A.J. and WRIGHT I.F. (1977) *Soil and Water: Water Retention, Porosity and Density of Field Soils*. Tech. Mon. 6, Soil Survey of England and Wales, Harpenden.

HANKS R.J. and ASHCROFT C.L. (1980) *Applied Soil Physics—Soil Water and Temperature Applications*. Springer, Berlin.

HANNA L.W. and SIAM N. (1980) The estimation of moisture content in the top 10 cm of soil using a neutron probe. *J. Agric. Sci.* (*UK*) **94**, 251–253.

HARKER A. (1962) *Petrology for Students*. Cambridge University Press, Cambridge.

HARRIS S. (1971) Index of structure: evaluation of a modified method of determining aggregate stability. *Geoderma* **6**, 155–162.

HARRISON A.F. and BOCOCK K.L. (1981) Estimation of soil bulk density from loss-on-ignition values. *J. Appl. Ecol.* **18**, 919–928.

HATCH F.H. (1971) *Petrology of the Sedimentary Rocks*. Hafner, New York.

HATCH F.H. (1972) *Petrology of the Igneous Rocks*. Hafner, New York.

HEPPLE S. and BURGES A. (1956) Sectioning of soil. *Nature* (*Lond.*) **177**, 1186.

HERBAUTS J. and DE BUYL E. (1981) The relation between spruce monoculture and incipient podzolization in ochreous brown earths of the Belgian Ardennes. *Plant and Soil* **59**, 33–49.

HESSE P.R. (1971) *A Textbook of Soil Chemical Analysis*. John Murray, London.

HILLEL D. (1971) *Soil and Water—Physical Principles and Processes*. Academic Press, New York.

HILLEL D. (1980) *Fundamentals of Soil Physics*. Academic Press, New York.

HILLEL D. (1982) *Introduction to Soil Physics*. Academic Press, New York.

HILLS R.C. (1970) *The Determination of the Infiltration Capacity of Field Soils using the Cylinder Infiltrometer*. Tech. Bull. 3, British Geomorphological Research Group.

HODGSON J.M. (Ed.) (1974) *Soil Survey Field Handbook*. Tech. Mon. 5, Soil Survey of England and Wales, Harpenden.

HODGSON J.M. (1978) *Soil Sampling and Soil Description*. Clarendon Press, Oxford.

HOLMES A. (1965) *Principles of Physical Geology*. Nelson, Walton on Thames.

INGRAM R.L. (1971) Sieve analysis. In *Procedures in Sedimentary Petrography* (ed. R.E. Carver), pp. 49–67. Wiley, New York.

INSTITUTE OF TERRESTRIAL ECOLOGY (1978) *Upland Land Use in England and Wales*. Publ. III, Countryside Commission, Cheltenham.

INTERNATIONAL SOCIETY OF SOIL SCIENCE (1962) *Transactions of Commissions IV and V joint meeting*. New Zealand.

INTERNATIONAL SOCIETY OF SOIL SCIENCE (1967) Proposal for a uniform system of soil horizon designations. *Bull. Int. Soc. Soil Sci.* **31**, 4–7.

JACKSON M.L. (1956) *Soil Chemical Analysis—Advanced Course*. Dept. of Soils, Madison, Wisconsin.

JEFFREY D.W. (1970) A note on the use of ignition loss as a means for the approximate estimation of bulk density. *J. Ecol.* **58**, 297–299.

JOHN M.K., LAVKULICH L.M. and ZOOST M.A. (1972) Representation of soil data for the computerized filing system used in British Columbia. *Can. J. Soil Sci.* **52**, 293–300.

JONES R.J.A. and COURT M.N. (1980) The measurement of mean temperatures in plant and soil studies by the sucrose inversion method. *Plant and Soil* **54**, 15–31.

KADDAH M.T. (1974) The hydrometer method for detailed particle size analysis. I. Graphical interpretation of hydrometer readings and test of method. *Soil Sci.* **118**, 102–108.

KERR P.F. (1959) *Optical Mineralogy*. McGraw-Hill, New York.

KINAKO P.D.S. and GIMINGHAM C.H. (1980) Heather burning and soil erosion on upland heaths in Scotland. *J. Env. Man.* **10**, 277–284.

KING C.A.M. (1966) *Techniques in Geomorphology*. Arnold, London.

KIRKBY M.J. and MORGAN R.P.C. (Eds) (1980) *Soil Erosion*. Wiley, New York.

KNAPP B.J. (1973) *A system for the field measurement of soil water movement*. Tech. Bull. 9, British Geomorphological Research Group.

KOOREVAAR P., MENELIK G. and DIRKSEN C. (1983) *Elements of Soil Physics*. Developments in Soil Science 13. Elsevier, Amsterdam.

KRAHN J. and FREDLUND D.G. (1972) On total, matric and osmotic suction. *Soil Sci.* **114**, 339–348.

KRAMER P.J. (1983) *Water Relations of Plants*. Academic Press, New York.

KRUMBEIN W.C. and PETTIJOHN F.J. (1938) *Manual of Sedimentary Petrography*. Appleton-Century Crofts, New York.

KUBIENA W.L. (1938) *Micropedology*. Collegiate Press, Ames, Iowa.

KUBIENA W.L. (953) *Soils of Europe*. Murby, London.

KUBIENA W.L. (1970) *Micromorphological Features of Soil Geography*. Rutgers University Press.

KUCHENBUCH R. and JUNGK A.A. (1982) A method for determining concentration profiles at the soil–root interface by thin slicing rhizospheric soil. *Plant and Soil* **68**, 391–394.

KUHNELT W. (trans. N. Walker) (1961) *Soil Biology*. Faber and Faber, London.

LELIEVRE F. (1982) [Testing several varieties of the cylinder method for the determination of soil bulk density] (in French). *Bull. Ass. Fr. Et. Sol* **3**, 245–253.

LIM H.S. (1982) An evaluation of the use of the Spex Mixer Mill unit for soil particle dispersion. *Soil Sci. Plant Nutr.* **28**, 131–140.

LIMBREY S. (1975) *Soil Science and Archaeology*. Academic Press, New York.

LINDSAY W.L. (1979) *Chemical Equilibria in Soils*. Wiley, New York.

LINES R. and HOWELL R.S. (1963) *The Use of Flags to Estimate the Relative Exposure of Trial Plantations*. Forest Record 51, Forestry Commission, HMSO, London.

LOW A.J. (1954) The study of soil structure in the field and in the laboratory. *J. Soil Sci.* **5**, 57–74.

MCCRONE W.C. (1982) Soil comparison and identification of constituents. *Microscope* **30**, 17–25.

MCLAREN A.D. and PETERSON G.H. (1967) *Soil Biochemistry*. Vol. 1. Arnold, London.

MCVEAN D.N. and RATCLIFFE D.A. (1962) *Plant Communities of the Scottish Highlands*. Monograph 1, The Nature Conservancy. HMSO, London.

MARRS R.H. and BANNISTER P. (1978) The adaption of *Calluna vulgaris* (L). Hull to contrasting soil types. *New Phytol.* **81**, 753–761.

MARRS R.H. and PROCTOR J. (1979) Vegetation and soil studies of the enclosed heathlands of the Lizard Peninsula, Cornwall. *Vegetatio* **41**, 121–128.

MARSHALL C.E. (1964) *The Physical Chemistry and Mineralogy of Soils*. Vol. I. *Soil Materials*. Wiley, New York.

MARSHALL T.J. (1959) *Relations between Water and Soil.*. Tech. Comm. 50, Commonwealth Agric. Bur., Harpenden.

MARSHALL T.J. and HOLMES J.W. (1979) *Soil Physics*. Cambridge University Press, Cambridge.

MEHLICH A. (1972) Uniformity of expressing soil test results. A case for calculating results on a volume basis. *Commun. Soil Sci. Plant Anal.* **3**, 417–424.

MEHLICH A. (1973) Uniformity of soil test results as influenced by volume weight. *Commun. Soil Sci. Plant Anal.* **4**, 475–486.

MIKHAIL E.H. and BRINER G.P. (1978) Routine particle size analysis of soils using sodium hypochlorite and ultra-sonic dispersion. *Aust. J. Soil Res.* **16**, 241–244.

MILES J. (1981) *Effects of Birch on Moorlands.* ITE, Cambridge.

MILES J. and YOUNG W.F. (1980) The effects on heathland and moorland soils in Scotland and Northern England following colonization by birch (*Betula spp.*). *Bull. Ecol.* **11**, 233–242.

MILNER H.B. (1962) *Sedimentary Petrography.* Allen and Unwin, London.

MIRREH H.F. and KETCHESON J.W. (1972) Influence of soil bulk density and matric pressure on soil resistance to penetration. *Can. J. Soil Sci.* **52**, 477–483.

MITCHELL C.W. (1973) Soil classification with particular reference to the Seventh Approximation. *J. Soil Sci.* **24**, 411–420.

MITCHELL C.W. (1973b) *Terrain Evaluation.* Longman, Harlow.

MOORE A.W., RUSSELL J.S. and WARD W.T. (1972) Numerical analysis of soils; a comparison of three soil profile models with field classifications. *J. Soil Sci.* **23**, 193–209.

MOSS D. (1985) An initial classification of 10-km squares in Great Britain from a land characteristic data bank. *Appl. Geog.* **5**, 131–150.

MOSS R.P. (Ed.) (1968) *The Soil Resources of Tropical Africa.* Cambridge University Press, Cambridge.

MÜCKENHAUSEN E. (1962) *Entstehung, Eigenschaften und Systematik der Boden der Bundesrepublik Deutschland,* DLG Frankfurt.

MÜCKENHAUSEN E. (1965) The soil classification system of the Federal Republic of Germany. *Pedologie,* Special issue 3, 57–89, Ghent, Belgium.

MUIR J.W. (1962) The general principles of classification with reference to soils. *J. Soil Sci.* **13**, 22–30.

MUIR J.W. and HARDIE H.G.M. (1962) A punched-card system for soil profiles. *J. Soil Sci.* **13**, 249–253.

MULLINS C.E. and HUTCHINSON J.B. (1982) The variability introduced by various sub-sampling techniques. *J. Soil Sci.* **33**, 547–561.

MUNN L.C., NIELSEN G.A. and MUEGGLER W.F. (1978) Relationship of soils to mountain and foothill range habitat types and production in Montana. *Soil Sci. Soc. Am. J.* **42**, 135–139.

NAKAYAMA F.S. and REGINATO R.J. (1982) Simplifying neutron moisture meter calibration. *Soil Sci.* **133**, 48–52.

NIELSEN D.R., BIGGAR J.W. and COREY J.C. (1972) Applications of flow theory to field situations. *Soil Sci.* **113**, 254–264.

NORRIS J.M. (1971) The application of multivariate analysis to soil studies. I. Grouping of soils using different properties. *J. Soil Sci.* **22**, 69–80.

NORRIS J.M. and LOVEDAY J. (1971) The application of multivariate analysis to soil studies II. The allocation of soil profiles to established groups: A comparison of soil survey and computer methods. *J. Soil Sci.* **22**, 395–400.

NYE P.H. and TINKER P.B. (1977) *Solute Movement in the Soil Root System.* Blackwell Scientific Publications, Oxford.

OERTLI J.J. (1973) The use of chemical potentials to express nutrient availabilities. *Geoderma* **9**, 81–95.

OLSON G.W. (1982) *Soils and the Environment.* Chapman and Hall, London.

OSMOND D.A. and BULLOCK P. (Eds) (1970) *Micromorphological Techniques and Applications.* Tech. Mon. 2, Soil Survey of England and Wales, Harpenden.

PARKINSON D., GRAY T.R.G. and WILLIAMS S.T. (1971) *Methods for Studying the Ecology of Soil Micro-organisms.* IBP Handbook 19. Blackwell Scientific Publications, Oxford.

PETERKEN G.F. (1967) *Guide to the Check Sheet for IBP Areas.* IBP Handbook 4. Blackwell Scientific Publications, Oxford.

PHILLIPSON J. (Ed.) (1977) *Methods of Study in Quantitative Soil Ecology.* IBP Handbook 18. Blackwell Scientific Publications, Oxford.

PITTY A.F. (1968) A simple device for the field measurement of hillslopes. *J. Geol.* **76**, 717–720.

PITTY A.F. (1969) *A Scheme for Hillslope Analysis.* Occ. Papers in Geog. 9, University of Hull.

PRITCHARD D.T. (1974) A method for soil particle-size analysis using ultrasonic disaggregation. *J. Soil Sci.* **25**, 34–40.

PYATT D.G., HARRISON D. and FORD A.S. (1969) *Guide to Site Types in Forests in North and Mid-Wales.* Forest Record 69, Forestry Commission. HMSO, London.

RACKHAM O. (1980) *Ancient Woodland: Its History, Vegetation and Uses in England.* Edward Arnold, London.

RAGG J.M. and CLAYDEN B. (1973) *The classification of some British soils according to the comprehensive system of the United States.* Tech. Mon. 3. Soil Survey of England and Wales, Harpenden.

RAWLS W.J. (1983) Estimating soil bulk density from particle size analyses and organic matter content. *Soil Sci.* **135**, 123–125.

READ H.H. and WATSON J. (1962) *Introduction to Geology, I. Principles.* Macmillan, London.

REEVE M.H., SMITH P.D. and THOMASSON A.J. (1973) The effect of density on water retention properties of field soils. *J. Soil Sci.* **24**, 355–367.

REYNOLDS E.R.C. (1966) The percolation of rainwater through soil demonstrated by fluorescent dyes. *J. Soil Sci.* **17**, 127–132.

REYNOLDS S.G. (1970) The gravimetric method of soil moisture determination. Pt. I. A study of equipment and methodological problems. *J. Hydrol.* **11**, 258–273.

REYNOLDS S.G. (1975) *Soil property variability in slope studies.* Z. Geomorph. **19**, 191–208.

RICE R.J. (1977) *Fundamentals of Geomorphology.* Longman, Harlow.

RICHARDS B.N. (1974) *Introduction to the Soil Ecosystem.* Longman, Harlow.

RIMMER D.L. (1982) Soil physical conditions on reclaimed colliery spoil heaps. *J. Soil Sci.* **33**, 567–579.

RODSKJER N. (1975) Thermoelectric thermometers for measurement of soil temperature in crops. *Swedish J. Agric. Res.* **5**, 23–26.

ROGERS J.S. (1974) Small laboratory tensiometers for field and laboratory studies. *Soil Sci. Soc. Amer. Proc.* **38**, 690–691.

RORISON I. (Ed.) (1969) *Ecological Aspects of the Mineral Nutrition of Plants.* Blackwell Scientific Publications, Oxford.

ROSE C.W. (1966) *Agricultural Physics.* Pergamon, Oxford.

ROWAN A.A. (1977) *Terrain Classification.* Forest Record 114, Forestry Commission. HMSO, London.

ROWE J.S. and SHEARD J.W. (1981) Ecological land classification: A survey approach. *Env. Man.* **5**, 451–464.

RUDEFORTH C.C. and WEBSTER R. (1973) Indexing and display of soil survey data by means of feature cards and Boolean maps. *Geoderma* **9**, 229–248.

RUSEK J. (1975) Die bodenbildende Funktion von Collembola und Acarina. *Pedobiologia* **15**, 299–308.

RUSSELL SIR E.J. (1971) *The World of the Soil.* Collins, London.

RUSSELL E.W. (1974) *Soil Conditions and Plant Growth.* Longman, Harlow.

RUTHERFORD G.K. (1974) *Soil Microscopy.* (Proc. 4th Int. Meeting on Soil Micromorphology, 1973). Limestone Press, Kingston, Ontario.

RUTTER A.J. (1955) The composition of wet heath vegetation in relation to the water table. *J. Ecol.* **43**, 507–543.

RUTTER A.J. and SANDS K. (1958) The relation of leaf water deficit to soil moisture tension in *Pinus sylvestris* L. I. The effect of soil moisture on diurnal changes in water balance. *New Phytol.* **57**, 50–65.

SAINI G.R. (1980) *Pedogenetic and Induced Compaction in Agricultural Soils.* Tech. Bull. Agric. Canada Res. Stat., Fredericton, New Brunswick.

SALTER P.J., BERRY G. and WILLIAMS J.B. (1966) The influence of texture on the moisture characteristics of soils. III. Quantitative relationships between particle size composition and available-water capacity. *J. Soil Sci.* **17**, 93–98.

SALTER P.J. and WILLIAMS J.B. (1965a) The influence of texture on the moisture characteristics of soils. I. A critical comparison of techniques for determining the available-water capacity and moisture characteristic curve of a soil. *J. Soil Sci.* **16**, 310–317.

SALTER P.J. and WILLIAMS J.B. (1965b) The influence of texture on the moisture characteristics of soils. II. Available-water capacity and moisture release characteristics. *J. Soil Sci.* **16**, 310–317.

SALTER P.J. and WILLIAMS J.B. (1967) The influence of texture on the moisture characteristics of soils. *J. Soil Sci.* **18**, 174–181.

SANDS C.R. and REID C.P.P. (1980) The osmotic potential of soil water in plant/soil systems. *Aust. J. Soil Res.* **18**, 13–25.

SATCHELL J.E. (1980) Soil and vegetation changes in experimental birch plots in a *Calluna* podzol. *Soil Biol. Biochem.* **12**, 303–310.

SAVAGE M.J. (1980) An inexpensive resistance thermometer for air temperature, soil temperature and relative humidity measurements. *Agrochemophysica* **12**, 1–3.

SCHMUGGE T.J., JACKSON T.J. and MCKIMM H.L. (1980) Survey of methods for soil moisture determination. *Water Resources Research* **16**, 961–979.

SCHNITZER M. and KHAN S.U. (1978) *Soil Organic Matter*. Elsevier, Amsterdam.

SCHOFIELD R.K. (1935) The pF of water in soil. *Trans. 3rd Int. Cong. Soil Sci.* **2**, 37–48.

SEDDON B. (1971) *Introduction to Biogeography*. G. Duckworth, London.

SHAW M.D. and ARBLE W.C. (1959) Bibliography on methods for determining soil moisture. *Engng. Res. Bull. Coll. Engng. Archit. Pa.* **3**, 13–78.

SHIPP R.F. and MATELSKI R.P. (1964) Bulk density and coarse fragment determinations on some Pennsylvania soils. *Soil Sci.* **99**, 392–397.

SLATYER R.O. and TAYLOR S.A. (1960) Terminology in plant- and soil-water relations. *Nature* **187**, 922–924.

SMITH C.J. (1980) *Ecology of the English Chalk*. Academic Press, London.

SMITH G.D. (1965) Lectures on soil classification. *Pédologie*, Special issue 4, Ghent, Belgium.

SNAYDON R.W. (1962) The growth and competitive ability of contrasting natural populations of *Trifolium repens* L. on calcareous and acid soils. *J. Ecol.* **30**, 439–447.

SOANE B.D. (1973) Techniques for measuring changes in the packing state and core resistance of soil after the passage of wheels and tracks. *J. Soil Sci.* **24**, 311–323.

SPANNER D.C. (1951) The Peltier effect and its use in measurement of suction pressure. *J. exp. Bot.* **2**, 145–168.

SPARKS B.W. (1972) *Geomorphology*. Longman, Harlow.

STRANGEWAYS I.C. (1983) Interfacing soil moisture gypsum blocks with a modern data logging system using a simple low-cost direct current method. *Soil Sci.* **136**, 322–324.

STRONG W.L. and LIMBIRD A. (1981) A key for classifying soils to the subgroup level of the Canadian system of soil classification. *Can. J. Soil Sci.* **61**, 285–294.

TANNER C.B. and JACKSON M.L. (1948) Nomographs of sedimentation times for soil particles under gravity or centrifugal acceleration. *Soil Sci. Soc. Amer. Proc.* **12**, 60–65.

TAYLOR N.H. and POHLEN I.J. (1962) *Soil Survey methods*. Soil Bur. Bull. 25. Dept. of Sci. Ind. Res., New Zealand.

TAYLOR N.H. and POHLEN I.J. (1968) Classification of New Zealand soils. In *Soils of New Zealand, Part I* pp. 15–46) Soil Bur. Bull. 26(1). Dept. of Sci. Ind. Res., New Zealand.

THOMAS A.M. (1966) *In situ* measurement of moisture in soil and similar substances by 'fringe' capacitance. *J. scient. Instrum.* **43**, 21–27.

THOMAS T.M. (1973) Tree deformation by wind in Wales. *Weather* **28**, 46–58.

THOMASSON A.J. (1971) Soil water regimes. *Report Welsh Soils Discussion Group* **12**, 96–105.

THOMASSON A.J. (1978) Towards an objective classification of soil structure. *J. Soil Sci.* **29**, 38–46.

THOMASSON A.J. and ROBSON J.D. (1967) The moisture regimes of soils developed on Keuper Marl. *J. Soil Sci.* **18**, 329–340.

THOMPSON T.R.E. (1979) A comprehensive system of soil description and classification. *J. Env. Man.* **9**, 247–253.

THORPE J. and SMITH G.D. (1949) Higher categories of soil classification: Order, Sub-order and Great Soil Groups. *Soil Sci.* **67**, 117–126.

TINKER P.B. (Ed.) (1980) *Soils and Agriculture: Critical Reports on Applied Chemistry*, Vol. 2. Blackwell Scientific Publications, Oxford.

TOLEMAN R.D.L. (1973) A peat classification for forest use in Great Britain. Paper 10, Proc. Int. Peat Soc. Symposium, Glasgow. In *Classification of Peat and Peatlands*. International Peat Society, Helsinki.

TOWNSEND J. (Ed) (1981) *Terrain Analysis and Remote Sensing*. Allen and Unwin, London.

TRUDGILL S.T. (1977) *Soil and Vegetation Systems*. Clarendon Press, Oxford.

TRUDGILL S.T. and BRIGGS D.J. (1981) Soil and land potential. *Progress in Phys. Geog.* **5**, 274–285.

UGOLINI F.C., REANIER R.E., RAU G.H. and HEDGES J.I. (1981) Pedological isotopic and geochemical investigation of soils at the boreal forest and alpine tundra transition in northern Alaska. *Soil Sci.* **131**, 359–374.

USDA (1938) *Soils and Men*. United States Dept. of Agriculture, Washington.

USDA (1951) *Soil Survey Manual*. Agric. Handbook 18. US Dept. of Agriculture, Washington.

USDA (1975) *Soil Taxonomy*. Agric. Handbook 436. US Dept. of Agriculture, Washington.

USDA (1981) *Soil Resource Inventories and Development Planning*. Tech. Mon. 1. Soil Management Services, United States Dept. of Agriculture, Washington.

VAN DEN BROEK M., VAN AMSTEL A., VERBAKEL A. and PEDROLI B. (1981) Variability of soil properties in a landscape ecological survey in the Tuscan Appenines, Italy. *Catena* **8**, 155–170.

VISVALINGAM M. and TANDY J.D. (1972) The neutron method for measuring soil moisture content—a review. *J. Soil Sci.* **23**, 499–511.

VISVALINGAM M. (1974) Well-point techniques and the shallow water-table in boulder clay. *J. Soil Sci.* **25**, 505–516.

WALKER N. (Ed.) (1975) *Soil Microbiology: A Critical Review*. Butterworth, London.

WALLWORK J.A. (1970) *The Ecology of Soil Animals*. McGraw-Hill, New York.

WATSON J.R. (1971) Ultrasonic vibration as a method of soil dispersion. *Soils Fertil., Harpenden* **34**, 127–134.

WEBSTER R. (1966) The measurement of soil water tension in the field. *New Phytol.* **65**, 249–258.

WEBSTER R. (1968) Fundamental objections to the 7th Approximation. *J. Soil Sci.* **19**, 354–366.

WEBSTER R. (1975) *Quantitative and Numerical Methods in Soil Classification and Survey*. Oxford University Press, Oxford.

WEBSTER R. and BECKETT P.H.T. (1970) Terrain classification and evaluation using air photography—a review of recent work at Oxford. *Photogrammetria* **26**, 51–75.

WELLS C.B. (1959) Core samplers for soil profiles. *J. Agric. Engng. Res.* **4**, 260–266.

WELLS T.C.E., SHEAIL J., BALL D.F. and WARD L. (1976) Ecological studies on the Porton Ranges: relationships between vegetation, soils and land-use history. *J. Ecol.* **64**, 589–626.

WHITE R.E. (1979) *Introduction to the Principles and Practice of Soil Science*. Blackwell Scientific Publications, Oxford.

WILLIAMS J.B. (1968) Measurements of total and matric suctions of soil water using thermocouple psychrometer and pressure membrane apparatus. *J. appl. Ecol.* **5**, 263–272.

WILLIAMS B.G., GREENLAND D.J., LINDSTROM G.R. and QUIRK J.P. (1966) Techniques for the determination of soil aggregates. *Soil Sci.* **101**, 157–163.

WILLIAMS T.H.L. (1978) An automatic scanning and recording tensiometer system. *J. Hydrol.* **39**, 175–183.

WILLIAMS T.H. (1980) An automatic electrical resistance soil-moisture measuring system. *J. Hydrol.* **46**, 385–390.

WOODWARD F.I. and SHEEHY J.L. (1983) *Principles and Measurements in Environmental Biology*. Butterworths, London.

WRIGHT M.J. (1971) The preparation of soil monoliths for the 9th International Congress of Soil Science, Adelaide, Australia. *Geoderma* **5**, 151–159.

YOUNG A. (1972) *Slopes*. Oliver and Boyd, Edinburgh.

YOUNG A. (1976) *Tropical Soils and Soil Survey*. Cambridge University Press, Cambridge.

6   Chemical analysis

S.E. ALLEN, H.M. GRIMSHAW and A.P. ROWLAND

1   Introduction

This chapter deals with the chemical analysis of soils, plant materials and waters that are of interest in plant ecology. The inclusion of waters is justified because certain types of waters such as rainwater, canopy through-fall and lysimeter waters are particularly relevant to terrestrial ecosystems. This chapter is designed to help those whose main speciality is not in analytical chemistry but a familiarity with basic laboratory practices is assumed. The procedures given for each method can be applied without clarification, although additional references are given where appropriate.

The procedures given here mainly concern the principal inorganic nutrient elements and physical analyses are largely dealt with elsewhere in the book. Organic constituents are omitted but further information about their estimation can be obtained from Allen (1974) and from other analytical works.

1.1   Laboratory practice and contamination

It is impossible in one short section to provide a comprehensive set of instructions on good analytical practice. All that can be done here is to highlight some of the precautions which need to be taken.

(a) All glass and plastic ware used in the procedures should be scrupulously clean. Washing with phosphorus-free detergent followed by several pure water rinses will suffice in most cases. More powerful cleaning agents such as chromic acid may occasionally be required. Borosilicate glass should be used where possible.

(b) Sources of contamination must be identified and eliminated. Some of the more common examples include:
  (i) external dust blown in from the surrounding environment—filtered air is desirable in these cases;
  (ii) internal dust arising from cleaning operations, rusty fittings, plaster and decorating materials;
  (iii) washing materials, particularly soap and scouring powders, and cosmetics;
  (iv) metallic dust from grinding and sieving operations;
  (v) laboratory reagents based on elements that have to be determined in other tests.

(c) Analytical grade reagents should always be used where possible. Many salts used in standards have to be dried in an oven at 105°C before weighing. Glass-distilled or deionized water should always be used. Deionized water should have a conductivity value of below 0·2 μS.

(d) In most procedures at least two blank determinations should be included to correct for any background introduced from reagents, filter papers, etc. Provided the two blanks agree the mean value can be subtracted from the sample result.

(e) It is good practice to include a reference material of known composition with each batch for analysis. This will allow any gross errors to be picked up and will provide a check on minor changes in analytical procedure which might occur over a long period.

1.2   Concentration levels

The information given in Table 6.1 has largely been derived from natural ecological materials which have been analysed in the author's laboratory. They cover plants and soils which are commonly found in the UK, but only data from the photosynthetic tissues of the plants have been

Table 6.1 Typical concentration ranges of elements in soil and plant materials (dry basis)

	Soil Extractable (mg 100 g^{-1})	Total (%)	Plant material (%)
Aluminium	10–200	1–10	0·005–0·025
Calcium	10–250*	0·4–4*	0·2–4
Carbon (organic)	–	0·5–15‡	44–46
Chlorine	–	0·004–0·04	0·05–0·5
Copper	2–30	0·001–0·01	0·0003–0·003
Iron	5–10	0·4–6	0·001–0·01
Magnesium	2–50	0·1–1†	0·1–0·4
Manganese	0·4–40	0·02–0·3	0·002–0·1
Nitrogen (organic)	–	0·1–1	1·5–5
(NH$_4$-N)	0·1–1	–	–
(NO$_3$-N)	0·05–1	–	–
Phosphorus (total)	0·2–2	0·01–0·2	0·03–0·5
Potassium	5–20	0·1–2	0·1–5
Silicon	–	5–40	0·05–0·5
Sodium	2–20	0·05–0·5	0·01–0·03
Sulphur	–	0·04–0·2	0·05–0·2
Zinc	10–400	0·003–0·03	0·002–0·01

*Higher in calcareous soils.
†Higher in dolomitic soils.
‡Higher in organic soils.

included. The soils are mostly brown earths but include some calcareous, podzolic and peaty-gley materials obtained from the A and B horizons.

1.3 Accuracy and precision

Results will invariably differ from the 'true' or 'absolute' value. If accuracy is defined as the closeness of the result to the 'true' value then it is desirable to have information about the size of the difference (systematic error or 'bias'). It is also important to know how much a value has been affected by random variation.

Errors therefore fall into two groups: random and systematic. Random errors are unpredictable and difficult to assign to a specific cause but are often associated with the operator and his equipment. Examples include minor variations in the use of balances, pipettes, volumetric flasks, etc. and in the behaviour of instruments. It should be noted that random errors arising in the field are usually greater than in the laboratory but an adequate field sampling programme will contain their effects (Section 2.1).

Systematic errors form a bias away from the true result, but as the latter is usually unknown this error is not easy to correct. Possible sources include failure to bring all forms of an element into solution during digestion, the use of inappropriate calibration curves and subtraction of faulty blank values. However, the most common source is chemical interference in the method. This can be suspected if the method fails to recover quantitatively known amounts of the element added to a portion of the sample solution ('standard addition' test). Attention is drawn to them in the text and recommendations made as appropriate.

Precision is the spread of results within a group of replicate determinations carried out at the same time. Reproducibility is better understood as the spread of results of replicate determinations carried out over a period of time. Both are governed by random errors associated with the operator and equipment. Their overall magnitude is easily expressed statistically as the 'standard deviation' whose calculation is described in most basic statistical texts. There are a number of publications that deal specifically with accuracy and precision in analytical chemistry and these should be referred to for further information.

2 Soils

Soil is an extremely complex medium and its variability causes many problems in analysis and in interpretation of analyical results. Only salient points can be mentioned here but further discussion on pedological aspects and physical analysis is given in Chapter 5. Reference books dealing with the analysis of soils include Hesse (1971), Allen (1974), Page *et al.* (1982) and Smith (1983).

Although there is no generalized technique for handling of soils in chemical analysis, certain basic principles underlie the variety of procedures in use. These will be discussed under each subheading as appropriate.

2.1   Collection

In general, soil sampling methods depend on the type of study being carried out. In experimental studies, sampling should be randomized to minimize the effect of any bias and to facilitate statistical assessment of the results. The number of replicate samples determines the error (precision) of the results. Large errors weaken conclusions so it is desirable to contain the precision within acceptable limits. The minimum number of replicates required to meet chosen limits is easily calculated if the variability (variance) of the site is known. This is usually not the case in field work so a small pilot study would be required to make an estimate. In practice, many experimental designs allow for five or ten replicates. However, it should be further noted that the removal of large numbers of samples from small experimental plots may itself become a factor influencing the experiment. In survey work sampling is usually systematic rather than random, but adequate replication is still required at each sampling point. The main aspects of sampling and design are covered in most elementary statistical texts which should be consulted for further details.

The depth of sampling is usually determined by the objective of the study but for many purposes rooting depth (up to 20 cm) will be sufficient. However, the study of a soil profile will require taking samples at specific depths or according to the soil horizon.

When results are to be expressed on a weight basis, sampling with a trowel or similar tool is sufficient. For analysis on a volume basis soil corers or augers are preferable but may not be suitable for stony soils. Methods for measuring bulk density of soils where augers cannot be used are described in Chapter 5. Care must be taken to guard against interhorizon contamination, particularly in sandy soils. This problem can become accentuated when using augers or corers.

Polythene bags are frequently used for the collection of soil samples. Although they are suitable for short-term use, there is a danger of incubation effects, particularly when left sealed in a warm atmosphere. Paper bags with waterproof liners were widely used before polythene become available. Plastic or aluminium containers are sometimes chosen but the possibility of metal contamination must not be overlooked.

2.2 Transport and storage

To minimize possible deterioration, transport times should be kept short and the samples cool during transit. Frozen 'cold packs' with the samples packed in insulated containers are effective for maintaining low temperatures. In the laboratory, prolonged storage of fresh soil samples should be avoided if possible. Even for brief storage the samples should be kept at a temperature just above freezing, although deep-freeze storage should be avoided as properties of a soil can be affected (Allen & Grimshaw, 1962). Labile constituents such as inorganic nitrogen fractions and values such as pH should be determined immediately on receipt of fresh material, but most other estimations can be made using the air-dried material.

2.3 Drying and moisture determination

For analytical purposes a soil is 'dry' when it has reached a constant weight at a given drying temperature. Complete dryness is difficult to achieve or even define, partly because it is not possible to remove all the water without loss of other volatile components or some breakdown of organic or mineral water. Such changes in soil may occur by heating at any elevated temperature, whilst at higher drying temperatures some of the structural water associated with organic compounds or held in soil mineral lattice structures will still be retained.

The most satisfactory compromise procedure for air-drying soils is to leave them thinly spread at 40°C in an air-circulated oven overnight. True air-drying by spreading on, say, a laboratory bench, is less satisfactory and the extra time needed to reach equilibrium provides ample opportunity for microbial and enzymic effects to occur. This may not be too serious for some soils but can be troublesome if the more labile materials are being studied. When estimates of extractable constituents are required then fresh or air-dried soils must be used rather than material that has been dried at high temperatures (*c.* 100°C) because the exchange properties may then be unduly affected. Recommendations are given in Table 6.2.

PROCEDURE

1 Dry a suitable container at 105°C for a few minutes. Cool in a desiccator.
2 Weigh the container empty and then with a convenient amount of fresh sample. Take enough to be representative of the sample but do not pack the container too tightly.
3 Dry at 105°C for 3 hours in an air-circulated oven. Cool in dessicator and reweigh.

Table 6.2 Procedures for the initial treatment of soils

Analyses required	Recommended treatment
Redox potential, pH	Field examination.
NO_3^-, NH_4^+, pH ⎫ Peat extractions ⎭	Storage for minimum period at + 1°C, then examine fresh.
Extraction of mineral soils ⎫ Mineralizable -N ⎬ Cation exchange capacity ⎭	Air-dry at not more than 40°C, then lightly crush through 2 mm sieve.
Loss-on-ignition ⎫ Total N, P, C ⎭	Dry as above, then grind finely.

4 Repeat drying for further 1-hour periods until constant weight is reached.

Calculate as percentage moisture if the figure is required as an analytical result, or as percentage dry matter if it is to be used to convert results from a fresh to a dry weight basis. The procedure is the same, except that 3 hours drying alone will be sufficient. This figure is normally used only for correction purposes.

2.4 Sieving and grinding

Fresh material is handled as received, although it is usual to separate large stones and roots before proceeding. Air-dry material should be passed through a 2 mm mesh sieve, aggregates being broken gently because some minerals are very soft. Machines are available to standardize this operation. A suitable model is marketed by D. Mackay of Britannia Works, Cambridge in which stainless steel bars inside a cylindrical cage drilled with 2 mm holes crush the soil as the cylinder rotates. Stainless steel sieves are recommended to minimize contamination. Brass should be avoided, particularly if copper and zinc are to be determined.

The soil will need to be well mixed after sieving and some bulk reduction may sometimes be necessary. Successive mixing, heaping and quartering with retention of opposite quarters is the conventional procedure, although there are also a number of mechanical aids which are suitable for the purpose.

Soil passed through 2 mm mesh should be used for the determination of extractable nutrients, but a further grinding is needed prior to using procedures requiring only a fraction of a gram of sample. In this case a representative fraction should be ground to pass 100 or even 250 mesh (Kleeman, 1967) using a mortar and pestle, ball or swing mill or an abrasive grinder. The grinding equipment should be made of agate, hardened steel or tungsten carbide.

2.5   Loss-on-ignition

This value is widely interpreted as a crude estimate of organic carbon in soil (see also Section 5.4.1). It is difficult to remove the last traces of carbon without loss of volatile inorganic components or 'bound-water', so there is a basic problem similar to that described for moisture. In some methods low ignition temperatures, even down to 375°C, have been suggested to minimize volatile loss but this involves a very long combustion period and traces of carbon often remain. In the procedure described below, 550°C is used. At this temperature errors due to volatile inorganic losses are minimal except when the organic content is low. In this case a loss-on-ignition value overestimates the organic matter.

PROCEDURE

1 Weigh a dry ignited porcelain crucible, add about 1 g of sieved dry (105°C) soil (or ground dry vegetation sample) and weigh again.
2 Ignite in a muffle furnace at 550°C and leave for about 2 hours. Peat and similar materials may only need about 1 hour. Allow the muffle furnace with the samples inside to come up to temperature from the cold, to avoid deflagration. Ensure adequate access of air.
3 When cool, remove the crucible, bring to room temperature in a desiccator and weigh.
4 Obtain net loss in weight for soils (or net ash weight for vegetation samples). Calculate both as a percentage of the dry weight.

In most cases loss-on-ignition or crude ash will be required along with a moisture determination and the sequence of events will then be: weigh fresh or air-dry sample; dry in oven; reweigh; ignite and reweigh.

2.6   pH

The pH values of aqueous extracts of soils are easily obtained (particularly with a pH meter) but are less easy to interpret. For example, hydrated aluminium influences the pH of soils and should not be overlooked (Coulter, 1969).

PROCEDURE

1 A conventional pH meter with glass and calomel electrodes and temperature compensation is suitable, although a dual glass-calomel electrode is more robust and is recommended for field use.
2 Standardize the meter with two buffers at different pH values.

3 Use a water to fresh soil ratio of about 2 : 1.
4 Mix the soil and water by stirring and allow to settle for about 20 minutes. Measure the pH of the supernatant liquid.
5 Ensure the temperature of the buffers and the sample solutions are the same and set the temperature compensator accordingly.

For further information consult Bohn (1971) and Hesse (1971).

2.7   Extractable fraction

The total concentrations of mineral elements in soils are generally of greater interest to pedologists than to ecologists and methods are not therefore dealt with here. This does not apply to carbon, nitrogen and phosphorus because large proportions of each are organically bound, resulting from biological processes and so a total (organic) content is often required in ecological studies.

A small proportion of mineral nutrient cation is present in true solution or held in the ionic form by negative organic and inorganic adsorption sites in the colloidal complex. In the simplest model all of this fraction is assumed to be readily available for plant uptake, so that an estimate of it should be much more informative for nutritional studies than the total content. The model assumes all adsorbed cations can be readily exchanged for another cation supplied in excess by a chemical extractant. The displaced cations are then measured in the filtered extract. The major cations to be estimated in the extract are potassium, calcium and magnesium, but traces of several others will be present.

However, this model has some limitations. It is found in practice that the exchange reaction of an extractant is accompanied by concurrent processes of solubilization of soil minerals and organic matter together with resorption from solution of ions already released. These features imply the model is too simple and that the assumption of 'availability' in itself needs qualification. Modern theories of cation exchange distinguish between the amount of a cation adsorbed by unit weight of soil (quantity factor) and the amount released to a leaching solution and, by assumption, to a plant (intensity factor). These and other aspects are summarized by Talibudeen (1981).

It has become accepted that the conventional single extraction measures mainly the intensity factor of an ion and values should be interpreted in this light. Meanwhile, extractants have long been used in agriculture where empirical relationships have been well established between values from specific extractants and crop uptake of cations and anions on various soil types. This data is used particularly for predicting fertilizer requirements. Ecologists do not normally have such information for native species, so extraction values are initially only a guide to the nutrient status of natural sites.

If meaningful data are to be obtained by an extraction the process of solubilization and resorption need to be minimized. This is covered below in the choice of extractant and conditions for extracting the major cations. Extractants for minor cations and for anions are listed separately later. Finally, it is recommended that all values obtained by these methods be recorded as 'extractable' cation rather than as 'exchangeable', and that the extractant and conditions employed should be specified.

2.7.1 Choice and conditions (for major mineral cations)

Clay minerals in the inorganic colloidal complex will readily adsorb ammonium ions (NH_4^+) when an excess is supplied and M ammonium acetate in pH 7 is probably the most widely used extractant. This extractant is also preferable to others when estimating cation exchange capacity (see below). However, the clay minerals will take up sodium and hydrogen ions only a little less readily and Morgan's reagent (10% sodium acetate in 3% acetic acid) has been used in agricultural advisory work in the UK. Another frequently used extractant is 2·5% acetic acid. Extractants based on divalent ions, such as Ba^{2+} are also sometimes employed but these are not described here.

Any extractant simulating 'exchange' should dissolve only a minimum of soil minerals and organic matter. The former is best achieved by extractants with a high pH value and the latter with a low pH level. Hence, neutral extractants at around pH 7 are widely preferred for most soil types. However, a high pH should in principle be retained for calcareous soils to minimize dissolution of carbonates. M ammonium acetate raised to pH 9 (by additional ammonia) will serve this purpose. 2·5% acetic acid is not suitable for calcareous soils. Organic soils such as peat are conventionally extracted or leached at pH 7 and not at lower pH levels.

In practice M ammonium acetate at pH 7 is suitable for a wide variety of non-saline soils whilst 2·5% acetic acid should be confined to non-calcareous and non-saline mineral soils. Both are described below but Morgan's reagent is omitted.

The other main difficulty (resorption) can be resolved by regulating the extraction conditions. These include the ratio of extractant to soil. If the proportion of soil is too high then resorption of released ions may become significant, particularly with clays and similar soils. If the proportion of extractant is excessive then the quantities extracted may be too low for satisfactory measurement. A ratio of 25 : 1 is suggested for air-dried sieved soil when cations are extracted. (Higher ratios are required for anions, see below.) Although ideally soil should be leached in columns it is standard practice to extract in bottles and use a shaker which avoids violent shaking.

A rotary end-over-end shaker is suitable. Leaching is retained for fresh peat but a funnel lined with filter paper can be used instead of a column.

Sufficient time should be allowed for the extraction to reach an equilibrium, although slow dissolution of minerals may continue after this stage has apparently been reached. The time should be standardized and periods of half to one hour are usual.

It is most important to keep all the conditions constant for samples whose results are to be subsequently compared.

2.7.2   Exchangeable constituents

Cation exchange capacity

The total amount of cations (including H^+) which can be displaced from the soil is still accepted as a measure of the cation exchange capacity (CEC) in spite of the limitations noted above. M ammonium acetate at pH 7 is generally used as the extraction solution. After filtering, excess ammonium ions are removed from the soil residue by washing. The adsorbed ammonium is then itself displaced by the cation of another extractant in which ammonium-nitrogen is later determined.

REAGENTS

1   Ammonium acetate, M (pH 7): add 575 ml of glacial acetic acid and 600 ml ammonium solution (0·880) to about 2 litres of water. Mix well and dilute to 10 litres. Check that the pH is 7·0 $\pm$ 0·1. If necessary adjust with acetic acid or ammonia solution.
2   Industrial spirit 60% (v/v).
3   Potassium chloride 5% (w/v).

PROCEDURE

1   Extract 10 g air-dry sieved soil (25 g fresh) with 250 ml ammonium acetate as given in Section 2.8.3 and filter. Reject filtrate unless required for analysis.
2   Wash the residue on the filter paper repeatedly with 50 ml portions of 60% industrial spirit until excess ammonium acetate has been removed. Add 10 ml 10% w/v NH_4Cl solution to the first portion of industrial spirit. Test filtrate for chloride (with $AgNO_3$ solution). When clear of chloride excess ammonium acetate is assumed to have been removed.
3   Leach the residue with about 10 ml of dilute KCl solutions (5% diluted about 5-fold) followed by several further portions of a 5% KCl solution, collecting the leachate in a vessel marked at 250 ml (500 ml for fresh samples).

Allow each portion to pass through before adding the next. Collect up to the mark.

4 Prepare blank leachates following the same procedure.

5 Determine NH_4^+-N in an aliquot of the lechate, by distillation and titration with M/140 HCl as given in Section 5.10.1 (or by the 'indophenol blue' method, see also Section 5.10.1).

CALCULATION

If 1 ml M/140 HCl = 0·1 mg NH_4^+-N

$$\text{then CEC (me 100 g}^{-1}) = \frac{\text{titre HCl (ml)} \times \text{leachate volume (ml)}}{1\cdot4 \times \text{aliquot (ml)} \times \text{sample wt (g)}}$$

Correct to dry weight basis.

Total exchangeable bases and exchangeable hydrogen

It is sometimes useful to know what fraction of the CEC is accounted for by mineral cations or 'bases' (TEB). The classical method involves ignition of the ammonium acetate extract, dissolution in excess standard acid and back titration with standard alkali (Bray & Willhite, 1929). However, summation of several metal cations estimated individually will give an approximate measure. If the TEB is expressed as a percentage of the CEC it is known as '% base saturation'.

Exchangeable hydrogen is sometimes measured, although it is difficult to define in terms of soil processes. Its role in soil acidity is partly bound up with that of aluminium. A rough estimate can be obtained simply as CEC-TEB. Direct methods seem to be of limited value and are normally based on a titration of the soil filtrate back to the initial pH (7·0) or use of the pH of the filtrate (Brown, 1943).

2.7.3 Extraction procedures

I Major nutrient cations (K, Ca, Mg)

EXTRACTANTS

1 Ammonium acetate M, pH 7: prepare as given in Section 2.7.2.
2 Acetic acid 2·5% v/v: dilute 250 ml glacial acetic acid to 10 litres with water.

PROCEDURE (AIR-DRIED, SIEVED SOIL)

1 Weigh 10 g soil into a 500 ml polythene bottle.
2 Add 250 ml selected extractant and shake for 1 hour on a rotary end-over-end shaker.
3 Filter through No. 44 filter paper rejecting first few ml of filtrate.
4 Prepare blank solutions following the same procedure.
5 Determine cations in the filtrate by the methods given in Section 5.

PROCEDURE (FRESH PEAT)

1 Weigh 25 g peat and transfer to funnel lined with No. 44 filter paper.
2 Leach with successive small portions of ammonium acetate (*not* acetic acid). Retain all filtrate in a container calibrated at 250 ml and allow each portion to pass through before adding the rest. Collect up to the mark.
3 Prepare blank solutions following the same procedure.

II Minor nutrient cations

Many extractants are used for particular cations and only brief comments are possible here. Some of the ions (e.g. ferrous iron) may exist in a reduced state in the soil and are subject to oxidation during extraction. Special precautions must then be taken to displace the air.

Manganese

Many extractants have been tried but M ammonium acetate at pH 7 has been found to be generally suitable.

Iron

Soil iron is of particular interest in pedology. Extractants based on oxalic acid, pyrophosphate or dithionite have all been used to study profile development. Ball & Beaumont (1972) recommend 3% oxalic acid. Soil iron is more often extracted in nutrition work when there is evidence of deficiency, e.g. iron chlorosis, and a variety of extractants have then been used.

Aluminium

3% oxalic acid is again useful in pedological work. In nutritonal studies M ammonium acetate at pH 4·8 has sometimes been used.

Copper and zinc

Mineral extractants such as 0·1 M hydrochloric acid have been used in the past although their action is largely dissolution. Better correlations with crop parameters have been obtained with complexiometric extractants such as EDTA.

III Nutrient non-metals

Many extractants have been tried and compared especially for the inorganic fractions of phosphorus. The organic fractions of both nitrogen and phosphorus have also been intensively studied in soil but require techniques which are not described here.

Nitrogen (NH_4^+-N, NO_3^--N)

The cation (NH_4^+-N) is included here for convenience. This has an exchange function in soil and both Na^+ and K^+ are used to displace it. 6% sodium chloride or potassium chloride are commonly used. Fresh soil is preferable with 10:1 ratio and intermittent stirring rather than shaking. Otherwise, extract air-dry at a ratio of not less than 25:1. Nitrate-nitrogen (NO_3^--N) is mobile in soil and cold water is a suitable extractant. However, wide ratios (up to 100:1) are desirable to reduce subsequent interference problems. Cleaning of extracts with adsorbents such as 'alumina cream' can give low recoveries.

Phosphorus (PO_4^{3-}-P)

Three extractants in common use are given here. The two acidic extractants have a solubilizing action and should be confined to non-calcareous soils but the main problem is the sensitivity of the phosphate ion to readsorption. High ratios are needed and Truog (1930) recommended 200:1. Calcareous soils are best extracted with Olsen's reagent at pH 8·5 (Olsen *et al.*, 1954). They accepted a fairly low ratio of 20:1 and this is still used. These extractants are all arbitrary in their action and methods for isotopic exchange, 'L' value and organic fractions may be needed in detailed studies of soil P but cannot be described here.

(a) *Truog's reagent.* 0·001 M H_2SO_4 buffered at pH 3 with $(NH_4)_2SO_4$. Use air-dried sieved material at a ratio of 200:1. A 30-minute shaking time is used. Use for acid and neutral soils.

(b) *Acetic acid 2·5% v/v.* Prepare as above, but because of the readsorption problem the extraction should be considered separately from that of the cations. For example a higher ratio (40 : 1) is desirable for some soils, especially clays. Use for acid and neutral soils.

(c) *Olsen's reagent.* 0·5 M $NaHCO_3$ adjusted to pH 8·5 with NaOH. Used for calcareous soils, taking air-dried sieved material and a ratio of 20 : 1. A 30-minute shaking time is used.

Sulphur (SO_4^{2-}-S)

Extractants have been less widely employed for this nutrient but Barrow (1968) and Sinclair (1973) are amongst those using 'calcium phosphate' for SO_4-S. The extractant is compatible with the turbidity method given in Section 5.

Note

All standard solutions must contain the same concentration of extractant as present in the sample solutions.

3   Plant materials

This section deals with the initial treatment of vegetation and preparation of the solution for subsequent determination of the individual elements. Litters and peats may be treated in the same way for a total analysis. General texts dealing with the analysis of plant materials include Chapman & Pratt (1961), AOAC (1970) and Allen (1974). Many of the methods used in the analysis of vegetation and also suitable for animal products, such as faeces or gut contents. In general, however, animal tissues are best considered separately from vegetation.

3.1   Collection

Some aspects of the sampling of plant communities are discussed elsewhere in this book. Several points must be taken into account.

(a) Choice of component for analysis: in general the green photosynthetic parts will reflect the nutrient status of the plant.

(b) Age of plant or component: for most purposes the current year's growth is adequate.

(c) Sampling time: in deciduous tissues the concentrations of the more mobile elements change throughout the growing season, although this tends to reach a maximum and remain relatively constant in the late July to mid August period.

(d) Site aspect and degree of shading, which have minor effects.

(e) Statistical aspects including randomization and replication.

(f) Field contamination by soil or animal droppings.

(g) Nutrient concentrations are governed by many factors and their interpretation is not simple. A clearer picture is obtained if the biomass of the tissue can also be estimated.

Methods of sampling are similar to those used in forestry and agriculture. High level pruners are available for tree leaves and shears or secateurs can be used for grassland and similar herbage.

3.2 Transport and storage

The fresh plant material should be transported back to the laboratory as soon as possible after collection. Avoid, if possible, using sealed polythene bags and containers because of respiration affects. If any delay is likely, early air-drying, by thinly spreading out the vegetation, should be considered. Most of the tests dealt with in this chapter can be carried out on dried plant material. Immediate drying on receipt is recommended unless labile constituents are being studied. If temporary storage is unavoidable a temperature just above freezing should be used. Deep-freeze storage may rupture the cell structure.

Contamination of vegetation by soil splash, airborne dust and animal materials can be troublesome. Removal of this contamination by brief agitation in water using ionic vibrators is sometimes recommended but such treatment can result in some nutrient loss. Steyn (1959) reports on the extent of elemental losses during cleaning.

3.3 Drying and grinding

Samples are finely ground to produce a sufficiently homogeneous material for a representative sample to be taken for analysis. Unlike soils, there is no standard mesh size, but between 0·4 and 0·7 mm is convenient for most purposes. Types of grinding machines available include the following.

(a) Beating mills: depend on the impact action of rotating arms or flails to reduce the sample size, which is then blown through a sieve. They are effective for bulky and woody materials but differential dust loss may occur.

(b) Shatter mills: depend on the shattering action of balls or swinging discs moving at high speed in an enclosed chamber. The sample can be reduced to a very fine powder but a long milling time may be needed.

(c) Cutting mills: depend on the cutting action of rotating blades. These mills produce little contamination but are not appropriate for harder materials because the cutting edges soon become blunt.

(d) Grinding mills: depend on a disc grinding action but they tend to overheat and it may be difficult to recover softer tissues which stick to the grinding discs.

Most mills—but especially those involving grinding and smashing processes—may contaminate the sample. This can be minimized through the use of hard steel, or better still, tungsten carbide, materials. Fresh samples are best treated in an homogenizer such as a Waring blender or even a kitchen model.

3.4   Moisture and ash

Most of the comments about moisture in soils (Section 2.3) also apply to plant materials. The main difference, which must be stressed, is the more labile nature of fresh plant material; so rapid drying is essential. The procedure given for the determination of moisture or dry weight of soils also applies to plant materials but for fresh moisture quite large weights of vegetation are needed to ensure that the sample is representative. The water holding capacity of different parts of the plant varies a great deal.

In the method for preparing a plant digest solution for analysis it is recommended that a dry, ground sample be weighed. This assumes that drying is carried out at 105°C. If the air-dried sample is taken then a moisture correction must be applied to the results.

The procedure given for loss-on-ignition in soils (Section 2.5) is basically the same for ash, except that the ash residue is calculated and not loss during combustion. As above, care is needed to ensure that the early stages are not too rapid, resulting in loss of ash.

3.4.1   Silica-free ash

By treating the residue obtained in a normal ash determination it is possible to obtain an approximate measure of the total non-silicate component of the plant minerals, which are essentially the nutrient elements. Dissolve the ash as given in Section 3.5.1 and retain the filter paper and residue from step 6. Ignite the paper in a weighed crucible, cool and calculate the residue as percentage silica. The silica-free ash can then be obtained by deduction of this result from the ash.

3.5   Preparation of solution for analysis

Analysis for total nutrient elements in plant materials requires the complete breakdown or oxidation of all organic matter. The two basic methods are dry ashing and acid digestion.

3.5.1 Dry ashing

Although not wholly convenient the method has two advantages:

(a) Relatively large sample weights can be handled.

(b) The residue is taken up in hydrochloric acid which is an excellent solvent and very suitable for subsequent analysis.

The main disadvantage of dry ashing is that non-metals are either lost completely in all samples (nitrogen) or suffer measurable volatile loss in some types of vegetation (phosphorus, sulphur, chloride and boron). Special dry-ashing procedures are required for the latter and relevant texts should be consulted. There may also be a slight loss of heavy metals or trace elements by retention on silica.

The usual temperature range from 450°C to 550°C for times varying from half an hour at 550°C to several hours at 450°C. The muffle furnace should not be preheated, but should be started from cold with the samples inside. This prevents sudden deflagration. Nitric acid added at the dissolution stage facilitates the oxidation of ferrous salts.

PROCEDURE

Suitable for metals including Al, Ca, Cu, Fe, K, Mg, Mn, Na and Zn but not for P or S unless calcium or magnesium salts are added.

1 Weigh 0·5 g dry ground sample into a dry acid-washed porcelain crucible.

2 Ignite at 550°C for 30 min. Make sure air has access to the muffle furnace and that the chimney vent is open.

3 Allow to cool, add 5 ml HCl (1 + 1) and warm to dissolve the residue.

4 Add 0·5 ml of conc HNO_3 and evaporate to dryness on a boiling water bath. If heating is continued for an hour the residue can be weighed after stage 6 to give a rough estimate of silica (as SiO_2).

5 Add 2 ml HCl (1 + 1), swirl, warm slightly to dissolve the residue and dilute.

6 Filter through a suitable paper into a 100 ml volumetric flask, wash the residue and dilute the filtrate to volume. This solution contains 1% HCl (v/v).

7 Prepare blank solutions following the same procedure.

NOTE

All standard solutions must contain 1% HCl (v/v) as present in the sample solutions.

3.5.2   Acid digestion

Acid digestion of organic material has the advantage that phosphorus, and with some digests, nitrogen, can be determined in the final solution along with other nutrients. Hence, in general, if suitable facilities are available, digestion is preferable to dry ashing. Two procedures are given below, namely the mixed acid (nitric–perchloric–sulphuric) digestion and the sulphuric–peroxide system. Both are suitable for a wide range of plant materials. However, certain points should be noted.

(a) The use of perchloric acid for oxidizing organic materials can be hazardous because of the explosive nature of some perchlorates if allowed to dry out. Recommendations for the use of perchloric acid are given by Steere (1971). In the procedure given here sulphuric acid is included in the digestion mixture to prevent drying out. Any interference in the subsequent analysis can be compensated for by inclusion of sulphuric acid in the standard solutions.

(b) As a general rule powerful oxidizing agents such as nitric and perchloric acids will complete the oxidation quickly but will drive off nitrogen. Less powerful oxidants such as hydrogen peroxide in sulphuric acid are slower and require a catalyst, but nitrogen is retained during the reaction.

(c) Wet digestion techniques are not very suitable for large sample weights and this applies in particular to the sulphuric–peroxide digestion. In this method the recommended weight (0·4 g) must not be exceeded. Low concentration elements can be estimated if the digest is run undiluted at the 5% acid level. It is permissible to exceed the weight recommended for the mixed acid digestion (0·2 g) but only if additional quantities of the nitric–perchloric mixture are added.

(d) Fatty materials, such as certain seeds, maybe difficult to oxidize. Prior digestion with nitric acid is then recommended but only the mixed acid digestion can be used subsequently.

The procedure given below should not lead to low recoveries of nutrients by the formation of insoluble salts such as certain perchlorates and sulphates even with nutrient-rich samples. A low recovery of iron due to ferric sulphate is possible but diluting the cold digest slightly and bringing to the boil will redissolve this precipitate. Manganese may precipitate in the most oxidized form if manganese-rich samples are digested, but this can be reduced by dilution and boiling in the presence of a few drops of sulphurous acid.

3.5.3   Digestion procedures

I Mixed acid

(Suitable for Al, Ca, Cu, Fe, K, Mg, Mn, Na, Zn, P)

PROCEDURE

1 Weigh 0·2 g of dry ground sample into a 50 ml Kjeldahl flask.
2 Add 1 ml 60% $HClO_4$;
        5 ml conc HNO_3;
        1 ml conc H_2SO_4.
3 Digest at moderate heat until white fumes are evolved.
4 Heat strongly for a few minutes to drive off most of the perchloric acid and allow to cool. The acid digest should now be colourless or occasionally pink (due to manganese). If yellow or brown reheat with a few drops of $HClO_4$.
5 Dilute the solution and boil for a few minutes if iron or aluminium are required. Filter through a No. 44 Whatman filter paper into 100 ml volumetric flask, wash paper and dilute to volume. (This will yield a 1% H_2SO_4 solution which is convenient for use in subsequent analysis.)
6 Prepare blank solutions following the same procedure.

II Sulphuric–peroxide mixture

(Suitable for Al, Ca, Cu, Fe, K, Mg, Mn, Na, Zn, P *and* N)

PROCEDURE

1 Prepare digestion mixture. Add 0·42 g Se and 14 g of $Li_2SO_4.H_2O$ to 350 ml of 100 vol H_2O_2. Mix well and add *with care* 420 ml of conc. H_2SO_4. Cool the mixture during the addition of the acid. Store at 2°C for not longer than 3 weeks.
2 Weigh 0·4 g of dry ground sample into a suitable Kjeldahl flask.
3 Add 4·4 ml of digestion mixture.
4 Digest at low heat until the initial reaction subsides avoiding loss of H_2SO_4 fumes.
5 Continue the digestion until a clear and almost colourless solution is obtained. For most samples this requires about 2 hours.
6 Dilute the solution and boil briefly if iron is to be determined.
7 Filter through a suitable paper and dilute to volume. Dilution to 50 ml will yield a 5% H_2SO_4 solution and this is normally used for Al, Cu and Zn. All other elements listed above require a further 5-fold dilution to give a 1% acid solution.
8 Prepare blank solutions following the same procedure.
    In both acid procedures recovery of the silica from the residue on the filter paper (by ignition) allows an approximate determination of silica (SiO_2).
    Solution preparation details for other elements, apart from those included above, are given in Section 5.

NOTE

All standard solutions must contain the same amount of acid and digestion reagents as present in the sample solution.

4   Waters

Some studies of terrestrial systems such as those of nutrient cycling must allow for nutrient pathways in various types of water. These include precipitation, canopy leachates, stem flows and soil waters such as run-off and lysimeter percolates. The methods given here are commonly used for fresh water, although limnological aspects of lakes or rivers and studies of wet and dry deposition from the atmosphere are outside the immediate scope of this book.

Further information on water analysis can be obtained from *Standard Methods for the Examination of Water and Wastewater* (American Public Health Association, 1975). Books catering for the needs of limnologists include Mackereth *et al.* (1978) and Golterman *et al.* (1978).

4.1   Collection and storage

Specialized collection procedures are needed to obtain samples of the waters mentioned above. Various aspects were discussed by Ciaccio (1971) whilst the particular problems of collecting rainwater are considered by Galloway & Likens (1978).

In all cases care is needed to avoid contamination, such as that due to bird droppings, soil splash or even fertilizers applied to nearby land. In nutrient cycling studies these sources may themselves need to be examined and allowed for.

Polythene containers are now almost universally used for collecting water samples. The high density polypropylene vessels are preferable. However, any plastic container must be washed regularly and thoroughly to prevent algal and bacterial growth. Heron (1962) suggested the impregnation of polythene collection and storage vessels with iodine to prevent bacterial and algal growth which particularly affects phosphate levels. There is also the possibility of other changes in chemical composition whilst standing in the field. Samples should be collected as frequently as possible but this is not always practicable. Addition of a mineral acid such as hydrochloric or sulphuric will deter bacterial growth and may also stabilize ammonium-nitrogen. Unfortunately, this approach introduces anions which are themselves of importance in water studies. Toxic chemicals such as mercuric chloride are effective, although the samples are then unsafe to be left unattended in the field. The chemical may also interfere with subsequent analytical procedures. Other chemicals which are sometimes recommended including chloroform and toluene are not fully effective. If only the dissolved fraction is being analysed then filtering in the field using membrane filters can be used to reduce microbial activity.

Once brought back to the laboratory the delay before analysis should be kept as short as possible. Freeze-storage of waters is not recommended, as the recovery of some elements, especially calcium, phosphorus and silicon appears to be affected. Perhaps the most effective short-term treatment is to keep the water at a temperature just above freezing. Storage in the dark is also recommended.

4.2 Preliminary treatment

Most natural waters include a suspended and dissolved component. Solids can be removed using membrane filters which are obtainable in different pore sizes. 0·4 μm is often used as a boundary size between dissolved and suspended matter. The dissolved fraction can be analysed after filtering using the methods described later with little modification. Glass fibre filters are faster than membrane but are variable in size and they must be well washed before use. Paper filters also have a variable pore size (0·5–1 μm) but it is possible to obtain grades that do not contaminate the sample (Whatman No. 44) although paper may remove minute traces of some ions. To estimate the suspended content the water must first be filtered and the residue then dried. The dissolved content may be determined on the filtrate which is then evaporated and dried. Total solids can be determined by evaporating the sample as received before further treatment.

Membrane filters will generally remove turbidity but coloured waters can be a problem. Some adsorbents (e.g. activated carbon and exchange resin) are effective for removing the colour from water but they must be carefully checked in case they retain the elements to be tested.

The concentration of some metal ions in solution may be too low to be conveniently measured. Concentration by evaporation is not always suitable because it may increase the viscosity and colour density. Ion exchange or complexing reagents (e.g. APDC) are more appropriate.

In many ecological studies there is an interest in the total income into the system so that the total concentrations of the nutrient elements in waters will be required. The treatment involves initial evaporation followed by a digestion stage. A suitable procedure for use when determining metals and phosphorus is given below. Nitrogen is treated separately as given in Section 5.10.

4.2.1   Total content

PROCEDURE

1   Measure a suitable aliquot of unfiltered water into a conical flask with small base (Taylor's pattern) or round-bottomed flask.
2   Add 0·5 ml conc. H_2SO_4.
3   Boil down carefully until white fumes appear but do not allow to dry.
4   Add 1 ml 60% $HClO_4$ and 2 ml conc. HNO_3.
5   Digest to white fumes and continue for a few minutes only. Take care not to allow the digest to boil dry.
6   Dilute, filter through a No. 44 Whatman filter and dilute to volume. (Dilution to 50 ml will yield a 1% H_2SO_4 solution which is convenient for use in subsequent analyses.)
7   Prepare blank solutions following the same procedure. (Blank determinations should not be based on the evaporation of deionized or distilled water in place of the water sample since these are not directly comparable. It is preferable to allow the blanks to commence at the digestion stage.)

If there is a need to analyse the particulate (filtered) fraction then the digestion procedure described for plant material should be followed. If the particulate fraction contains refractory inorganic residues then fusion may be needed to bring the sample into solution. The main problem in looking at the particulate components is that of acquiring sufficient for analysis.

4.3   Initial tests

4.3.1   Conductivity

The specific conductance, or conductivity of a water sample is a measure of the total ionic concentration. Specific conductivity, the reciprocal of specific resistance, is measured in a conductivity cell by a low voltage AC meter. A number of commercial instruments are available. Samples are not filtered or treated before measurement.

The conductivities of most fresh waters are low and are expressed in μS cm^{-1}. A conductivity cell is calibrated against a suitable potassium chloride solution (0·01 M KCl = 1412 μS cm^{-1} or 0·001 M KCl = 147 μS cm^{-1} both at 25°C). The cell constant (C) can then be calculated from the measured resistance of the standard solution in ohms (R_{KCl}) and the specific conductance of the same solution (K_{KCl}) at 25°C.

$$C = R_{KCl} \times K_{KCl}$$

Conductivity varies with temperature which should be standardized usually at $20°$ or $25°C$. This can be done by bringing the sample to the standard temperature, by a compensating circuit in the conductivity meter, or by calculation.

In acid water the hydrogen ions can contribute a major proportion of the total conductivity, and it is often ecologically more meaningful to subtract the conductivity due to hydrogen ions leaving a corrected conductivity (K_{corr}) that is more representative of the plant nutrients in solution.

$$K_{corr} = K_{20} - K_{(H^+)_{20}}$$

This approach has been described by Sjörs (1950) and is used as a convenient method of investigating overall nutritional relationships of peat ecosystems under field conditions.

4.3.2 pH

pH is a function of the hydrogen ion concentration and may be measured by a colorimetric or an instrumental method. The colorimetric method is simple to use but lacks precision and is unsuitable for coloured waters. Measurement of pH using a pH meter is always preferable because it is more sensitive and precise. The meter should be standardized using two buffer solutions. The sample should be at the same temperature as the buffer solution (see also Section 2.6).

4.3.3 Alkalinity

Alkalinity is a measure of the anions of weak acids, notably bicarbonate with carbonate, hydroxide and traces of others. Total alkalinity is estimated by titration to pH $4·5$ with standard acid and a suitable indicator as described below. Dissolved carbon dioxide gives a small error which should not be overlooked if the alkalinity is low (Mackereth et al., 1978). The three main anions can be distinguished qualitatively if this procedure is carried out in two stages. The method described below would then begin with a titration to pH $8·3$ using phenolphthalein as an indicator (American Public Health Association, 1975; Golterman et al., 1978). Alkalinity is widely expressed in terms of calcium carbonate although milli-equivalents are to be preferred. Rose (1983) defined alkalinity solely in terms of bicarbonate and carbonate dissolved from limestone rock. He further advocated that the usual method should include a blank titration to allow for carbon dioxide.

Strong acid anions such as chloride, nitrate and sulphate are usually estimated separately but if required collectively the ion exchange method of Mackereth et al. (1978) may be used.

The acidity of waters can be estimated by titration with 0·01 M sodium hydroxide but the actual pH at the equivalence point depends on the type of acid present. Dissolved carbon dioxide is the main form present in unpolluted waters and titration to pH 8·3 with phenolphthalein is usually recommended in these cases.

Total alkalinity

REAGENTS

1 Hydrochloric acid (0·01 M).
2 Mixed indicator;
0·02% methyl red and 0·1% bromocresol green in 95% industrial spirits.

PROCEDURE

1 Measure an aliquot of unfiltered water into a titration flask.
2 Add 2 or 3 drops of mixed indicator and titrate with the acid from blue to the first trace of pink.

CALCULATION

If V ml 0·01 M hydrochloric acid was required

$$\text{total alkalinity} = \frac{V(\text{ml}) \times 0.5005 \times 10^3}{\text{aliquot (ml)}} \text{ mg}\,l^{-1} \text{ as CaCO}_3$$

$$= \frac{V(\text{ml}) \times 10}{\text{aliquot (ml)}} \text{ me}\,l^{-1}$$

5   Individual elements

5.1   Introductory notes

5.1.1 Standards

In both flame and colorimetric methods the standards must be run at the same time as the samples and should always contain equivalent levels of the background components (acid, soil extractant, etc.). Any sample solutions which have to be diluted for specific purposes should include sufficient background components to restore the original levels. Low level standards can be unstable and should be replaced regularly from stock standard solutions. They must be stored in a cool, dark place.

5.1.2 Calculations

The procedures described in this chapter can be split into three groups: the flame methods, the colorimetric methods and those with a titration as the last step. The first two, which are instrumental procedures, depend on the use of standard solutions to prepare a calibration curve from which a concentration value can be obtained. Many of the procedures used in this book have a linear relationship between the scale readings and concentrations. In the case of the titration method the titrant has to be first calibrated against a standard solution. It is usually preferable to work in terms of molarity, but in routine instrumental work it is often more convenient to calculate in terms of elemental concentration.

To avoid repetition the calculation stages are not detailed for most methods and general formulae are given for the first two of the three groups referred to above. However, a number of points should be noted.

(a) The standard solutions used in the flame and colorimetric procedures must contain the background components (acid, digestion reagents, soil extractant) in the same concentrations as the sample solutions (Table 6.3).

(b) Factors for any dilution and concentration stage must be taken into account. For example, there will always be a dilution factor for calcium and magnesium by atomic absorption because of the need to introduce a lanthanum salt.

(c) Blank values must also be read from the calibration graph as an elemental concentration and deducted from the sample concentration.

(d) The final result should be corrected to a dry weight basis if necessary.

CALCULATION FORMULAE

I Flame emission and atomic absorption

(Al, Ca, Cu, Fe, K, Mg, Mn, Na, Zn)

If $C = \mathrm{mg\,l}^{-1}$ element is obtained from the calibration curve then for plant materials:

$$\text{total element (\%)} = \frac{C(\mathrm{mg\,l}^{-1}) \times \text{solution volume (ml)}}{10^4 \times \text{sample wt (g)}}$$

soils:

$$\text{extractable element (mg } 100\,\mathrm{g}^{-1}) = \text{ as above} \times 10^3$$

waters:

$$\text{dissolved element (mg\,l}^{-1}) = C(\mathrm{mg\,l}^{-1})$$

II Colorimetry

(Cl, Fe, P, NO_3, S, SO_4-turbidimetric)

If C = mg element obtained from the calibration curve then for plant materials:

$$\text{total element (\%)} = \frac{C(\text{mg}) \times \text{solution volume (ml)}}{10 \times \text{aliquot (ml)} \times \text{sample wt (g)}}$$

soils:

$$\text{Extractable element (mg 100 g}^{-1}) = \text{as above} \times 10^3$$

waters:

$$\text{Dissolved element (mg l}^{-1}) = \frac{C(\text{mg}) \times 10^3}{\text{aliquot (ml)}}$$

III Other calculations

These are given separately for particular elements.

5.1.3 Expression of results

The SI system of units is now in almost universal use. The only exceptions to its use in this text are that the litre (1) has been used in place of dm^3 and millilitre (ml) instead of cm^3. Soil extraction data are quoted as mg $100 g^{-1}$ but can be expressed in terms of equivalents, i.e. me $100 g^{-1}$ by dividing mg $100 g^{-1}$ by the equivalent weight of the element in question. The term 'parts per million' (ppm) should not be used when quoting results and if employed at all it should be confined to internal use in the laboratory.

5.2 Aluminium

Atomic absorption is the preferred method for estimating aluminium in soil, plant material and aqueous solutions. However, a high temperature nitrous oxide–acetylene flame is required and both soil extracts and plant digests may present problems. Some ionization also occurs at this higher temperature so that fewer atoms are in the ground state, but this effect is lessened in the presence of 0·1% potassium chloride. Iron interferes and in running oxalic acid extracts of soils the iron level in the standards should match that of the samples.

The preparation of plant digests resembles the procedure used for iron (Section 5.7) in that the digests are diluted and boiled before making up to volume to ensure that all aluminium is in solution.

1 Stock standard solution ($100\,\mathrm{mg\,l^{-1}}$).
Dissolve $1{\cdot}759\,\mathrm{g}$ aluminium potassium sulphate ($\mathrm{AlK(SO_4)_2.12H_2O}$) in water, add $4\,\mathrm{ml}$ conc. $\mathrm{HNO_3}$ and dilute to 1 litre.

2 Working standards.
Prepare a range from 0 to $5\,\mathrm{mg\,l^{-1}}$ Al by dilution of the stock solution. Include acids and extractants to match sample solutions.

PROCEDURE

1 Prepare the sample solutions as described in Sections 2.7, 3.5 or 4.2 but for all plant digests boil as for iron (Note 1 in Section 5.7). It is also preferable to include KCl to give a concentration of $0{\cdot}1\%$ in each standard.

2 Set up the instrument as in Section 6.2.

3 Select the $309{\cdot}3\,\mathrm{nm}$ wavelength.

4 Use a rich nitrous oxide–acetylene flame ensuring the correct burner is fitted.

5 Construct a calibration curve and use it to obtain the concentration of Al in solution. Subtract the blank value. Calculate Al concentrations in the original material (Section 5.1.2).

5.3   Calcium

Atomic absorption is a rapid and sensitive method for the determination of calcium. Other procedures such as titration using EDTA have been widely used and an EDTA method is outlined as a less sensitive alternative.

5.3.1   Atomic absorption method

Atomic absorption is the preferred method for estimating calcium in soils, plant materials and waters. The air–acetylene flame is widely used for this element even though it is not hot enough to prevent refractory compounds of aluminium, phosphorus and other elements interfering. However, this is controlled by the addition of lanthanum which is effective in quantities lower than usual when sulphuric acid is present. The hotter nitrous oxide–acetylene flame is almost free of interferences so that lanthanum need not be added, but the ionization buffer mentioned for aluminium (Section 5.2) is desirable for calcium. The sample dilutions recommended below to incorporate lanthanum can be used for magnesium with dual element standards but are not suitable for other elements.

The detection limit for this method is about $0{\cdot}05\,\mathrm{mg\,l^{-1}}$ and reproducibility $\pm1\%$.

REAGENTS AND STANDARDS

1 Stock standard solution ($1000\,mg\,l^{-1}$ Ca).
Dissolve $2 \cdot 4973$ g dry $CaCO_3$ in about 200 ml water containing 5 ml conc. HCl, boil to drive off CO_2, cool and dilute to 1 litre in a volumetric flask with water.
2 Lanthanum chloride solutions.
  (a) $2000\,mg\,l^{-1}$ La (for plant material and water):
dissolve $10 \cdot 6939$ g $LaCl_3 . 7H_2O$ in water with aid of $1 \cdot 6$ ml 2 M HCl and dilute to 2 litres.
  (b) $4000\,mg\,l^{-1}$ La in 5% H_2SO_4 (for soil extracts):
dissolve $21 \cdot 388$ g $LaCl_3 . 7H_2O$ in water, add 100 ml conc. H_2SO_4 and dilute to 2 litres.
Alternatively, take 80 ml 10% La solution, add 100 ml conc. H_2SO_4 and dilute to 2 litres.
3 Working standards.
Prepare a range of standards up to $40\,mg\,l^{-1}$ by dilution of the stock solution. Lower ranges may be necessary for some waters and soil extracts. Include acids and extractants to match the sample solutions, and also $400\,mg\,l^{-1}$ La (plant and water standards) and $800\,mg\,l^{-1}$ La in 1% H_2SO_4 (soil standards).
4 $2 \cdot 5$% sulphuric acid (for water analysis only).

PROCEDURE

1 Prepare the sample solution as described in Sections 2.7, 3.5 or 4.2.
2 Dilute the sample solutions as given in Table 6.3 to include lanthanum solution at $400\,mg\,l^{-1}$ for plant digests and waters and $800\,mg\,l^{-1}$ for soils in 1% H_2SO_4.
3 Set up the instrument as outlined in Section 6.2.
4 Select the absorption line at $422 \cdot 7$ nm.
5 Use an air–acetylene flame and adjust the acetylene flow rate until maximum transparency is just obtained.
6 Construct a calibration curve to obtain the concentration of Ca in solution. Subtract the blank value. Calculate concentration of Ca in the original material (Section 5.1.2).

5.3.2 EDTA method

The method involves the use of EDTA (the di-sodium salt of ethylene diamine tetra-acetic acid) which forms stable complexes with many elements at specific pH values. Various indicators are available for calcium but in many cases the end-point is not sharp and a reference end-point should be

Table 6.3 Dilution volumes (ml) required to provide 15 ml of a five times diluted solution containing lanthanum and 1% H_2SO_4 and other components as required. All acid strengths as v/v (for Ca and Mg estimation only)

Initial solution	Diluted solution	Sample volume (ml)	Lanthanum (2000 mg l^{-1})	Lanthanum (4000 mg l^{-1}) in 5% H_2SO_4	Water	Compensating solution
Vegetation						
Dry ashing (1% HCl)	1% HCl 1% H_2SO_4	3·0	3·0	—	6·0	3·0 (5% HCl + 5% H_2SO_4)
Mixed acid (1% H_2SO_4)	1% H_2SO_4	3·0	3·0	—	6·0	3·0 (5% H_2SO_4)
Sulphuric-peroxide (5% H_2SO_4)	1% H_2SO_4	3·0	3·0	—	9·0	—
Water						
Water	1% H_2SO_4	3·0	3·0	—	6·0	3·0 (5% H_2SO_4)
Soil						
MNH$_4$OAc	MNH$_4$OAc 1% H_2SO_4	3·0	—	3·0	6·0	3·0 (5M NH$_4$OAc)
2·5% HOAc	2·5% HOAc 1% H_2SO_4	3·0	—	3·0	6·0	3·0 (12·5% HOAc)

prepared. Otherwise they are best used in conjunction with a photoelectric titrator. Three alternative indicators are described below.

Interference from heavy metals is not serious with most organic materials, but with soils it can be significant and can cause difficulty with the end-point. The high dilution recommended in the method minimizes end-point interference but limits the value of the procedure if the calcium level is low.

About 10 μg calcium can be detected but reproducibility at normal levels is unlikely to be better than $\pm 5\%$ for ecological materials.

REAGENTS AND STANDARDS

1   Calcium standard (100 mg l^{-1} Ca).
Dissolve 0·2497 g dry CaCO$_3$ in water containing approximately 1 ml conc. HCl. Boil to drive off the CO$_2$, cool and dilute to 1 litre.
2   EDTA solution (1 ml = 0·1 mg Ca).
Dissolve 0·931 g of di-sodium ethylene diamine tetra-acetate in 1 litre of water and standardize the solution by titrating against the Ca standard (see above).
3   Sodium hydroxide, M (40·01 g l^{-1}).
4   Indicators:
(a) Murexide: grind together 0·1 g murexide and 50 g NaCl in a mortar and store in a dark bottle.
(b) Calcon: dissolve 20 mg calcon in 50 ml methanol. Prepare fresh weekly.
(c) Glyoxal: dissolve 0·20 g glyoxal-bis-(2-hydroxyanil) in 50 ml methanol.

PROCEDURE

1   Prepare the sample solution as described in Sections 2.7, 3.5 or 4.2.
2   Add 5 ml M sodium hydroxide and indicator (0·1 g murexide, 5 drops calcon or glyoxal) to water, mix and dilute to about 100 ml. This gives a bluish reference end-point.
3   Standardize the EDTA solution as follows: pipette 10 ml calcium standard into a titration flask and add water, sodium hydroxide and indicator as above. Titrate with EDTA solution until the colour matches that of the reference end-point.
4   Pipette an aliquot (usually up to 5 ml) of the sample solution into a titration flask, add water, sodium hydroxide and the indicator. Titrate with the EDTA solution as above.
5   Carry out blank determinations following the same procedure and subtract from the sample values. Calculate the calcium content of the original material.

CALCULATION

If 1 ml EDTA solution $= 0.1$ mg Ca and V ml are required then for plant materials:

$$Ca(\%) = \frac{V(ml) \times \text{solution volume (ml)}}{100 \times \text{aliquot (ml)} \times \text{sample wt (g)}}$$

soil extracts:

$$Ca(mg\ 100\ g^{-1}) = \text{as above} \times 10^3$$

waters:

$$Ca(mg\ l^{-1}) = \frac{V(ml) \times 100}{\text{aliquot (ml)}}$$

5.4 Carbon

5.4.1 Organic carbon

The methods available for the determination of organic carbon in soils and plant materials fall into two main categories, absolute methods and 'rapid' methods. The absolute methods, which are not described here, depend on combustion or acid oxidation to generate carbon dioxide which is then absorbed and weighed. Combustion systems are available from laboratory suppliers.

Rapid methods are also based on wet oxidation but rely on titration of unused oxidant rather than the trappings of carbon dioxide. Unfortunately, earlier rapid methods such as that of Walkley & Black (1934) often gave poor recoveries and are not now recommended. However, the rapid method of Tinsley (1950) gives acceptable values for plant material but is less suitable for soils (Bremner & Jenkinson, 1960). A modification of this method proposed by Kalembasa & Jenkinson (1973) gives better recoveries for soil and is described below. It can be readily adapted for plant material.

If desired, an approximate estimate of organic carbon can be obtained from the loss-on-ignition values which is a crude measure of organic matter. Conversion factors used for this purpose in the past include one which assumes soil organic matter contains 58% carbon. However, Ball (1964) and Howard (1966) have queried the validity of this approach.

REAGENTS

1 Potassium dichromate (0·0833 M).
Dissolve 24·52 g pure $K_2Cr_2O_7$ in water and dilute to 1 litre.

2   Sulphuric–phosphoric acid mixture (5:1).

Mix 1500 ml conc. H_2SO_4 (SG 1·84) with 300 ml conc. H_3PO_4 (SG 1·75).

3   Ferrous ammonium sulphate (0·5 M).

Dissolve 196·0 g ferrous ammonium sulphate $((NH_4)_2SO_4.FeSO_4.6H_2O)$ in water, add 20 ml conc. H_2SO_4 and dilute to 1 litre. Make up fresh daily.

4   Indicator solution.

Dissolve 0·20 g N-phenylanthranilic acid in 100 ml 0·2% Na_2CO_3 solution.

PROCEDURE

1   Weigh not more than 0·5 g air-dry sieved sample into a 250 ml round-bottomed flask with a ground glass joint. This should contain between 5 and 15 mg organic carbon.

2   Add 20·0 ml dichromate solution and 30 ml acid mixture, fit a condenser and reflux for 20 minutes.

3   Rinse the condenser with water, add 5 drops indicator solution and titrate the unused dichromate with ferrous ammonium sulphate solution until the violet colour changes to a dark green end-point (T ml).

4   Treat the blank (dichromate and acid only) in the same way and record titre.

CALCULATION

If T ml of 0·5 M ferrous ammonium sulphate are used in the sample titration and X ml in the blank titration then:

$$\text{Organic-C (\%)} = \frac{(X - T) \times 0·15}{\text{sample wt (g)}}$$

Correct to dry weight as necessary.

5.4.2   Carbonate–carbon

Inorganic carbon is usually present as the carbonates of calcium or magnesium. A number of methods of analysis are available including the gravimetric method of Allison (1960) and that described by Bascomb (1961) involving the reaction of dilute hydrochloric acid with the carbonate in the soil and the measurement of the evolved carbon dioxide using a calcimeter. For soils with a low carbonate content the simple apparatus described by Pittwell (1968) is quite effective. A more rapid method described below uses dilute hydrochloric acid but estimates unused acid rather than carbon dioxide evolved. The method is not, therefore, specific for carbonates but still gives a useful though approximate value for calcareous soils. The method titrates excess acid with standard alkali but the end-point is not easy to detect by indicator if the extract is brown. In these cases it may be better to gauge the end-point from a titration curve obtained with a pH meter as the alkali is added.

REAGENTS

1 Hydrochloric acid, 0·5 M (can be prepared from ampoules).
2 Sodium hydroxide, 0·5 M (can be prepared from ampoules).
3 Phenolphthalein, 1% w/v in 95% industrial spirit.

PROCEDURE

1 Weigh up to 2 g air-dry, sieved sample into a 250 ml beaker flask (see Note 1).
2 Add 40·0 ml of 0·5 M hydrochloric acid.
3 Swirl to mix and then leave for 1 hour.
4 Filter the solution through No. 44 paper and wash the residue (see Note 2).
5 Titrate the excess acid in the filtrate against 0·5 M sodium hydroxide ($= B$) using phenolphthalein solution as indicator.
6 Standardize sodium hydroxide against 40·0 ml hydrochloric acid ($= A$).

CALCULATION (as $CaCO_3$)

$$CaCO_3 \ (\%) \ = \ \frac{(A - B) \ (\text{ml}) \ \times \ 2·502}{\text{sample wt (g)}}$$

Correct to dry weight.

NOTES

1 Take not more than 1 g of calcareous materials.
2 Filtering is not necesary if a titration curve is to be obtained.

5.5 Chloride

Both total and extracted chloride can be readily determined in soils and plant materials, but these forms are only rarely needed in ecology and are not described here. The thiocyanate method is widely used for estimating chloride ions in waters. It is not particularly sensitive but is adequate for the relatively high levels present in most natural waters. The method is an indirect procedure whereby chloride liberates thiocyanate ions which react with excess ferric ions to produce a coloured complex. Some interference from nitrate is possible.

The detection limit for this method is about $0.5\,mg\,l^{-1}$ and reproducibility $\pm 2\%$.

REAGENTS AND STANDARDS

1   Stock standard ($1000\,mg\,l^{-1}$ Cl).
Dissolve $1.6484\,g$ dry NaCl in water and dilute to 1 litre.
2   Sodium acetate–acetic acid buffer.
Dissolve $30\,g$ sodium acetate trihydrate in $500\,ml$ water. Add $1\,ml$ glacial acetic acid and dilute to 1 litre.
3   Ferric alum–nitric acid.
Dissolve $30\,g$ $FeNH_4(SO_4)_2.12H_2O$ in $350\,ml$ water. Add $95\,ml$ conc. HNO_3. Bring to boil, cool and filter. Dilute to $500\,ml$.
4   Mercuric thiocyanate solution.
Suspend $0.5\,g$ $Hg(SCN)_2$ in $250\,ml$ water. Stir for 12 hours at room temperature and filter.

PROCEDURE

1   Pipette 0–5 ml stock standard into 50 ml graduated flasks to give a range from 0 to 5 mg Cl.
2   Measure not more than 15 ml sample into 50 ml flasks.
3   Add 20 ml buffer and mix.
4   Add 10 ml acid ferric alum and mix.
5   Add 4 ml mercuric thiocyanate, dilute to volume and mix.
6   Measure the optical density at 460 nm using water as a reference.
7   Construct a calibration curve and use it to obtain mg Cl in sample aliquot. Calculate Cl concentration in original material (Section 5.1.2).

5.6   Copper

Atomic absorption is suitable for copper and is preferable to colorimetric or other methods. No serious interferences occur in the air–acetylene flame. Levels in most materials are low (Table 6.1) but the solutions prepared as given earlier will usually be adequate. The sulphuric–peroxide digests are run undiluted at the 5% acid level (Section 3.5). Levels in waters may sometimes be difficult to measure but copper can be concentrated using an APDC-MIBK extraction as described in Cresser (1978).

The detection limit for this method is about $0.02\,mg\,l^{-1}$ and reproducibility $\pm 3\%$.

1 Stock standard solution ($100 \, mg \, l^{-1} \, Cu$).
Dissolve $0.3930 \, g \, CuSO_4.5H_2O$ in water and dilute to 1 litre.
2 Working standards.
Dilute the stock solution to give an intermediate standard of $10 \, mg \, l^{-1}$ and from this produce a range from 0 to $1.0 \, mg \, l^{-1} \, Cu$. Include acids or extractants to match sample solution.

PROCEDURE

1 Prepare the sample solutions as described in Sections 2.7, 3.5 or 4.2.
2 Set up the instrument as in Section 6.2.
3 Select the 324·8 nm wavelength.
4 Use a non-reducing lean air–acetylene flame.
5 Construct a calibration curve and use it to obtain the concentration of Cu in solution. Subtract the blank value. Calculate Cu concentrations in original material (Section 5.1.2).

5.7 Iron

Colorimetric methods are still widely used for iron although atomic absorption is becoming more popular. Both are described here and are comparable in sensitivity.

5.7.1 Atomic absorption method

Allan (1959) found atomic absorption to be satisfactory for the determination of iron in biological materials. The sulphuric acid present in the digests will depress the absorption reading (Curtis, 1969) but this is compensated for by the addition of sulphuric acid to the standards. For soil extracts checks should be made on each soil type for interference by sulphate, and if necessary standards should be compensated accordingly.

The detection limit for this method is about $0.05 \, mg \, l^{-1}$ and the reproducibility $\pm 3\%$.

REAGENTS AND STANDARDS

1 Stock standard solution ($100 \, mg \, l^{-1} \, Fe$).
Dissolve $0.1 \, g$ clean untarnished Fe wire in about $10 \, ml$ warm $10\% \, H_2SO_4$. When cool dilute to 1 litre.
2 Working standards.

Prepare a range from 0 to $5 \, mg \, l^{-1}$ Fe by dilution of the stock solution. Include acids or reagents to match sample solution.

1   Prepare the sample solutions as described in Sections 2.7, 3.5 or 4.2 (see Note 1 for digests and Note 2 for oxalic acid extractions).
2   Set up the instrument as outlined in Section 6.2.
3   Select the wavelength setting of 248·3 nm (there are two less sensitive lines close to this one).
4   Use a non-reducing air–acetylene flame.
5   Construct a calibration curve and use it to obtain the concentration of Fe in solution. Subtract blank value. Calculate Fe concentration in the original material (Section 5.1.2).

NOTES

1   Digests which contain sulphuric acid require the addition of 10–15 ml water and bringing to the boil before filtering and diluting to volume. This is due to the fact that ferric sulphate is insoluble in anhydrous sulphuric acid and precipitates from solution during the final stages of the digestion.
2   3% oxalic acid soil extracts are rich in iron and usually require a 50-fold dilution. Compensate the standards by inclusion of the extractant at this diluted level to reduce interference and burner carbonization.

5.7.2   Colorimetric method

The method described below uses a sulphonated form of bathophenanthroline and was described by Quarmby & Grimshaw (1967) from the procedure of Riley & Williams (1959).

The detection limit for this method is about $0 \cdot 1 \, mg \, l^{-1}$ and reproducibility $\pm 2\%$.

REAGENTS AND STANDARDS

1   Stock standard solution ($100 \, mg \, l^{-1}$ Fe).
Prepare as described in Section 5.7.1.
2   Working standard ($1 \, mg \, l^{-1}$).
Prepare weekly from the stock solution.
3   Sulphonated bathophenanthroline reagent.
Add 4·0 ml fuming H_2SO_4 (20% SO_3) to 0·4 g bathophenanthroline (4:7-diphenyl-1:10 phenanthroline). Stir until dissolved and allow to stand for 30 min. Pour into about 400 ml water.

Neutralize with NH_4OH to between pH 4 and 5 and finally dilute to 1 litre.

4 Sodium acetate hydrate, 33% w/v (see Note 1).

5 Hydroxylamine hydrochloride 2·5% w/v.

6 Combined reagent.

Mix 33% sodium acetate solution, sulphonated bathophenanthroline reagent and 2·5% hydroxylamine hydrochloride in the ratio 4:3:1.

PROCEDURE

1 Prepare the sample solutions as described in Sections 2.7, 3.5 or 4.2 (for soil extracts see Note 2).

2 Measure suitable sample aliquots not exceeding 20 ml into 50 ml volumetric flasks (see Note 3).

3 Pipette aliquots of 0–30 ml of the working standard solution into 50 ml volumetric flasks to give a standard range from 0 to 0·03 mg Fe.

4 Add acid or soil extractant to the standards to match the sample aliquots.

5 From this point treat standards and samples in the same way.

6 Add 16 ml of the combined reagent and dilute to volume (see Note 1).

7 Measure the optical density at 536 nm or with a yellow filter using water as a reference.

8 Construct a calibration curve and use it to obtain mg Fe in the sample aliquots. Subtract blank value. Calculate concentrations of Fe in the original material (Section 5.1.2).

NOTES

1 The sodium acetate buffer is sufficient to control acid levels up to the equivalent of 20 ml of 1% (v/v) H_2SO_4.

2 Refer to the Note 1 under Section 5.7.1 concerning the treatment of plant digests.

3 3% oxalic acid extracts are rich in iron and a 1 ml aliquot will be sufficient (see also Note 2 under Section 5.7.1).

Particular care is needed in the determination of iron because of the possibility of contamination from laboratory materials.

5.8   Magnesium

The atomic absorption method is preferable to any other procedure for this element. The EDTA method, in which magnesium is determined by difference is less sensitive and subject to interferences. Colorimetric methods in general are less satisfactory.

The atomic absorption method given here is basically the same as that used to determine calcium. Oxyacids of aluminium and phosphorus interfere in an air–acetylene flame and necessitate the inclusion of lanthanum as a releasing agent. The hotter nitrous oxide–acetylene flame is almost free of interferences, but it is preferable to include potassium chloride to reduce ionization.

The detection limit for this method is about $0.005\,\mathrm{mg\,l^{-1}}$ and reproducibility $\pm 1\%$.

REAGENTS AND STANDARDS

1   Stock standard solution ($100\,\mathrm{mg\,l^{-1}}$ Mg).
Dissolve $1.0136\,\mathrm{g}$ $MgSO_4.7H_2O$ in water containing about $1\,\mathrm{ml}$ H_2SO_4. Dilute to 1 litre.
2   Lanthanum chloride solutions.
Prepare 2000 or acid $4000\,\mathrm{mg\,l^{-1}}$ La stock solutions as given in Section 5.3.
3   2.5% sulphuric acid (for water analysis only).
4   Working standards.
Prepare a range from 0 to $4\,\mathrm{mg\,l^{-1}}$ Mg by dilution of the stock solution. Include acids and extractants to match the sample solutions and include $400\,\mathrm{mg\,l^{-1}}$ La (plant and water standards) and $800\,\mathrm{mg\,l^{-1}}$ La in 1% H_2SO_4 (soil standards).

PROCEDURE

1   Prepare the sample solution as described in Sections 2.7, 3.5 or 4.2.
2   From this point follow the procedure outlined in Section 5.3 (starting at step 2).
3   The most sensitive wavelength setting for magnesium is $285.2\,\mathrm{nm}$.
4   Construct a calibration curve to obtain concentration of Mg in solution. Subtract the blank value. Calculate concentration of Mg in the original material (Section 5.1.2).

5.9   Manganese

In the past, colorimetric procedures have been used for this element, and the formaldoxine method is sensitive enough to determine manganese in most samples (Allen, 1974). However, atomic absorption is a sensitive and convenient method and is relatively free from interferences, but it is preferable that the standards and samples contain manganese in the same valency state. This may be variable in soil extracts and if perchloric is not driven off in the mixed acid digestion a reduction stage may be needed (Thompson & Reynolds, 1978). Also, after mixed acid digestion manganese dioxide may be precipitated, in which case it should be redissolved with sulphurous acid. The application of atomic absorption to the determination of manganese is discussed by Christian & Feldman (1970) and air–acetylene is the most suitable flame.

The detection limit for this method is about $0.01\,\mathrm{mg\,l^{-1}}$ and reproducibility $\pm 2\%$.

REAGENTS AND STANDARDS

1 Stock standard solution $(100\,\mathrm{mg\,l^{-1}\,Mn})$.
Dissolve $0.4060\,\mathrm{g}$ $MnSO_4.4H_2O$ in water containing 1 ml conc. H_2SO_4 and dilute to 1 litre.
2 Working standards.
Prepare a range from 0 to $2\,\mathrm{mg\,l^{-1}}$ Mn by dilution of the stock standard. Include acids or extractants to match the sample solution.

PROCEDURE

1 Prepare the sample solutions as in Sections 2.7, 3.5 or 4.2.
2 Set up the instrument as outlined in Section 6.2.
3 The wavelength setting for manganese is 279·5.
4 Adjust the air–acetylene flame to give maximum transparency.
5 Construct a calibration curve to obtain the concentration of Mn in solution. Subtract the blank value. Calculate concentration of Mn in the original material (Section 5.1.2).

5.10 Nitrogen

5.10.1 Organic nitrogen

The classical Kjeldahl digestion remains the standard procedure for converting organic-nitrogen to ammonium-nitrogen which can then be readily estimated. The sample is digested with sulphuric acid in the presence of a salt and a catalyst. The purpose of the salt, generally potassium or sodium sulphate, is to raise the digestion temperature. Many different catalysts have been shown to be effective but mercuric oxide is used in the method given here.

Levels of inorganic nitrogen are relatively low in soils and plant materials. Nitrate and nitrite-nitrogen are not readily converted to the ammonium form by the normal digestion procedure but the inclusion of salicylic acid will facilitate this reduction. 'Fixed' ammonium nitrogen, that is the fraction held in the soil mineral lattice, may not be fully recovered.

The sulphuric–peroxide digestion procedure given in Section 3.5 was introduced to enable phosphorus and nitrogen to be determined in the same digest solution. A lithium salt is used to elevate the digestion temperature and selenium is added as a catalyst. Neither interferes with the determinations given in the text. The method is effective for organic nitrogen in soils and plant materials (although not mineral elements in soils).

Usually ammonia is distilled off from the digest solution after reaction with excess alkali. It is then measured by titration (as described here). The detection limit for this method is about 10 μg and reproducibility $\pm 2\%$. A colorimetric method can also be used to measure ammonium nitrogen directly in the digest.

In waters, levels of inorganic nitrogen may be comparable to, or exceed, those of the organic form so the former must be first removed before the organic nitrogen can be determined.

REAGENTS AND STANDARDS

1   Standard ammonium chloride ($100\,mg\,l^{-1}\,NH_4$-N).
Dissolve 0·1910 g dry NH_4Cl in water and dilute to 500 ml.
2   Hydrochloric acid, M/140 (1 ml = 0·1 mg NH_4-N).
Prepare 0·1 M HCl and standardize it against 0·05 M Na_2CO_3 solution using 0·1% bromo-phenol blue as indicator. When standardized dilute to exactly M/140.
3   Potassium sulphate–mercuric oxide mixture.
Mix K_2SO_4 and HgO (NH_3-free grades) in the ratio of 20:1 (2 g tablets of this preparation are available*).
4   Sulphuric acid, conc. (NH_3-free grade).
5   Alkali mixture.
Dissolve 500 g NaOH and 25 g sodium thiosulphate in water with care, cool and dilute to 1 litre.
6   Devarda's alloy powder.
The commercially available Devarda's alloy should be finely ground before use. It contains 50% Cu, 45% Al and 5% Zn.
7   Magnesium oxide, powder, ignited at 550° C for 1 hour.
8   Boric acid-indicator solution. Place 20 g of pure boric acid (H_3BO_3) in a 2-litre beaker, add about 900 ml water, heat and swirl until the H_3BO_3 is dissolved. Cool and solution, and add 20 ml of mixed indicator solution prepared by dissolving 0·099 g of bromocresol green and 0·066 g of methyl red in 100 ml of industrial methylated spirit. Then add 0·1 M NaOH cau-

*Available from Thompson & Capper, 3 Goddard Road, Ashmoor Industrial Estate, Runcorn, Cheshire.

tiously until the solution assumes a reddish purple tint (pH *c*. 5·0). Dilute to 1 litre and mix.

I Kjeldahl digestion

1 (a) Plant materials and soil.
Weigh a suitable quantity of dried ground sample into a 50 ml round-bottomed Kjeldahl flask. (Take 0·1 g of organic samples but 0·2 g of mineral soil low in organic matter.)
  (b) Water.
Measure up to 250 ml into a Taylor flask and add 0·1 g Devarda's alloy plus 0·05 g magnesium oxide. Boil down carefully to a few ml but do not allow to dry out.

2 Add 2 g of potassium sulphate–mercuric oxide mixture followed by 3 ml concentrated sulphuric acid. Run the acid slowly down the neck of the flask whilst rotating the flask.

3 Heat the bulb of the flask gently on a digestion rack while frothing continues.

4 When the frothing has stopped increase heat until sulphuric acid refluxes down the neck of the flask.

5 Continue heating until the solution becomes colourless or pale yellow/green and continue for a further period ranging from 15 min for vegetation, 30 min for peat, litter and animal tissue and up to 1 hour for soils.

6 Allow the flask to cool and dilute to 50 ml with water unless the entire solution is to be taken for distillation.

7 Prepare blank solutions following the same procedure.

II Distillation

8 Set up a steam distillation apparatus as shown in Fig. 6.1. Use a steam generator containing NH_3-free water (see Note).

9 Pass steam through apparatus for 30 min. Check the steam blank by collecting 20–30 ml distillate and titrating with M/140 HCl as given below. The steam blank should not require more than 0·1 ml acid.

10 Transfer the entire sample solution or an aliquot to the reaction chamber and add 12 ml of alkali mixture.

11 Commence distillation immediately and collect 50 ml distillate in a suitable receiver containing 5 ml of boric acid-indicator solution.

12 Titrate the distillate with M/140 hydrochloric acid to a pale yellow end-point using a microburette.

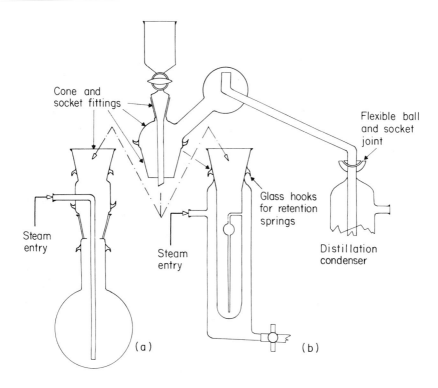

Figure 6.1 Apparatus suitable for (a) macro, and (b) semi-micro distillation of solutions containing ammonium-nitrogen.

13   Occasionally check that the distillation recovery is satisfactory by taking an aliquot of the standard ammonium chloride solution in place of the sample.

14   Subtract blank value from the sample titrations.

CALCULATION

If 1 ml M/140 HCl = 0·1 mg NH_4^+-N

then total N (%) $= \dfrac{\text{titre (ml)} \times \text{solution volume (ml)}}{10^2 \times \text{aliquot (ml)} \times \text{sample wt (g)}}$

NOTE

Deionized water is preferable to distilled water since the latter often gives a high blank value. Ammonium ions can be removed from distilled water by shaking with a strong cation exchange resin. (Mains tap water may have a lower NH_4-N content than distilled water in some areas.)

III Colorimetry

Ammonium-nitrogen may also be determined in the distillate using a colorimetric method instead of the titration step just described. In some ways a more convenient alternative is to omit the distillation stage and to determine ammonium-nitrogen in the original digest. The direct procedure generally favoured is based on the formation of the indophenol blue complex using a phenolic compound, hypochlorite and a catalyst. There are many procedures in the literature, some of which are coupled with digestions to determine total nitrogen (Verdouv *et al.*, 1977; Hinds & Lowe, 1980), whilst others have been extensively used for water analysis (APHA *Standard Methods of Water and Wastewater*, 1975; DOE Standing Committee of Analysts, 1981). In using these methods it is important to note that an elevated temperature will be required to speed up the reaction, the timing will need to be strictly controlled and the pH of the reaction is critical. The critical conditions that have to be maintained make the method rather more suitable for automated than manual operation. A recent review of the indophenol method evaluated the role of all the reagents commonly used in this reaction (Krom, 1980).

5.10.2 Inorganic nitrogen fractions

It is often of value to determine inorganic nitrogen in soils and waters. Levels in soils are usually low compared with organic nitrogen but both ammonia and nitrate ions are important because they are the forms available for plant uptake. Sodium or potassium chloride are used for extracting ammonium-nitrogen from soils whilst a water extract is sufficient for nitrate-nitrogen.

Ammonium-nitrogen

To determine extractable ammonium-nitrogen in soil, first follow the extraction procedure discussed in Section 2.7.3.

I Distillation

1 Use the procedure outlined in Section 5.10.1 taking 50 ml of soil extract or up to 250 ml water in a distillation flask.
2 Add about 0·2 g MgO and proceed from step 11.
3 Run reagent blanks for soil extracts.

II Colorimetry

See discussion in Section 5.10.1.

Nitrate-nitrogen

To determine extractable nitrate-nitrogen in soil, first follow the extraction procedure discussed in Section 2.7.3. Water should preferably be unfiltered.

I Distillation

Initially, the ammonia in solution is removed by distillation (and estimated separately if required). Nitrate is then reduced by Devarda's alloy and the resultant ammonia distilled into boric acid and titrated with hydrochloric acid as before.

1   Use the procedure outlined in Section 5.10.1 taking 50 ml of soil extract or up to 250 ml water in a distillation flask.
2   Add about 0·2 g MgO, collect 50 ml distillate and discard.
3   Remove the flask, add about 0·2 g Devarda's alloy and proceed from step 11 in Section 5.10.1.

II Colorimetry

Direct colorimetric methods for nitrate estimation in waters are generally less reliable and sensitive than indirect procedures. The latter are based on the reduction of nitrate to nitrite, followed by complex formation. A suitable method for waters has been recommended by Mackereth et al. (1978). The method can also be used for nitrate-nitrogen if the reduction stage is omitted.

The detection limit for this method is about $0·01 \, mg \, l^{-1}$ and reproducibility $\pm 3\%$.

REAGENTS AND STANDARDS

1   Stock standard ($100 \, mg \, l^{-1} \, NO_3\text{-}N$).
Dissolve 0·7216 g dry KNO_3 in water and dilute to 1 litre.
2   Cadmium reductant.
Stand zinc rods in 20% w/v cadmium sulphate solution overnight. Scrape the cadmium from the rods and break into small particles. Treat before and after use with 2% HCl and then wash until acid free and store under water.
3   Ammonium chloride, 2·6% w/v.

4 Borax, 2·1% w/v.
5 Sulphanilamide, 1% w/v in 10% v/v HCl
6 N-1-napthylethylenediamine dihydrochloride, 0·1% w/v (NNED).

PROCEDURE

1 Prepare a range of working standards up to $1 \text{ mg} \text{l}^{-1}$ and transfer 10·0 ml to a suitable container.
2 Add 3·0 ml ammonium chloride solution, 1·0 ml borax solution and approximately 0·5 g cadmium reductant.
3 Seal the container and shake for 20 min.
4 Allow to settle and pipette 7·0 ml supernatant into a 50 ml graduated flask.
5 Add 1·0 ml sulphanilamide and mix.
6 After 5 min, add 1·0 ml NNED, mix and dilute to volume.
7 Measure the absorbance at 543 nm.
8 Prepare a calibration curve and estimate nitrate levels and calculate final results as in Section 5.1.2.

5.10.3 Mineralizable soil nitrogen

The amount of soil organic nitrogen mineralized and consequently available for plant uptake is dependent on the activity of micro-organisms and is a key stage in the nitrogen cycle. It is an important characteristic of a soil but no ideal method has been suggested for its estimation. The method given here is designed to stimulate microbial activity, and although arbitrary, is widely used. The inorganic nitrogen which is released is most conveniently extracted and estimated by distillation. Acceptable results may be obtained from soils if total nitrogen levels exceed 0·5%. See Page *et al.* (1982) for further discussion.

REAGENTS AND STANDARDS

1 Standard ammonium chloride solution ($100 \text{ mg} \text{l}^{-1} \text{ NH}_4^+$ -N).
2 Standard hydrochloric acid solution, M/140.
3 Boric acid-indicator solution.
4 Devarda's alloy powder.
5 Magnesium oxide.
6 Sand, acid washed, 30–60 mesh.
7 Potassium chloride, 6% w/v.

Prepare as in Section 5.10.1.

PROCEDURE

I Incubation

1 Weigh 5 g air-dry sieved soil into a 250 ml screw cap polythene bottle for zero time determination. Weigh a separate 5 g portion into a 250 ml beaker flask for incubation.
2 Add 15 g sand to each and mix.
3 Add 6 ml water to each.
4 Cover the mouth of the flask with a porous polythene film and incubate at 30°C for 14 days. (The zero time samples must be extracted on the same day that incubation is started.)
5 Include two sand blanks each for zero time and for incubation.

II Extraction

1 Add 125 ml 6% KCl to zero time samples prepared as above.
2 Use 125 ml 6% KCl to wash the incubated sample into 250 ml polythene bottles.
3 Shake for 1 hour on an end-over-end shaker.
4 Filter through Whatman No. 44 paper.

III Distillation

1 Use the procedure outlined in Section 5.10.1 using 50 ml of soil extract.
2 Add about 0·2 g Devarda's alloy and about 0·2 g MgO and proceed from step 11 in Section 5.10.1.
3 Run a reagent blank.

CALCULATION

If 1 ml M/140 HCl = 0·1 mg NH_4^--N

then mineralizable N (mg 100 g^{-1})

$$= \frac{\text{titre (ml)} \times (125 + 6) \text{ ml} \times 2}{\text{aliquot (ml)}}$$

Mineralizable N is obtained by subtracting the zero time value from the incubation result. Correct to dry weight.

NOTE

Extracts of incubated soil should be stored just above 0°C and distilled within 24 hours.

5.11 Phosphorus

A common method for estimating phosphate-phosphorus in solution is based on the formation of the 'molybdenum blue' complex and is described here. The reducing agent is stannous chloride and is slightly more sensitive than ascorbic acid which is also widely used. Under optimum conditions amounts as low as 1 μg phosphate-phosphorus can be measured. No method for organic phosphorus in soil is given here but two basic methods are widely used and are reviewed by Black & Goring (1953) and Williams *et al.* (1970).

Total phosphorus in soils is preferably determined following a hydrofluoric acid plus perchloric acid digestion. However, the mixed acid digestion (Section 2.7) is an acceptable substitute for many soil types taking 50 mg dry ground soil.

The detection limit for this method is about $0\cdot1$ mg l^{-1} and reproducibility $\pm1\%$.

REAGENTS AND STANDARDS

1 Stock standard solution (100 mg l^{-1} P).
Dissolve $0\cdot4393$ g dry KH_2PO_4 in water and dilute to 1 litre.
2 Working standard (2 mg l^{-1} P).
Dilute from the stock standard solution. Make up fresh weekly.
3 Ammonium molybdate–sulphuric acid reagent.
Dissolve 25 g ammonium molybdate $(NH_4)_6Mo_7O_{24}.4H_2O$ in about 200 ml water in a beaker. It may be necessary to warm slightly to dissolve.
Carefully add 280 ml conc. H_2SO_4 to about 400 ml water with mixing and cooling.
Filter the molybdate solution into the acid solution, mix and cool.
4 Stannous chloride reagent.
Dissolve $0\cdot2$ g $SnCl_2.2H_2O$ in 100 ml of 2% v/v HCl. Prepare immediately before use.

PROCEDURE

1 Prepare the sample solution as described in Sections 2.7, 3.5 or 4.2.
2 Pipette a suitable aliquot of sample solution into a 50 ml volumetric flask (up to 40 ml soil extract and 5 ml plant digest) (see Notes 1 and 2).
3 Pipette aliquots of 0–15 ml of working standard solution into separate 50 ml volumetric flasks. This gives a standard range from 0 to $0\cdot03$ mg phosphorus. Include in each flask either digest acid or soil extractant to match the sample aliquots.
4 From this point treat standards and samples in the same way.

5 Add water to each flask until it is about two-thirds full.

6 Add 2 ml ammonium molybdate reagent and mix.

7 Add 2 ml stannous chloride reagent, mix and dilute to volume.

8 After 30 min measure the optical density at 700 nm using water as a reference.

9 Construct a calibration curve for the standards and use it to determine the mg phosphorus in the sample aliquots. Subtract blank values as necessary. Calculate the phosphorus content in the original samples (see Section 5.1.2).

NOTES

1 Aliquots of plant digests (1% acid) should not exceed 10 ml but up to 40 ml of neat water and soil extracts (2·5% acetic acid, Truog's reagent) may be taken. Aliquots of Olsen's extracts should not exceed 10 ml and must be neutralized with dilute sulphuric acid until yellow with 0·1% p-nitrophenol indicator. If a brown colour masks the indicator then add a pre-determined amount of acid.

2 For samples in which the phosphorus level is very low (e.g. natural waters) the above procedure can be modified by including an extraction procedure in which the phospho-molybdate is extracted into n-butanol and the colour developed in the organic phase (see Allen, 1974).

5.12   Potassium

Potassium has an intense emission line which makes flame photometry a very suitable method, although atomic absorption is also sometimes used for this element. Flame photometry has the advantage that dual channel instruments for sodium and potassium can be used although care is required in preparing the standards (see Note). Some photometers allow for flame fluctuation by incorporating an internal lithium standard but models without this feature are satisfactory if flame conditions are stabilized.

The detection limit for this method is about $0·05 \, mg \, l^{-1}$ and reproducibility $\pm 2\%$.

REAGENTS AND STANDARDS

1 Stock standard solutions ($1000 \, mg \, l^{-1} \, K$).

Dissolve 1·9067 g dry KCl in deionized water and make up to 1 litre. Check the Na content (see Note): commercial stock solutions contain significant levels of Na.

2 Working standard solutions.

For most ecological materials the three separate ranges $0–5 \, mg \, l^{-1}$,

0–25 mg l^{-1} or 0–100 mg l^{-1} will be advised. Include acid or soil extractants to match sample solutions. Dual standards with sodium can be used but see Note.

PROCEDURE

1 Prepare the sample solutions as described in Sections 2.7, 3.5 or 4.2.
2 Set up the instrument as in Section 6.2.
3 Select the emission line at 766 nm or use a potassium filter.
4 Use an air–acetylene or air–propane flame.
5 Construct a calibration curve and use it to obtain the concentration of K in solution. Subtract blank values. Calculate K concentrations in the original material (Section 5.1.2).

NOTE

If dual sodium–potassium standards are required then sources of cross-contamination must be minimized. The stock solutions must be prepared separately and neither should contain more than 1 mg l^{-1} of the other element. Commercial stock solutions are suspect in this respect and unbalanced combinations must be avoided.

5.13   Silicon

The most suitable method is the colorimetric procedure based on the formation of a 'heteropoly blue' colour. The method given below for waters follows that of Morrison & Wilson (1963). It will measure levels down to about 5 μg silicon under these conditions. The reaction is similar to that used for phosphate-phosphorus but interference from the latter can be controlled by adding tartaric acid.

Total silicon in plant material and soil required a fusion technique not described here but see Page et al. (1982) and other texts. An approximate estimate as silica (SiO_2) in vegetation can be obtained by weighing the residue remaining after the dry ashing or wet oxidation procedures described earlier. However, quantitative removal of this residue from a digestion flask may be difficult.

The detection limit for this method is about 0·2 mg l^{-1} and reproducibility ±2%.

REAGENTS AND STANDARDS

1 Stock standard (100 mg l^{-1} Si).
Fuse 0·2139 g dry SiO_2 powder with 1 g anhydrous Na_2CO_3 in a platinum

crucible at 950°C until a clear melt is obtained. Cool, and immerse the crucible and contents in water contained in a polypropylene beaker. Warm to dissolve, cool and dilute to 1 litre, washing the crucible carefully. Store in a polythene bottle.

2   Working standard solution ($5\,mg\,l^{-1}$ Si).

Dilute the stock standard solution. Store in a polythene bottle.

3   Ammonium molybdate–sulphuric acid reagent.

Dissolve 89 g ammonium molybdate $(NH_4)_6Mo_7O_{24}.4H_2O$ in about 800 ml water. Dilute 62 ml conc. H_2SO_4 to about 150 ml by adding it carefully to water. Allow to cool and add the acid to the molybdate solution. Dilute to 1 litre.

4   Tartaric acid, 28% w/v.

5   Reducing solution.

Dissolve 2·4 g $Na_2SO_3.7H_2O$ and 0·2 g 1-amino-2-napthol-4-sulphonic acid in about 70 ml water. Add 14 g $K_2S_2O_5$, shake to dissolve and dilute to 100 ml. Prepare fresh weekly.

6   Hydrochloric acid, 1% v/v.

PROCEDURE

1   Pipette 1–20 ml of the water into a 50 ml flask.

2   Pipette 0–6 ml of working standard solution into volumetric flasks to give a range of standards from 0 to 0·03 mg silicon.

3   Neutralize the standard aliquots with 1% HCl.

4   From this point treat standard and sample aliquots in the same way.

5   Add 1·25 ml acid molybdate reagent. Mix and leave for 10 min.

6   Add 1·25 ml tartaric acid. Mix and leave for 5 min.

7   Add 1·0 ml reducing solution. Mix and dilute to volume with distilled water.

8   Allow the flask to stand for 15 min to ensure maximum colour development.

9   Measure the optical density of each solution at 810 nm using water as a reference.

10   Construct a calibration curve and use it to obtain mg Si in the sample aliquots. Subtract blank value. Calculate concentration of Si in the original material (Section 5.1.2.).

NOTE

Distilled water is preferred to deionized water (which in some cases contains significant levels of silica) and should be used for all solution preparation.

5.14 Sodium

Sodium may be easily determined by flame emission photometry using an air–propane flame and the technique is the same as for potassium (Section 5.12). Atomic absorption does not have any marked advantage over the emission method except that the high levels of calcium sometimes encountered in soil extracts do not interfere. This interference may be eliminated in the flame photometric method by dilution or inclusion of 0·2% sulphuric or phosphoric acid. It is convenient to determine sodium and potassium together on a dual channel flame photometer but see Note 1 below.

The detection limit for this method is about $0·05\,mg\,l^{-1}$ and reproducibility $\pm 2\%$.

REAGENTS AND STANDARDS

1 Stock standard solutions ($1000\,mg\,l^{-1}\,Na$).
Dissolve 2·5420 g dry NaCl in deionized water and make up to 1 litre.
2 Working standard solutions.
Prepare a range of standards by suitable dilution of the stock standard.
For ecological materials two ranges of standards will be found convenient, namely $0–5\,mg\,l^{-1}$ or $0–25\,mg\,l^{-1}$. Reproducibility for routine work should be of the order of $\pm 2\%$. Include acids or extractants to match the sample solutions. Dual standards can be prepared (see Note 1).

PROCEDURE

The steps set out for potassium (Section 5.12) should be followed with the exception that the 589 nm line or appropriate filter are used for sodium.

NOTES

1 If dual sodium–potassium standards are required care is necessary in their preparation, particularly from commercial stock solutions (see Note under potassium (Section 5.12)).
2 Due to the wide spread of sodium salts in laboratory reagents and in products outside the laboratory, blank values will often be higher than for other elements and particular care should be taken to ensure that contamination is minimized.

5.15 Sulphate-sulphur

Increase in sulphur levels in the atmosphere has renewed the interest in sulphur cycling in the environment. The element generally occurs as the sulphate ion in natural waters. No ideal chromogenic reagent has yet been developed for sulphate-sulphur, although thoron is sometimes used. Many turbidimetric methods have been described, but the one given here appears to be more precise than most and is taken from that of Butters & Chenery (1959) who modified an earlier method of Chesnin & Yien (1950). The conditions for the development of turbidity need to be carefully controlled to get the best results. A fairly uniform size of barium chloride crystals is desirable and violent shaking at any stages is avoided.

The method here is applied to waters. It can also be used for solutions of vegetation (total sulphur) and soil (extractable sulphate-sulphur) but the preparations of these solutions requires special dry ashing and extraction procedures respectively. Details are given in the texts already quoted.

The detection limit for this method is about $0.2 \, mg \, l^{-1}$ and reproducibility $\pm 5\%$.

REAGENTS AND STANDARDS

1   Sulphur standard (1 ml $= 0.05 \, mg \, S$).
Dissolve $0.3844 \, g$ of $MgSO_4.7H_2O$ in water and make up to 1 litre.
2   Nitric acid, 25% v/v.
3   Acetic acid, 50% v/v.
4   Ortho-phosphoric acid, SG 1.75.
5   Barium chloride crystals.
Grind $BaCl_2.2H_2O$ crystals to pass between 15 and 40 BS mesh.
6   Gum acacia, 0.5% w/v.
Dissolve $0.5 \, g$ gum acacia in 100 ml warm water. This reagent should be freshly prepared.

PROCEDURE

1   Pipette 0–10 ml of the standard into 50 ml volumetric flasks to give a range of standards from 0 to 0.50 mg S.
2   Transfer a suitable sample aliquot into a 50 ml volumetric flask.
3   From this point treat standards and samples in the same way.
4   Add 5 ml nitric acid, 5 ml 50% acetic acid and 1 ml H_3PO_4 and swirl the solution to mix. Dilute to the base of the neck and mix again.
5   Add 1 g $BaCl_2.2H_2O$ crystals without mixing, and leave to stand for 10 min.
6   Invert the flasks twice and stand for 5 min.
7   Invert twice again and leave for a further 5 min.
8   After about ten inversions, add 1 ml gum acacia solution and make up to 50 ml.

9 Invert several times more and stand for 1·5 hours.

10 Measure the turbidity as optical density at 470 nm or with a blue green filter using water as a reference.

11 Construct a calibration curve and use it to obtain mg S in the sample aliquot. Calculate SO_4-S concentration in the original material (Section 5.1.2).

5.16 Zinc

Atomic absorption is a very suitable procedure for zinc and is preferable to other methods. No serious interferences occur using the air–acetylene flame. Zinc is a trace nutrient but the method is sensitive and the solutions prepared as given earlier are all suitable. In general they should not require dilution and the sulphuric-peroxide digests are run undiluted at the 5% acid level.

The detection limit for this method is about $0·01\,mg\,l^{-1}$ and reproducibility $\pm 2\%$.

REAGENTS AND STANDARDS

1 Stock standard solution ($100\,mg\,l^{-1}$ Zn).
Dissolve 0·4398 g $ZnSO_4.7H_2O$ in water and dilute to 1 litre. Commercially prepared stock standards are available.

2 Working standards.
Dilute the stock solution ten times to give an intermediate standard of $10\,mg\,l^{-1}$ and from this prepare a range from 0 to $1·0\,mg\,l^{-1}$ Zn. Include acids or extractants to match the sample solutions.

PROCEDURE

1 Prepare the sample solutions as described in Sections 2.7, 3.5 or 4.2.

2 Set up the instrument as in Section 6.2.

3 Select the wavelength of 213·9 nm for zinc.

4 Use an air–acetylene flame.

5 Construct a calibration curve and use it to obtain the concentration of Zn in solution. Subtract blank value. Calculate Zn concentrations in the original material (Section 5.1.2).

NOTE

Particular care is needed in the determination of zinc because of the possibility of contamination from laboratory materials.

6   Instrumental procedures

The two analytical techniques used for the determination of most of the elements in Section 5 are atomic absorption (with flame emission) and colorimetry. They are recommended as being rapid and technically easy to apply, and more sensitive than older methods, although not always free from interferences. Some gravimetric and volumetric methods are also given but the theory and practice on these techniques are well covered in standard texts such as Vogel (1962) and are not repeated here.

6.1   Colorimetry

The colorimetric methods in Section 5 require only a basic spectrophotometer. In these methods the test solution is treated with a chromogenic reagent to produce a colour which is proportional to the amount of element being tested in the solution. The methods given here have been adapted to the normal levels of nutrients in soil, vegetation and waters, but concentration techniques such as evaporation or solvent extraction may be necessary for determining low concentrations.

A spectrophotometer consists of four main parts: namely, a radiation source, a wavelength selector, a test solution compartment and a detection system. In visible spectrophotometry the radiation source is usually a tungsten filament bulb. Monochromatic light is obtained by use of a filter, or by a prism or grating. Sample cells (cuvettes) are housed in a black enclosure and are made from optical glass. The cells normally have a path length of 1 cm but longer path lengths may be used for increasing the sensitivity of a method. Most modern spectrophotometers use a photocell and photomultiplier detection system, usually with digital readout, although older instruments employed a null-point meter. Double beam instruments are widely available, and utilize a reference cell containing only water or solvent.

Colorimetry is based on the absorption of monochromatic light by the solution under standard conditions. The concentration of ions in the coloured solution is related to the absorption by the Beer–Lambert Law given by:

$$\log_e \frac{I_0}{I} = kcl$$

where I_0 = incident light, I = emergent light, c = concentration, k = constant, and, l = solution thickness. A plot of optical density against concentration should be linear for the ranges specified in the methods. Deviations from linearity may, however, occur due to chemical and physical effects. A few practical points are listed below.

1 Although electronic components now require very little warm up time the instrument itself should be allowed to reach a temperature equilibrium after the lamp is switched on.

2 Select the wavelength appropriate for the method.

3 The optical surfaces of cuvettes should be cleaned carefully with a soft tissue and the cuvettes must be matched before use.

4 The instrument should be zeroed with water in both the reference and sample cells. This setting should be checked occasionally during a run.

5 Use the sample cell to obtain the absorbance (optical density) value of standards and samples after colour development. Some waters are themselves coloured. To allow for this, obtain a colour blank by omitting the chromogenic reagent. Later subtract from the sample when the concentration has been obtained.

6.2 Flame spectroscopy

Atomic absorption and flame photometry are sensitive and complementary techniques and, in general, for metallic elements, the flame spectroscopy methods outlined in Section 5 are preferred to colorimetry when the instrumentation is available.

6.2.1 Atomic absorption

Atomic absorption spectroscopy (AAS) is a sensitive and specific technique enabling many mineral elements to be estimated in solution with relatively few interferences. Only a basic description can be included here but the methodology is well documented by various workers including Christian & Feldman (1970), Ramirez-Munoz (1968), Thompson & Reynolds (1978) and Price (1979). Electrothermal techniques are well covered by Fuller (1977).

AAS involves three separate units, notably the light source, the sample volatilization unit and the detector system. The light source is a hollow cathode lamp which is now available for most mineral elements. The volatilization unit normally includes a nebulizer and a burner with a long path length. Gas mixtures commonly used for this purpose include air–acetylene and nitrous oxide–acetylene. For trace nutrients this system can be replaced by an electrothermal atomizer which enables very low levels to be detected. The detection system comprises a monochromator, detector and amplifier very similar to that found in colorimetric instruments. Single beam and double beam instruments are available. Cheaper single beam instruments are less stable but later models are much improved and have the edge in stability. Most models are equipped with a digital display which in some cases is in concentration units. Facilities for attaching a recorder are usually available and enable the procedure to be automated if a sampler is also attached.

When a solution of the element is sprayed into the flame an atomic vapour is formed. Radiation from the hollow cathode lamp passing through the flame will be absorbed by atoms of the same element, thereby resulting in a decrease in the signal at the detector. This decrease in energy is proportional to the concentration of the element in solution. Interferences can occur at this stage but are relatively minor except for calcium and magnesium where lanthanum is included as given in Section 5. Even for these elements the hotter nitrous oxide–acetylene flame is relatively free from interferences. It is important, however, that standards and samples are comparable regarding reagents and extractants as stressed elsewhere.

For flame work generally it is important that the manufacturers' instructions be followed, particularly with regard to gas flow rates and ignition procedures. A few additional hints are included here to supplement these instructions and other details given in Section 5.

1   Allow a few minutes for the hollow cathode lamp temperature to stabilize. This is most important for single beam instruments.

2   Adjust the lamp current to a setting suitable for the sensitivity required. However, this should preferably be not more than two-thirds of the maximum current indicated on the lamp if its life is not to be significantly shortened.

3   Select the appropriate wavelength and optimize the wavelength setting to give maximum reading on the energy meter.

4   Ensure the lamp is aligned correctly.

5   Check the correct burner is in place. Nitrous oxide–acetylene flames use a shorter burner slot. Use of an incorrect burner may result in an explosion.

6   Ignite the flame and adjust the fuel and oxidant levels to give the most appropriate flame conditions. Low fuel pressure will result in increased noise.

7   Optimize the burner settings by aspirating a standard. A slow response indicates a blocked nebulizer, which can be cleared with a fine wire.

8   Select the appropriate slit width and integration time. Set instrument readings using zero and top standard solutions to obtain a suitable response. If necessary expand the scale or conversely rotate burner to reduce the signal.

9   Periodically recheck calibrations to detect any drift. Obtain a plot of reading against standard concentration and use this to obtain sample concentrations. For the ranges given in Section 5 this is usually linear or slightly curved. Rich samples should be diluted rather than run on an abnormally wide standard range.

10   Note that in extended use high acid concentrations will gradually corrode the burner and 1% acid is recommended in the methods. Soil extracts may result in carbon being deposited on the burner. Flush regularly with water.

6.2.2 Flame photometry

In general, although atomic absorption is more sensitive than flame emission the latter technique is particularly suitable for the determination of sodium and potassium. There is little difference in sensitivity for either of these elements. Flame photometers are relatively inexpensive and simple to use and basically comprise an aspirator–burner unit and a detection measuring system. The nebulizer in most cases is similar to that used in AAS but an air–propane flame is normally used. Characteristic emission in the flame is isolated using an optical filter and the transmitted energy is collected on a photocell for amplification and output. Dual channel instruments are also available and some of these use an internal standard to compensate for changes in flame conditions.

In emission methods, atoms in the flame absorb the heat energy and then release it at a wavelength characteristic for that element. If the operating conditions are constant then the intensity of this light is proportional to the concentration of the element in the flame. The control of flame conditions is most important in flame emission methods as the emission is related to the flame temperature and, therefore, variations in fuel flow or aspiration rate through blockages in the nebulizer must be avoided. This type of variation is overcome by using an instrument with a lithium internal standard where the signal from sodium (or potassium) is compared to that from a constant lithium concentration. Interferences can also occur by overlap of molecular band spectra and this is most important for sodium estimation in calcareous soil extracts. At higher levels of sodium or potassium, e.g. up to $100\,mg\,l^{-1}$, some self-absorption may occur when atoms reabsorb emitted radiation resulting in reduced sensitivity and a curvilinear calibration.

A few practical points for operating a flame photometer are given below. They also apply to an AAS instrument operated in the emission mode with the correct burner in position.

1   Select the appropriate wavelength (AAS equipment) or filter (most flame photometers). See also Note 3 in Section 6.2.1.

2   Use only an air–acetylene or air–propane flame, usually the latter. Adjust flow to obtain the most appropriate flame conditions.

3   Adjust the scale reading to a suitable value with the top standard after setting the baseline with zero standard.

4   Check both standard readings for stability. Then aspirate the full range of standards to prepare the calibration curve.

5   Aspirate samples and check standards periodically to detect drift or nebulizer blockage.

6   Flush regularly with water.

7   References

AMERICAN PUBLIC HEATH ASSOCIATION (1975) *Standard Methods for the Examination of Water and Wastewater.* 14th Edn. New York.

ALLAN J.E. (1959) Determination of iron and manganese by atomic absorption. *Spectrochim. Acta* **10**, 800–806.

ALLAN J.E. (1961) The determination of copper by atomic absorption spectrophotometry. *Spectrochim. Acta* **17**, 459–466.

ALLEN S.E. (Ed.) (1974) *Chemical Analysis of Ecological Materials.* Blackwell Scientific Publications, Oxford.

ALLEN S.E. and GRIMSHAW H.M. (1962) Effect of low-temperature storage on the extractable nutrient ions in soils. *J. Sci. Fd Agric.* **13**, 525–529.

ALLISON L.E. (1960) Wet-combustion apparatus and procedure for organic and inorganic carbon in soil. *Proc. Soil Sci. Soc. Am.* **24**, 36–40.

ASSOCIATION OF OFFICIAL ANALYTICAL CHEMISTS (1970) *Official Methods of Analysis.* Washington.

BALL D.F. (1964) Loss-on-ignition as an estimate of organic matter and organic carbon in non-calcareous soils. *J. Soil Sci.* **15**, 84–92.

BALL D.F. and BEAUMONT P. (1972) Vertical distribution of extractable iron and aluminium in soil profiles from a brown earth–peaty podzol association. *J. Soil Sci.* **23**, 298–308.

BARROW N.J. (1968) Determination of elemental sulphur in soils. *J. Sci. Fd Agric.* **19**, 454–456.

BASCOMB C.L. (1961) A calcimeter for routine use on soil samples. *Chemy. Ind.* **45**, 1826–1827.

BLACK C.A. and GORING C.A.I. (1953) Organic phosphorus in soils. In *Soil and Fertilizer Phosphorus in Crop Nutrition* (Eds. W.H. Pierre and A.G. Norman), pp. 123–152. Academic Press, New York.

BOHN H.L. (1971) Redox potentials. *Soil Sci.* **112**, 39–45.

BRAY R.H. and WILLHITE F.M. (1929) Determination of total replaceable bases in soils. *Ind. Engng. Chem. Analyt. Edn.* **1**, 144.

BREMNER J.M. and JENKINSON D.S. (1960) Determination of organic carbon in soil. II. Effect of carbonized materials. *J. Soil Sci.* **11**, 403–408.

BROWN I.C. (1943) A rapid method of determining exchangeable hydrogen and total exchangeable bases of soils.. *Soil Sci.* **56**, 353–357.

BUTTERS B. and CHENERY E.M. (1959) A rapid method for the determination of total sulphur in soils and plants. *Analyst (Lond.)* **84**, 239–245.

CHAPMAN H.D. and PRATT P.F. (1961) *Methods of Analysis for Soils, Plants and Waters.* University of California, California.

CHESNIN L. and YIEN C.H. (1950) Turbidimetric determination of available sulfates. *Proc. Soil Sci. Soc. Am.* **5**, 149–151.

CHRISTIAN G.D. and FELDMAN F.J. (1970) *Atomic Absorption Spectroscopy: Applications in Agriculture, Biology and Medicine.* Wiley-Interscience, Chichester.

CIACCIO L.L. (Ed.) (1971) *Water and Water Pollution Handbook.* Dekker, Basle.

COULTER B.S. (1969) The chemistry of hydrogen and aluminium ions in soils, clay minerals and resins. *Soils Fertil.* **32**, 215–223.

CRESSER M.S. (1978) *Solvent Extraction in Flame Spectroscopic Analysis.* Pergamon Press, Oxford.

CURTIS K.E. (1969) Inteferences in the determination of iron by atomic-absorption spectrophotometry in an air-acetylene flame. *Analyst (Lond.)* **94**, 1068–1071.

DEPARTMENT OF THE ENVIRONMENT. Standing Committee of Analysts (1981) *Ammonia in Waters.* (Methods for the examination of water and associated materials). HMSO, London.

FULLER C.W. (1977) *Electrothermal Atomization for Atomic Absorption Spectrometry.* Chemical Society, Letchworth.

GALLOWAY J.N. and LIKENS G.E. (1978) The collection of precipitation for chemical analysis. *Tellus* **30**, 71–82.

GOLTERMAN H.L., CLYMO R.S. and OHNSTAD M.A.N. (1978) *Methods for Physical and Chemical Analysis of Fresh Waters.* Blackwell Scientific Publications, Oxford.

HERON J. (1962) Determination of phosphate in water after storage in polythene. *Limnol. Oceanog.* **7**, 316–321.

HESSE P.R. (1971) *A Textbook of Soil Chemical Analysis.* Murray, London.

HILLEBRAND W.F. and LUNDELL G.E.F. (1953) *Applied Inorganic Analysis.* John Wiley, New York.

HINDS A.A. and LOWE L.E. (1980) Application of Berthelot reaction to determination of ammonium-nitrogen in soil extracts and soil digests. *Commun. Soil Sci. Plant Anal.* **11**, 469–475.

HOWARD P.J.A. (1966) The carbon-organic matter factor in various soil types. *Oikos* **15**, 229–236.

KALEMBASA S.J. and JENKINSON D.S. (1973) A comparative study of titrimetric and gravimetric methods for the determination of organic carbon in soil. *J. Sci. Fd Agric.* **24**, 1085–1090.

KLEEMAN A.W. (1967) Sampling error in the chemical analysis of rocks. *J. Geol. Soc. Aust.* **14**, 43–48.

KROM M.D. (1980) Spectrophotometric determination of ammonia: a study of a modified Berthelot reaction using salicylate and dichloroisocyanurate. *Analyst (Lond.)* **105**, 305–316.

MACKERETH F.J.H., HERON J. and TALLING J.F. (1978) *Water Analysis.* Scient. Publs. Freshwat. biol. Ass. No. 36.

MORRISON I.R. and WILSON A.L. (1963) The absorptiometric determination of silicon in water. Part II. Method for determining 'reactive' silicon in power-station waters. *Analyst (Lond.)* **88**, 100–104.

OLSEN S.R., COLE C.V., WATANABE F.S. and DEAN L.A. (1954) Estimation of available phosphorus in soils by extraction with sodium bicarbonate. *Circ. U.S. Dep. Agric.* No. 939.

PAGE A.L., MILLER R.H. and KEENEY D.R. (Eds.) (1982) *Methods of Soil Analysis. Part 2, Chemical and Microbiological Properties.* American Society of Agronomy and Soil Science Society of America, Madison.

PITTWELL L.R. (1968) Apparatus for the analysis of small quantities of carbonates. *Mikrochim. Acta* 903–904.

PRICE W.J. (1979) *Spectrochemical Analysis by Atomic Absorption.* Heyden, London.

QUARMBY C. and GRIMSHAW H.M. (1967) A rapid method for the determination of iron in plant material with application of automatic analysis to the colorimetric procedure. *Analyst (Lond.)* **92**, 305–310.

RAMIREZ-MUNOZ J. (1968) *Atomic Absorption Spectroscopy.* Elsevier, Amsterdam.

RILEY J.P. and WILLIAMS H.P. (1959) Micro-analysis of silicate and carbonate minerals. III. Determination of silica, phosphoric acid and metal oxides. *Mikrochim. Acta* 804–824.

ROSE L. (1983) Alkalinity—its meaning and measurement. *Cave Science* **10**, 21–29.

SINCLAIR A.G. (1973) An autoanalyser method for determination of extractable sulphate in soil. *New Zealand J. Agric. Res.* **16**, 287–292.

SJÖRS H. (1950) On the relation between vegetation and electrolytes in north Swedish mire waters. *Oikos* **2**, 241–258.

SMITH K.A. (Ed.) (1983) *Soil Analysis—Instrumental Techniques and Related Procedures.* Dekker, Basle.

STEERE N.V. (Ed.) (1971) *Handbook of Laboratory Safety.* Chemical Rubber Co, Cleveland.

STEYN W.J.A. (1959) *A statistical study of the errors involved in the sampling and chemical analysis of soils and plants with particular reference to citrus and pineapples.* Ph.D. Thesis, Rhodes University, South Africa.

TALIBUDEEN O. (1981) Cation exchange in soils. In *The Chemistry of Soil Processes* (Eds. D.J. Greenland and M.H.B. Hayes), pp. 115–177. Wiley, Chichester.

THOMPSON K.C. and REYNOLDS R.J. (1978) *Atomic Absorption, Fluorescence and Flame Emission Spectroscopy.* Griffin, London.

TINSLEY J. (1950) The determination of organic carbon in soils by dichromate mixtures. *Trans. 4th Int. Cong. Soil Sci.* **1**, 161–164.

TRUOG E. (1930) The determination of the readily available phosphorus of soils. *J. Am. Soc. Agron.* **22**, 874–882.

VERDOUV H., VAN ECHTELD C.J.A. and DEKKERS E.M.J. (1977) Ammonia determination based on indophenol formation with sodium salicylate. *Water Research* **12**, 399–402.

VOGEL A.I. (1962) *A Textbook of Quantitative Inorganic Analysis*, 3rd edn, Longmans, London.

WALKLEY A. and BLACK C.A. (1934) An examination of the Detjareff method for determining soil organic matter and a proposed modification of the chromic acid titration method. *Soil Sci.* **37**, 29–38.

WILLIAMS J.D.H., SYERS J.K., WALKER T.W. and REX R.W. (1970) A comparison of methods for the determination of soil organic phosphorus. *Soil Sci.* **110**, 13–18.

7 Data analysis

S.D. PRINCE

1 Introduction

Plant ecology has a longer tradition of observation and description than of experimentation and inference. This is partly a result of its youth—all sciences start with a descriptive or phenomenological phase—but there are even younger developments in biology which have largely passed beyond this, and so one is forced to recognize that it is partly the nature of plant ecology itself and in particular the intrinsic complexity of vegetation that ensures a continuing place for description, at least at the start of each new investigation. The foregoing chapters have introduced the methods appropriate for describing and analysing the particular phenomena with which they are concerned and it is not the purpose of this chapter to reiterate these; rather it is to proceed to the methods which are available for manipulating the measurements in order to gain insight into relationships and to test hypotheses.

Although analysis of ecological data can be regarded as progress beyond description—in the sense of being a later stage of the scientific method—the newcomer may be surprised to discover that many of the methods of data analysis in common use by plant ecologists are by means peculiar to plant ecology; instead they are very general techniques, widely used throughout the observational and experimental sciences. Plant ecology and agriculture in particular have contributed some new methods which have later been adopted by other disciplines but the majority of methods employed have their origins outside the subject. Many of the methods with which we are concerned are those of applied statistics. These methods must be distinguished from what are commonly called 'statistics' which are really measurements, often presented in tables and more or less misleading graphs or histograms, and sometimes subject to tactical arrangement to emphasize or obscure an argument; the popular jibes against statistics generally refer to these edited data sets and not to statistical methods.

The difficulties of data analysis in plant ecology often necessitate the cultivation of applied mathematicians and statisticians, although the ideal is to be competent in mathematics at least to the level needed to conduct an intelligent conversation with a professional. Unfortunately, most of us have been unable to exercise much conscious choice in our own training and so, for the foreseeable future, there will be a need to introduce methods of data

345

analysis to practising plant ecologists. Many excellent elementary texts are available (see below) but most attempt a logical development from first principles which, although essential at some stage, can be a formidable deterrent to the busy, working ecologist or to the beginner who is only half convinced of the relevance of mathematical methods. The enforced brevity of this chapter allows a quite different approach to be adopted, i.e. that of the market-stall in which the goods are laid out for inspection rather than the formal, text book method of progressive revelation.

This, then, is the main aim: to provide a brief catalogue of methods of data analysis—some of which may have escaped the notice of the working ecologist—and to whet the appetite of the newcomer with a description of the potential of each method without attempting to provide all the details necessary for their actual application. There are dangers in this approach, as there are in any other, but these should be minimized if the references to the literature on any particular method are carefully checked before irrevocable commitment to any one technique, however appropriate or attractive it may have been made to appear.

Data analysis is an integral part of any investigation and should be considered soon after the idea for an investigation has started to take form; in fact, many good research programmes have grown out of the realization that a particular type of analysis could be applied. Failure to plan the investigation with the method of data analysis in mind generally makes the project less efficient; either the same conclusion could have been reached with less effort or the purpose of the investigation cannot be realized from the data. For example, an ecologist planning a vegetation survey might decide to describe a number of sites, or relevés, by making a list of the species present in each one. Time may be limited and so he may decide to list only the major species at each site. Pragmatic decisions of this sort are often forced on an ecologist but the consequences need to be carefully thought through. In this example the decision to list major species would be appropriate for a subjective comparison of sites and would probably yield more information in the time available than if fewer sites were to be examined and the extra time spent in searching for minor species. However, time could also have been saved if a pilot survey had been carried out and a restricted species list compiled of just those species which discriminate between the sites, then the methods of objective site comparison and multivariate data analysis could be used since all the sites have been scored for all the species on the restricted list. The relationship of experimental design and data analysis is more explicit in some types of data analysis, e.g. analysis of variance, but no matter what the purpose of the investigation only good fortune will save wasted effort if the appropriate method of analysis has not been determined at the start.

The beginner will find a plethora of elementary books on statistical methods and a good number written specifically for the biologist (e.g. Bailey, 1981; Bishop, 1983; Campbell, 1974; Clarke, 1980; Finney, 1980; Mead & Curnow, 1983; Parker, 1979; Sokal & Rohlf, 1973). Bailey's *Statistical Methods in Biology* has stood the test of time and includes a useful summary of statistical formulae, but the choice is very much a matter of personal taste and background; each of these books provides a suitable way into the subject for someone with no previous knowledge of statistics and only an elementary mathematical background, and will enable them to move on to more comprehensive works providing a wide range of methods, e.g. *Statistical Methods* (Snedecor & Cochran, 1967; Cochran & Snedecor, 1980), *Biometry: The Principles and Practice of Statistics in Biological Research* (Sokal & Rohlf, 1981) and *Statistics in Biology* (Bliss, 1967). *Quantitative and Dynamic Plant Ecology* (Kershaw & Looney, 1985) introduces some methods of data analysis applied specifically to plant ecology with special emphasis on elementary statistical considerations in the measurement and sampling of plant communities. Without doubt the most important work on the quantitative analysis of vegetation is *Quantitative Plant Ecology* (Greig-Smith, 1983). General books on quantitative methods in ecology, e.g. *An Introduction to Quantitative Ecology* (Poole, 1974) and *Mathematical Ecology* (Pielou, 1977), and others with a bias towards animal ecology, e.g. *Ecological Methods* (Southwood, 1978) and *Introduction to Experimental Ecology* (Lewis & Taylor, 1974), provide the plant ecologist with at least an introduction to methods of data analysis in common use. Nevertheless, methods which are new to plant ecology are constantly being added and no comprehensive catalogue can be made. Reviews of some more advanced methods are available written specifically for biologists (e.g. Jeffers, 1978, 1982a; Gauch, 1982; Hewlett & Plackett, 1979; Causton & Venus, 1981), but others are only to be found in technical form in the mathematical literature. Finally, a variety of check lists (Natural Environment Research Council, 1978, 1979, 1980, 1982) and pro-forma work sheets (Dawkins, 1975) are helpful to some, although the latter is likely to be superceded by the publication of statistical programs for microcomputers (see Section 7).

Introductions to mathematical methods which may be needed for ecological data analysis are available at various levels of complexity. Machin (1976) and Causton (1983) give brief accounts of a variety of topics including matrix algebra, whereas Smith (1966, 1969) and Batschelet (1979) are more comprehensive.

Technical terms and names of methods are here given in italics wherever they first appear. It is neither appropriate nor possible to include much detail in this short overview, but proper names are given so that terms and methods can be traced in the reference works listed above and in Section 8.

Before embarking on the discussion of statistical and mathematical techniques it may be helpful to disclose some of the private activities of the professional data analyst which are rarely mentioned in public presentations of results. This silence can give the impression that it is good practice to proceed directly from data acquisition to a specific and often complex analytical technique, but cautionary tales abound of advanced statistical methods wrongly applied to inappropriate data, with absurd results. The problem is that only the products of the data analysis are presented, not the routes taken in reaching these conclusions. In reality the more skilled the analyst, generally the more time is spent in a preliminary and private contemplation of the data. When these preliminary activities are revealed nothing particularly spectacular emerges, merely careful plotting of scatter diagrams of one variable against another, plotting frequency distributions and comparison with model frequency distributions and application of transformations to the measurements, possibly using logarithmic graph paper or one of the other special purpose rulings which are available. Because they are so simple, little is said about them but this initial exploration of the data is often extremely informative.

2   Sampling

Studies involving observation require just as careful design as those involving experimental treatments. Generally a *sample* is measured rather than the whole population, but most elementary statistical formulae are intended for samples from an infinite population whereas, in reality, populations are finite. So long as the sample does not exceed about 10% of the total population no modification of the usual formulae is needed. It should not be thought, however, that there is some fixed proportion of a population which ought to be measured in order to obtain an acceptable degree of accuracy; the precision depends primarily on the total number of measurements made. The size of sample needed to measure a mean value can be calculated if the size of error in the mean acceptable in, say, 95% of cases is known. This allowable error is expressed as a *confidence limit* and the number of measurements is the product of the square of the *normal deviate* (1.96^2) and the *variance* (which can usually be estimated to an adequate degree of accuracy for this purpose), divided by the square of the allowable error. If the allowable error cannot be stated or no estimate of the variance is available, a rather simpler way of fixing a sample size is to plot a graph, as the measurements are made, of the cumulative mean (or the variance) against a number of measurements which have been made. The graph generally settles down after a certain number of measurements and this is often an adequate sample size (Greig-Smith, 1983).

The technique of *randomization* is stressed in ecological sampling where the aim is to ensure that every member of the population has an equal chance of being included in the sample. It is appropriate when there is no information about the type of variation present, but the same accuracy can be achieved with far less effort if the population is divided into strata, within which the variation is reduced; such *stratified samples* are more precise because the variation between the stratum means does not contribute to the population variance. The analogue of stratification in experimental studies is *blocking*, pairing of subjects or the use of permanent quadrats in studies of vegetation changes. The most appropriate method of randomization—either in the whole population or within the strata—depends on the nature of the investigation. Coordinates drawn from a set of random numbers are often used in spatial sampling. The method can be adapted for rectangular areas by setting limits on the ranges of the random coordinates or by transforming the coordinates to maintain equal probabilities. For example, circular sample areas are often convenient since only one reference point (the centre) need be marked; if a random number is selected between 0 and 360 and is used as a compass direction, and if n is the number of potential positions required along a radius of length x, a second random number r (between 0 and $n/10$) can be transformed to give a radial coordinate $R(R = x\sqrt{r/n})$ which maintains the requirements of equal probability. Random numbers can be found in published sets of statistical tables (e.g. Rohlf & Sokal, 1981; Fisher & Yates, 1974) and many inexpensive pocket calculators have a random number generator of sufficient quality for this purpose.

It is often stated that randomization of quadrats or measurement locations is essential to measure the precision of the mean. Certainly any bias in the choice of samples must be eliminated, but it is questionable if it is always necessary to randomize in the sense of selecting locations using random coordinates. There is often strong correlation between neighbouring measurements in the natural environment, and this accounts for the common observation that *systematic (regular) sampling* is more accurate than a random sample of the same size. Clearly, care must be exercised to avoid regular samples coinciding with any periodic variation in the vegetation, but this is very unlikely in practice and checks for pattern can be made (see Section 6.2) if it is thought necessary. Since the most certain form of dependence is between neighbouring samples, it could be argued that the distance between them should be maximized, i.e. by using a regular grid to place the quadrats; the sample sites are much easier to locate and the data is just as likely to be random in the sense required for normal calculation of the precision of the mean. Moreover, a regular grid of measurements carries spatial information which a randomized sample lacks.

General introductions to the principles of sampling are given in Cochran & Snedecor (1980) and Mead & Curnow (1983). A more advanced account can be found in Cochran (1977).

3   Purpose of an investigation

The advantages of objective measurement over subjective judgement in ecological or, for that matter, in any type of scientific investigation are obvious. However, measurements can be made for quite different purposes in different investigations and it is worth reviewing these before proceeding.

The first purpose we need to consider is that of measurement for its own sake, where the aim is to obtain the value of a specific variable; for instance, the mean stomatal aperture of leaves at a particular water potential, the biomass of plankton per unit volume of stream water, the cover of grass on a plot or the proportion of plants killed by frost. The role of data analysis in this case, apart from the trivial one of calculating the mean value, is in determining the accuracy of the measurements (see Section 3.1).

In ecology absolute measurement is uncommon. Generally, the variability of a single measurement is calculated only as a prelude to comparing two or more measurements; mean stomatal apertures at several leaf water potentials or the proportion of plants killed in different degrees of frost are likely to be compared with one another (see Section 3.2).

All the remaining methods of data analysis can be regarded as more or less explicit forms of mathematical modelling (see Section 6). The more familiar forms of mathematical models are equations which specify the functional relationships between one or more controlling factors and the phenomenon which is the object of study. For example, the density of habitable sites for a plant, the density of those which are actually occupied and the rates of occupation and extinction of sites can be modelled as an epidemic process (Carter & Prince, 1981). The exercise of expressing a phenomenon in the symbolic logic of a mathematical model is valuable in several ways: (i) it is a parsimonious description of the functional relationship; (ii) it promotes clear thinking in the investigation; (iii) it directs attention to the relevant relationships for future examination; (iv) it exposes irrelevancies; (v) it may enable responses to, as yet, untried treatments to be predicted and (vi) it may reveal counter-intuitive relationships (Section 6.2). Normally, any model of this type will be more or less limited in its application to the range of phenomena for which it was designed, although epidemic models are an example of a case in which two analogous but quite distinct phenomena (spread of disease and the dynamics of plant metapopulations) can be modelled by homologous mathematical relationships.

By far the most frequent type of data analysis, however, involves not the use of special purpose models but of completely general mathematical or statistical models which are either not limited to any particular phenomenon at all or are only very loosely so. It is easy to overlook the model underlying many of these more general methods, but the dangers of inappropriate use of a technique will be reduced if it is kept in mind; for example, the linear models implicit in the methods of analysis of variance and regression are quite restrictive and it is important to select one which is realistic for each particular application (see Section 6.1). Even calculation of a standard deviation requires the model to be specified, i.e. whether the frequency distribution is considered to be approximately normal or binomial or of some other form (see Section 3.1).

3.1 Accuracy of measurements

It has been suggested that biologists, physicists and chemists can be distinguished by their behaviour when faced with a set of repeated measurements, the successive values of which vary slightly: the chemist is likely to repeat the experiments in the hope of minimizing the variation, the physicist will throw away the apparatus and design a better one while the biologist will simply take the average of the original values. Each behaves according to the type of variation to which he or she has become accustomed. What they will all do is to calculate a measure of the degree of variation to present alongside the original or improved mean value.

Standard error measures the variation of a mean value whereas *standard deviation* and the *coefficient of variation* measure the degree of variation of the individual values which contribute to the mean. Both standard error and standard deviation are in the units of the original measurement whereas the *variance*, from which they are calculated, is in squared units. For continuous measurements such as length, weight, number (when the mean is much greater than 1) experience shows that successive values form a particular type of *frequency distribution* called the *normal distribution* and the standard deviation and standard error can be calculated from the observations. Discontinuous measurements such as point quadrat measurements of the cover of a plant or the proportions of each morph in a population of a heterostylic plant generally fit a *binomial distribution* for which the standard error can be calculated using a rather simpler formula. A *Poisson distribution*, for which the variance equals the mean, describes the circumstance in which an individual is unlikely to fall into a particular sample but since there are a large number of individuals most samples contain a few, e.g. the numbers of individual plants in a small quadrat or the number of pollen grains in each cell of a haemocytometer slide. The appropriate distribution for any set of measurements can often be assumed using past experience but there are objective tests available. Certain types of deviation from these distributions such as *skewness* and *kurtosis* are important. In many cases a *normal approximation* is possible: for example, when the mean proportion (p) of a binomial distribution lies between 0.7 and 0.3 and the number of observations (n)

exceeds about 30, then there is little error introduced by treating it as a normal distribution with mean p and variance $p(1 - p)/n$; similarly, a Poisson distribution with a high mean (x) estimated from a reasonably large number of observations (n) tends to normality when nx exceeds 30. The *arcsine transformation* is often successful in normalizing proportions and percentages and a *square root transformation* is used before using a normal approximation for observations which follow a Poisson distribution. *Corrections for continuity* allow for the difference between these discrete distributions and a continuous frequency distribution such as the normal.

The variation of the mean can be specified by its standard error or by *confidence limits*, which give the range of values (the *confidence interval*) within which the mean is expected to lie at the chosen level of probability. In this instance, *probability* refers to the frequency with which successive values of the mean will be within the range of values specified by the confidence interval. This probability can be calculated without having to make repeated measurements of the mean. The confidence interval is useful when the mean has been calculated from few observations because it takes into account the degree of repetition of the measurement. The higher the level of probability chosen and the fewer the number of observations, the wider will be the interval between the upper and lower confidence limits.

These pieces of information are required to convey the result; the mean itself with its units and either the standard error and the number of observations or the confidence limits and their associated probability.

3.2   Comparison of measurements

A clear distinction must be made between these measures of accuracy of the mean and a *test of significance* of the difference between two or more means (Jeffers, 1982b). Rarely does an ecologist make a single measurement, e.g. of the herbaceous biomass in one site. Generally, several sites are to be compared and the question arises whether a difference between two sites is because of a real, underlying difference between them or—since there are bound to be differences between repeated measurements on one site— because, by chance, different values were obtained whereas the same values might be obtained in another sample or on another occasion. The appropriate test of significance under these circumstances would be to calculate the *standard error of the difference* between the mean biomasses in the two sites and to compare it with the actual difference. The probability of obtaining such a difference by chance can be estimated and, if this is very low, the *null hypothesis* that the means do not differ can be rejected at the appropriate probability and the *experimental hypothesis*, that there is a real difference, provisionally accepted. In this context probability refers to the frequency with which such a difference would arise in repeated observations. The *exact probabiltiy* can be reported but it is more usual to state whether it is less than 5, 1 or 0·1%. By convention only probabilities less than or equal to 5% are commonly accepted as indicative of a significant difference between means and the lower probabilities are used to indicate greater significance.

Two sets of observations which cannot be assumed to be normally distributed can nevertheless be compared using tests intended for normal data if they can be transformed to normality (see Section 3.1). Student's *t-distribution* is used instead of the normal distribution when the number of observations in either sample is small; 30 is often used as the number below which it is preferred.

Binomial proportions and percentages can sometimes be transformed sufficiently for a test based on the normal distribution but values near the extremes cannot and, in this case, the proportions are tested directly using a χ^2 (*chi-squared*) test criterion. The number of observations required to demonstrate the significance of differences in proportions at a stated probability can be calculated and this is often a salutary exercise since it will show that very large numbers of observations are often needed to achieve the desired degree of significance.

A more common problem is how to make comparisons between all the values of, say, biomass in a survey of more than two sites. Here the variance of each mean would be pooled, possibly after a *test of homogeneity*, in order to obtain a better estimate of the overall population variance. The difference between each individual mean and the overall mean can then be tested using the *pooled variance* in much the same way as for just two sites. Since the criterion of significance which is adopted is whether or not the particular mean exceeds a value which only occurs at a chosen, low probability, this value can be calculated and used as a yard-stick for comparing means. This measure is called the *least significant difference* (LSD). Objections to the indiscriminate use of LSD have often been made since a certain proportion of comparisons will always erroneously be declared significant. Cochran & Snedecor (1980) examine the alternatives and conclude, nevertheless, that thoughtful use of LSD or the *Newman–Keuls* method should not introduce serious errors.

Much more powerful methods are available if the different means to be compared relate to plants, populations, sites, etc., which have received different amounts or *levels* of the same treatment, e.g. the herbaceous biomass on plots to which different quantities of an inorganic fertilizer have been added. The effect of the fertilizer overall would be examined using a one-way analysis of variance and then the effects of individual levels could be examined using a *response curve* which uses *orthogonal polynomial partitioning*. Some of the absurdities which arise from the application of LSD and the other methods mentioned in the previous paragraph to this type of data have been described by Dawkins (1981).

4   Types of measurement

Not all measurements useful to ecologists can be expressed as numbers; sometimes it can only be said in what order the observations fall or to what class they belong without, in either case, being able to say anything about the numerical differences between positions in the rank or between the classes. For example, the degree of damage caused to trees by large herbivores such as elephants cannot easily be given a numerical value but trees can be classed as 'damaged' or 'undamaged', or perhaps in a rank of a few classes of increasing damage such as 'none', 'slight', 'medium' and 'extensive'. The statistical methods dealt with so far generally involve the estimation of population parameters, i.e. numbers which, when inserted into the appropriate equations, adjust a general-purpose model to fit the particular data set under consideration (standard deviation and mean are examples of population parameters). For this reason they are known as *parametric* tests. The diversity of population distributions from which ranked and classified data are drawn makes it much more convenient to use tests which do not depend on the identification of the appropriate distribution and, for this reason, these tests are referred to as *non-parametric*; they are also called *rank* or *order statistics*, which stresses the type of measurements for which they are appropriate. Sometimes they are called 'short-cut' methods, which emphasizes another characteristic, i.e. the simplicity and rapidity with which many of the tests can be applied.

Numerical measurements such as length, weight or number are continuous in the sense that any value can be specified rather than being limited to a few discontinuous classes or ranks. Whereas continuous, numerical measurements can later be expressed as a rank or placed in fewer classes than the original measurements, with some loss of information, discontinuous measurements cannot be converted into continuous ones. It might be worthwhile ranking numerical measurements, e.g. to compare two sets of measurements made with instruments which had not been calibrated against each other or were not adjusted to an absolute scale. There exists, therefore, a hierarchy of information content in which continuous, numerical measurements contain the most, followed in order by ranks, classifications and, least of all, binary or two-state classifications. Clearly, a non-parametric test of continuous data reduced to a rank will be less powerful than a parametric test which utilizes all the information in the original measurements.

The techniques considered so far are mostly parametric tests of significance. In the sections that follow some examples of non-parametric methods are described (Section 4.1) and also those appropriate for classified data (Section 4.2). Siegel (1956) gives a comprehensive introduction to non-parametric statistics and at least some methods are described in most text-books of applied statistics (see Section 1). Binary measurements are par-ticularly important in toxicology and biological assay, for which the standard experiment involves measuring how many individuals succumb and how many survive a particular treatment. Such is the importance of this type of experiment in applied biology that parametric tests have been developed to analyse these *quantal responses*. The tests are applicable not only when the observations are the number of deaths but also such 'all-or-nothing' phenomena as seed germination (see Section 4.3).

4.1 Non-parametric tests

The speed of execution, simplicity and lack of restriction to a particular type of frequency distribution which characterize many non-parametric tests is illustrated by the *Wilcoxon signed ranks test* which is used to compare the performance of a set of entities under two different treatments. This test only requires the differences between the entities to be ranked whereas the para-metric equivalent—a *t*-test of a difference—demands completely numerical data which is normally distributed, and much more calculation is involved. The cost of the non-parametric advantages in this case is a loss in efficiency of about 5% compared with the *t*-test. The *Mann–Whitney U test* has similar advantages and is used for the comparison of two, independent, unpaired groups which need not be equal in size.

The *one-sample runs test* enables the sequence of two classes of individuals in a sample to be tested to determine whether the sequence is random. It has no parametric equivalent and it can be applied to spatially ordered data such as the sequence of two species along a line transect or to temporal data such as the occurrence of heavy fruiting (masting) in forest trees in a run of years.

The type of measurement used assumes greater importance when select-ing a non-parametric test because methods exist for classified, ranked, interval and numerical measurements. Many of the methods for classified data are ingenious variations of contingency tables, while comparisons of positions in an overall ranking of the data is a common strategem in tests of ranked measurements. Maximum deviations between two or more cumula-tive distributions can be used to test for homogeneity without any assump-tions about the nature of the distributions to be compared.

Non-parametric tests are usually inferior to parametric tests in situations where either could be used but in these cases they are useful as rapid checks before a more formal parametric test is undertaken. Their special value lies in their distribution-free nature and their ability to test classificatory and ranked data which has inherently less information content than numerical data and the opportunity to compensate for the reduced power of some tests by an increase in the sample size.

4.2   Classified data

Perhaps the most common type of analysis of classified data is concerned with comparisons of the numbers of individuals which fall into each class in two separate samples. It might reasonably be supposed that the degree of large herbivore damage to trees depends on the species of the tree and this hypothesis could be tested by counting the numbers of individual trees of two species which fall into each of four classes of damage. The two sets of numbers would be compared using a χ^2 test criterion. Similarly, χ^2 can be used to test a hypothesis, not only about the equality of proportions in two samples as in this example, but also about some theory which predicts a particular proportion of the total number of individuals which are expected in each class. Many experiments in genetics which are concerned with segregation of one or a few genes are analysed in this way. Spatial pattern is often tested against a Poisson model of randomness by calculating the numbers of samples which are expected to contain a particular number of individuals using an expansion of the Poisson expression; observations and expectation are then compared using χ^2. A *correction for continuity* is recommended in the calculation of χ^2 since a continuous distribution is being used to approximate a set of discrete classes.

In many applications the units of the investigation—individuals, sites, years, etc.—are simultaneously classified according to more than one set of criteria. This sort of data is conveniently arranged in *contingency tables*. For instance, plants which have been presented to two different insects may be attacked by both, by only one or by neither; this is the simplest case in which all the subjects are classified according to two attributes which are either present or absent and a *2 × 2 contingency table* is formed. The probability that the proportion of plants attacked by insect A is unaffected by insect B, can be tested by χ^2. The much more complex contingency tables which result from *multiple* classifications can also be tested using χ^2. Data of many sorts can be cast in the form of contingency tables and a great variety of hypotheses can be tested in this way (see Siegel, 1956; Greig-Smith, 1983). Often a general hypothesis is examined first and then further contingency tables are constructed for more detailed examination of the data. It is normally recommended to *combine neighbouring classes* if their expected numbers fall below 5, but values as low as 1 are permissible. In cases with few observations or small numbers expected *Fisher's exact test of significance* is helpful.

4.3 Quantal responses

The relationship between increasing amounts of a treatment and a proportion which is observed at each level of the treatment is suitable for *probit or logit analysis* (Hewlett & Plackett, 1979; Finney, 1971, 1978) if the distribution of the logarithms of the levels of treatment to which individuals respond in the experimental population is approximately normal. Common uses are in the determination of LD_{50} values (the dose which kills 50% of the population) for pollutants and extreme environmental factors, but it is also appropriate for studies of the effects of environmental factors on such quantal responses as the rate of seed or spore germination.

Essentially, probit analysis involves a transformation of the dose–response curve to produce a linear relationship between the logarithm of the dose and the response expressed as standard deviations. In order to keep the response scale positive, by convention, 5 standard deviation units are added to each transformed, measured response. A straight line is fitted by an iterative, weighted regression procedure rather than a simple linear regression since the points differ in weight according to the level of response and number of individuals counted. Once linearized, the level of treatment most likely to cause any chosen level of response in the 10–90% range can be interpolated and errors calculated. Two-dimensional *probit planes* are used where the treatment can be regarded as consisting of two components.

5 Number of variables

So far we have considered the analysis of measurements of just one sort made on each individual, except for the case of binomial data where contingency tables for multiple classification were mentioned. It is a very common practice to make two different measurements on each individual; for example, measurements of heights or girths and fruit production of trees might be used (i) to examine the dependence of fruit production on the size of a tree, (ii) to predict fruit production of additional trees for which only the size is known, or (iii) to examine the variation in fruit production having first removed any influence of differences in size between trees. This type of data is *bivariate*, whereas measurements of girth alone, which might be required for a study of the tree's population ecology, are *univariate*. Where more than two measurements are made on each individual the data are said to be *multivariate*.

Whenever measurements of more than one characteristic are available the possibility arises of improving the description of the units under consideration by using all the characteristics simultaneously. Alternatively, we may wish to examine the characteristics themselves, perhaps with the intention of selecting the more informative ones and discarding the rest. This is the province of multivariate analysis (see Morrison, 1967).

5.1   Bivariate data

A *scatter diagram*, with one axis for each variable, gives a visual impression of the relationship, if any, between two sets of measurements. A common occurrence is a more-or-less linear scatter of points either ascending or descending across the diagram. This linear relationship can be summarized by drawing a straight line through the points to represent the trend. Such a line, if fitted by eye, can only be an approximate summary and it is more usual to calculate the *line of best fit* by *linear regression*. The line will always pass through the point which corresponds to the mean of both variables; it can also be arranged to pass through the point which corresponds to a zero value for both variables, but this requires special care to avoid unjustified extrapolation and anyway it is rarely necessary. The slope of the regression line and its location with respect to the vertical axis are referred to as the *regression coefficient* and *regression constant*, respectively. The errors associated with these parameters can be estimated and confidence limits can be calculated. A particularly clear format for the comparison of several regression lines is provided by Dawkins (1975). The errors in the predicted values of one variable increase in both directions away from the mean value of the other and for this reason, as well as uncertainty about its shape, using the relationship to predict values outside the range of measurements can be very misleading.

The assumptions which underlie the technique of linear regression are: for every value of the independent variable (x), there is a normal distribution of values of the dependent variable (y) from which the measured value is drawn; the means of all the y distributions all lie on a straight line with respect to x and the standard deviations of all the y distributions are equal. When the two variables are normally distributed, not necessarily with the same standard deviations, the mutual relationship can be described using a *correlation coefficient* (sometimes called Pearson's product moment correlation, r). r has no units and takes values between $+1$ and -1 (perfect positive and perfect negative correlations respectively); $r = 0$ indicates no relationship. The regression coefficient and r are closely related and r^2 is a measure of the proportion of the variance of a dependent variable that can be attributed to its linear regression on the independent variable. Some tests of r can be carried out directly and tables are available which can be used to test the null hypothesis that $r = 0$. All other tests of hypotheses require r to be transformed into a normally distributed measure known as z. Regression is more widely used than the correlation coefficient because, apart from the less stringent demands on the shape of the frequency distribution of the independent variable, r only estimates the closeness to a linear relationship of the two variables, whereas regression tells us how much y changes for a given x, the shape of the relationship, and the accuracy with which one variable can be predicted by the other.

Many bivariate data sets in which the two variables are not linearly related can, nevertheless, be analysed using linear regression if a *transformation* can be found which renders the relationship linear. All the properties demanded by linear regression must of course be met by the transformed data. Replacement of counts by their logarithms, especially when low values are common, often enables a linear regression on some continuous variable such as temperature or altitude. Another situation in which transformation prior to linear regression is commonly used is when the data are to be examined to see if they are consistent with some law which can be expressed as a non-linear relationship between the two variables; for example, data on the growth of a population might be expected to be described by either an *exponential* or a *logistic growth curve* of numbers against time and a convenient test is to undertake a mathematical manipulation of the model such that functions of the numbers and time will be linearly related if they obey the appropriate law. Linear regression then enables the goodness of fit of the data to the law to be examined.

In a few cases *curvilinear regression* is attempted because all reasonable transformations are ineffective or a predicting equation in the original units is considered essential. The procedure is similar to multiple linear regression in that several functions of the independent variable are regressed. *Second degree polynomials* are widely used and, if the values of the independent variable are equally spaced, the lengthy computations are made more easy using tables of *orthogonal polynomials*. Quite general methods of *curve fitting* are also available.

The non-parametric equivalents of the Pearson product moment correlation coefficient are particularly useful in ecological surveys when subjective scores are used and the restrictive assumptions of the bivariate normal distribution are unlikely to apply. The *Spearman* and the *Kendall rank correlation coefficients* are widely used and Kendall's coefficient can be extended to a measure of *partial rank correlation*. *Kendall's coefficient of concordance* tests for agreement between two or more rankings of the same entities; a useful test to check, for example, the comparability of different observer's subjective estimates of vegetation characteristics.

5.2 Multivariate data

Most of the more important multivariate methods were developed long before they came to be widely used owing to the complexity of the calculations they involve but, now that most ecologists have access to a computer, these methods are used routinely and have taken their place alongside those of the uni- and bi-variate statistics already considered.

The observations are usually grouped into a *data matrix*, or table, which has columns and rows for each object and each characteristic. Typical data sets may consist of (i) morphometric measurements of a number of individual organisms, (ii) measurements of several environmental factors in a number of locations, (iii) lists of species found in a set of vegetation samples or (iv) species of pollen grains found in a set of peat samples. The measurements may be binary, multistate, ranked, or continuous but not all multivariate methods are applicable to all types of data and only a few can handle more than one type at once. For ranked and continuous data, scatter diagrams of the objects can be plotted using the values of two of the characteristics as the axes and, since each of the remaining characteristics can be envisaged as an additional axis, this type of data is sometimes described as 'multidimensional'. The various objectives of multivariate analysis can be summarized as: *structure simplification* in which a set of interdependent variables is transformed to independence or the number of variables is reduced; *classification* where the distribution of the objects in *variable space* is examined, in particular to see if they fall into groups; *grouping of variables* in which we attempt to detect cognate groups of variables; *analysis of interdependence* between the variables; *analysis of dependence* of one variable on the rest; *hypothesis construction and testing* of multivariate relationships (Kendall, 1975).

Some multivariate methods can be regarded as extensions of univariate analyses but this is far from being generally true since most multivariate analyses do not involve considerations of probability. It is important also to realize that many methods assume that the data can be described by the *multivariate normal distribution*. Unfortunately, the range of alternative frequency distributions used in univariate statistics is not available and there are few distribution-free multivariate methods. The ecologist must therefore be careful when considering a multivariate analysis to ensure that the method being contemplated does not make unrealistic assumptions about the nature of the data.

Sometimes several independent variables may be available, all of which give information about the dependent variable. Under these circumstances *multiple linear regression* is used to summarize and express the multiple relationship. Prediction of the dependent variable with greater accuracy than could be achieved with just one independent variable is one way to use this technique but the tests and measures of significance of the *partial regression coefficients* enable it also to be used to help select the best set of independent variables, if several are available, or to rate the independent variables in order of importance of their relationship with the dependent variable. The calculations are extensive for more than two independent variables and computers are essential especially for large data sets.

The standard error of *deviations from the regession* measures how closely the calculated regression fits the data and it is the measure of primary interest when multiple regression is used to predict the dependent variable, especially if there is a choice of ways to do this. Comparison of the *deviations mean square* with the *total mean square* indicates how much of the variation of the dependent variable has been accounted for by the regression. This is often very much less than 50% even when all the partial regression coefficients are highly significant. The significance of each additional independent variable is measured by the reduction its addition causes in the deviations from the regression. It is common for the reduction to change if the order of inclusion of independent variables is changed, since it is quite usual to find the independent variables are correlated with one another as well as with the dependent variable.

Multiple regression is often used in observational studies where causative relationships are sought amongst a set of possible controlling factors. Whereas useful hypotheses may be generated in this way, only a controlled experiment can remove the effects of correlations of variables which are unknown, or cannot be measured, or are wrongly considered to be unimportant by the observer (Hauser, 1974). It is all too easy for the naïve investigator to be misled into believing that multiple regression is a panacea, or can solve the problem of identifying causative factors like a sieve to trap causative factors from amongst all those which are thoughtlessly thrown into it. Without an experiment even the most significant coefficient only indicates a correlation.

Ordination is a frequently used multivariate method and *principal components analysis* has proved to be the most popular single technique. A complete multivariate data set is necessary in which there are no missing values but this method is not restricted to normally-distributed data. It may be undertaken for one or more of the following reasons; (i) to simplify the observations by reducing the number of variables needed to describe the relationships amongst a set of individual objects or locations; (ii) to examine the relationships among the variables, perhaps to identify those which account for the greatest amount of variation and which might therefore be considered to be the most important; (iii) to display groups of individuals which are similar, or to detect trends of variation. In general, principal components analysis finds the principal axes of a multidimensional configuration of individuals and determines the coordinates of each individual relative to the principal axes. In this way it carries out a convenient transformation to new axes which enables the individuals to be displayed in a few dimensions, in a configuration which is as faithful as possible to the original multidimensional one. Principal components analysis does not itself reduce dimensionality, rather it concentrates the information on the first few components leaving the user to discard the later ones if, as is usually the case, they are of no interest. The new axes or principal components are often rotated using *Kaiser's Varimax criterion* which facilitates identification of the principal components in terms of combinations of the original measurements. Variables are normally *standardized* to equalize their variance, otherwise those having the greatest variance are likely to dominate the first principal components. Data matrices have an inherent symmetry and rows can be ordinated in column space or vice versa; similarly, standardizations can be carried out on both rows and columns. The final ordination can be affected very much by these manipulations and this can best be appreciated by studying examples (e.g. Jeffers, 1978; Sneath & Sokal, 1973; Blackith & Reyment, 1971; Gittins, 1969). *Reciprocal averaging* and a modification of it called *detrended correspondence analysis* are forms of principal components analysis which have the advantage of providing, simultaneously, equally valid ordinations of the individuals and their measured characteristics (see Gauch, 1982).

In a principal components ordination the arrangements of the individuals relative to each other is an approximation of the *Euclidean distances* between them, i.e. the physical distance between individuals in a multidimensional version of a scatter diagram. However, there are many other *distance measures*, and *principal coordinates analysis* can be used to ordinate several of these. Unfortunately, identification of the original variables with the principal axes is more difficult than in principal components analysis and principal coordinates analysis requires a larger computer to process the same amount of data.

General dissatisfaction with the performance of similarity measures in those cases when the real, underlying differences between individuals happens to be known has led to the use of *non-parametric* (or *non-metric*) *multidimensional scaling* which seeks to preserve only the original ranking of dissimilarities between individuals in the final ordination (Greig-Smith, 1983).

Factor analysis (Seal, 1964) is a multivariate technique which depends upon an explicit mathematical model of the variation. The number of common factors which are allowed to contribute to the observed response is defined by the user according to some prior hypothesis. It therefore has more in common with the aims of linear modelling than it has with principal components and coordinate analysis (Blackith & Reyment, 1971).

Methods for creating objective classifications using multivariate data, sometimes called *cluster analysis*, are available for most types of data from binary to quantitative. The objective is to allocate each unit to one of a set of classes or clusters, the members of which are more similar to one another than they are to members of any other class. The method may associate the classes themselves into groups to form higher orders of classes, the final product of which is a *hierarchical classification*. The hierarchy reveals more about the structure of the multivariate data than does a set of unrelated classes, and a hierarchy can sometimes be used to allocate new individuals to the most appropriate class more efficiently than can be done if the relationships of the classes are undefined. The results are often presented diagrammatically in the form of a tree. Useful reviews of the many methods by which a hierarchy may be constructed are provided by Cormack (1971), Greig-Smith (1983) and Gordon (1981). All techniques which are compatible with the data will produce the same classification if there are clear-cut discontinuities between the groups of individuals, but where there are no clear-cut groups the classes are formed by *dissection* of the trends of similarity and different methods are unlikely to do this in the same positions. Since most real data is to some extent like this, the final classes are at least partly dependent on the method used to detect them.

Hierarchical classification can start by forming groups out of pairs of similar individuals or by subdivision of the whole population into two internally more homogeneous groups. The first strategy is called *agglomerative* and the second *divisive* and there are advantages to both. In a divisive classification the most fundamental divisions are made first but, if we are more interested in obtaining the best groupings of individuals, agglomerative methods are to be preferred. Clearly, the possibility of misclassification is reduced if more than one characteristic of the individuals is used to divide or fuse groups and the terms *monothetic* and *polythetic* are used to distinguish these single and multiple-character classifications. Modern computers have made it possible to undertake all monothetic divisions of binary data and then to select the best of these but, as the number of characteristics increases, so do the number of polythetic divisions and this intuitively satisfying approach is soon unmanageable. Various strategies such as *indicator species analysis* use one-dimensional ordinations to direct the search for division criteria. Agglomerative methods are necessarily polythetic and are more flexible than divisive methods. The first step is to calculate a *similarity matrix* (or its complement the *dissimilarity matrix*) using a suitable *similarity measure* (Sneath & Sokal, 1973; Greig-Smith, 1983; Cormack, 1971). Individuals, and later the groups, are joined according to some carefully defined *sorting strategy* of which *single-link*, *centroid* or *average-link*, *furthest neighbour* or *complete link*, *median*, *group-average*, and *minimum variance* or *optimal agglomeration* are the main types. Lance & Williams (1967) introduced a *flexible* strategy which combines some of the desirable features of the others. A *minimum spanning tree* can be derived from a similarity matrix and is a useful means of representing the relationships between individuals which complements ordination and, although it is not a classificatory technique, a single-link classification can easily be generated from it (Gower & Ross, 1969). The tree is plotted on a scatter diagram of individuals using a pair of measured or derived variables such as two principal components.

There are several multivariate techniques which aim to define the differences between two or more individuals or between two or more distinct sets of measurements which relate to the same individuals. They are known collectively as *discriminant analysis* or *canonical analysis*. Ecologists often collect data which consists of two sets of multivariate data, e.g. the species composition and a set of environmental measurements for quadrats. *Canonical correlations* between these sets of properties are analogous to the correlation coefficient between say, the abundance of a species and the mineralizable nitrogen content of the soil. *Canonical variates* are linear combinations of the original measurements which are uncorrelated with each other. *Multiple discriminant analysis* is mathematically identical to canonical variate analysis, only it is performed on measurements of the same set of characteristics made on different individuals. The aim is to derive a linear combination of the original measurements or *discriminant function* which maximizes the differences amongst the set means. Unfortunately, these techniques generally assume linear data structures which are not always appropriate for ecological data. Introductions to the methods are given in Jeffers (1978), Pielou (1977) and Kendall (1975) and an example is provided by Barkham & Norris (1970) who use them to analyse measurements of woodland vegetation and soils.

6   Mathematical and statistical models

6.1   Linear statistical models

A number of references have already been made to *analysis of variance*, a powerful class of methods introduced by Fisher in the 1920s, initially for the analysis of agricultural field trials. It was soon realized that analysis of variance was capable of quite general application and it is now one of the most important methods of applied statistics. The essence of analysis of variance is a linear model or equation which is used to describe the components of the variation in a dependent variable. In the case of a fertilizer trial the dependent variable would be the yield or performance of the plants and the components of variation would be the various fertilizers used. Linear models are of many types but they all have the same general form which they share with regression analysis (see Section 5.1).

In analysis of variance, as the name implies, the variation in a set of observations or measurements can be analysed into more fundamental components. In the simplest case of a *one-way classification* the data are measurements of experimental items, each of which has been subjected to one or a number of treatments; for example, the areas of leaves which have been consumed by a selection of different species of mollusc, say ten individuals of each of five species of snails and slugs. The leaf areas consumed in a given period form the observations. Now the variation in areas of leaf consumed may be due partly to any of the following: (i) differences between the species of mollusc used in the experiment, (ii) variation between the appetites of individuals of the same species, (iii) undetected differences between the supposedly identical shoots on which they fed and (iv) errors in measurement of the holes in the leaves after the meal. A suitable design of experiment followed by an analysis of variance will separate the effect of the mollusc species from these other causes of variation. The place of the mollusc in other experimental designs of this sort may be taken by a chemical or physical treatment of which different quantities are applied, and in such a case the meaning of the term *levels* is obvious. In more descriptive investigations the 'treatments' may be locations or years or some other passive condition rather than deliberate treatments.

The results of an analysis of variance are presented in an *analysis of variance table* which, in the case of a one-way classification, is composed of the *error mean square*, the *treatment mean square*, and the result of a *variance ratio* or *F-test* by which the significance of the treatments can be tested. The calculations are performed on the *sums of squares* of the deviations from the means, which become *mean squares* or variances when divided by the appropriate number of *degrees of freedom*. The degrees of freedom are closely related to the number of observations which contribute to the sums of deviations. Even in this, the simplest of analysis of variance designs, the elegance of the technique is clear; moreover it makes explicit the essence of good scientific method, in that it shows an experiment must be executed sufficiently well to enable any real effect to emerge above the background or error variation but that excessive precision is wasteful (Heath, 1970).

A common practice in analysis of variance which can lead to problems is the desire to make comparisons between individual pairs of means derived from the data. To be valid these must be *orthogonal comparisons* in that they are planned and satisfy certain rules of additivity. *Ad hoc* tests of pairs of means which unexpectedly turn out to be very different are best avoided but, if made, must satisfy more rigorous criteria (e.g. *Sheffés test*). The validity of making unplanned comparisons amongst all pairs of means has already been discussed in connection with significance tests.

Analysis of variance is a versatile technique but there are strict requirements which, if not satisfied, invalidate the method. For a one-way classification the observations within each class must be normally distributed with the same variance as in all the other classes. Typical cases when the conditons are not met are when the observations consist of small whole numbers, when proportions or percentages are involved, especially if these are near 0 or 100, and when an increase in the level of a treatment does not have a linear, additive effect.

There are many types of analysis of variance. If the means of the classes themselves differ it is called a *fixed-effects model*. *Random effects models* are appropriate when the class means are random samples from a normal distribution, as might occur in replicate sets of chemical analyses of leaves. *Mixed models* are also used. The one-way classification refers to the fact that the observations all belong to two or more levels of one treatment. *Two-way classifications* are a natural extension to two treatments and more treatments can be added, although above a three-way classification the numbers of observations needed to maintain an adequate number of degrees of freedom often becomes excessive. *Nested* or *split-plot* classifications are used when observations are naturally grouped in some way, e.g. measurements of the chemical composition of individual leaves of several plants or application of the same set of treatments to a few large areas of vegetation. A two-way classification is implicit in the type of experimental design where a set of treatments is replicated in a number of *randomized blocks* in order to minimize unwanted variation; each observation then belongs to a treatment and block. Blocks are sometimes further divided into two classes if it is expected that two sources of background variation are present: a *Latin square* design is an example, and these can be used to control variation in time as well as in space; for example, randomizing the order of a sequence of treatments may be advisable, when the responses of the same set of individuals to a sequence of treatments form the data, if there could be residual effects of the earlier treatments.

Analysis of covariance combines the methods of linear regression with analysis of variance and is useful to control unwanted variation which is continuous and which could only be dealt with very crudely by blocking.

Data which do not conform to the requirements of the analysis variance model can sometimes be made suitable by *transformations* of the observations; *square root, angular or arcsine* and *logarithmic* transformations are commonly used. *Tukey's test* can help to decide whether certain types of transformation are justified and which one to adopt. *Missing data* due to accidents, errors or mislaid records, but not due to real treatment effects, destroy the symmetry of the analysis of variance and this can be awkward especially in two-way and higher-order classifications and methods exist for *estimating the missing value*. Modifications of the subsequent analysis prevent any undue effect on the results.

In some investigations there is no symmetry implicit in the data and this is especially common in survey data: a set of conditions are chosen and a search made for all the individuals which fall in the resulting classes. There is no logic in rejecting individuals simply to obtain equal numbers in each class or *cell*, especially when one cell has very few members. In such cases methods for *unequal numbers* of observations in the cells of a multiple classification are used.

Once it was generally believed that scientific study of the effects of treatments or factors had to be carried out by holding all factors constant except one, which was then allowed to vary and its effect studied. However, *factorial analysis of variance* has rendered this approach obsolete. If two or more factors are to be studied, measurements are made of the effects of all combinations of factor levels, as in the classical fertilizer trial in which two levels of nitrogen, phosphorus and potassium fertilizers are combined in all eight possible ways. The results are no less informative than the 'one-factor-at-a-time' approach and in most circumstances are more so. This advantage arises (i) because every observation is used to furnish information on all the factors, thus reducing the number of observations needed, and (ii) because of a new feature of multifactorial experiments, that of *interaction effects* as well as *main effects*. Main effects in a fertilizer trial are the overall effects of, say, phosphorus irrespective of what other fertilizer was present. Interaction effects can arise in this example because phosphorus often has a different effect at one level of nitrogen from that at another. Interactions can be between two or more factors and the analysis of variance models used enable main effects and two-factor interactions to be separated from higher-order interactions. Factorial forms of the *split-plot design* exist and these are particularly appropriate if the precise effect of one factor and its interaction with a second factor is required but there is less interest in the second factor itself. For example, a major treatment such as different times of burning might only be carried out in large plots, which can therefore not be replicated more than a few times in the area available, but many individuals of a set of species can be observed in each plot. This could be analysed as a split-plot in which the different species and any interactions with burning are examined with great precision but the overall effects of types of burning are less precisely measured.

A standard format is used to indicate the number of factors and levels in a factorial design; for example, a 3 × 4 factorial has two factors, one at three levels and the other at four. If there are equal numbers of levels of each factor this is abbreviated to the number of levels raised to a power equal to the number of factors (the fertilizer trial was a 2^3 factorial, i.e. 2 × 2 × 2).

Response curves are particularly appropriate to express the relationship between the level of a factor and the observations. They may be linear or more complex in form. The calculations are facilitated by published tables (e.g. Fisher & Yates, 1974). *Response surfaces* summarize the simultaneous effects of the levels of two factors.

For multiple classifications non-parametric analogues of the simpler forms of analysis of variance exist, e.g. the *Kruskal–Wallis one-way analysis of variance by ranks*; otherwise χ^2 can be used if the data is classified.

6.2   Mathematical models

Many techniques of data analysis involve adapting and fitting standard mathematical relationships to the particular requirements of a set of experimental or observational data; *growth analysis* (Hunt, 1982), *matrix models* of population growth (Jeffers, 1982a), *pattern analysis* (Greig-Smith, 1983; Pielou, 1977) and soil mapping (Webster, 1978) are examples of this. The properties of mathematical models of this sort are often very well known, in which case the existing theory surrounding them can be applied to the specific problem which confronts the ecologist, thereby freeing him from the need to analyse the relationship from first principles. Clearly his main concerns are (i) to recognize which type of model is appropriate and (ii) to adapt it to the specific problem. On other occasions it is preferable to elaborate a novel mathematical relationship which simulates the phenomenon of interest rather than adapt an existing one. There are many different reasons why a model of this sort may be useful: (i) the mental exercise of expressing the phenomenon in the symbolic logic of mathematics is worthwhile in itself, in that it concentrates attention on the relevant aspects and shows up irrelevant details for what they are; (ii) it may help to expose the nature of the fundamental biological mechanism which is involved, and (iii) it may indicate which critical measurements would enable a hypothesis to be tested. There is no end to the variety and variations in models which have been used to analyse and describe ecological phenomena and it is not possible to discuss even a representative selection here. Useful introductions to the concept of modelling and to some of the more common types of model are given by Jeffers (1972, 1978, 1982a) and by Southwood (1978).

7   Computers and data analysis

The volume of arithmetic involved in most statistical tests was once an important consideration, and many types of ecological investigations which are now commonplace could not be contemplated for this reason alone. The advent of microelectronic calculators and computers effectively removed this obstacle. Even a very modest calculator can now simultaneously add up a list of numbers and their squares, and inexpensive 'scientific' calculators often have keys for standard deviations, linear regression, and correlation coefficients, and standard mathematical functions such as logarithms, powers, square roots and factorials. Calculators can be used outdoors for initial data analysis, which has the advantage that doubtful values can be identified and checked at source. Because the number of significant figures is limited on any machine, it is important to take care to avoid loosing precision in the result. A common cause of this is in the use of a *correction factor* when calculating sums of squares; wherever possible the sum of square deviations should be calculated from the differences directly, and it is good practice to subtract the mean or some other offset from every value in order to reduce the magnitude of the numbers used in the calculations.

Until the late 1970s only professional ecologists employed by large institutions had easy access to computers, but now a microcomputer of considerable power can be purchased for about the price of a domestic television receiver and some are much less expensive. Although the speed and capacity to work on large sets of data of these *microcomputers* are limited in comparison with *mini* and *mainframe computers*, they nevertheless enable a vast increase in complexity of operations, flexibility and speed compared with a calculator. All manual tasks, apart from entering the data and any instructions necessary to start the program, can be taken over by the computer leaving the user virtually free of the repetitive, elementary arithmetic and of the need for constant vigilence for arithmetical errors.

For various reasons the mainframe computer is likely to be the workhorse of statistical analysis in large institutions for the foreseeable future. Unfortunately, there is often a direct relationship between the size of a computer installation and the difficulty and delay experienced by an individual user owing to the increase in administration and number of users. There is much to be said for a configuration which enables data to be entered or even recorded directly by a microcomputer, the data to be organized and the early stages of analysis undertaken before sending it automatically to a mini or mainframe computer to which the microcomputer can be connected.

Now that the initial euphoria engendered by computers has died down, it has become clear that their capacity to increase the volume of data which can be analysed and the complexity of analytical techniques which are practicable necessitates a reassessment of the whole activity of data analysis. The early statistical computer programs, many of which are still widely used, were designed to solve the arithmetical problems of routine statistical analysis. They are often efficient and very fast but rarely can a user who has not written the program find out exactly what they are doing and what assumptions are being made about the data. Early programs in effect simulated the key sequence which was followed on a calculator once the analytical procedure

Table 7.1 Statistical and data analysis software: some of the more widely used packages which are suitable for the analysis of ecological data using mainframe, mini and microcomputers

Package name	Program language	Suitable for hardware (*some, **most) Micros	Minis	Mainframes	Description	Source
BMDP	Fortran	*	**	**	Collection of 41 separate programs for statistical analyses ranging from simple data description to advanced techniques. Suitable for casual user of individual programs.	BDMP Statistical Software Inc., 1965 Westwood Blvd, Suite 202, Los Angeles, CA 90025, USA
Genstat	Fortran and Assembler	*	*	**	Powerful statistical package which incorporates statistical procedures and data handling routines and a language which permits user to write algorithms. Not suitable for the unaided non-expert.	Nag Central Office, Mayfield House, 256 Banbury Road, Oxford OX2 7DE, UK
Glim-3	Fortran and Assembler	*	*	**	Interactive package for generalized linear modelling—a statistical technique which subsumes many standard statistical methods such as regression, analysis of variance, contingency tables, etc. Good facilities for data display and plots. Manual demands considerable knowledge of statistics.	Nag Central Office, Mayfield House, 256 Banbury Road, Oxford OX2 7DE, UK
Minitab	Fortran	**	**	*	General purpose statistical package for users with little computing experience. Interactive and good for data exploration. Suitable for smaller data sets.	Minitab Data Analysis Software, 1124 Edge Hill Drive, Madison, WI 53705, USA
P-Stat	Fortran	*	**	**	General purpose statistics and data handling. Can be run in interactive mode using conversational command language.	Timberlake Clark Ltd, 40B Royal Hill, Greenwich, London SE10 UK or P.O. Box Att. Princeton, NJ 08540, USA

Program	Language				Description	Supplier
SAS	PL/1 and Assembler		IBM		General statistics and data handling program. Also available on DEC-VAX, Prime and DG-ECLIPSE machines	SAS Software Ltd, 68 High Street, Weybridge, Surrey, UK or SAS Institute, Box 8000, Cary, NC 27511-8000, USA
SCSS	Fortran		**		Interactive, simpler version of SPSS.	See below
SPSS	Fortran		**	**	Very popular general purpose statistical and data-handling package. Documentation explains statistical methods as well as program. Uses a control language.	Pakhoed Computer Services B, P.O. Box 863, 3000 AW Rotterdam, The Netherlands
Microstat	Basic	**			Very popular general purpose statistical and data-handling package for microcomputers. Includes analysis of variance.	Ecosoft Inc., P.O. Box 68602, Indianapolis, IN 46268, USA
NWA Statpak	Basic	**			General purpose statistical and data-handling package for microcomputers.	Northwest Analytical Inc., 1532 SW Morrison St, Portland, OR 97205, USA
SAM	Basic	**			General purpose statistical analysis and data-handling package for microcomputers. Includes some multivariate techniques.	International Software (UK), P.O. Box 160, Welwyn Garden City, Herts., UK
Cornell Ecology Programs	Fortran		**	**	A widely used set of programs designed for multivariate analysis of plant community data and preliminary data handling.	H.G. Gauch Jr, Ecology & Systematics, Cornell University, Ithaca, NY 14850, USA
NAG Library	Fortran and Algol compatible versions		**	**	A library of subroutines for numerical and statistical problems which can be built into the user's own programs.	Nag Central Office, Mayfield House, 256, Banbury Road, Oxford OX2 7DE, UK.
Microtab	BBC Basic	*			A new general-purpose data handling and statistics package designed specifically for the BBC microcomputer. Includes parametric and non-parametric methods. Disc only.	Edward Arnold (Publishers) Ltd, 41 Bedford Square, London WC1B 3DQ

had been chosen. What is now slowly emerging is a new type of program, which makes full use of interactive computing, in which the user is involved in the choice of each new step in the analysis. This effectively restores the degree of control which was implicit in the careful choice of a procedure before a manual analysis was undertaken but was lost when the very existence of a program was the reason for its use; all too often, appropriateness was traded for virtually instantaneous results. An ecologist who has a dim awareness that a multiple regression is needed could easily be seduced by the apparent ease of using a computer program which someone else has written, only to discover later or, worse, never to discover at all that the programmed technique was not the appropriate one and the results are nonsense!

A valuable feature of most microcomputers and the more modern mainframes is the visual presentation of results on a screen (visual display units and graphics terminals) and this has restored another important characteristic of data analysis which was neglected in the early days of computers, namely preliminary examination of the data. It is a simple matter to instruct a computer which has graphics capability to plot scatter diagrams of one measurement against another, to apply a transformation and replot, to plot frequency distributions with expected values—tasks which used to be very time-consuming but which can be invaluable even when the analytical technique was decided upon at the planning stage and the investigator feels that his duty has been done!

There are opposing views on the subject of using ready-written programs for data analysis. Clearly some data manipulations are so trivial that it would be a waste of time to find, comprehend, and adapt an existing program, but in other cases the technique is too complex or the programming skills available are inadequate for the method to be applied unless an existing program is used. Under these circumstances the user must ensure the program does only what he wants it to do. To some extent the danger of ignoring this task is receding as more programs become available and the user, once again, has a choice of methods and is therefore reminded of the need carefully to consider which is the most informative statistical technique to use in the circumstances.

Mainframe computers used for statistical analysis are usually equipped with *packages* or libraries containing complete programs, *subroutines* and *algorithms* which have been written and tested by professional programmers. The contents of these packages can be used directly or can be used as building blocks to construct a new program for a special purpose. Similar packages are available for microcomputers although the complexity, speed and capacity of these programs will always be more limited than on larger computers. Computer manufacturers often supply lists of *software* which will run on their machine but the variety and the volume of software being published in books, home-computer magazines or offered for sale on magnetic tapes or discs has led to independent publications of lists of products; current ones include *Software for Statistical and Survey Analysis* (Study Group on Computers in Survey Analysis), *Microcomputer Software Directory* (Computer Publications Ltd.), *Statistical Packages on Microcomputers* (Woodward & Elliott, 1983) and the *Small Computer Program Index* (ALLM Books). Cooke *et al.* (1982) and Lee & Lee (1982) provide listings of programs in BASIC for microcomputers and the former is helpfully cross-referenced to the textbook on elementary statistics by Clarke & Cook (1983). Davies (1971) gives a more extensive set of programs in FORTRAN. Some of the widely available packages used in the analysis of ecological data are listed in Table 7.1. The choice of the large, expensive packages for mini and mainframe computers is generally in the hands of professional statisticians and computer staff from whom users can find out what is available locally. Before approaching the computer service organization it is wise to have a clear idea of what analytical technique is required and the size of the data set to be processed. An elementary piece of advice when considering the purchase of software is not to rely on a general description but, if at all possible, to ask the supplier to run the programs with the prospective user's own data to enable a more realistic and thorough evaluation. Comprehensive microcomputer statistics packages could be obtained for approximately £200 ($300) in 1984.

8 References

ALLM BOOKS. *Small Computer Program Index* (ed. A. Pritchard). 21 Beechcroft Rd, Bushey, Watford, Herts. WD2 2JU.

BAILEY N.T.J. (1981). *Statistical Methods in Biology* (2nd edn). London. Hodder & Stoughton.

BARKHAM J.P. and NORRIS J.M. (1970) Multivariate procedures in an investigation of vegetation and soil relations of two beech woodlands, Cotswold Hills, England. *Ecology* **51**, 630–639.

BATSCHELET E. (1979) *Introduction to Mathematics for Life Scientists* (3rd edn). Springer-Verlag, Berlin.

BISHOP O.N. (1983) *Statistics for Biology*. Longman, Harlow.

BLACKITH R.E. and REYMENT R.A. (1971) *Multivariate Morphometrics*. Academic Press, London.

BLISS C.I. (1967) *Statistics in Biology*. Vols 1 and 2. McGraw Hill, New York.

CAMPBELL R.C. (1974) *Statistics for Biologists* (2nd edn). Cambridge University Press, Cambridge.

CARTER R.N. and PRINCE S.D. (1981) Epidemic models used to explain biogeographical distribution limits. *Nature* **293**, 644–645.

CAUSTON D.R. (1983) *Biologist's Basic Mathematics*. E. Arnold, London.

CAUSTON D.R. and VENUS J.C. (1981) *Biometry of Plant Growth*. E. Arnold, London.

CLARKE G.M. (1980) *Statistics and Experimental Design* (2nd edn). E. Arnold, London.

CLARKE G.M. and COOKE D. (1983) *A Basic Course in Statistics* (2nd edn). E. Arnold, London.

COCHRAN W.G. (1977) *Sampling Techniques* (3rd edn). Wiley, New York.

COCHRAN W.G. and SNEDECOR G.W. (1980) *Statistical Methods* (7th edn). Iowa State University Press, Ames.

COMPUTER PUBLICATIONS LTD (1983) *Microcomputer Software Directory* (ed. K. Floyd). Computer Publications Ltd., Evelyn House, 62 Oxford Street, London W1A 2HG.

COOKE D., CRAVEN A.H. and CLARKE G.M. (1982) *Basic Statistical Computing*, E. Arnold, London.

CORMACK R.M. (1971) A Review of Classification. *Jl. R. Statist. Soc.* A. **134**, 321–67.

DAVIES R.G. (1971) *Computer Programming in Quantitative Biology*. Academic Press, London.

DAWKINS H.C. (1975) *Statforms: 'Pro-Formas' for the Guidance of Statistical Calculations*. E. Arnold, London.

DAWKINS H.C. (1981) The misuse of *t*-tests, LSD and multiple range tests. *Bulletin of the British Ecological Society* **12**(4), 112–115.

FINNEY D.J. (1971) *Probit Analysis* (3rd edn). Cambridge University Press, Cambridge.

FINNEY D.J. (1978) *Statistical Methods in Biological Assay* (3rd edn). Charles Griffin, London.

FINNEY D.J. (1980) *Statistics for Biologists*. Chapman and Hall, London.

FISHER R.A. and YATES F. (1974) *Statistical Tables for Biological, Agricultural and Medical Research* (6th edn). Longman, Harlow.

GAUCH H.G. (1982) *Multivariate Analysis in Community Ecology*. Cambridge University Press, Cambridge.

GITTINS R. (1969) The application of ordination techniques. In *Ecological Aspects of Mineral Nutrition in Plants*, (ed. I.H. Rorison), pp. 37–66. Blackwell Scientific Publications, Oxford.

GORDON A.D. (1981) *Classification—Methods for the Exploratory Analysis of Multivariate Data*. Chapman and Hall, London.

GOWER C. and ROSS G.J. (1969) Minimum spanning trees and single linkage cluster analysis. *Appl. Statist.* **18**, 54–64.

GREIG-SMITH P. (1983) *Quantitative Plant Ecology* (3rd edn). Blackwell Scientific Publications, Oxford.

HAUSER D.P. (1974) Some problems in the use of stepwise regression techniques in geographical research. *Can. Geogr.* **18**, 148–158.

HEATH O.V.S. (1970) *Investigation by Experiment*. E. Arnold, London.

HEWLETT P.S. and PLACKETT R.L. (1979) *An Introduction to the Interpretation of Quantal Responses in Biology*. E. Arnold, London.

HUNT R. (1982) *Plant Growth Curves. The Functional Approach to Plant Growth Analysis*. E. Arnold, London.

JEFFERS J.N.R. (Ed.) (1972) *Mathematical Models in Ecology*. Blackwell Scientific Publications, Oxford.

JEFFERS J.N.R. (1978) *An Introduction to Systems Analysis: With Ecological Applications*. E. Arnold, London.

JEFFERS J.N.R. (1982a) *Modelling*. Chapman and Hall, London.

JEFFERS J.N.R. (1982b) Presentation of data. *Bulletin of the British Ecological Society* **13**(1), 2–3. 2–3.

KENDALL M.G. (1975) *Multivariate Analysis*. Charles Griffin, London.

KERSHAW K.A. and LOONEY J.H.H. (1973) *Quantitative and Dynamic Plant Ecology* (3rd edn). E. Arnold, London.

LANCE G.N. and WILLIAMS W.T. (1967) A general theory of classificatory sorting strategies. I. Hierarchical systems. *Computer Journal* **9**, 373–80.

LEE J.D. and LEE T.D. (1982) *Statistics and Numerical Methods in BASIC for Biologists*. Van Nostrand Reinhold, New York.

LEWIS T. and TAYLOR L.R. (1974) *Introduction to Experimental Ecology*. Academic Press, London.

MACHIN D. (1976) *Biomathematics: an Introduction*. Macmillan, London.

MEAD R. and CURNOW, R.N. (1983) *Statistical Methods in Agriculture and Experimental Biology*. Chapman and Hall, London.

MORRISON D.F. (1967) *Multivariate Statistical Methods*. McGraw-Hill, New York.

NATURAL ENVIRONMENT RESEARCH COUNCIL (1978, 1979, 1980, 1982) *Statistical Checklists*. 1. *Design of Experiments*, 2. *Sampling*, 3. *Modelling*, 4. *Plant Growth Analysis*. Institute of Terrestrial Ecology, Cambridge.

PARKER R.E. (1979) *Introductory Statistics for Biology*. E. Arnold, London.

PIELOU E.C. (1977) *Mathematical Ecology*. Wiley, New York.

POOLE R.W. (1974) *An Introduction to Quantitative Ecology*. McGraw-Hill, New York.

ROHLF F.J. and SOKAL R.R. (1981) *Statistical Tables* (2nd edn). W.H. Freeman, San Francisco.

SEAL H. (1964) *Multivariate Statistical Analysis for Biologists*. Methuen, London.

SIEGEL S. (1956) *Non-parametric Statistics for the Behavioural Sciences*. McGraw-Hill, New York.

SMITH C.A.B. (1966, 1969) Biomathematics. *The Principles of Mathematics for Students of Biological and General Science*. Vol. 1, *Algebra, Geometry, Calculus*. Vol. 2, *Numerical Methods, Matrices, Probability, Statistics*. Charles Griffin, London.

SNEATH P.H.A. and SOKAL R.R. (1973) *Numerical Taxonomy: The Principles and Practice of Numerical Classification*. W.H. Freeman, San Francisco.

SNEDECOR G.W. and COCHRAN W.G. (1967) *Statistical Methods* (6th edn). Iowa State University Press, Ames.

SOKAL R.R. and ROHLF F.J. (1973) *Introduction to Biostatistics*. W.H. Freeman, San Francisco.

SOKAL R.R. and ROHLF F.J. (1981) *Biometry: the Principles and Practice of Statistics in Biological Research* (2nd edn). W.H. Freeman, San Francisco.

SOUTHWOOD T.R.E. (1978) *Ecological Methods with Particular Reference to the Study of Insect Populations* (2nd edn). Chapman and Hall, London.

STUDY GROUP ON COMPUTERS IN SURVEY ANALYSIS. *Software for Statistical and Survey Analysis* (eds D. Cable and B. Rowe). Central Statistical Office, London.

WEBSTER R. (1978) *Quantitative and Numerical Methods in Soil Classification and Survey*. Oxford University Press, Oxford.

WOODWARD W.A. and ELLIOTT A.C. (1983) Statistical Packages on Microcomputers, In *Computer Science and Statistics: the Interface*, (ed. Gentle), pp. 63–70. North-Holland Publishing Co., Amsterdam.

8 Plant population biology

M. J. HUTCHINGS

1 Introduction

Demography, the study of populations, is defined in the Oxford English Dictionary as 'that branch of anthropology which deals with the life conditions of communities of people as shown by statistics of births, deaths, diseases, etc.' In two respects this definition departs from the way in which ecologists are accustomed to use the word nowadays. First, the demography of any species may be the object of study, and thus demography extends beyond anthropology. Secondly, whereas the term community, as used in the definition in connection with human beings, will offend few readers, it must be dropped in ecological studies, since for the ecologist it is applied to assemblages of coexisting populations belonging to many species. A demographer, however, will usually confine his attention to the life conditions of populations of a single species. Populations are thus the components from which communities are constructed. (A population may be defined as a group of organisms of the same species occupying a particular space at the same time: Krebs, 1972.)

Development of the study of populations of plants has lagged well behind the counterpart study of animals. The reasons for this, and the history of the subject have been reviewed several times (e.g. Harper, 1967; Harper & White, 1974; White, 1979). Apart from the difficulties associated with studying organisms which can have an extremely plastic form and which in some cases can multiply vegetatively—difficulties regarded less and less these days as major barriers to progress—it is now widely recognized that the study of plant populations is greatly complicated because of the variety of levels from which it can be approached. Plant populations can be studied at the level of genetical units, termed genets, at the level of vegetatively-produced repeating structures termed ramets, such as shoots or tillers in clonal species, or at the level of still more fundamental modules such as leaves or the twigs of a tree. Plant populations also possess dynamic properties in the seed phase as well as in the growing phase. Despite these complexities there has been an enormous increase of interest in plant population biology in recent years, reflected in an explosion of publications on different aspects of the subject. Unfortunately, few of these studies have been detailed enough to enable compilation of full life-tables for different species.

The writing of this chapter has been governed by the conviction that plant population biology must be primarily a field-based science which should develop from the study of natural populations, and thus the methods dealt with are mainly those appropriate to field-based investigations. (This view may be considered narrow since it apparently relegates laboratory and green-house experimentation to a level of secondary importance. Such experimentation is considered of great value, however, since it enables significant reduction in the level of uncontrolled variation in the objects of study, and is therefore more likely than field-based studies to provide definitive tests of hypotheses.) No attempt is made in the chapter to deal comprehensively with the study of plant competition, population genetics or evolution, and thus certain concepts and theories which are cornerstones of modern plant population ecology (e.g. energy allocation patterns and reproductive effort (van Andel & Vera, 1977; Thompson & Stewart, 1981), r- and K-selection (Gadgil & Solbrig, 1972), $-3/2$ self-thinning (Hutchings & Budd, 1981) have not been discussed. Even so, the theory and methods which *are* dealt with reflect the author's selection. As more plant species have been studied and more questions have been asked about plant populations, so appropriate new methods and techniques have been devised to enable acquisition of information. This proliferation of methods is necessary and inevitable, given the diversity of plant species and of their behaviour. It is bound to be the case that some valuable, but perhaps less widely usable methods will be omitted in a review of limited length. However, it is hoped that the information provided here will lead the reader to the more useful of these methods, and that familiarity with the species being studied will inspire the investigator to develop new techniques, where necessary.

The chapter is divided into three main sections dealing with, respectively, methods for studying the seed phase of populations, methods for studying growing plants, and analytical techniques.

2   Population biology of the seed phase

Following Sagar & Mortimer (1976) the term seed is used in this chapter in such a way as to include structures which should more accurately be referred to as fruits (van der Pijl, 1972). The unit that is actually dispersed should strictly be termed a diaspore (van der Pijl, 1972; Howe & Smallwood, 1982). The seed is a dormant or resting stage in the life of a plant. Seeds survive adverse conditions better than growing plants, and thus plants can 'ride out' difficult environmental circumstances in the seed state with low levels of metabolic activity, and resume active growth when more favourable conditions return. The survival value of these resting stages in the life-cycles of plants has led to the evolution of different types of seed dormancy states, which can persist for a long time and be difficult to break (Harper, 1957;

Roberts, 1972; Cook, 1980), and to the evolution of certain germination syndromes which are characteristic of species occupying different habitats with distinct types of mortality risk (Angevine & Chabot, 1979). These germination syndromes promote resumption of active growth under more favourable conditions. An ultimate consequence of dormancy and germination syndromes, both of which display great variation in details from species to species, is the variety of types of seed bank—the store of viable seeds in the soil—which have evolved. These have been subjected to broad classification (Thompson & Grime, 1979; Grime, 1979, 1981) and attempts have been made to assess their ecological significance.

Many studies have demonstrated that dormant seeds in the soil usually outnumber the growing plants in a habitat. Watkinson (1975) has stressed the importance of including seeds in plant population studies, particularly when the seeds are capable of long-term survival; the growing plants are probably a genetically-biased sample of the whole population, whereas the seed bank in the soil may contain a very large number of genotypes which have as yet been untried as growing plants. Thus, the seed bank may be a store containing genotypes formed over a long period in the past—a fund of genetic diversity with great adaptive and evolutionary potential. The seed fraction of a plant population also exhibits its own population dynamics (Fig. 8.1), and thus merits investigation; consideration of this subject therefore constitutes the first part of a review of data collection techniques.

2.1 Seed production

Interphase a (Fig. 8.1), strictly represents the probability of a mature plant producing seeds; this can be determined by direct counting. The number of viable seeds produced, B, is more difficult to determine, since seed maturation and release from the parent plant may take place over a long period of time, and many seeds may abort during maturation. In addition, the frequency distribution of the number of seeds produced by individual plants in a population is usually highly skewed, with a few plants producing a large number of seeds and most plants producing only a few (Levin & Wilson, 1978). In some species it is possible to estimate seed production accurately by counting the number of mature capsules and multiplying this by the average number of seeds produced per capsule (Leverich & Levin, 1979; Waite & Hutchings, 1982).

2.2 Pre-dispersal seed population biology

The period of time during which the seed phase of a plant's life-cycle can be censused for population studies commences shortly after fertilization

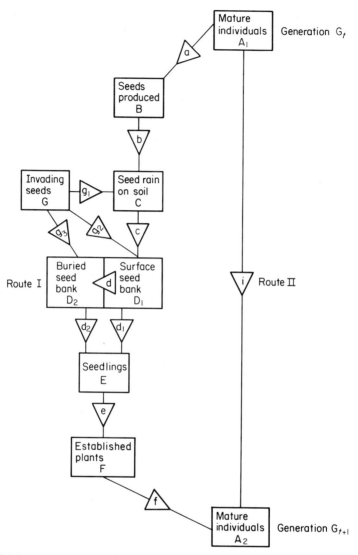

Figure 8.1 A generalized life-table for a species which does not have ramet reproduction. Route I is genet reproduction, or plant establishment from seed. Route II indicates that some plants in some species may be able to survive from one generation to the next. Capitals are used to represent the number at each phase through the life-table. Small letters represent the probability of passage from one phase to the next. (From Sagar & Mortimer, 1976.)

has taken place; the new genetical unit at this stage becomes a target for plant predation and plant parasites (Sagar & Mortimer, 1976; Fig. 8.1, interphase b). The subject of herbivory has been reviewed several times (Ehrlich & Birch, 1967; Harper, 1969, 1977; Janzen, 1969, 1971; Crawley, 1983), and examined experimentally on a number of occasions. A recent example is the field investigation conducted by Louda (1982), in which the level of predation on flowers and seeds was controlled by the application of insecticides; this reduced the loss of fruits of *Haplopappus squarrosus* caused by flower- and seed-feeding insects by between 41 and 94%. Louda went on to show that the impact of herbivores on populations subject to flower- and seed-predation was also apparent later in the life-cycle from depressed numbers of seedlings and young plants, and that insect predation was the proximate factor limiting recruitment and abundance in this species.

2.3 Dispersal of seeds: the seed rain

Dispersal of seeds from the area where they were produced can account for loss of some seeds between their production and reaching the soil surface. This loss is part of interphase c in Fig. 8.1. Immigration of seeds into the area (phase G in Fig. 8.1) may make good or even exceed this reduction in numbers. It may be difficult in practice, however, to determine the numbers of seeds involved in these processes.

2.3.1 Measuring dispersal distances

Those seeds which escape pre-dispersal predation contribute to the seed rain falling on an area (phase C, Fig. 8.1). Most seeds, even light, wind-dispersed seeds, are deposited close to the parent plant, and in many cases there is a negative exponential decline in density of deposited seeds (as with dispersing pollen grains) with increasing distance from the parent source (e.g. Werner, 1975a) (Fig. 8.2). In order to characterize the seed shadow within which dispersing seeds are deposited around a parent plant, it is usually necessary to count seeds as they land, rather than to record those already at the soil surface; this is difficult to do accurately and it does not enable the observer to distinguish between seeds deposited in different years or at different times of the year.

Seeds with structures which enable them to become attached to dispersal agents and to the soil or vegetation where they land can be collected in cheesecloth traps as they disperse (Levin & Kerster, 1969). Bullock & Primack (1977) simulated animal dispersal of seeds by trailing denim, muslin and cotton cloths through vegetation. These cloths removed ripe seeds from inflorescences, and the distances moved before they were re-deposited could be measured. Platt (1975) and Platt & Weis (1977) followed wind-dispersed seeds visually from source to landing site to measure distance of dispersal.

Distances moved by the dispersal units of *Vulpia fasciculata* were monitored by Watkinson (1978b) by spraying them with aerosol paint before dispersal and searching for them after dispersal. Similar methods have been

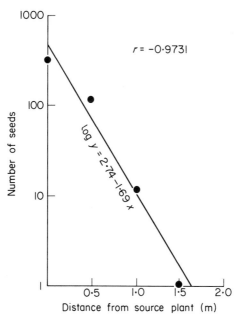

Figure 8.2 Pattern of seed deposition in teasel, *Dipsacus fullonum*. The number of seeds deposited as a function of distance from the parent plant exhibits a negative exponential function ($r = -0.9731$). (From Werner, 1975.)

devised using dyes and radioactive tracers (Mortimer, 1974; Watkinson, 1978a; see also Colwell, 1951, for studies of spore dispersal), but recovery may only be at an acceptable level if the majority of dispersal distances are short. For example, Watkinson (1978b) recovered 97·3% of paint-sprayed dispersal units; overall, 79% of dispersal units fell within 10 cm of the parent plants, and the maximum recorded dispersal distance was only 36·0 cm. Naturally, such methods can only be used when dispersal is not impaired by addition of the marker.

The experimental approach which has been used most frequently in measuring diaspore movements in wind-dispersed species is to determine the rate of fall, or terminal velocity of seeds in a column of still air—usually a long, transparent plastic pipe arranged vertically (e.g. Sheldon & Burrows, 1973; Werner & Platt, 1976; Platt & Weis, 1977; Rabinowitz & Rapp, 1981). Prior to measuring it may be necessary to treat the pipe to remove static, which can affect the fall rate by attracting seeds to the sides of the plastic pipe (Rabinowitz & Rapp, 1981). The time taken by seeds to fall through a standard distance is timed by stop-watch; measurements may need considerable replication before stable mean values for fall rates are obtained, and must be rejected if the seed touches the sides of the pipe. Estimates of terminal velocities vary between seeds of the same species, between measurements made on the same seed, and under different relative humidities.

Results obtained using this method must be treated with caution. Seeds in the field will rarely be dispersed in still air conditions similar to those found within a vertical plastic pipe. To introduce greater realism, Sheldon & Burrows (1973) measured dispersal distances of single seeds under constant wind speeds in a wind-tunnel. H. Ford (personal communication) has, however, calculated that the wind speeds used in their experiments on *Taraxacum officinale* would have been insufficient to dislodge even completely ripe seeds from the receptacle. Given a requirement for faster wind speeds to initiate dispersal, the subsequent distances which seeds would travel in such winds would often be greater than the estimated seed dispersal ranges given by Sheldon & Burrows (1973).

In studies of seed dispersal on salt marshes, W.G. Beeftink (personal communication) has measured dispersal fluxes caused by the tides both at and below the sea surface, using fine-meshed net traps. The problem with such methods is likely to be that the net mesh size may need to be so fine to ensure efficient trapping of seeds of many species (particularly *Juncus* spp., of which the seeds are very small and numerous) that the rate of flow of water through the trap may be seriously reduced, leading to unrealistically low rates of capture of dispersing seeds. Ryvarden (1971) described a rudimentary trap (with considerable disadvantages in its use) for collecting seeds dispersed in streams, and also a trap suitable for collecting seeds at ground level at the fringe of a glacier, where the vegetation cover was very sparse. His technique would be inefficient in most types of vegetation (Rabinowitz & Rapp, 1980).

Watkinson (1978b) distinguished two aspects of seed dispersal: Phase I dispersal involves movement of seeds from the infructescence to the soil and Phase II dispersal is subsequent movement of seeds along the soil surface. Phase II dispersal could result in the seed returning to a position closer to the parent plant than its resting place at the end of Phase I dispersal. Both Watkinson (1978b) and Mortimer (1974) have measured Phase II dispersal distances, using the techniques referred to earlier of labelling individual seeds with radioactive tracers, paints or dyes. Phase II movement of seeds by inorganic dispersal agents only appears to be a major factor on smooth soil surfaces with an absence of vegetation. (Harper *et al.* (1965) describe a point quadrat technique for quantifying the roughness of the soil surface.) On the other hand, organic agents of dispersal may have a major influence upon the positions of seeds at the end of Phase II dispersal.

2.3.2 Density of the seed rain

In addition to measuring and predicting dispersal distances, it may be of importance to determine the density of seeds making up the seed rain. Werner (1975a) devised a simple, inexpensive seed trap using filter paper sprayed with the non-drying sticky substance 'Tanglefoot'. (Such traps placed at different distances from an isolated seed source also provide information on dispersal distances.) Seeds landing on this surface become tightly held. Periodically the filter paper can be replaced, and the old filter paper, together with its trapped sample of seeds, can be returned to the laboratory where the seeds can be identified and counted. Thus, not only can the density of the seed rain at a point on the ground be estimated, but temporal patterns in the seed rain can also be determined. Rabinowitz & Rapp (1980) have used this technique; they caution that the seed rain exhibits spatial variations, and thus a valid sampling design must be established for accurate assessment of seed density. However, while total seed rain showed a variable spatial density (see also Falińska (1968) for trapping of dispersing seeds in woodland), few *species* showed spatial heterogeneity in seed rain density, and few seeds were dispersed in clumps. (This is not always the case. See, for example, Waite & Hutchings, 1978.) Other problems which arose in the use of this technique were firstly that no seeds could be collected when traps were frozen and snow-covered, although seeds were still being dispersed throughout the winter (Rabinowitz & Rapp, 1980), and secondly that the efficiency of the traps declined as vegetation grew above them and intercepted seeds. As yet, death of seeds intercepted by vegetation and failing to reach the soil surface has not been adequately studied.

2.4 Post-dispersal seed population biology

The fates which may befall seeds landing on the soil have been categorized in broad terms by Sagar & Mortimer (1976) as: (i) to remain where they land on the soil surface; (ii) to move along the soil surface, thus altering their final distance of dispersal from the parent plant (Phase II dispersal); (iii) to become buried (interphase d, Fig. 8.1); (iv) to be predated on the soil surface; (v) to be removed from the ground and carried long distances by dispersal agents (Phase II dispersal); (vi) to be killed; (vii) to die, or to germinate (interphases d_1, d_2, Fig. 8.1) on or beneath the soil surface. Some seeds may experience more than one of these fates, whereas predation, death and germination are terminal from the point of view of the seed phase. The lengths of time for which seeds remain on or in the soil, and their ultimate fates should be quantified in any complete study of the population biology of a plant. Much depends on whether seeds are in a dormant or non-dormant state, and on the length of time for which seeds can remain viable.

The subject of seed dormancy has been extensively reviewed (e.g. Taylor-son & Hendricks, 1977; Cook, 1980), and Harper (1957, 1977) has categorized three types of dormancy state.

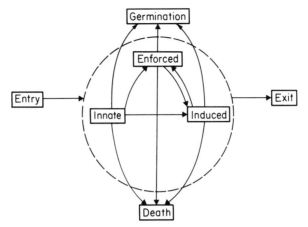

Figure 8.3 The states and fates of seeds in the soil. Within the broken circle, seeds are in a state of dormancy. (Re-drawn from Sagar & Mortimer, 1976, after Roberts, 1970.)

(a) Innate dormancy, which prevents the germination of seeds while they are still on the parent plant. Seeds are dispersed in this condition.

(b) Induced dormancy, a state of inability to germinate which the seed enters as a result of being subjected to some experience after ripening.

(c) Enforced dormancy, a state in which germination will not take place until some environmental limitation or constraint has been removed.

All three categories of dormancy have been discussed at length by Harper (1977, pp. 65–79). Most dispersing seeds arrive at the soil surface in a state of dormancy from which in order to germinate they must be released by the application of appropriate stimuli or environmental conditions. However, additional states of dormancy may be imposed upon seeds before the correct conditions to promote germination are experienced (Fig. 8.3).

It appears likely that in many species a high proportion of the seeds in a dormant state will die before germination, and of those which do germinate, successful establishment is far from ensured (Sagar, 1970). Failure to germinate indicates that the correct stimulus to break dormancy has not been applied, or that the correct environmental conditions to constitute a 'safe-site' (Sagar & Harper, 1961) for germination have not been experienced by the seed. Different species of plants adapted to grow under certain broadly-defined environmental conditions can sometimes be classified together as having the same germination characteristics (Grime *et al.*, 1981) or germination syndromes (Angevine & Chabot, 1979), although it is an emerging fact that diversity of adaptations to a particular set of environmental conditions seems to be the case rather than the exception for the species found together in any habitat.

While seeds of some species, particularly some grasses, have short-term viability, and most of their seeds either germinate very soon after release from the parent plant or die, seeds of other species can persist in a viable, dormant state on or in the soil for considerable periods of time (see e.g. Cook, 1980). It has even been demonstrated that certain species which had long been considered locally extinct could still be recovered as seedlings from seed which has remained dormant in the soil at sites they once occupied (e.g. Moore, 1983); thus the seed bank may be of importance in conservation.

Natural selection will have operated on all species to favour strategies of germination and viability which are a compromise between conflicting benefits and disadvantages. The possible benefits of rapid germination include increasing the potential rate of population increase, (i) by reducing mortality in the seed phase by shortening its duration, and (ii) by shortening the generation time. In addition, greater competitiveness may be conferred in comparison with plants germinating later. A major disadvantage is that competition between seedlings of the same species may be intense because of almost simultaneous germination. Moreover, short-term viability necessitates early, simultaneous germination even though conditions may be unfavourable, and therefore could lead to high rates of mortality. By contrast, longer viability may enable germination of the species to be protracted, thus reducing competition; individual seeds may remain in a dormant state until some highly specific safe-site requirements for germination are satisfied. Cohen (1966, 1967) and Jain (1979) have modelled the ecological and evolutionary benefits which may accrue from different strategies of germination from the seed bank and prolonged dormancy, and have made predictions concerning the types of environmental conditions under which different strategies will be advantageous.

2.5   The seed bank

Soil seed banks can have great economic and agricultural significance. They are also of considerable academic interest because of their demographic and evolutionary consequences for plant species (Cohen, 1966; Grime, 1979). It is therefore of no surprise that they have been the subject of much research which has been extensively reviewed (e.g. Harper, 1977; Cook, 1980; Roberts, 1981). Harper (1977) has defined the seed bank as the store of seeds buried in the soil, whereas Sagar & Mortimer (1976) use the term in a broader sense, distinguishing between a surface seed bank and a buried seed bank. The density of viable seeds in the seed bank normally decreases with increasing successional age of the site, with length of time since the last cropping or disturbance, and with increasing latitude and altitude (Johnson, 1975; Thompson, 1978).

In the small number of studies which have been conducted upon the fates of seeds on the soil surface (interphase d_1, Fig. 8.1), predation appears to have been a major cause of seed loss. Pathogens are assumed to account for the deaths of many seeds, and some seeds are lost by dispersal out of the study area. In intensive studies of the fates of seeds of *Plantago lanceolata* introduced into three habitat types by Mortimer (1974), there was an alarming rise as time passed in the proportion of seeds for which the ultimate fate was unknown. Similarly Waite (1980) and Russell (1983) were unable to account for the loss of many seeds between the seed rain and the soil seed bank (interphase c, Fig. 8.1) in a range of types of coastal habitat; in some of the sites they studied, water transport of seeds out of the area may have caused significant losses. Rabinowitz & Rapp (1980) reported a 71% fall in seed density from the seed rain to the buried viable seed bank (interphases c and d, Fig. 8.1). Detailed information is lacking in all but a small number of cases on the main causes of such losses.

The majority of seeds which do not germinate at the soil surface and which escape predation and pathogens, become part of the buried seed bank. This occurs slowly by burial beneath litter, by incorporation during natural soil movement, or more rapidly because of the activities of soil animals; for example, after ingestion by earthworms and voiding in the cast, and by falling into cracks and burrows (McRill & Sagar, 1973; McRill, 1974). In many cases the viability of the seeds is not affected. When the land is ploughed, dug, or disturbed in other ways, incorporation of seeds into the soil is more rapid, and this results in deeper burial because of soil inversion. The effects of burial upon the subsequent fates of seeds have been reviewed by many authors (Cook, 1980; Roberts, 1981). All of these buried seeds may become combined with a residue of viable seeds added to the seed bank from previous seed generations, particularly when the species has potential longevity in the seed phase.

2.5.1 Quantifying the seed bank and determining its composition

Methodology for determining the size of seed banks in soils has been reviewed by Roberts (1981). Surface seed banks can sometimes be recovered by directing vacuum pumps over a specified area, although in dense vegetation, recovery of seeds lying on the surface may be incomplete. This technique may also collect much unwanted material (e.g. soil, plant litter), from which the seeds must be sorted. Seeds of some species, such as those of the genus *Plantago*, present problems using this method, because on contact with moisture they produce copious mucilage by means of which they become firmly attached to the substrate. To recover the buried seed bank, soil samples of the desired depth are normally removed by means of a soil corer; the diameter of the corer can be selected to suit the volume of soil required.

The aim of the investigator will usually be not only to identify the species of seeds in the seed bank, but to determine their density, either as number per unit area of ground or per unit volume of soil, and the way this alters throughout the soil profile. The distribution of seeds within the soil is not uniform, either in the horizontal or the vertical plane, and thus several soil samples must usually be collected in order to overcome the variation in seed content between samples and to obtain a stable estimate of density. Roberts (1981) states that it is not possible to generalize about the number of samples to be taken or their spatial frequency, but recommends, in common with Kropáč (1966) that a large number of small samples should be taken, rather than a small number of large samples. Ideally a preliminary sampling study should be conducted in which the investigator determines the number of soil samples necessary to provide stable mean values of the parameters being measured with an acceptably small standard error. A suitable technique is described in Greig-Smith (1983, pp. 31–2). While ample replication of samples should always be striven for, the logistic problems of dealing with many soil samples may place strict sampling considerations beyond the means of many workers. Roberts (1981; see also Roberts, 1958) suggests suitable spatial arrangements for sampling points to estimate seed density, and recommends sampling along transects whenever seed banks are to be investigated along environmental gradients.

Just as replication of soil samples for accurate estimation of the seed bank has been minimal in some studies rather than ample, so the volume of soil investigated has sometimes been small. There is little published information on the volume which should be used, although Thompson & Grime (1979), Roberts (1981) and Graham (1983) have provided some guidelines. In cases where the emphasis is on determining the species of seeds in the seed bank rather than their number, a curve relating number of species recovered to volume of soil sampled can be constructed (Fig. 8.4). This is used to estimate the minimum volume of soil which must be sorted to reveal the majority of species present. Hayashi (1975) recommends that twice the minimum volume of soil should be sampled. The minimum volume was estimated to be $400\,cm^3$ under early succession vegetation (Numata et al., 1964), $500\text{–}600\,cm^3$ under grassland (Hayashi & Numata, 1971) and $4000\text{–}6000\,cm^3$ beneath climax forest (Hayashi & Numata, 1968); the smaller the buried seed bank is likely to be, the larger the volume of soil recommended for sampling. This method of estimating the volume of soil to use is analogous to the use of species–area curves for determining areas to sample in quadrat surveys of plant communities, and it suffers from the same problems of subjectivity in its determination (Hopkins, 1957).

Apart from horizontal and vertical variations in density and composition, the seed bank may exhibit marked seasonal fluctuations. Thus, a single sample will rarely characterize the seed bank adequately, particularly where

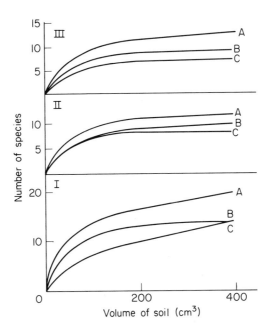

Figure 8.4 Relationship between number of species recovered from the seed bank, and volume of soil sampled, in three seral stages. I = *Erigeron* stage, II = *Imperata* stage, III = *Pinus* stage. A = soil from depth 0–1 cm, B = soil from depth 1–2 cm, C = soil from depth 2–5 cm. (From Numata *et al.*; 1964).

seasonal inputs of seed, or germination flushes are marked. Thompson & Grime (1979), for example, have documented marked temporal changes in the density of seeds and in the germinable species in the soil from ten ecologically different sites in the Sheffield region. Some species, which were extremely short-term components of the seed bank, germinated too rapidly after release from the parent plants to be recovered from the soil, given the rate of sampling (6-week intervals) imposed in this study. Thompson & Grime (and Grime 1979, 1981) distinguish between such species, which have transient seed banks, none of the seed output remaining in the soil for more than a year, and species with persistent seed banks, in which some of the seeds are at least one year old. Grime (1981) subdivides these classes further, and provides examples of species with each type of seed bank.

   Although the presence of large numbers of viable seeds in the soil has long been recognized (Darwin, 1859; Brenchley, 1918; Brenchley & Warington, 1930, 1933; Chippindale & Milton, 1934), the characterization and quantification of the buried seed bank poses major technical and logistic problems. The least energy-intensive method of quantifying the seed bank is to count and identify seedlings as they emerge. While this is usually carried out in the laboratory on collected soil samples, it can be done *in situ*. If such studies are conducted in the field, seed density can only be reported on a unit area, rather than volume of soil, basis. Care must also be taken to prevent further seeds arriving on the study area. Seedling identification guides have been published by Chancellor (1966), Hanf (1970), Häfliger & Brun-Hool (1973) and Muller (1978). Conditions favouring high rates of germination for many species can be created by removing living vegetation and disturbing the underlying soil, which may release seeds from dark-enforced dormancy (Champness & Morris, 1948; Budd *et al.*, 1954; Harper, 1957). Remaining innate or induced dormancy may be overcome as a result of exposure to light (Wesson & Wareing, 1967), greater temperature fluctuations (Thompson *et al.*, 1977; Thompson & Grime, 1983), or to stratification or high temperatures.

In large-scale studies of seed banks, particularly when species with small seeds are involved, it may be impractical to quantify the seed bank completely. In the extensive studies conducted by Thompson & Grime (1979) a method was devised which was intended to estimate only the 'readily-germinable' component of the seed bank. Because such methods underestimate the size of the seed bank, caution must be exercised in comparing results with those obtained in studies where the seed bank has been completely determined. Methods for complete determination of the seed bank are described below (see p. 392). The method used by Thompson & Grime was to collect soil from the field, air-dry it, pass it through a coarse mesh sieve to remove large stones, and to spread it out in a thin layer in trays in the greenhouse. A regime of 16 hours of light at 20°C and 8 hours of dark at 15°C was used for seed germination because many native species had previously been found to germinate readily under these conditions. Trays were regularly watered from below. Germinated seedlings were recorded and identified at frequent intervals.

In Thompson & Grime's study most seedlings germinated during the first 3 weeks of testing and few emerged after 5 weeks. However, the soil was not disturbed in these experiments, so that further germination, which would have been mostly of seeds being released from enforced dormancy after exposure to light, did not occur. In similar experiments Graham (1983) recommended 5 weeks as a minimum period of time over which to record germination, although in his experiments the soil was regularly turned. However, Roberts (1958) kept soil samples in similar conditions for 3 years; few seedlings emerged in the third year. He concluded that in quantification of the seed bank 'there would be little advantage in extending the period (of observation) beyond 2 years'. About 5% of the seedlings of *Anagallis arvensis*, *Plantago major* and *Matricaria* spp. emerged in the third year. Similarly, 83% of all seedlings emerged from samples of soils from tropical regions within 1 year, 99% within 2 years and only 1% in the third year (Vega & Sierra, 1970). Froud-Williams *et al.* (1983) retained soil samples for 2 years, turning them every month during the first year to encourage germination, and every 3 months in the second year. Watkinson (1978a) retained soil samples for 12 months, stirring them at intervals, while Rabinowitz (1981) kept samples of prairie soils for 145 days, stirring them occasionally. Germination had virtually ceased at the end of this time; 74% of germination took place within the first month.

In addition to disturbing the soil at intervals to improve germination, subjecting it to fluctuating temperatures has been found to stimulate germination of many species (e.g. Warington, 1936; Roberts, 1958). Different responses to fluctuating temperature have been recognized in ecologically distinct groups of species (Thompson et al., 1977; Thompson & Grime, 1983). Mott (1972) observed different germination responses to identical amplitudes of temperature fluctuations in monocotyledonous and dicotyledonous species, with germination of the former predominating under a 30°/25°C temperature regime, and of the latter when the fluctuations were 20°/15°C. It is clear therefore, that the time of year at which germination experiments are conducted in the field, or the environmental regime applied to germinate seeds in soil samples in the laboratory, may have a great influence on the species of seedlings which emerge. In addition, if the time of year over which germination is studied is less than a year, this can also significantly alter results because of the seasonality of the seed rain and the seasonality of germination and dormancy (Angevine & Chabot, 1979).

Three further problems may arise. Firstly, many viable seeds may die without first germinating during the period of study, and thus be omitted from the estimation of the viable seed bank. Secondly, some seedlings may germinate *and* die in the time interval between two recording dates. This may happen even when consecutive recording dates are close, for it is a well established fact that many newly-germinated seedlings have very short lives. The underestimation of seed bank size from this cause may be considerable. Thirdly, some seeds may have prolonged dormancy which is very difficult to break (e.g. some members of the Leguminosae, in which a high proportion of 'hard' seeds are produced); as already mentioned, seeds of many species can remain viable but dormant for very long periods (see e.g. Cook, 1980). These seeds may remain outside the 'readily-germinable' component of the seed bank throughout the study, and thus never be recorded.

The methods of Thompson & Grime (1979) described above were most effective in detecting persistent seed banks, even for species which were fairly uncommon in the vegetation. They were less effective for recording seedlings of species which were released from the parent plant in a state of innate dormancy, requiring chilling to attain a germinable state. After chilling, seeds of these species germinated rapidly over a short time interval. Because sampling frequency was too low to coincide with the brief period between the attainment of germinability and the onset of germination (Thompson &

Grime, 1979), seedlings of these species were never recorded from the soil. Finally, the danger of extrapolating results obtained in one year, which may be unrepresentative, to other years, must not be ignored.

While methods such as those employed by Thompson & Grime (1979) can only provide a measure of the seeds which are 'available' for germination at the time of soil disturbance, total enumeration of the seeds in the soil—both viable and non-viable, active and dormant (Harper, 1977)—requires their physical separation from the soil. The techniques involved are laborious and extremely time-consuming. They have been reviewed in detail by Roberts (1981), and include wet sieving, dispersion by flotation, separation by air-flow methods (Fig. 8.5) and hand-sorting. The reader is directed to Roberts' review and the papers by Hyde & Suckling (1953), Kropáč (1966), Malone (1967), whose flotation technique is reported as unsatisfactory for seeds of *Ranunculus* spp. by Sarukhán (1974), Jones & Bunch (1977), Fay & Olson (1978), Roberts & Ricketts (1979) and Standifer (1980). Identification of seeds can be achieved with the use of suitable reference works (e.g. Martin & Barkely, 1961; Corner, 1976; Montgomery, 1977).

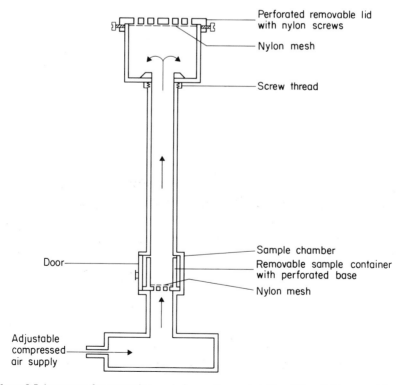

Figure 8.5 Apparatus for separating seeds from other material by differential transport in an air current. Compressed air is fed into the inlet, and rises through the vertical tube to deposit lighter material in the top chamber. The apparatus is made from perspex.

2.5.2 Determination of seed viability

Determination of the viable fraction of the seed bank requires further painstaking labour. Hand-sorting of obviously dead, decaying and unfilled seeds may be carried out, and the embryos of seeds can be examined. Alternatively, optimal conditions can be provided to stimulate seed germination, although this is far easier to achieve when interest is confined to one species rather than the total seed bank.

Chemical tests for seed viability have been described by Freeland (1976). The best known of these is staining of living tissues with tetrazolium salts, a technique which can be slow and tedious, particularly with small seeds. The tissues of seeds of different species exhibit different vital staining patterns with tetrazolium, however, and some familiarity with the species under investigation may be necessary in order to diagnose viability correctly. Unfortunately, even non-viable seeds may exhibit tetrazolium staining. (See also Sarukhán, 1974; Roberts, 1972.)

2.5.3 Classifying seeds in the seed bank

A practical classification of the fates of seeds arriving on the soil surface which may be used as a framework for further studies is that utilized by Sarukhán (1974) in a study of the seed population dynamics of three species of *Ranunculus*. (Other more detailed but perhaps less practical schemes are those of Schafer & Chilcote (1969) and Mortimer (1974).) For one year Sarukhán recorded the fates of seeds sown in batches on the soil surface in small, circumscribed areas from which they could be recovered after different intervals of time. The categories recognized were: (i) germinated (G), the total number of seedlings observed in the field from the starting point of the investigation when seeds were sown, up to the date of collection of the sample under investigation; (ii) 'dormant-enforced' (DE), the seeds which germinated from a soil core when favourable conditions were applied; (iii) 'dormant-induced' (DI), the seeds recovered from the soil cores which failed to germinate under favourable conditions but showed a positive reaction to the tetrazolium test for viability; (iv) the dead seeds (D), the empty and non-living seeds recovered from the soil, plus those which were found to be dead during the germination and viability tests. From this final fraction the number of germinated seeds was subtracted since all seedlings shed their seed coats, and these were included in the total number of empty and dead seeds. Thus, the total number of seeds, S, consisted of the total of the categories $G + DE + DI + D$. Using this scheme Sarukhán analysed changes in the distribution of seeds into each of these categories throughout the year.

The schematic diagram of Sagar & Mortimer (1976, see also Begon &

Mortimer, 1981) illustrates the passage of seeds from one stage to another along the path towards becoming established plants (Fig. 8.1, Route I). The probability of transition from each stage to the next is indicated by inter-phases. Determination of the values of such transition probabilities, and of the scale of losses of seeds between each stage should be a major objective in investigations into the population dynamics of the seed phase of different species. Few studies appear to have followed this scheme, however, (but see Waite, 1980; Russell, 1983), although Sarukhán (1974), Mortimer (1974) and Watkinson (1978b) have documented the fates of populations of seeds in the field. As a general conclusion, it seems clear that, in the absence of distur-bance of the vegetation or soil, the majority of seeds in the seed bank are destined to die *in situ*.

2.5.4   Sizes of seeds

One of the least variable measures of plant performance, even under highly competitive conditions (Harper *et al.* 1970; Palmblad, 1968), is mean seed weight. (Individual seed weight within seed populations does show variations, however, and species may exhibit seed polymorphism; see Harper *et al.*, 1970.) The advantages of maintaining seed weight under adverse conditions include the following: (i) much of the seed weight consists of endosperm—food reserves which are needed to sustain the growing seedling before it is capable of supporting itself entirely by photosynthesis; (ii) seed-ling weight and subsequent performance are often correlated with seed weight (Black, 1957, 1958). The disadvantages associated with increased size include the following: (i) fewer seeds can be produced per unit investment of resources in reproduction; (ii) larger seeds will disperse shorter distances; (iii) larger seeds may be more apparent and valuable to herbivores; (iv) the problem of gaining moisture from the substrate faster than it is lost to the atmosphere during germination is more acute for larger seeds, in which a greater proportion of the surface will be out of contact with the substrate (Harper & Benton, 1966).

However, although seeds generally exhibit a marked lack of plasticity in mean size and weight within a species, there are circumstances under which variations in the weights of seeds of a species have been observed. These include variations following experimental disturbance inflicted upon plants (Weaver & Cavers, 1980), systematic variation in seed weight within indi-vidual plants (see Harper *et al.*, 1970 for a review), and weight variations both between (Hodgson & Blackman, 1957) and within seasons (Falińska, 1968; Rorison, 1973; Fuller *et al.*, 1983). Such variations have not been fully documented, and are as yet imperfectly understood, and further work in this field, especially at the level of measuring individual seed weights rather than mean seed weight, would be valuable.

2.6 The bud bank and pollen grains

Other types of banks of plant parts exist in the soil and these may require investigation in population studies. Just as seeds should not be overlooked in plant population studies because of their bearing on the numbers, and fluxes in numbers, of genets in populations, so other dormant meristems should not be ignored, since in their various forms as bulbs, bulbils, corms, tubers and rhizome buds, they reflect the extent of growth and success of genets which are already established. Such meristems may either break dormancy each year as the growing season approaches or, in the case of clonal, connected systems of such meristems, may be subject to correlative inhibition controlling which buds develop and which remain dormant, although such control may be disrupted by disturbance. (Similarly, populations of buds on a tree can be treated as a bank of meristems which can also be subjected to demographic analysis; Maillette, 1982a,b,c).

Finally, the accumulated pollen grains in the soil can be of use in population studies. Pollen in the soil does not constitute a bank in the same way as seeds, meristems and buds, since its presence and density are irrelevant to future population sizes. However, the use of pollen grains in the soil for interpreting the history of vegetation is well known (e.g. Moore, p. 529). Bennett (1983) has used measurements of accumulation rates of pollen grains of different species as input for equations of population growth to determine postglacial rates of increase of tree populations in Norfolk, England.

3 Population biology of growing plants

While population studies have been conducted on an extremely wide variety of plant species, strictly demographic data have been collected for very few species and a relatively small number of these possess sufficient features of life history, behaviour or morphology in common to enable straightforward comparisons of their demography. Comparative demographic studies can be carried out on species which possess at least some of the characteristics listed by Sarukhán & Harper (1973). These are that they should be (i) closely related, (ii) very common, (iii) living in the same area within an extensive and stable ecosystem, (iv) representative of contrasting life-cycle strategies. They should also produce (i) well-defined, discrete vegetative propagules which are connected by above-ground structures, readily distinguishable from the parent plant and other propagules, and which quickly become independent of the parent, (ii) conspicuous seeds which do not migrate extensively and present no serious dormancy problems, (iii) seedlings and vegetative propagules at discrete periods in the year, and (iv) adult plant bodies with markedly different phenology and some morphological character that permits the age of the plant to be determined. These characteristics are also worth considering when the practicality of working on any single species is being assessed.

While it is freely admitted that ideal sets of species satisfying all of these criteria will not be found, some species may possess many of these attributes. An alternative to comparing the demography of different species, clearly, is to compare the demographic behaviour of a single species under a variety of environmental conditions or experimental treatments.

3.1   Sample sizes and sampling frequency

No firm guidelines have been established concerning the size of samples of plants which should be utilized in demographic studies, or the frequency with which the condition of plant populations should be recorded. It will probably prove impossible to develop a single set of rules because of the wide differences which species may exhibit in their structure and behaviour, and in the rate at which significant demographic events occur; for example, in an annual species, recordings must be made frequently during the growing season whereas for tree species, after the first few seasons of growth, some types of data will require collection no more than once a year. Difficulties over the number of individuals to be used in investigations are more complex to solve. For practical reasons populations are nearly always sampled rather than completely recorded. Since there is always variation between the demographic behaviour of the individual plants in a population, a sampling scheme must be devised which will produce accurate estimates of population parameters with acceptable confidence intervals. Law (1981) has been justly critical of population studies where inadequate sampling schemes or sample sizes have been employed, and he cautions against inferring too much about plant population biology from such data. In particular, he warns of the effects of spatial heterogeneity of biotic and abiotic factors within plant populations on the behaviour of the population and its constituent parts. This warning cannot be reiterated too strongly, for in recent years it has become clear that generalizations about the population biology of a single species over wide categories of space and time can be very misleading; plants in different sampling plots may behave very differently (e.g. Whigham, 1974; Barkham, 1980b; Waite, 1980; Mack & Pyke, 1983), and populations may change their demographic patterns markedly through time (Law, 1981; Mack & Pyke, 1983; Waite, 1984). At the present time no simple rules can be offered for sampling plant populations apart from the considerations normally applying in sampling.

A further complication emerging from recent studies is the fact that, not surprisingly, there are marked differences between the demographic behaviour expressed by populations of a given species under stable and changing conditions. That such differences might exist has often not been recognized in studies of plant population biology. In most investigations there is either an explicit statement or an implicit assumption that populations have been studied under 'stable' conditions, even though this may be doubtful. Stability cannot, in any case, be readily measured, nor can its continued existence be forecast when a demographic study is initiated. Both Law (1981) and Waite (1984) have demonstrated marked changes in demography of a given species in a given site as early succession proceeds. Our knowledge about this subject is still very incomplete, and further comparative studies of population biology under changing conditions would be welcome.

Finally, the performance (e.g. size or reproductive effort) of individual plants in populations also displays temporal changes. Changes occur at a variety of time scales, including seasonal changes related to the phenology of growth and stages in the life cycle, directional changes which are succession-based (Law, 1981; Waite, 1984) and characterized by the increase and subsequent decrease of species, and, in some species, cyclic changes, in which plants of a given species occupying different patches of the same habitat may be displaying different population properties (e.g. Watt, 1947a,b; Thomas & Dale, 1974).

3.2 The population units to be studied

In any study of plant population dynamics a primary consideration is the level at which the study is to be conducted. While the geneticist or evolutionist may be most interested in the demographic behaviour of genetical individuals (the genet population, each member of which is the product of a single seed), the ecologist may be more interested in behaviour at the level of functional parts (the ramet population) which includes tillers, shoots or rosettes in clonally-growing species, or at the level of modules (units of construction such as leaves, which may be repeated within genets and ramets). These different levels of organization as bases for study have been discussed by Harper & White (1974), Hutchings & Barkham (1976), Bazzaz & Harper (1977), Hunt (1978a), Harper (1977, 1978, 1981), and White (1979, 1980).

3.3 Censusing populations

Regardless of the unit under study, data must be collected at sufficiently frequent intervals to enable the investigator to build up a picture of the changes in numbers of plants, and their status at different times during the study period. Counts of density alone are of little value here, since they do not establish any link between the plants recorded at one census and those recorded at subsequent censuses. Mapping or tagging to enable reliable re-identification of population units at different census dates is imperative if useful data for demographic analysis is to be obtained; without this information, total or partial turnover of the population at a site may have occurred but the investigator will be unable to analyse these events. Repeated mapping or tagging can be very time-consuming, tedious and tiring. Given the investment of effort which must be devoted to such work it is imperative that careful planning of sampling schemes, data collection routines and methods of analysis is carried out. In studies where demographic analysis is not of primary interest, collection of such data will probably be an unaffordable luxury.

The more regularly mapping or tagging is carried out, the closer the observer will approach a measurement of the *true* flux of plants through the population; particularly in populations reproducing freely by seed, many seedlings suffer mortality after extremely short periods of time, and thus will come and go between censuses without ever being recorded. The ultimate causes of death of seedlings are only broadly known or guessed at by most investigators; determination of the proportions dying as a result of predation, fungal invasion, dislodging from the soil, rain-splash, etc. may be a worthwhile aim in some studies.

Various methods exist which can facilitate accurate re-identification of plants at different census dates. It is, of course, essential that the method chosen does not affect the plant or its population dynamics, or alter external factors such as grazing intensity.

3.3.1   Tagging

For woody shrubs and trees, metal tree tags, each with a unique number inscribed, can be either nailed to the trunk of the tree, or fastened around a suitable limb with non-corroding wire. In the case of herbaceous species, aluminium plant tags with numbers written indelibly can be tied loosely around the stem so as not to interfere with growth (Hutchings & Barkham, 1976). A soft graphite pencil on metal tags is sufficiently weatherproof. In the study quoted there was no evidence of plants suffering from aluminium poisoning. Plastic tags can also be obtained on which numbers can be written in waterproof ink.

While trees and shrubs are often spaced widely enough for identification to pose few problems—in some cases even without the use of labels—this is often not the case with herbaceous species. Here the problem may also arise of plants, ramets or modules emerging over an extended period of time during each growing season. Much demographic information may be contained in each cohort of population units, a cohort being defined as the set of units commencing growth within a certain time interval. Different cohorts can be tagged in different ways; favourite methods of plant demographers have been to use colour-coded plastic rings, plastic-coated wires, toothpicks or sections of drinking straws. However, identification of different plants or units within cohorts usually necessitates numbering of tags.

3.3.2 Photography

Law (1981) was able to identify individual plants of *Poa annua* from photographs of his study plots. This technique, though labour saving, can usually only be applied unambiguously for small areas where the vegetation is sparse and essentially single-layered. However, large-scale aerial photography has also been successfully used in woodland surveys, where individual trees can be identified at different dates (Pigott & Wilson, 1978).

3.3.3 Mapping

More sophisticated methods of recording population units involve use of the following.

(a) Pantographs (Sagar, 1959; Sarukhán & Harper, 1973; Hawthorn & Cavers, 1976), from which maps showing the positions and sizes of plants can be produced at a convenient scale. However, pantographs are bulky to transport to field sites, time-consuming to set up and use and only practical in short vegetation.

(b) Mapping tables and quadrats which fit onto fixed points in the ground marking the corners of the plots being studied (e.g. Waite, 1984) or some other reference point. At each recording date a transparent mapping table, usually made from a perspex sheet, is arranged at a height above the vegetation so as not to cause damage or interfere with its growth. A sheet of clear acetate or tracing paper is fixed onto the mapping table, and a map of the outlines of the plants in the plots is drawn in indelible ink. A cross-wire sighting device should be used to avoid errors caused by parallax when mapping the positions of plants, particularly when the mapping table is more than a few centimetres above ground level and when plant density is high, otherwise identification of closely-positioned plants may be ambiguous. A range of complexities of design can be devised to suit different conditions.

(c) Simple bar plotters from which rectangular coordinates can be read (e.g. Cullen *et al.*, 1978). This method is a direct development from that of recording plants by their positions along line transects (e.g. Antonovics, 1972).

(d) Automatic field digitizers (Mack & Pyke, 1979) which can feed data directly to a logging device such as a tape recorder. Such automation, while saving time and labour, is regrettably expensive.

The techniques described above are generally suitable for use on a small scale, i.e. where the quadrat size in use is no larger than about 1 m². For larger square or rectangular quadrats a useful technique is to attach measuring tapes to adjacent corner posts of the sampling plot and to measure the distances from the corners to each plant in turn. The measurements can be converted to rectangular coordinates within the plot using the cosine rule (Fig. 8.6). This technique has been successfully used to record the positions of orchids in a 20 × 20 m plot with enough accuracy to leave little doubt in interpreting the presence or absence of individual plants over a period of 11 years (M.J. Hutchings, unpublished). In species such as orchids, in which, because of growth habit, some alteration in the position of the emerging shoot can be expected each year, there can be no absolute position assigned to any plant. Despite this, by careful application of the method the coordinates of given plants emerging in different years differed in most cases by much less than 5 cm. This method is, however, unlikely to be satisfactory for accurate identification when plants are very densely packed.

The data obtained by these techniques can be compared from one census to the next. In the case of maps, visual comparisons of consecutive sets of records may be enough to determine whether new plants have established,

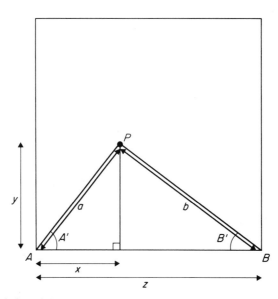

Figure 8.6 Calculation of the rectangular coordinates, x, y, of plant P from measurements a and b, taken respectively from corners A and B of a study plot.
$\cos A' = (z^2 + a^2 - b^2/2za)$; $x = a \cos A'$; $y = \sqrt{a^2 - x^2}$.

and whether previously-recorded plants have survived or died, in the time interval between recordings. Some subjectivity, using knowledge of the behaviour of the plants and their density is normally necessary during this process because it is unlikely that all records of a given plant will appear in exactly the same place on maps made at each census, or have exactly the same coordinates. This will be especially true when species which perennate below ground are being recorded (e.g. many bulbous species and orchids) and when the study plot has large dimensions, resulting in greater likelihood of error. In the case of computer-held files of coordinates, automatic searches for the same plants appearing at different census dates are unlikely to save labour unless *approximately* identical positions are sought, rather than identical coordinate pairs. However, the greater the approximation, the more cases of error in identification are likely to occur. Comparisons of certain aspects of the plant's performance at different censuses may help confirm or deny the presence of the same plant when its recorded position has changed slightly, e.g. does the plant's height increase from one census to the next?, if the plant is in flower at the second census, was it already in flower, or at least in bud, at the first?, etc. Another complication when attempting accurate re-identification with young plants is movement of germinated seedlings on the ground surface (Liddle *et al.*, 1982); this may be caused, particularly in open communities, by rain-splash, frost heaving and movement by soil fauna. A final problem in recording the survival of plants from a series of censuses is that certain species which perennate over winter do not reappear in some seasons but re-emerge in succeeding years (Wells, 1967, 1981; Tamm, 1972a; Epling & Lewis, 1952). In such cases the difficulty of matching records of the same plant made at different dates becomes greater as the length of the absence of the plant increases.

3.4 Measurements on growing plants

Where possible in plant population studies, it is desirable to obtain several sets of measurements on the same population without interfering with its behaviour, rather than collecting data on a single date using methods which involve destroying or mutilating the object of study. Unfortunately while it is true that most plants, unlike most animals 'stand still and wait to be counted' (Harper, 1977), the majority, particularly herbaceous species, are extremely sensitive to interference from external agencies. This can be seen in the response of increased respiration rates after even gentle handling (Evans, 1972); this increase can be considerable and prolonged. Repeated handling has an accumulative effect which will eventually be reflected in poorer growth; thus, any interference with growing plants must be kept to a minimum during the study.

3.4.1 Classification by age states

The measurements which can be made without inflicting unacceptable damage are mostly descriptive, and amount to classifying the state in which the plant is censused. Plants can be classified as being in a vegetative state, in flower, setting seed, senescing or multiplying vegetatively. In some cases, records of morphological features (e.g. number and size of leaves, flowers, stems or branches) can add information which can be useful in describing the population and modelling its behaviour (see below), and different morphological categories can be defined for different types of plants. The most detailed development in this area is the classification of plants by age states (Rabotnov, 1978; Gatsuk et al., 1980), or life-cycle stages (Sharitz & McCormick, 1973). Each age state can be characterized by a certain combination of quantitative and qualitative features. Thus, the plant is not characterized by its age, but by its state of development, which has been shown to be of greater usefulness for a number of demographic purposes, in a number of species (e.g. Fortanier, 1973; Kawano, 1975; Kawano & Nagai, 1975; Werner, 1975b; Barkham, 1980a; Solbrig et al., 1980; Newell et al., 1981). Individuals of a given species may achieve given states at very different ages as a result, particularly, of local microhabitat influences. Reversal of status of individual plants, for example, from generative (reproductive) in one year to vegetative in the next, or from senile to generative, is not uncommon in many species (Rabotnov, 1978). Gatsuk et al. (1980) claim two major advantages of a census of a plant population based on age states in comparison with calendar age based censuses. These are, firstly, that similar states, which will be characterized by quantitatively and qualitatively similar morphology, may be attained by individuals at very different calendar ages—age and developmental state are often very poorly correlated. Thus, at different ages, two plants in the same age state may be ecologically equivalent in terms of size, reproduction, competitive influence and resource-capturing ability. Conversely, two plants of the same age may be extremely different in terms of these attributes, and thus occupy very different age states in the population. Secondly, they point out that definition of calendar age is very difficult for many species of herbs and woody plants, whereas classification by age states is usually feasible. In modelling populations of the short-lived monocarpic perennial Dipsacus fullonum, Werner & Caswell (1977) produced far more satisfactory predictions of population behaviour using models based upon classification by developmental stage (various categories of seeds, plants of different rosette size, flowering plants) rather than by age; such findings may support the conclusions of Gatsuk et al. (1980). Nevertheless, Gatsuk et al. recommend, where possible, the simultaneous definition of both calendar ages and life states for plants in populations.

3.4.2 Quantitative measurements

With care, some simple quantitative measurements can also be made on living populations of many plants. The technique of dimensional analysis has been widely applied in studies of standing crop and primary production of woody plants, particularly in forests (e.g. Whittaker & Woodwell, 1968, 1971; Newbould, 1967). It involves establishing predictive regression relationships between certain easily measurable dimensions, such as height and the diameter of the trunk at a fixed height, and the weight of the tree or a part of the tree. Goodall (1945) was one of the first workers to apply these methods to follow the changes in dry weight of plant organs in herbaceous species, because they avoided the need for mutilation or destruction of the plant under study. Similar methods have been used to estimate various shoot weights for the woodland herb *Mercurialis perennis*. Monthly measurements of stem height and stem basal diameter were made on growing populations in the field, and used in various equations to predict fresh and dry weights, although there were many potential sources of error (Hutchings, 1975). Similar techniques can be devised to predict flowering and seed output from measurements of morphological size.

Leaf area can also be recorded in the field using the technique of optical planimetry; with the key measurements of plant weight, weight of plant parts and leaf area at different dates, even if these are only estimations, the investigator is in a position to undertake growth analysis (Evans, 1972; Hunt, 1978b), although it must be recognized that the potential for errors may be high.

3.5 Destructive measurements on plants

3.5.1 Historical reconstruction

Natural stands of vegetation also contain much historical detail which can be utilized in reconstructing a picture of the past status of the vegetation. By examining the live trees and fallen dead trees in forest plots, Henry & Swan (1974) and Oliver & Stephens (1977) were able to reconstruct the history and past structure of species populations and vegetation.

3.5.2 Ageing plants from morphological markers

The techniques involving direct examination of perennial plants to assess their ages (Roughton, 1962) are based on the ability to recognize morphological features laid down at fixed intervals of time—in most cases yearly intervals. In many trees and other woody plants, ageing can be achieved directly from counts of annual rings which are either exposed by cutting the trunk or a disc from the trunk (or branch), or by removing a core sample using a Pressler borer (Newbould, 1967, p. 15). These techniques, however, have pitfalls which must be avoided. Firstly, it is essential to count rings at a standard height, because the wood added to a tree each year is accumulated as a tall, narrow, hollow cone placed on top of the previous years' accumulated timber. As a result, the higher up the trunk the annual rings are counted, the fewer of them will be intercepted by the section through the trunk, or by the core borer. The usual solution is to standardize the resulting error by sampling at a standard height. Diameter at breast height, DBH, which is equivalent to 1·3 m above ground level, is a common measure in forestry investigations. Other problems include the occasional formation of more than one growth ring, or no rings, in a year, and the complexity of structure of the rings where branches and branch initials arise from the trunk, making interpretation difficult. Finally, the anatomical differences between spring wood and winter wood, which create the timber density differentials which can be seen as growth rings, vanish in tropical regions where climate lacks seasonality. Additional valuable information can be obtained by measuring the width of annual rings, which can be related to past historical events and climatic conditions (e.g. Henry & Swan, 1974; Fritts, 1976; Hughes *et al.*, 1982).

Many plants with herbaceous above-ground parts develop repeated subterranean markers or woody structures which offer accurate clues to an individual's age. Examples of studies where ageing of herbaceous plants has relied on such markers include those of Linkola (1935) in Finnish meadow communities, Wager (1938) in arctic fjaeldmark in Greenland, Pigott (1968) on *Cirsium acaulon* and Kerster (1968) on *Liatris aspera*. The lively debate (Werner, 1978; Levin & Kerster, 1978) on the accuracy of ageing *Liatris* corms from what appeared to be annual growth rings illustrates the fact that the meaning of morphological markers can confuse even experienced observers, and that errors are potentially devastating for accurate interpretation of population structure. Further use of anatomical structures to interpret age has been made for *Carex arenaria* (Noble *et al.*, 1979), *Carex bigelowii* (Callaghan, 1976) and the moss *Polytrichum alpestre* (Collins, 1976).

Estimation of time since germination of perennial herbs with extensive clonal growth presents a greater problem for the population ecologist. Many clonal species possess virtually unlimited life-spans, particularly in those cases where vegetative proliferation of ramets is followed by the decay of the link between parent and offspring (Harper, 1977; Noble *et al.*, 1979), resulting in proliferation over a wide area of many fragments with the same genetic origin. Two techniques have been used to age the genet in such cases. In the first, pioneered by Harberd (1961, 1962, 1963, 1967), samples of living plants were collected from positions established in the field on a grid basis, and subjected to detailed scoring for a range of morphological characteristics. Plants with the same characteristics which were thus the products of the same genetic parent could then be mapped, so that the spatial extent of the clone could be determined. The age of the clone could then be estimated, albeit with wide confidence intervals, from a knowledge of the probable rate of spread of the clone in the habitat. Recent documentation of clones with relatively phalanx and guerilla growth forms (Lovett Doust, 1981; clones with a phalanx growth form have slow radial expansion in comparison with clones with a guerilla growth form) within genet populations of a given species (Burdon, 1980; Bülow-Olsen *et al.*, 1984) increases the uncertainty associated with estimates of rates of radial expansion, and thus increases the confidence intervals associated with age estimations.

The second method, used by Oinonen, does not necessarily involve destruction of the plants. Again, it relies on being able to map the extent of clones from morphological markers. Oinonen (1967a,b, 1969) estimated ages of clones of *Pteridium aquilinum*, *Lycopodium complanatum*, *Convallaria majalis* and *Calamagrostis epigeios* from the relationship between clone diameter and the length of time since major disruptive events—in this case fires—in the habitat. The clones established immediately after fires. Thus, rates of expansion of such clones could be estimated and used to predict the age of further clones.

Use of morphological markers can also enable determination of the number or density of genets occupying a site. In some cases, the extent of clones can be mapped using visible markers, such as the presence in a single habitat of different leaf morphs of the same species. In the case of *Trifolium repens*, 3–6 genetic individuals persisted per dm^2, with none being permanently eliminated from an old pasture habitat, although all showed changing distribution patterns as time passed (Cahn & Harper, 1976; see also Burdon, 1980).

3.5.3 Electrophoresis

Genetic diversity in populations of plants can also be revealed by means of electrophoretic techniques (e.g. Hubby & Lewontin, 1966; Scandalios, 1969; Wilson & Hancock, 1978). Recent examples of the use of these methods in plant population studies are provided by Gottlieb (1975, 1977a,b).

4   Analysis

4.1   Within-population variation in demography

It was stated earlier that many plant population studies can be criticized on the grounds that the number of plants which have been studied is inadequate, that little attention has been paid to the heterogeneous nature of the habitat as experienced by individual plants in populations and that consequently the population parameters which have been calculated are often poorly defined, having wide confidence intervals (Law, 1981). The demographic properties of populations result from the integrated behaviour of their members; however, these members may have very different properties which are at least partly a result of the fact that each plant exists in a unique local microenvironment. Within-population variation in demographic properties contains information which is worthy of closer investigation. The spatial pattern of the individuals in a population—the size, arrangement (or angular dispersion; see Mack & Harper, 1977), proximity and identity of neighbours—and habitat heterogeneity, differences in relative time of germination and genetically-based differences in growth rate, can all affect performance and life history of the individual plants in a population. In this area the field of plant population biology closely approaches those of plant competition, population genetics and evolution. While there are no clear divisions between these subjects, and it is not the brief of this chapter to cover the material pertinent to them all, their interrelations should be borne in mind by the investigator. Some papers which discuss the importance of the variables listed above in connection with population biology include Pielou (1960, 1961, 1962), Mead (1966), Ross & Harper (1972), Yeaton & Cody (1976), Yeaton et al. (1977), Mack & Harper (1977), Antonovics & Levin (1980), Waller (1981), Liddle et al. (1982), Bülow-Olsen et al. (1984). Fitness, in terms of number of seeds or descendants produced, has often been shown to depend on size or other measures of plant performance which themselves depend strongly on the variables listed above.

Analyses such as those used in the publications listed above can reveal the forms of response exhibited by individual plants in response to their local microenvironments. In contrast, various ways can be suggested of subdividing populations into groups of plants with some feature in common, in order to enable analysis of the properties of the groups to be undertaken. For example, several authors have demonstrated differences in the demographic behaviour of cohorts of plants initiating growth at different times of year or at different stages during secondary succession (e.g. Law, 1981; Weaver & Cavers, 1979; Waite, 1984), and in different years (Antonovics, 1972). Plants of different ages, sizes and stages of development within populations have been shown to differ in a range of demographic properties (e.g. Werner & Caswell, 1977; Waite, 1984) and, in dioecious species, plants of different sexes may have different demographic properties (Meagher & Antonovics, 1982; Hutchings, 1983) and may segregate spatially within the habitats they occupy (Freeman et al., 1976).

4.2 Equation of population flux

The most basic demographic relationship links the number of plants (or ramets or modules) present at time $t + 1$ to the number which were present at some earlier time, t, via the number of 'births' or recruitment events (B), deaths (D), immigrations (I) and emigrations (E) which have taken place in the interval between t and $t + 1$. The form of this relationship is

$$N_{t+1} = N_t + B - D + I - E.$$

This equation can be used to summarize the flux in a population at any level of organization, be it genets, ramets or modules, where N represents the number of units present at the chosen level of organization. In the majority of cases, particularly when seeds are not being considered, the immigration and emigration terms in the equation may be small enough to be ignored. However, when clonally-spreading organisms are being studied, immigration and emigration of ramets at the edges of study sites may be an important component of overall population flux.

4.2.1 Analysis of survival

Despite the recruitment of large numbers of seedlings in many plant communities, the appearance of much vegetation only alters at a very slow rate (Darwin, 1859). Thus, although the flux of individual plants into and out of populations may be substantial, the numbers of plants present at different censuses may remain relatively constant over considerable periods of time (Fig. 8.7). This is because the majority of seedlings only survive for very short periods, and thus have little impact upon population processes.

Two closely-related methods are available for analysing the survival of plants in populations. Both involve plotting the logarithm of the number or proportion of plants (or ramets, or modules such as leaves) surviving against an arithmetic time axis (Figs 8.8, 8.9). In the first approach, the survival of a cohort of plants, i.e. a uniformly-aged set of plants, is plotted to produce a survivorship curve (Deevey, 1947; Harper, 1977) which illustrates the magnitude of the age-specific mortality risks experienced by the cohort (Fig. 8.8); the steeper the gradient of the survivorship curve, the greater the mortality risk suffered by the cohort. Using the second approach, which produces a depletion curve (Harper, 1977), the survival of all the plants present on a given census date is plotted through time. In this case the integrated mortality risk suffered by a multi-aged population of plants is illustrated. Deevey (1947) classified survivorship curves in broad terms on the basis of the magnitude and timing of mortality risks during life (Fig. 8.9).

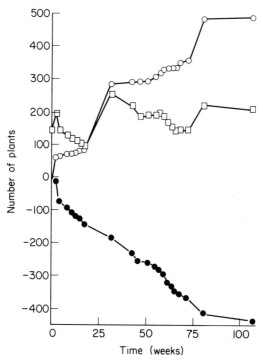

Figure 8.7 Changes in size of a population of *Ranunculus repens*: (○) cumulative gains; (●) cumulative losses; (□) net population size (from Sarukhán & Harper, 1973).

Note that interpretations of mortality risks based on this classification can not be applied in the same way to cohorts and to the plants for which depletion curves are constructed, since depletion curves illustrate the behaviour of plants with a wide range of ages. Depletion curves will thus not reflect the age-specific mortality risks these plants suffer, although they will reflect the intensity of time-based mortality risks to which the whole population is being subjected. Survivorship curves were classified as three major types.

(a) Type I: the plants have a low mortality risk early in life and a heavy mortality risk when maximum life span is approached.

(b) Type II: there is a constant, age-independent mortality risk through-out the life of the cohort, shown by a linear survivorship graph. Such a graph illustrates a constant proportion of the surviving population dying in each time interval; decline in *numbers* of survivors is exponential.

(c) Type III: young plants suffer a heavy mortality risk. The mortality risk declines with age, and therefore a few individuals will normally achieve long life-spans.

Type I survivorship curves have been observed in cohorts of several annual species which die after a single flowering episode (Mack, 1976; Watkinson, 1981), in cohorts of leaves of *Linum usitatissimum* (Bazzaz &

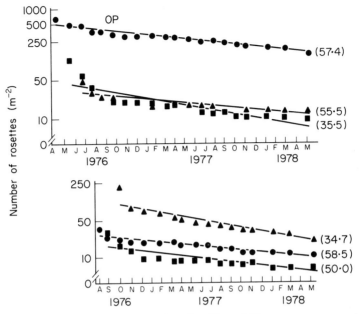

Figure 8.8 Survivorship curves for five cohorts of seedlings of *Plantago coronopus*, and depletion curves for the original population (OP) at Seaford Cliffs, Sussex. Half-life values, in weeks, are given in brackets. All fitted regressions are significant at $P < 0.001$. (From Waite, 1984.)

Harper, 1977; Harper & Bell, 1979) and in some cohorts of perennial range-land grasses (e.g. data of Canfield (1957) re-analysed by Sarukhán & Harper, 1973). In the case of Canfield's data, censuses to record survivors were made at widely-spaced intervals, and the survivorship curves may not have been

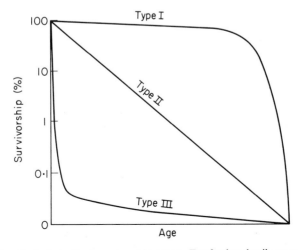

Figure 8.9 Hypothetical standard survivorship curves. For further detail, see p. 408.

adequately defined, particularly over the first year of life, when there is reason to believe that in some of the species he studied mortality risks were far greater than in subsequent seasons.

A large number of investigations have revealed Type II survivorship or depletion curves in plant populations. These include not only studies of cohorts and mixed age populations of a wide range of species, mostly herbaceous perennials (Sagar, 1959; Harper, 1967; Antonovics, 1972; Tamm, 1972a,b; Sarukhán & Harper, 1973; Waite, 1984), but also investigations into the decrease in numbers of viable seeds in soil seed banks (Roberts, 1962; Roberts & Dawkins, 1967). The decline in numbers of survivors in a population exhibiting a Type II curve can be fitted by an equation of the form

$$Y_t = Y_0 e^{-b'x}$$

or, transforming to logarithms,

$$\log_e Y_t = \log_e Y_0 - b'x$$

where Y_t is the number of survivors at time t, Y_0 is the initial population size, b' is a constant instantaneous rate of mortality and x is the age of the population. In populations exhibiting such a decline a useful statistic for comparative purposes is the half-life, which is the length of time which passes before half of the population has died. This is analogous to the half-life of a radioisotope. Before calculating half-life values it is necessary to establish that the linear regression model provides a good fit to the logarithmically transformed data on survival. If it does, the time required for half of the population to die can be calculated by substitution in the regression equation.

In spite of the predominantly linear nature of many survivorship and depletion curves, close inspection may reveal regular seasonal changes in the mortality risk. These can only be observed when censuses have been made at regular intervals. Normally, a higher mortality risk is observed during the time of most active growth in spring and early summer, and mortality is lower during winter (Sagar, 1959; Harper & White, 1971; Tamm, 1972a,b; Sarukhán & Harper, 1973; Hawthorn & Cavers, 1976). In addition, survivorship curves may differ in slope and therefore, half-life for cohorts establishing at different dates during the growing season, at different times after a disturbance and in different years (Antonovics, 1972; Weaver & Cavers, 1979; Law, 1981; Waite, 1984). Different rates of decay may result from different degrees of adaptedness of the genotypes in the cohort to environmental conditions pertaining at the specific time when they were initiated (Waite, 1980).

In addition to seasonal departures from a constant death risk, repeated closely-spaced censuses conducted soon after cohorts are initiated may reveal a high mortality rate for very young seedlings compared with their subsequent mortality rate. As already stated, many seedlings entering populations have extremely short lives which widely-spaced censuses would fail to record. While fungal attack and herbivory may account for many of these losses, there may be a period of loss of poorly-adapted genotypes from the population. Additionally, many seeds are likely to germinate in sites which subsequently prove unfavourable for seedling survival. Certainly, the spatial distribution patterns of newly-germinated seedlings often differ markedly from those of the plants which survive to maturity. Thus, early in life, cohorts may display mortality patterns characteristic of Type III survivorship curves.

The best documentation of Type III survivorship curves comes from work on populations of tree species (Hett & Loucks, 1971, 1976; van Valen, 1975; Sarukhán, 1980). The pattern of decreasing age-specific mortality risks in such species can often be described by fitting a power function equation to the data:

$$Y_t = Y_0 x^{-b''}$$

or

$$\log_e Y_t = \log_e Y_0 - b'' \log_e x.$$

Symbols are as for the exponential decay model, but b'' represents a mortality rate which falls with age (Hett, 1971).

Many species fail to conform to any of these three classes of survivorship curves, while others show behaviour more complex than can be described using one of the types alone, e.g. *Tragopogon orientalis* quoted in Harper & White (1971), ramets of *Ranunculus repens* (Sarukhán & Harper, 1973) some populations of *Primula veris* and orchid species (Tamm, 1972a,b).

Survivorship curves can be used to measure the cost to plants, in terms of survival, of engaging in certain activities. For example, in most perennial species, reproduction results in a reduction in post-reproduction survival, sometimes caused by a reduction in growth and competitive ability of reproducing plants relative to non-reproducing plants. Separate analysis of the survival of reproducing and non-reproducing plants can enable quantification of the risk attached to reproduction to be undertaken. Comparison of the mortality risk suffered by sexual progeny and vegetative propagules may also be instructive (Sarukhán & Harper, 1973).

Finally, it is worth recalling that, as Harper (1977, p. 528) has pointed out, valid comparisons of genet survivorship curves for different species must commence at the same developmental point. He reminds us that, whereas the survivorship of zygotes is of the greatest interest to geneticists and evolutionists, ecologists have usually recorded survivorship only from the time that seeds germinate, thus omitting the period from zygote formation to seed germination. This period is admittedly difficult to study, but, as pointed out in an earlier section of this chapter, mortality risks during this time may be considerable, and their study and quantification therefore have a place in any complete demographic study.

4.2.2   Analysis of population flux

Survivorship and depletion curves are used to analyse the rate of loss of plants in populations. Demography is also concerned with additions to populations. Analysis of the flux of plants in regularly-censused populations can be presented using a tabular format and diagrams similar to those presented by Sarukhán & Harper (1973), Noble *et al.* (1979) and Waite (1984) (see Table 8.1, Figs 8.7 and 8.10). These studies are notable for allowing comparisons to be made either of the fluxes exhibited by closely-related species, or by populations of one species growing under a range of conditions. Fluxes are normally calculated on the basis of behaviour of the population over a period at least as long as one year, but within-season analyses of population birth and death rates have also been calculated for *Ranunculus* spp. (Sarukhán & Harper, 1973), *Carex arenaria* (Noble *et al.*, 1979) and *Plantago coronopus* (Waite, 1984) (Fig. 8.10).

An interesting extension of the population flux analyses referred to above is the use of the data to produce a scale flow diagram of the life cycle of the species under study, relating census data on the number of plants present to

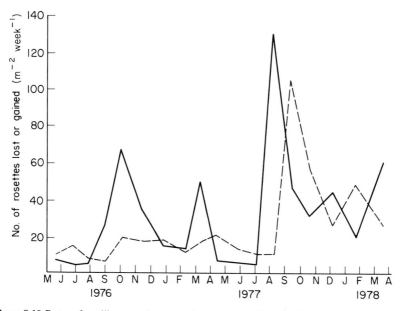

Figure 8.10 Rates of seedling recruitment and rosette mortality of *Plantago coronopus* at Seaford Cliffs, Sussex: (——) rate of seedling recruitment, plants m^{-2} week^{-1}; (----) rate of rosette mortality, plants m^{-2} week^{-1} (from Waite, 1984).

Table 8.1 Analysis of population flux for *Plantago coronopus* at Seaford. Values refer to the number of rosettes recorded per m² (From Waite, 1984)

	May 1976 (t) May 1977 (t + 1)	May 1977 (t) May 1978 (t + 1)	May 1976 (t) May 1978 (t + n)
(a) Number of rosettes present in May, year t	394	759	394
(b) Number of rosettes present in May, year t + n	759	1103	1103
(c) Number of rosettes present in May, year t, surviving until May, year t + n	181	324	88
Net change in rosette density per m² (b − a)	+365	+344	+709
Percentage net change in rosette density (b − a)/a × 100	+92.6%	+45.3%	+179.9%
(d) Number of seedlings recruited between May in year t and May in year t + n	1147	2235	3382
(e) Number of vegetatively-derived rosettes recruited between May in year t and May in year t + n	19	21	40
(f) Total number of rosettes recruited (d + e)	1166	2256	3422
Percentage of recruited rosettes which arose vegetatively (e/f) × 100	1.6%	0.9%	1.2%
(g) Number of seedling mortalities recorded between May in year t and May in year t + n	585	1477	2062
(h) Number of vegetatively-produced rosette mortalities recorded between May in year t and May in year t + n	3	13	16
(i) Total number of rosettes lost	801	1912	2713
Percentage survival of seedling-recruited rosettes (d − g/d) × 100	49.0%	34.0%	39.0%
Percentage survival of vegetatively-recruited rosettes (e − h/e) × 100	84.2%	38.1%	60.0%
Percentage survival of rosettes present in May in year t. (= j) j = (c/a) × 100	45.9%	42.7%	22.3%
Expected time for the complete turnover of the population present in May in year t = (n/100 − j) × 100, where n = time in years	1.85 years	1.73 years	2.57 years
(k) Total number of rosettes recorded	1560	3015	4575
Percentage mortality of all recorded rosettes (i/k) × 100	51.3%	63.4%	59.3%

the time of year at which the data were collected. This has been carried out by Sarukhán (1971) for the three species of *Ranunculus* which he studied (see Harper, 1977, pp. 582–4) (Fig. 8.11). Such diagrams can become very complicated. They can be used to present information both on the seed phase and the growing plant population: they also provide a visual indication of the periods during the life of the species, and the times of year, when mortality and different modes of reproduction are occurring, and show when dormancy, germination and seed death are most important. Using a different diagrammatic format, Barkham (1980a) has also constructed flow diagrams for the bulbous perennial *Narcissus pseudonarcissus*, illustrating the yearly losses and gains to and from each of three age-state categories (juveniles, sub-adults, adults) of plants in natural populations.

4.3   Analysis of population structure

The failure of analysing population flux to reveal differences in behaviour of different categories of plants can be partially overcome by subjecting the population to analysis of its structure, based on plant ages, sizes, states, etc. Usually, such analysis involves the construction of frequency distribution histograms. One of the most obvious manifestations of the differences between individuals in a population is in their sizes (although size differences are far less apparent within populations of ramets or modules). Even in widely-spaced populations of plants where competition may be of

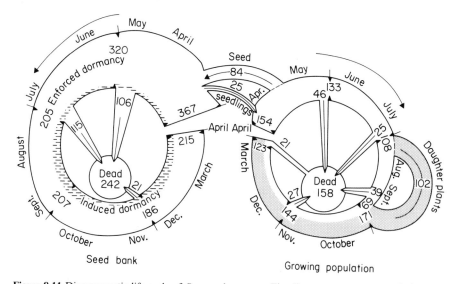

Figure 8.11 Diagrammatic life-cycle of *Ranunculus repens*. The diagram represents population fluxes within average quadrats of 1 m² of permanent grassland in a site in North Wales. (From Harper, 1977, after Sarukhán, 1971.)

marginal importance in restricting growth, a range of weights usually develops, referred to as a size hierarchy (White & Harper, 1970). Expressed as a histogram, the weight distribution, which is often symmetrical or normal for seeds and seedlings, becomes positively skewed, the degree of skewness increasing through time and with competition between plants (Obeid *et al.*, 1967). Some authors have proposed that there is an upper limit to the level of skewness reached when the weight frequency distribution becomes lognormal (i.e. if all weights were logarithmically transformed the resulting values would be normally distributed), e.g. Koyama & Kira (1956), and that at this point the tendency for the population to develop further skewness by differential growth of large and small plants is offset by the progressive death of small plants (White & Harper, 1970). Other authors have argued that a bimodal weight frequency histogram is a good indication that competition has caused the hierarchy of plant weights (Ford, 1975). Ford & Newbould (1970) provide some rules for the construction of frequency histograms, and the calculation of skewness has been discussed by Hutchings (1975). Frequency histograms can be constructed for other quantifiable aspects of plant performance, and also to reveal age and state structures of natural populations (e.g. Kerster, 1968; van Andel, 1975). With the exception of seed weight in many species, frequency histograms illustrate the fact that measures of plant performance are rarely symmetrically, let alone normally, distributed around the mean value, and thus parametric statistical tests are often inappropriate for making comparisons of performance between populations. Additionally, mean values are often poor summary statistics to provide in describing plant populations (see Chapter 7).

4.4 Life-tables and fecundity schedules

The primary aim of population biology is to achieve an understanding of the numbers and changes in numbers in populations from place to place and from time to time (Harper, 1977). A more ambitious task than analysing flux in populations is to produce a life-table which relates mortality risks and fecundity to discrete categories of plants in a population. The basis for these categories could be age, size or state. Although a number of such tables have been produced for animals, life-tables for plants are rare, and to date virtually restricted to annual species with non-overlapping generations, one of the least complicated types of life-cycle.

4.4.1 Diagrammatic life-tables

While demographic data for actuarial purposes are normally presented using a more or less standard format, plant ecologists have been more flexible in their interpretation of the term life-table, producing both conventional tabular forms (Sharitz & McCormick, 1973; Leverich & Levin, 1979), and a diagrammatic form (Sagar & Mortimer, 1976; Barkham, 1980a). The more recent diagrammatic format was introduced by Sagar & Mortimer (1976), and its construction is fully described by Begon & Mortimer (1981). These authors have used variants of their diagrammatic life-table to illustrate demographic processes in a number of plant species with a range of types of life cycle (Fig. 8.12). The format can be expanded to accommodate complex life cycles of perennial species with overlapping generations (Sagar & Mortimer, 1976; Barkham, 1980a), and the fates of seeds between their production and germination (Waite, 1980), although unless such studies are very carefully planned, many of the numbers at different steps during the seed phase, and the values of many interphases from one step to another may be either speculative or unknown.

4.4.2 Tabular life-tables

A simple introduction to the construction of conventional, tabular life-tables is provided by Begon & Mortimer (1981). In particular they make clear the difference between cohort (or 'dynamic' or 'horizontal') life-tables and static (or 'time-specific' or 'vertical') life-tables. The former follows the fate of a cohort from its recruitment until the death of its last member, whereas the latter is constructed from observations of the population age-structure at, usually, a single point in time (occasionally two). Although the demographic behaviour of two cohorts in the same population may differ (and thus a cohort life-table is strictly accurate only for the cohort which has been recorded), the cohort life-table has the advantage of being based on observations made on a number of different occasions. Construction of cohort life-tables is the only way in which life-tables can be produced for species with non-overlapping generations. In long-lived species it may prove impossible to follow a cohort from recruitment to its ultimate loss from the population and in such cases, preparation of a static life-table will be the only way of carrying out a full demographic analysis. Static life-tables can generate considerable errors as a result of their failure to allow for age-specific changes in birth and death rates, and year-to-year changes in these values. The observed age structure of the population must be assumed to be the product of stable birth and death rates, an assumption which is often unjustified.

However, despite the practical difficulties which may be associated with collection of data suitable for life-table compilation, the calculation of life-tables presents few problems. It should be remembered that values in a life-table may differ for populations of the same species in different habitats and in different seasons.

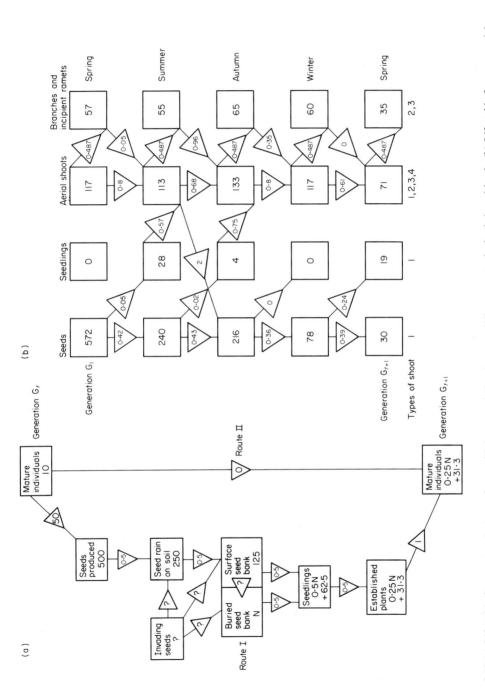

Figure 8.12 (a) Life-table for an annual, *Avena fatua*, with low population increase. '*N*' represents the buried seed bank. (b) Life-table for a perennial, *Ranunculus repens*. (Both from Sagar & Mortimer, 1976.)

Table 8.2. Life-table for *Phlox drummondii* at Nixon, Texas (Adapted from Leverich & Levin, 1979)

Age interval (days) $x - x'$	Length of interval (days) D_x	Number surviving to day x, $= N_x$	Survivorship l_x	Number dying during interval d_x	Average mortality rate per day q_x	Mean expectation of life (days) E_x	Killing power during interval k_x
0–63	63	996	1·0000	328	0·0052	122·87	0·1735
63–124	61	668	0·6707	373	0·0092	104·73	0·3550
124–184	60	295	0·2962	105	0·0059	137·59	0·1911
184–215	31	190	0·1908	14	0·0024	137·05	0·0322
215–231	16	176	0·1767	2	0·0007	115·72	0·0050
231–247	16	174	0·1747	1	0·0004	100·96	0·0025
247–264	17	173	0·1737	1	0·0003	85·49	0·0025
264–271	7	172	0·1727	2	0·0017	68·94	0·0051
271–278	7	170	0·1707	3	0·0025	62·71	0·0077
278–285	7	167	0·1677	2	0·0017	56·78	0·0052
285–292	7	165	0·1657	6	0·0052	50·42	0·0161
292–299	7	159	0·1596	1	0·0009	45·19	0·0027
299–306	7	158	0·1586	4	0·0036	38·46	0·0111
306–313	7	154	0·1546	3	0·0028	32·36	0·0085
313–320	7	151	0·1516	4	0·0038	25·94	0·0117
320–327	7	147	0·1476	11	0·0107	19·55	0·0338
327–334	7	136	0·1365	31	0·0325	13·85	0·1123
334–341	7	105	0·1054	31	0·0422	9·90	0·1520
341–348	7	74	0·0743	52	0·1004	5·58	0·5268
348–355	7	22	0·0221	22	0·1428	3·50	
355–362	7	0	0·0000				

Table 8.3. Glossary of symbols used in the life-table and fecundity schedule. (Adapted from Leverich & Levin, 1979)

x	Age in days. (This designates the first day of an interval.)
x'	The first day of the interval following the interval of x.
D_x	Length in days of the interval beginning with day x.
N_x	Number of individuals surviving to day x.
l_x	Survivorship: the probability that an individual of age zero will survive to day x.
d_x	Number of individuals dying during the interval beginning with day x.
q_x	Average mortality rate per day during the interval beginning with day x.
E_x	Mean expectation of life on day x, in days.
k_x	Killing power, or rate of mortality during the interval beginning with day x.
B_x	Total number of progeny produced during the interval beginning with day x. Superscript indicates the basis of measurement (B_x^{seed}).
b_x	Average number of progeny per individual during the interval beginning with day x. Superscript indicates the basis of measurement (b_x^{seed}).
$l_x b_x$	Contribution to R_0 during the interval beginning with day x.
R_0	Net reproductive rate. $R_0 = \Sigma l_x b_x$.

The cohort life-table produced by Leverich & Levin (1979) for the annual herb *Phlox drummondii* is the most detailed life-table produced to date for a plant species. It consists of two parts (Tables 8.2, 8.4); in the first (Table 8.2), the mortality risks suffered at different ages by the members of a cohort are documented, and in the second (Table 8.4), the fecundity schedule—a summary of the age-specific ability to produce seeds—is presented. In view of the completeness of the analysis and its clear presentation, this example is used to illustrate the steps involved in calculating the life table and fecundity schedule. The definitions of the symbols used are shown in Table 8.3. (A good illustration of the construction of a static life-table—though not one for a

Table 8.4. Fecundity schedule for *Phlox drummondii* at Nixon, Texas, based on seed production (Adapted from Leverich & Levin, 1979)

$x - x'$	B_x^{seed}	N_x	b_x^{seed}	l_x	$l_x b_x$
0–299	0·000	996	0·0000	1·0000	0·0000
299–306	52·954	158	0·3352	0·1586	0·0532
306–313	122·630	154	0·7963	0·1546	0·1231
313–320	362·317	151	2·3995	0·1516	0·3638
320–327	457·077	147	3·1094	0·1476	0·4589
327–334	345·594	136	2·5411	0·1365	0·3470
334–341	331·659	105	3·1587	0·1054	0·3330
341–348	641·023	74	8·6625	0·0743	0·6436
348–355	94·760	22	4·3073	0·0221	0·0951
355–362	0·000	0	0·0000	0·0000	0·0000
				$R_0 = \Sigma l_x b_x =$	2.4177

plant species—is provided by Begon & Mortimer (1981, pp. 12–14), and they also present a cohort life-table for *Poa annua*, adapted from Law, 1975.)

Phlox drummondii is strictly annual, seeds germinate over a short period and virtually no seeds survive beyond one year. Thus, seeds produced in one season rarely contribute to any generation except the next, and consecutive generations of plants are non-overlapping. Moreover, each generation can be treated as a cohort since all germination takes place over a short period of time in late November and early December.

Leverich & Levin recorded numbers of seeds in the soil prior to germination at irregular and widely-spaced intervals. The intervals, $x - x'$, are shown in the first column of the life-table, and the length of each interval, D_x, is calculated in the second column. Both the density of the seed population in the soil and of the germinated seedlings were recorded from equivalent areas of ground, so that direct comparisons of both phases of the population could be made. The third column of the life-table shows the number, N_x, surviving to the start, on day x, of each recording interval. It can be seen that for this species, numbers fall rapidly during the seed phase, (first three figures in column 3); this was apparently due to predation rather than to loss of viability.

Altogether, 190 growing plants were recorded (fourth figure in column 3 of the life-table); the positions of each were determined to allow their re-identification at subsequent censuses, most of which were at 7-day intervals. Survivorship, l_x, is calculated in column 4. Conventionally, it is the probability of an individual of age zero (in practice those recorded at the very first census) surviving until the start of each recording period. If a graph of survivorship is plotted through time, it can be seen that for *P. drummondii*, survivorship falls far more rapidly during the seed phase than during the phase of establishment of seedlings (Fig. 8.13). This may appear surprising, given the usual high death rate of seedlings. However, censuses during this period may have been too irregular (the intervals between censuses at this time were over 2 weeks) to allow recording of many short-lived seedlings, and thus the mortality risks during this period may not have been adequately estimated.

The number dying during each time interval, d_x, is determined (column 5) by subtracting the number still surviving at each time, x', from the number surviving at each preceding time x. The age-specific mortality rate, q_x, for the population at different dates is calculated by dividing the number dying during the time interval in question, d_x, by the number N_x, surviving to the start of the time interval (i.e. dividing column 5 by column 3; see also Sarukhán & Harper, 1973). In the table for *Phlox* these values have been standardized to a daily rate, by dividing by the length, in days, of the time interval in question, to produce the values tabulated in column 6.

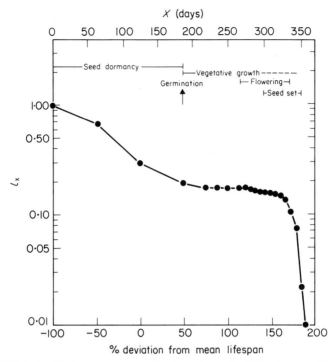

Figure 8.13 Relationship between survivorship, l_x, and time for *Phlox drummondii* at Nixon, Texas, in 1974–75. Age is indicated both in days and as percentage deviation from mean life span (= 122·87 days).

The final column in the life-table presented by Leverich & Levin is a calculation of the mean expectation of life, E_x, in days. Consider calculation of E_x for the 136 plants continuing to survive after 327 days. Of these, 31 died in the 7-day interval between 327 and 334 days; their mean life expectation was thus 3·5 days. Of the 105 still surviving, 31 died between 334 and 341 days. Their mean life expectation after 327 days was therefore 10·5 days. Of the 74 still surviving, 52 died between 341 and 348 days. Thus, after 327 days their mean life expectation was 17·5 days. The final 22 plants died between 348 and 355 days, and thus, after 327 days their mean expectation of life was 24·5 days. For the whole population of 136 plants still surviving after 327 days therefore, the overall average expectation of life was $(31 \times 3·5 + 31 \times 10·5 + 52 \times 17·5 + 22 \times 24·5)/136 = 13·85$ days, as tabulated in column 7. Similar calculations will reveal the mean expectation of life of the surviving population on all census dates.

In addition to the life-table values calculated by Leverich & Levin it is possible to calculate values, k_x, which reflect the 'killing-power', or rate of mortality during any period in the life of the species (Varley & Gradwell, 1970; Begon & Mortimer, 1981). k_x-values can thus be summed to measure the killing-power between any pair of census dates. k_x-values are calculated for each time interval between censuses as $\log_{10} N_x - \log_{10} N_{x+1}$ (see column 8). Calculation of killing-power values at different stages during a species' life cycle is the basis for k-factor analysis, a technique which has been extensively used in analysis of the population dynamics of animals, particularly insects (Varley & Gradwell, 1960; Varley et al., 1973). There are two major reasons why this technique has not been widely used to analyse the population dynamics of plants; (i) it is difficult to apply when successive generations overlap, and (ii) it requires data collected over several years, a luxury rarely available to the plant demographer. However, Silvertown (1982, pp. 113–116) has described its use in analysing the population dynamics of the sand dune annual *Androsace septentrionalis*, both to determine the key factor contributing to generation mortality and to reveal the stages during the life cycle at which mortality factors, as measured by the values of k_x, regulate the population in a density-dependent manner.

Density-dependent regulation of plant populations can also be investigated either by observation of natural populations (e.g. Lovett Doust, 1981; Waite, 1984) or by recording the response of populations to manipulation of their density (Watkinson & Harper, 1978; Keddy, 1981). Caution is necessary in interpreting results from such studies, however, since the intensity of density-dependent regulation strongly depends on the intensity of interactions between neighbouring plants. This itself depends on the species of plant involved, their sizes and the distances between them (Mack & Harper, 1977). Thus, although density-dependent regulation may be apparent at the level of the whole population, such regulation acts with differential intensity on a local scale. Therefore, the use of mean density when investigating population regulation mechanisms only provides a crude measure of the intensity with which such mechanisms may act (Law, 1981; Waite, 1984). Moreover, statistical evidence of density-dependent regulation does not necessarily prove causality (Royama, 1977).

Fecundity schedules

Table 8.4 presents the fecundity schedule for *P. drummondii* calculated from its seed production. It can be seen that no seeds were produced before the 299th day of recording the population. However, between days 299 and 306, and between all subsequent census dates apart from the last pair, seeds were produced. In column 2 of the fecundity schedule, B_x, the total number of progeny produced during each successive interval starting with day x, is calculated. In the study this was achieved by counting the capsules matured within the time interval in question, and multiplying by the average number of seeds per capsule (= 2·787). Thus, in the first time interval, 19 capsules were matured, resulting in an output of roughly 53 seeds. The fourth column, b_x, tabulates the mean seed production within each time interval by the plants surviving at the start of each time interval. The final column measures the contribution within each time interval to the net reproductive rate, R_0. This is calculated by multiplying the terms l_x and b_x for each time interval. The net reproductive rate (i.e. the average number of offspring produced per individual alive at day 0 of the study) is equal to $\Sigma l_x b_x$ (Table 8.4).

In the study of *Phlox drummondii* schedules of births and deaths are based upon the age of plants in the population. The life-tables constructed by Sharitz & McCormick (1973) for the winter annuals *Sedum smallii* and *Minuartia uniflora* are based on life-cycle stages, each of which is a period during the life cycle when some significant advance in population development takes place. These life-cycle stages are related to plant age. However, age may be an unsatisfactory basis for calculation of life-tables, since, as already stated, plants of a given age may have grown to very different sizes and stages of development. Life-tables calculated on the basis of plant age as described above provide only mean measures of the performance of the whole population, and indeed Leverich & Levin (1979) illustrate this by re-calculating several life-table parameters separately for the large and small plants in their populations; the life-tables and fecundity schedules for these different classes of plants are naturally very different. More detailed analysis is necessary to document such within-population variation and to enable us to fully assess its evolutionary significance.

4.5 Modelling population behaviour

4.5.1 Transition matrix models

While the results obtained in demographic studies are of intrinsic interest to ecologists, evolutionists and geneticists, one of the major additional uses to which they may be applied is the modelling of populations to enable predictions of their future size and behaviour to be made. The most common modelling technique used in predicting the behaviour of plant populations is the Leslie matrix (Leslie, 1945) which, by commencing with populations subdivided into different classes on the basis of age, and incorporating the probabilities of transition from one age-class to another and the fecundities specific to each age, enables the future population size and age composition to be predicted through discrete time steps. This is not the place to describe the technique in detail; clear and simple introductions have been written by Williamson (1967) and Begon & Mortimer (1981), and several more advanced texts are available (e.g. Searle, 1966; Usher, 1972). Many examples of use of the technique in plant population biology have been published (e.g. Sarukhán

& Gadgil, 1974; Callaghan, 1976; Werner & Caswell, 1977; Maillette, 1982b; Fetcher & Shaver, 1983).

Begon & Mortimer (1981) use results of Law (1975) for *Poa annua* to illustrate the use of Leslie matrix models. They demonstrate that, when the transition probabilities fed into the matrix remain constant at each iteration, the projected population age distribution becomes stable as the population's size increases. They also offer an insight into the potential for sophistication of the model, firstly by incorporating seasonal changes in fecundity and survival and secondly by subjecting birth and death rates to density-dependent control. Such alterations, if they realistically model the behaviour of populations in the field, may control the tendency for the modelled population to increase in size indefinitely.

The value of age-based classifications for plant populations is, as already stated, limited in many species, and in such cases the use of size or stage of development for classification may be more informative (see Werner & Caswell, 1977). Hubbell & Werner (1979) and Law (1983) have drawn attention to the fact that whereas the Leslie matrix is an appropriate model for populations classified by age, use of the stage projection matrix of Lefkovitch (1965) is more appropriate when populations have been classified by size or state (although the Leslie matrix model has been used in such cases). This method has the drawback, however, of necessitating the assumption that the population has a stable age distribution before the numbers in each size-class can be predicted, an assumption which is often unjustified (Vandermeer, 1975). Hubbell & Werner (1979) and Law (1983) have provided additional methods of population projection designed to overcome this problem.

4.5.2   Population models with a spatial component

A different approach to modelling plant populations has been pioneered by Barkham & Hance (1982). They modelled the effects of changing the probabilities of mortality, clonal growth and reproduction from seed on the development of populations of *Narcissus pseudonarcissus*. The probabilities of mortality, clonal growth and seed reproduction used in the model were based on field measurements of these variables. The model simulated the behaviour of the population in a square containing a large number of small units of space, within each of which a single plant or vegetative propagule could establish and grow. The whole modelled area was considered large enough to allow for the occurrence of all of the processes relevant to the demography of *N. pseudonarcissus*. A starting population of 100 adult plants was distributed randomly within the modelled area, and, within each year of the simulation each plant was allowed to function in one of the eight ways in which it might behave under field conditions; the manner of functioning of each plant was selected at random. (The eight ways of functioning were (i) to continue to survive without reproducing, (ii) to multiply vegetatively, (iii) to multiply by seed, (iv) to multiply vegetatively *and* (v) to reproduce by seed, (vi) to multiply vegetatively *or* reproduce sexually and die, (vii) to multiply vegetatively *and* reproduce sexually and die, (viii) to die without reproducing. The possible ways of functioning could be adapted to suit other species.) Realistic constraints were built into the model for the positions in which vegetative propagules and seedlings would establish in the modelled area relative to prior-established plants, and for the chances of either successfully establishing when both entered the same unit of space in the same year.

The value of this model was that in addition to enabling long-term predictions of population behaviour and of the impact of management changes on this behaviour to be made, much information about the spatial organization of the population, and its genetic and clonal composition could be obtained. While this type of modelling may necessitate development of new computer programs for different species (whereas matrix modelling involves adaptation of an existing framework), acquisition of the information pertaining to spatial and genetical structure in populations requires this type of approach.

5   Conclusion

Now that plant population biology has become a fashionable field of research, it is perhaps timely to warn of the danger of the subject becoming an area in which documentation is the ultimate aim, rather than problem-solving and synthesis. Many plant population studies have provided infor-mation which has so far not been compared with what is known about closely-related species, species of the same growth form or life cycle, or with what is known about the same species growing in different conditions. Nevertheless, in the majority of cases it is not the intention of the ecologist to study a given species solely for its intrinsic interest; a species will very often be chosen for study because it is the most suitable tool to use to answer a specific question. The results obtained from investigations conducted on one species should be integrated, as early as possible, into broader statements and hypotheses about the expected behaviour of certain categories of plants, or of plants under certain growing conditions. This has been done, for example, by Watkinson (1981) for winter annuals, by Hutchings (1979) and Noble *et al.* (1979) for clonal perennial herbs and by White (1980) for a range of species. Another approach is to collect sound data on which to build models which can be used to predict the behaviour of plant populations when conditions are changed (e.g. Barkham & Hance, 1982). Invariably, the initial formulation of clear questions to be answered (see Jeffers, 1978, 1979, 1980) is the source from which all such synthesis will spring.

6   References

ANGEVINE M.W. and CHABOT B.F. (1979) Seed germination syndromes in higher plants. In *Topics in Plant Population Biology* (eds O.T. Solbrig, S. Jain, G.B. Johnson and P. H. Raven), pp. 188–206, Macmillan, New York.

ANTONOVICS J. (1972) Population dynamics of the grass *Anthoxanthum odoratum* on a zinc mine. *J. Ecol.* **60**, 351–365.

ANTONOVICS J. and LEVIN D.A. (1980) The ecological and genetic consequences of density-dependent regulation in plants. *Ann. Rev. Ecol. Syst.* **11**, 411–452.

BARKHAM J.P. (1980a) Population dynamics of the wild daffodil (*Narcissus pseudonarcissus*). I. Clonal growth, seed reproduction, mortality and the effects of density. *J. Ecol.* **68**, 607–633.

BARKHAM J.P. (1980b) Population dynamics of the wild daffodil (*Narcissus pseudonarcissus*). II. Changes in number of shoots and flowers, and the effect of bulb depth on growth and reproduction. *J. Ecol.* **68**, 635–664.

BARKHAM J.P. and HANCE C. E. (1982) Population dynamics of the wild daffodil (*Narcissus pseudonarcissus*). III. Implications of a computer model of 1000 years of population change. *J. Ecol.* **70**, 323–344.

BAZZAZ F. A. and HARPER J.L. (1977) Demographic analysis of the growth of *Linum usitatissimum*. *New Phytol.* **78**, 193–208.

BEGON M. and MORTIMER M. (1981) *Population Ecology: A Unified Study of Animals and Plants.* Blackwell Scientific Publications, Oxford.

BENNETT K. D. (1983) Postglacial population expansion of forest trees in Norfolk, UK. *Nature (Lond.)* **303**, 164–167.

BLACK J. N. (1957) The early vegetative growth of three strains of subterranean clover (*Trifolium subterraneum* L.) in relation to size of seed. *Aust. J. agric. Res.* **8**, 1–14.

BLACK J.N. (1958) Competition between plants of different initial seed sizes in swards of subterranean clover (*Trifolium subterraneum* L.) with particular reference to leaf area and the light microclimate. *Aust. J. agric. Res.* **9**, 299–318.

BRENCHLEY W.E. (1918) Buried weed seeds. *J. agric. Sci.* **9**, 1–31.

BRENCHLEY W.E. and WARINGTON K. (1933) The weed seed population of arable soil. II. Numerical estimation of viable seeds and observations on their natural dormancy. *J. Ecol.* **18**, 235–272.

BRENCHELY W.E. and WARINGTON K. (1933) The weed seed population of arable soil. II. Influence of crop, soil, and methods of cultivation upon the relative abundance of viable seeds. *J. Ecol.* **21**, 103–127.

BUDD A.C., CHEPIL W.S. and DOUGHTY J.L. (1954) Germination of weed seeds. III. The influence of crops and fallow on the weed seed population of the soil. *Can. J. agric. Sci.* **34**, 18–27.

BULLOCK S.H. and PRIMACK R.B. (1977) Comparative experimental study of seed dispersal on animals. *Ecology* **58**, 681–686.

BÜLOW-OLSEN A., SACKVILLE HAMILTON N.R. and HUTCHINGS M.J. (1984) A study of growth form in genets of *Trifolium repens* L. as affected by intra- and interplant contacts. *Oecologia* **61**, 383–387.

BURDON J.J. (1980) Intraspecific diversity in a natural population of *Trifolium repens*. *J. Ecol.* **68**, 717–735.

CAHN M.G. and HARPER J.L. (1976) The biology of leaf mark polymorphism in *Trifolium repens* L.: I. Distribution of phenotypes at a local scale. *Heredity* **37**, 309–325.

CALLAGHAN T.V. (1976) Growth and population dynamics of *Carex bigelowii* in an alpine environment. Strategies of growth and population dynamics of tundra plants 3. *Oikos* **27**, 402–413.

CANFIELD R.H. (1957) Reproduction and life span of some perennial grasses of southern Arizona. *J. Range Mgmt.* **10**, 199–203.

CHAMPNESS S.S. and MORRIS K. (1948) The population of buried viable weed seeds in relation to contrasting pasture and soil types. *J. Ecol.* **36**, 149–173.

CHANCELLOR R.J. (1966) *Identification of Weed Seeds of Farm and Garden*. Blackwell Scientific Publications, Oxford.

CHIPPINDALE H.G. and MILTON W.E.J. (1934) On the viable seeds present in the soil beneath pastures. *J. Ecol.* **22**, 508–531.

COHEN D. (1966) Optimizing reproduction in a randomly varying environment. *J. Theor. Biol.* **12**, 119–129.

COHEN D. (1967) Optimizing reproduction in a randomly varying environment when a correlation may exist between the conditions at the time a choice has to be made and the subsequent outcome. *J. Theor. Biol.* **16**, 1–14.

COLLINS N.J. (1976) Growth and population dynamics of the moss *Polytrichum alpestre* in the maritime Antarctic. Strategies of growth and population dynamics of tundra plants 2. *Oikos* **27**, 389–401.

COLWELL R. (1951) The use of radioactive isotopes in determining spore distribution patterns. *Am. J. Bot.* **38**, 511–523.

COOK R.E. (1980) The biology of seeds in the soil. In *Demography and Evolution in Plant Populations* (ed O.T. Solbrig) pp. 107–129, Blackwell Scientific Publications, Oxford.

CORNER E.J. (1976) *The Seeds of Dicotyledons*, Vols I and II. Cambridge University Press, Cambridge.

CRAWLEY M.J. (1983) *Herbivory: The Dynamics of Animal–Plant Interactions*. Studies in Ecology, Vol. 10. Blackwell Scientific Publications, Oxford.

CULLEN J.M., WEISS P.W. and WEARNE G.R. (1978) A plotter for plant population studies. *New Phytol.* **81**, 443–448.

DARWIN C. (1859) *The Origin of Species*. Murray, London.

DEEVEY E. S. (1947) Life tables for natural populations of animals. *Q. Rev. Biol.* **22**, 283–314.

EHRLICH P.R. and BIRCH L.C. (1967) The 'Balance of Nature' and 'Population Control'. *Am. Nat.* **101**, 97–108.

EPLING C. and LEWIS H. (1952) Increase of the adaptive range of the genus *Delphinium*. *Evolution* **6**, 253–267.

EVANS G.C. (1972) *The Quantitative Analysis of Plant Growth*. Studies in Ecology, Vol. I. Blackwell Scientific Publications, Oxford.

FALIŃSKA K. (1968) Preliminary studies on seed production in the herb layer of the *Querco–Carpinetum* association. *Ekologia Polska, Seria A*, **19**, 395–409.

FAY P.K. and OLSON W.A. (1978) Technique for separating weed seed from soil. *Weed Science* **26**, 530–533.

FETCHER N. and SHAVER G.R. (1983) Life histories of tillers of *Eriophorum vaginatum* in relation to tundra disturbance. *J. Ecol.* **71**, 131–147.

FORD E.D. (1975) Competition and stand structure in some even-aged plant monocultures. *J. Ecol.* **63**, 311–333.

FORD E.D. and NEWBOULD P. J. (1970) Stand structure and dry weight production through the sweet chestnut (*Castanea sativa* Mill.) coppice cycle. *J. Ecol.* **58**, 275–296.

FORTANIER H. (1973) Reviewing the length of the generation period and its shortening, particularly in tulips. *Scientia Horticulturae* **1**, 107–116.

FREELAND P.W. (1976) Tests for the viability of seeds. *J. Biol. Ed.* **10**, 57–64.

FREEMAN D.C., KLIKOFF L.G. and HARPER K.T. (1976) Differential resource utilisation by the sexes of dioecious plants. *Science* **193**, 597–599.

FRITTS H.C. (1976) *Tree-Rings and Climate*. Academic Press, London.

FROUD-WILLIAMS R.J., CHANCELLOR R.J. and DRENNAN D.S.H. (1983) Influence of cultivation regime upon buried weed seeds in arable cropping systems. *J. appl. Ecol.* **20**, 199–208.

FULLER W., HANCE C.E. and HUTCHINGS M.J. (1983) Within-season fluctuations in mean fruit weight in *Leontodon hispidus* L. *Ann. Bot.* **51**, 545–549.

GADGIL M. and SOLBRIG O.T. (1972) The concept of *r*- and *K*-selection: evidence from wild flowers and some theoretical considerations. *Am. Nat.* **106**, 14–31.

GATSUK E., SMIRNOVA O.V., VORONTZOVA L.I., ZAUGOLNOVA L.B. and ZHUKOVA L.A. (1980) Age-states of plants of various growth forms: a review. *J. Ecol.* **68**, 675–696.

GOODALL D.W. (1945) The distribution of weight change in the young tomato plant. I. Dry weight changes of the various organs. *Ann. Bot.* **9**, 101–139.

GOTTLIEB L.D. (1975) Allelic diversity in the outcrossing annual plant *Stephanomeria exigua* ssp. *carotifera* (Compositae). *Evolution* **29**, 213–225.

GOTTLIEB L.D. (1977a) Phenotypic variation in *Stephanomeria exigua* ssp. *coronaria* (Compositae) and its recent derivative species 'Malheurensis'. *Am. J. Bot.* **64**, 873–880.

GOTTLIEB L.D. (1977b) Genotypic similarity of large and small individuals in a natural population of the annual plant *Stephanomeria exigua* ssp. *coronaria*. *J. Ecol.* **65**, 127–134.

GRAHAM D.J. (1983) *The role of the seed bank in the establishment of vegetation on a disturbed grassland.* M. Phil. Thesis, University of Sussex.

GREIG-SMITH P. (1983) *Quantitative Plant Ecology.* Studies in Ecology, Vol. 9, Blackwell Scientific Publications, Oxford.

GRIME J.P. (1979) *Plant Strategies and Vegetation Processes.* Wiley, Chichester.

GRIME J.P. (1981) The role of seed dormancy in vegetation dynamics. *Ann. Appl. Biol.* **98**, 555–558.

GRIME J.P., MASON G., CURTIS A.V., RODMAN J., BAND S.R., MOWFORTH M.A.G., NEAL A.M. and SHAW S. (1981) A comparative study of germination characteristics in a local flora. *J. Ecol.* **69**, 1017–1059.

HÄFLIGER E. and BRUN-HOOL J. (1973) *Weed Communities of Europe.* CIBA-GEIGY, Basle.

HANF M. (1970) *Weeds and their Seedlings.* BASF, UK.

HARBERD D.J. (1961) Observations on population structure and longevity of *Festuca rubra* L. *New Phytol.* **60**, 184–206.

HARBERD D.J. (1962) Some observations on natural clones in *Festuca ovina*. *New Phytol.* **61**, 85–100.

HARBERD D.J. (1963) Observations on natural clones of *Trifolium repens* L. *New Phytol.* **62**, 198–204.

HARBERD D.J. (1967) Observations on natural clones of *Holcus mollis*. *New Phytol.* **66**, 401–408.

HARPER J.L. (1957) The ecological significance of dormancy and its importance in weed control. *Proceedings of the 4th International Congress on Crop Protection*, Hamburg 415–420.

HARPER J.L. (1967) A Darwinian approach to plant ecology. *J. Ecol.* **55**, 247–270.

HARPER J.L. (1969) The role of predation in vegetational diversity. In *Diversity and Stability in Ecological Systems, Brookhaven Symposium in Biology* **22**, 48–62.

HARPER J.L. (1977) *Population Biology of Plants.* Academic Press, London.

HARPER J.L. (1978) The demography of plants with clonal growth. In *Structure and Functioning of Plant Populations* (eds A.H.J. Freysen and J.W. Woldendorp), pp. 27–48. North-Holland Publishing Company, Amsterdam.

HARPER J.L. (1981) The concept of population in modular organisms. In *Theoretical Ecology: Principles and Applications* (ed. R.M. May), 2nd edn, pp. 53–77. Blackwell Scientific Publications, Oxford.

HARPER J.L. and BELL A.D. (1979) The population dynamics of growth form in organisms with modular construction. In *Population Dynamics* (eds R.M. Anderson, B.D. Turner and L.R. Taylor), 20th Symposium of the British Ecological Society, pp. 29–52, Blackwell Scientific Publications, Oxford.

HARPER J.L. and BENTON R.A. (1966) The behaviour of seeds in soil. II. The germination of seeds on the surface of a water-supplying substrate. *J. Ecol.* **54**, 151–166.

HARPER J.L., LOVELL P.H. and MOORE K.G. (1970) The shapes and sizes of seeds. *A. Rev. Ecol. Syst.* **1**, 327–356.

HARPER J.L. and WHITE J. (1971) The dynamics of plant populations. In *Proceedings of the Advanced Study Institute on Dynamics and Numbers in Populations* (eds P.J. den Boer and G.R. Gradwell), pp. 41–63. Oosterbeek, 1970. Centre for Agricultural Publication and Documentation, Wageningen.

Harper J.L. and WHITE J. (1974) The demography of plants. *A. Rev. Ecol. Syst.* **5**, 419–463.

HARPER J.L., WILLIAMS J.T. and SAGAR G.R. (1965) The behaviour of seeds in the soil. I. The heterogeneity of soil surfaces and its role in determining the establishment of plants from seed. *J. Ecol.* **53**, 273–286.

HAWTHORN W.R. and CAVERS P.B. (1976) Population dynamics of the perennial herbs *Plantago major* L. and *P. rugelii* Decne. *J. Ecol.* **64**, 511–527.

HAYASHI I. (1975) The special method of inventory of buried seed population of weeds. *Workshop on Research Methodology in Weed Science, Bandung* 1, paper 4.

HAYASHI I. and NUMATA M. (1968) Ecological studies on the buried seed population of the soil as related to plant succession. V. From overmature pine stand to climax *Shiia* stand. *Ecological Studies of Biotic Communities in the National Park for Nature Study* **2**, 1–7.

HAYASHI I. and NUMATA M. (1971) Viable buried-seed populaton in the *Miscanthus*- and *Zoysia*-type grasslands in Japan—ecological studies on the buried-seed population in the soil related to plant succession VI. *Jap. J. Ecol.* **20**, 243–252.

HENRY J.D. and SWAN J.M.A. (1974) Reconstructing forest history from live and dead plant material—an approach to the study of forest succession in southwestern New Hampshire. *Ecology* **55**, 772–783.

HETT J.M. (1971) A dynamic analysis of age in sugar maple seedlings. *Ecology* **52**, 1071–1074.

HETT J.M. and LOUCKS O.L. (1971) Sugar maple (*Acer saccharum* Marsh.) seedling mortality. *J. Ecol.* **59**, 507–520.

HETT J.M. and LOUCKS O.L. (1976) Age structure models of balsam fir and eastern hemlock. *J. Ecol.* **64**, 1029–1044.

HODGSON G.L. and BLACKMAN G.E. (1957) The analysis of the influence of plant density on the growth of *Vicia faba*. Part I. The influence of density on the pattern of development. *J. Exp. Bot.* **7**, 147–165.

HOPKINS B. (1957) The concept of the minimal area. *J. Ecol.* **45**, 441–449.

HOWE H.F. and SMALLWOOD J. (1982) Ecology of seed dispersal. *A. Rev. Ecol. Syst.* **13**, 201–228.

HUBBY J.L. and LEWONTIN R.C. (1966) A molecular approach to the study of genic heterozygosity in natural populations. I. The number of alleles at different loci in *Drosophila pseudoobscura*. *Genetics* **54**, 577–594.

HUBBELL S.P. and WERNER P.A. (1979) On measuring the intrinsic rate of increase of populations with heterogeneous life histories. *Am. Nat.* **113**, 277–293.

HUGHES M., KELLY M., LAMARCHE V. and PILCHER J. (Eds) (1982) *Climate from Tree-Rings.* Cambridge University Press, Cambridge.

HUNT R. (1978a) Demography versus plant growth analysis. *New Phytol.* **80**, 269–272.

HUNT R. (1978b) *Plant Growth Analysis.* Studies in Biology No. 96. Arnold, London.

HUTCHINGS M.J. (1975) Some statistical problems associated with determinations of population parameters for herbaceous plants in the field. *New Phytol.* **74**, 349–363.

HUTCHINGS M.J. (1979) Weight-density relationships in ramet populations of clonal perennial herbs, with special reference to the $-3/2$ power law. *J. Ecol.* **67**, 21–33.

HUTCHINGS M.J. (1983) Shoot performance and population structure in pure stands of *Mercurialis perennis* L., a rhizomatous perennial herb. *Oecologia*, **58**, 260–264.

HUTCHINGS M.J. and BARKHAM J.P. (1976) An investigation of shoot interactions in *Mercurialis perennis* L., a rhizomatous perennial herb. *J. Ecol.* **64**, 723–743.

HUTCHINGS M.J. and BUDD C.S.J. (1981) Plant competition and its course through time. *BioScience* **31**, 640–645.

HYDE E.O.C. and SUCKLING F.E.T. (1953) Dormant seeds of clovers and other legumes in agricultural soils. *N. Z. J. Sci. Technol.* A34, 375–385.

JAIN S. (1979) Adaptive strategies, polymorphism, plasticity and homeostasis. In *Topics in Plant Population Biology* (eds O.T. Solbrig, S. Jain, G.B. Johnson and P.H. Raven), pp. 160–187. MacMillan, New York.

JANZEN D.H. (1969) Seed-eaters versus seed size, number, toxicity and dispersal. *Evolution* **23**, 1–27.

JANZEN D.H. (1971) Seed predation by animals. *A. Rev. Ecol. Syst.* **2**, 465–492.

JEFFERS J.N.R. (1978) *Design of Experiments*. Statistical Checklist 1. Institute of Ecology, Cambridge.

JEFFERS J.N.R. (1979) *Sampling*. Statistical Checklist 2. Institute of Ecology, Cambridge.

JEFFERS J.N.R. (1980) *Modelling*. Statistical Checklist 3. Institute of Ecology, Cambridge.

JOHNSON E.A. (1975) Buried seed populations in the subarctic forest east of Great Slave Lake, Northwest Territories. *Can. J. Bot.* **53**, 2933–2941.

JONES R.M. and BUNCH G.A. (1977) Sampling and measuring the legume seed content of pasture and cattle faeces. *CSIRO Tropical Agronomy Technical Memorandum*, Number 7.

KAWANO S. (1975) The productive and reproductive biology of flowering plants. II. The concept of life history strategy in plants. *J. Coll. Lib. Arts, Toyama Univ.* (*Nat. Sci.*) **8**, 51–86.

KAWANO S. and NAGAI Y. (1975) The productive and reproductive biology of flowering plants. I. Life history strategies of three *Allium* species in Japan. *Bot. Mag., Tokyo*, **88**, 281–318.

KEDDY P.A. (1981) Experimental demography of the sand-dune annual *Cakile edentula*, growing along an environmental gradient in Nova Scotia. *J. Ecol.* **69**, 615–630.

KERSTER H.W. (1968) Population age structure in the prairie forb, *Liatris aspera*. *BioScience* **18**, 430–432.

KOYAMA H. and KIRA T. (1956) Intraspecific competition among higher plants. VIII. Frequency distribution of individual plant weight as affected by the interaction between plants. *J. Inst. Polytech., Osaka City Univ.* D7, 73–94.

KREBS C.J. (1972) *Ecology: The Experimental Analysis of Distribution and Abundance*. Harper & Row, New York.

KROPÁČ Z. (1966) Estimation of weed seeds in arable soil. *Pedobiologia* **6**, 105–128.

LAW R. (1975) *Colonisation and the evolution of life histories in* Poa annua. Ph.D. Thesis, University of Liverpool.

LAW R. (1981) The dynamics of a colonizing population of *Poa annua*. *Ecology* **62**, 1267–1277.

LAW R. (1983) A model for the dynamics of a plant population containing individuals classified by age and size. *Ecology* **64**, 224–230.

LEFKOVITCH L.P. (1965) The study of population growth in organisms grouped by stages. *Biometrics* **21**, 1–18.

LESLIE P.H. (1945) On the use of matrices in certain population mathematics. *Biometrika* **33**, 183–212.

LEVERICH W.J. and LEVIN D.A. (1979) Age-specific survivorship and reproduction in *Phlox drummondii*. *Am. Nat.* **113**, 881–903.

LEVIN D.A. and KERSTER H. (1969) Density-dependent gene dispersal in *Liatris*. *Am. Nat.* **103**, 61–74.

LEVIN D.A. and KERSTER H.W. (1978) Rings and age in *Liatris*. *Am. Nat.* **112**, 1120–1122.

LEVIN D.A. and WILSON J.B. (1978) The genetic implications of ecological adaptations in plants. In *Structure and Functioning of Plant Populations* (eds A.H.J. Freysen and J.W. Woldendorp), pp. 75–100. North-Holland Publishing Company, Amsterdam.

LIDDLE M.J., BUDD C.S.J. and HUTCHINGS M.J. (1982) Population dynamics and neighbourhood effects in establishing swards of *Festuca rubra*. *Oikos* **38**, 52–59.

LINKOLA K. (1935) Über die Dauer und Jahresklassenverhältnisse des Jungenstadiums bei einigen Wiesenstauden. *Acta for. Fenn.* **42**, 1–56.

LOUDA S.M. (1982) Limitation of the recruitment of the shrub *Haplopappus squarrosus* (Asteraceae) by flower- and seed-feeding insects. *J. Ecol.* **70**, 43–53.

LOVETT DOUST L. (1981) Population dynamics and local specialization in a clonal perennial (*Ranunculus repens*). I. The dynamics of ramets in contrasting habitats. *J. Ecol.* **69**, 743–755.

McRILL M. (1974) *Some botanical aspects of earthworm activity*. Ph.D. Thesis, University of Wales.

McRILL M. and SAGAR G.R. (1973) Earthworms and seeds. *Nature* (*Lond.*) **243**, 482.

MACK R.N. (1976) Survivorship of *Cerastium atrovirens* at Aberffraw, Anglesey. *J. Ecol.* **64**, 309–312.

MACK R.N. and HARPER J.L. (1977) Interference in dune annuals: spatial pattern and neighbourhood effects. *J. Ecol.* **65**, 345–363.

MACK R.N. and PYKE D.A. (1979) Mapping individual plants with a field-portable digitizer. *Ecology* **60**, 459–461.

MACK R.N. and PYKE D.A. (1983) The demography of *Bromus tectorum*: variation in time and space. *J. Ecol.* **71**, 69–93.

MAILLETTE L. (1982a) Structural dynamics of silver birch I. The fates of buds. *J. Appl. Ecol.* **19**, 203–208.

MAILLETTE L. (1982b) Structural dynamics of silver birch. II. A matrix model of the bud population. *J. appl. Ecol.* **19**, 219–238.

MAILLETTE L. (1982c) Needle demography and growth pattern of Corsican pine. *Can. J. Bot.* **60**, 105–116.

MALONE C.R. (1967) A rapid method for enumeration of viable seeds in soil. *Weeds* **15**, 381–382.

MARTIN A.C. and BARKELY W.J. (1961) *Seed Identification Manual.* University of California Press, Berkeley.

MEAD R. (1966) A relationship between individual plant spacing and yield. *Ann. Bot.* **30**, 301–309.

MEAGHER T.R. and ANTONOVICS J. (1982) The population biology of *Chamaelirium luteum*, a dioecious member of the lily family: life history studies. *Ecology* **63**, 1690–1700.

MONTGOMERY F.A. (1977) *Seeds and Fruits of Plants of Eastern Canada and Northeastern United States.* University of Toronto Press, Toronto.

MOORE P.D. (1983) Seeds of thought for plant conservationists. *Nature (Lond.)* **303**, 572.

MORTIMER A.M. (1974) *Studies of germination and establishment of selected species with special reference to the fates of seeds.* Ph.D. Thesis, University of Wales.

MOTT J.J. (1972) Germination studies on some annual species from an arid region of Western Australia. *J. Ecol.* **60**, 293–304.

MULLER F.M. (1978) *Seedlings of the North West European Lowland. A Flora of Seedlings.* Dr. W. Junk and Co-edited with PUDOC, Wageningen.

NEWBOULD P.J. (1967) *Methods for Estimating the Primary Production of Forests.* IBP Handbook No. 2. Blackwell Scientific Publications, Oxford.

NEWELL S.J., SOLBRIG O.T. and KINCAID D.T. (1981) Studies on the population biology of the genus *Viola* III. The demography of *Viola blanda* and *Viola pallens*. *J. Ecol.* **69**, 997–1016.

NOBLE J.C., BELL A.D. and HARPER J.L. (1979) The population biology of plants with clonal growth. I. The morphology and structural demography of *Carex arenaria*. *J. Ecol.* **67**, 983–1008.

NUMATA M., HAYASHI I., KOMURA T. and ÔKI K. (1964) Ecological studies on the buried-seed population in the soil as related to plant succession. I. *Jap. J. Ecol.* **14**, 207–215.

OBEID M., MACHIN D. and HARPER J.L. (1967) Influence of density on plant to plant variation in Fiber Flax, *Linum usitatissimum* L. *Crop Science* **7**, 471–473.

OINONEN E. (1967a) The correlation between the size of Finnish bracken (*Pteridium aquilinum* (L.) Kuhn) clones and certain periods of site history. *Acta for. Fenn.* **83**, 1–51.

OINONEN E. (1967b) Sporal regeneration of ground pine (*Lycopodium complanatum* L.) in southern Finland in the light of the size and age of its clones. *Acta for. Fenn.* **83**, 76–85.

OINONEN E. (1969) The time-table of vegetation spreading of the lily-of-the-valley (*Convallaria majalis* L.) and the wood small-reed (*Calamagrostis epigeios* (L.) Roth) in southern Finland. *Acta for. Fenn.* **97**, 1–35.

OLIVER C.D. and STEPHENS E.P. (1977) Reconstruction of a mixed species forest in central New England. *Ecology* **58**, 562–572.

PALMBLAD I.G. (1968) Competition in experimental studies on populations of weeds with emphasis on the regulation of population size. *Ecology* **49**, 26–34.

PIELOU E.C. (1960) A single mechanism to account for regular, random and aggregated populations. *J. Ecol.* **48**, 575–584.

PIELOU E.C. (1961) Segregation and symmetry in two-species populations as studied by nearest-neighbour relationships. *J. Ecol.* **49**, 255–269.

PIELOU E.C. (1962) The use of plant-to-neighbour distances for the detection of competition. *J. Ecol.* **50**, 357–367.

PIGOTT C.D. (1968) *Cirsium acaulon* (L.) Scop. *J. Ecol.* **56**, 597–612.

PIGOTT C.D. and WILSON J.F. (1978) The vegetation of North Fen at Esthwaite in 1967–9. *Proc. R. Soc. Lond.* B 200, 331–351.

PLATT W.J. (1975) The colonization and formation of equilibrium plant species associations on badger disturbances in a tall-grass prairie. *Ecol. Monogr.* **45**, 285–305.

PLATT W.J. and WEIS I.M. (1977) Resource partitioning and competition within a guild of fugitive prairie plants. *Am. Nat.* **111**, 479–513.

RABINOWITZ D. (1981) Buried viable seeds in a North-American tall grass prairie: the resemblance of their abundance and composition to dispersing seeds. *Oikos* **36**, 191–195.

RABINOWITZ D. and RAPP J.K. (1980) Seed rain in a North American tall grass prairie. *J. appl. Ecol.* **17**, 793–802.

RABINOWITZ D. and RAPP J.K. (1981) Dispersal abilities of seven sparse and common grasses from a Missouri prairie. *Am. J. Bot.* **68**, 616–624.

RABOTNOV T.A. (1978) On coenopopulations of plants reproducing by seeds. In *Structure and Functioning of Plant Populations* (eds A.H.J. Freysen and J.W. Woldendorp), pp. 1–26. North-Holland Publishing Company, Amsterdam.

ROBERTS E.H. (1972) Dormancy: a factor affecting seed survival in soil. In *Viability of Seeds* (ed. E.H. Roberts), pp. 321–359, Syracuse University Press, Syracuse.

ROBERTS H.A. (1958) Studies on the weeds of vegetable crops. 1. Initial effects of cropping on the weed seeds in the soil. *J. Ecol.* **46**, 759–768.

ROBERTS H.A. (1962) Studies on the weeds of vegetable crops. II. Effect of six years of cropping on the weed seeds in the soil. *J. Ecol.* **50**, 803–813.

ROBERTS H.A. (1981) Seed banks in soils. *Adv. appl. Biol.* **6**, 1–55.

ROBERTS H.A. and DAWKINS P.A. (1967) Effect of cultivation on the numbers of viable weed seeds in soil. *Weed Research* **7**, 290–301.

ROBERTS H.A. and RICKETTS M.E. (1979) Quantitative relationships between the weed flora after cultivation and the seed population in the soil. *Weed Res.* **19**, 269–275.

RORISON I.H. (1973) Seed ecology-present and future. In *Seed Ecology* (ed W. Heydecker), pp. 497–519, Butterworths, London.

ROSS M.A. and HARPER J.L. (1972) Occupation of biological space during seedling establishment. *J. Ecol.* **60**, 77–88.

ROUGHTON R.D. (1962) A review of literature on dendrochronology and age determination of woody plants. *Colo. Dept. Game Fish. Tech. Bull.* 15, 99pp.

ROYAMA T. (1977) Population persistence and density dependence. *Ecol. Monogr.* **47**, 1–35.

RUSSELL P.J. (1983) *A study of vegetation composition and species interactions in emergent salt marsh communities.* D. Phil. Thesis, University of Sussex.

RYVARDEN L. (1971) Studies in seed dispersal I. Trapping of diaspores in the alpine zone at Finse, Norway. *Norw. J. Bot.* **18**, 215–226.

SAGAR G.R. (1959) *The biology of some sympatric species of grassland.* D. Phil. Thesis, University of Oxford.

SAGAR G.R. (1970) Factors controlling the size of plant populations. *Proceedings of the 10th British Weed Control Conference*, pp. 965–979.

SAGAR G.R. and HARPER J.L. (1961) Controlled interference with natural populations of *Plantago lanceolata, P. major* and *P.media. Weed Research* **1**, 163–176.

SAGAR G.R. and MORTIMER A.M. (1976) An approach to the study of the population dynamics of plants with special reference to weeds. *Appl. Biol.* **1**, 1–47.

SARUKHÁN J. (1971) *Studies on plant demography.* Ph.D. Thesis, University of Wales.

SARUKHÁN J. (1974) Studies on plant demography: *Ranunculus repens* L., *R. bulbosus* L. and *R. acris* L. II. Reproductive strategies and seed population dynamics. *J. Ecol.* **62**, 151–177.

SARUKHÁN J. (1980) Demographic problems in tropical systems. In *Demography and Evolution in Plant Populations* (ed. O.T. Solbrig), pp. 161–188, Blackwell Scientific Publications, Oxford.

SARUKHÁN J. and GADGIL M. (1974) Studies on plant demography: *Ranunculus repens* L., *R. bulbosus* L., and *R.acris* L. III A mathematical model incorporating multiple modes of reproduction. *J. Ecol.* **62**, 921–936.

SARUKHÁN J. and HARPER J.L. (1973) Studies on plant demography: *Ranunculus repens* L., *R. bulbosus* L. and *R. acris* L. I. Population flux and survivorship. *J. Ecol.* **61**, 675–716.

SCANDALIOS J.G. (1969) Genetic control of multiple molecular forms of enzymes in plants: a review. *Biochem. Genetics* **3**, 37–79.

SCHAFER D.E. and CHILCOTE D.O. (1969) Factors influencing persistence and depletion in buried seed populations. I. A model for analysis of parameters of buried seed persistence and depletion. *Crop Science* **9**, 417–419.

SEARLE S.R. (1966) *Matrix Algebra for the Biological Sciences*, Wiley, New York.

SHARITZ R.R. and McCORMICK J.F. (1973) Population dynamics of two competing annual plant species. *Ecology* **54**, 723–740.

SHELDON J.C. and BURROWS F.M. (1973) The dispersal effectiveness of the achene-pappus units of selected Compositae in steady winds with convection. *New Phytol.* **72**, 665–675.

SILVERTOWN J.W. (1982) *Introduction to Plant Population Ecology*, Longmans, London.

SOLBRIG O.T., NEWELL S.J. and KINCAID D.T. (1980) The population biology of the genus *Viola*. I. The demography of *Viola sororia*. *J. Ecol.* **68**, 521–546.

STANDIFER L.C. (1980) A technique for estimating weed seed populations in cultivated soil. *Weed Science* **28**, 134–138.

TAMM C.O. (1972a) Survival and flowering of some perennial herbs. II. The behaviour of some orchids on permanent plots. *Oikos* **23**, 23–28.

TAMM C.O. (1972b) Survival and flowering of perennial herbs. III. The behaviour of *Primula veris* on permanent plots. *Oikos* **23**, 159–166.

TAYLORSON R.B. and HENDRICKS S.B. (1977) Dormancy in seeds. *A. Rev. Pl. Physiol.* **28**, 331–354.

THOMAS A.G. and DALE H.M. (1974) Zonation and regulation of old pasture populations of *Hieracium floribundum*. *Can. J. Bot.* **52**, 1452–1458.

THOMPSON K. (1978) The occurrence of buried viable seeds in relation to environmental gradients. *J. Biogeog.* **5**, 425–430.

THOMPSON K. and GRIME J.P. (1979) Seasonal variation in the seed banks of herbaceous species in ten contrasting habitats. *J. Ecol.* **67**, 893–921.

THOMPSON K. and GRIME J.P. (1983) A comparative study of germination responses to diurnally-fluctuating temperatures. *J. appl. Ecol.* **20**, 141–156.

THOMPSON K., GRIME J.P. and MASON G. (1977) Seed germination in response to diurnal fluctuations of temperature. *Nature (Lond.)* **267**, 147–149.

THOMPSON K. and STEWART A.J.A. (1981) The measurement and meaning of reproductive effort in plants. *Am. Nat.* **117**, 205–211.

USHER M.B. (1972) Developments in the Leslie matrix model. In *Mathematical Models in Ecology* (ed. J.N.R. Jeffers), 12th Symposium of the British Ecological Society, pp. 29–60, Blackwell Scientific Publications, Oxford.

VAN ANDEL J. (1975) A study on the population dynamics of the perennial plant species *Chamaenerion angustifolium* (L.) Scop. *Oecologia* **19**, 329–337.

VAN ANDEL J. and VERA F. (1977) Reproductive allocation in *Senecio sylvaticus* and *Chamaenerion angustifolium* in relation to mineral nutrition. *J. Ecol.* **65**, 747–758.

VAN DER PIJL L. (1972) *Principles of Dispersal in Higher Plants*, 2nd edn. Springer, Berlin.

VAN VALEN L. (1975) Life, death and energy of a tree. *Biotropica* **7**, 260–269.

VANDERMEER J.H. (1975) On the construction of the population projection matrix for a population grouped in unequal stages. *Biometrics* **31**, 239–242.

VARLEY G.C. and GRADWELL G.R. (1960) Key factors in population studies. *J. Anim. Ecol.* **29**, 399–401.

VARLEY G.C. and GRADWELL G.R. (1970) Recent advances in insect population dynamics. *A. Rev. Entomol.* **15**, 1–24.

VARLEY G.C., GRADWELL, G.R. and HASSELL M.P. (1973) *Insect Population Ecology*. Blackwell Scientific Publications, Oxford.

VEGA M.R. and SIERRA J.N. (1970) Population of weed seeds in a lowland rice field. *Philipp. Agric.* **54**, 1–7.

WAGER H.G. (1938) Growth and survival of plants in the Arctic. *J. Ecol.* **26**, 390–410.

WAITE S. (1980) *Autecology and population biology of* Plantago coronopus *L. at coastal sites in Sussex.* D.Phil. Thesis, University of Sussex.

WAITE S. (1984) Changes in the demography of *Plantago coronopus* at two coastal sites. *J. Ecol.* **72**, 809–826.

WAITE S. and HUTCHINGS M.J. (1978) The effects of sowing density, salinity and substrate upon the germination of seeds of *Plantago coronopus* L. *New Phytol.* **81**, 341–348.

WAITE S. and HUTCHINGS M.J. (1982) Plastic energy allocation patterns in *Plantago coronopus.* *Oikos* **38**, 333–342.

WALLER D.M. (1981) Neighbourhood competition in several violet populations. *Oecologia* **51**, 116–122.

WARINGTON K. (1936) The effect of constant and fluctuating temperature on the germination of the weed seeds in arable soil. *J. Ecol.* **24**, 185–204.

WATKINSON A.R. (1975) *The population biology of a dune annual,* Vulpia membranacea. Ph.D. Thesis, University of Wales.

WATKINSON A.R. (1978a) The demography of a sand dune annual: *Vulpia fasciculata* II. The dynamics of seed populations. *J. Ecol.* **66**, 35–44.

WATKINSON A.R. (1978b) The demography of a sand dune annual: *Vulpia fasciculata* III. The dispersal of seeds. *J. Ecol.* **66**, 483–498.

WATKINSON A.R. (1981) The population ecology of winter annuals. In *The Biological Aspects of Rare Plant Conservation* (ed. H. Synge), pp. 253–264, Wiley, Chichester.

WATKINSON A.R. and HARPER J.L. (1978) The demography of a sand dune annual: *Vulpia fasciculata* I. The natural regulation of populations. *J. Ecol.* **66**, 15–33.

WATT A.S. (1947a) Contributions to the ecology of bracken (*Pteridium aquilinum*). IV. The structure of the community. *New Phytol.* **46**, 97–121.

WATT A.S. (1947b) Pattern and process in the plant community. *J. Ecol.* **35**, 1–22.

WEAVER S.E. and CAVERS P.B. (1979) The effects of date of emergence and emergence order on seedling survival rates in *Rumex crispus* and *R. obtusifolius. Can. J. Bot.* **57**, 730–738.

WEAVER S.E. and CAVERS P.B. (1980) Reproductive effort of two perennial weed species in different habitats. *J. appl. Ecol.* **17**, 505–513.

WELLS T.C.E. (1967) Changes in a population of *Spiranthes spiralis* (L.) Chevall. at Knocking Hoe National Nature Reserve, Bedfordshire, 1962–65. *J. Ecol.* **55**, 83–99.

WELLS T.C.E. (1981) Population ecology of terrestrial orchids. In *The Biological Aspects of Rare Plant Conservation* (ed. H. Synge), pp. 281–295. Wiley, Chichester.

WERNER P.A. (1975a) A seed trap for determining patterns of seed deposition in terrestrial plants. *Can. J. Bot.* **53**, 810–813.

WERNER P.A. (1975b) Predictions of fate from rosette size in teasel (*Dipsacus fullonum* L.). *Oecologia* **20**, 197–201.

WERNER P.A. (1978) On the determination of age in *Liatris aspera* using cross-sections of corms: implications for past demographic studies. *Am. Nat.* **112**, 1113–1120.

WERNER P.A. and CASWELL H. (1977) Population growth rates and age *versus* stage-distribution models for teasel (*Dipsacus sylvestris* Huds.). *Ecology* **58**, 1103–1111.

WERNER P.A. and PLATT W.J. (1976) Ecological relationships of co-occurring goldenrods (*Solidago*: Compositae). *Am. Nat.* **110**, 959–971.

WESSON G. and WAREING P.F. (1967) Light requirements of buried seeds. *Nature (Lond.)* **213**, 600–601.

WHIGHAM L.D. (1974) An ecological life history study of *Uvularia perfoliata* L. *Am. Midl. Nat.* **91**, 343–359.

WHITE J. (1979) The plant as a metapopulation. *A. Rev. Ecol. Syst.* **10**, 109–145.

WHITE J. (1980) Demographic factors in populations of plants. In *Demography and Evolution in Plant Populations* (ed. O.T. Solbrig), pp. 21–48. Blackwell Scientific Publications, Oxford.

WHITE J. and HARPER J.L. (1970) Correlated changes in plant size and number in plant populations. *J. Ecol.* **58**, 467–485.

WHITTAKER R.H. and WOODWELL G.M. (1968) Dimension and production relations of trees and shrubs in the Brookhaven Forest, New York. *J. Ecol.* **56**, 1–25.

WHITTAKER R.H. and WOODWELL G.M. (1971) Measurement of the primary production of forests. In *Productivity of Forest Ecosystems* (ed. P. Duvigneaud), pp. 159–175. Proceedings of the Brussels Symposium, 1969. IBP/UNESCO.

WILLIAMSON M.H. (1967) Introducing students to the concepts of population dynamics. In *The Teaching of Ecology* (ed. J.M. Lambert), 7th Symposium of the British Ecological Society, pp. 169–176. Blackwell Scientific Publications, Oxford.

WILSON R.E. and HANCOCK J.F. (1978) Comparison of four techniques used in the extraction of plant enzymes for electrophoresis. *Bull. Torrey bot. Club* **105**, 318–320.

YEATON R.I. and CODY M.L. (1976) Competition and spacing in plant communities: the northern Mojave desert. *J. Ecol.* **64**, 689–696.

YEATON R.I., TRAVIS J. and GILINSKY E. (1977) Competition and spacing in plant communities: the Arizona upland association. *J. Ecol.* **65**, 587–595.

9 Description and analysis of vegetation

F. B. GOLDSMITH, C. M. HARRISON and
A. J. MORTON

1 Introduction

Many ecologists view vegetation as a component of ecosystems which displays the effects of other environmental conditions and historic factors in an obvious and easily measurable manner. The careful analysis of vegetation is therefore used as a means of revealing useful information about other components of the ecosystem. In this way the research worker's study frequently proceeds from the description of vegetation in the field to the subsequent analysis of these records in the laboratory. The two phases of study reflect fundamentally different concepts of vegetation. The description is of 'real' vegetation while the analysis is concerned with forming 'abstractions' or 'generalized types' of vegetation which are simplifications of the complexities of the 'real' world. At the outset of the study the worker therefore has to decide which method of description and analysis best suits his purposes. The choice is a difficult one and the decision is affected by the purpose of study, the scale of enquiry, the botanical knowledge of the worker and the nature of vegetational variation itself (Webb, 1954).

Ecologists with a botanical training and studying a botanically-oriented problem generally prefer an approach based upon floristics, i.e. upon plant species composition. This necessitates a knowledge of the flora and may require considerable time and/or expertise, especially if working in an unfamiliar area. Their approach also depends upon whether the problem is autecological (single species) or synecological (communities), and orientated towards a production study or the identification of causality. Autecologists usually require a measure of abundance or performance of a species which can be easily supplied whereas synecologists, dealing with communities, become blighted by the problems associated with the nature of vegetational variation. An introductory text in plant ecology is provided by Willis (1973). Production ecologists require data about dry weight and calorific content which is very time-consuming to collect and destructive. Also they are often concerned about the number of samples necessary and their most informative spatial arrangement.

Zoologists, on the other hand, being interested in vegetation as a matrix in which animals live and feed, are usually more interested in its structure, usually the degree of stratification, and habitat diversity (Elton & Miller, 1954; Elton, 1966).

Vegetation description is also an integral part of much resource survey work, especially in the preparation of inventories of timber and the assessment of range-land carrying capacity. Soil scientists, and to a lesser extent, geologists and climatologists, are interested in vegetation as an expression of the factors they study. Their research is often conducted on more general scales than that of the ecologists and requires classifications that produce mappable units (Birse & Robertson, 1967).

Foresters frequently use an assessment of species composition to indicate site potential and to help in selecting species for planting. In practice, however, this usually involves the establishment of indicator species which identify the main characteristics of the site (*Juncus, Sphagnum, Pteridium*, etc.) and objective methods for describing vegetation are usually restricted to research projects (van Groenewoud, 1965).

The objective of this chapter is to review the range of methods available to the ecologist and to discuss the theoretical and practical problems involved with each and to indicate the kind of situation in which its use is appropriate. Full descriptions will not be given where reference to a readily available book or journal can provide the theoretical background and examples of the application of a method.

The primary objective in selecting a method should be its informativeness, but no ecologist has yet succeeded in establishing an objective test for comparing the information return from a range of techniques. Other criteria which should be used in selecting a method include relevance to the aims of the project, speed, accuracy, objectivity (reproducibility) and non-destructiveness (Moore *et al.*, 1970). In practice, the research worker usually has to make a compromise because the most objective method may not be the fastest, or the most accurate may be destructive. However, if these criteria are considered prior to sampling and analysing vegetation the final outcome should be proportionally more useful.

The first decision is to identify the objective of the study. Although this sounds unnecessarily obvious it determines a series of decisions that must subsequently be made. Fig. 9.1 summarizes the kind of decision-making process the research worker should adopt. However, it must be emphasized that these are only guidelines. In practice, every problem is unique, and the selection of a method should be based on a detailed consideration of the characteristics of the problem. It is hoped, however, that Fig. 9.1 will provide an indication of the kinds of questions that should be asked and the decisions that are likely to be made.

Four textbooks are of general relevance to this chapter and should be consulted for reference (Greig-Smith, 1983; Kershaw & Looney, 1985; Kuchler, 1967; Shimwell, 1971).

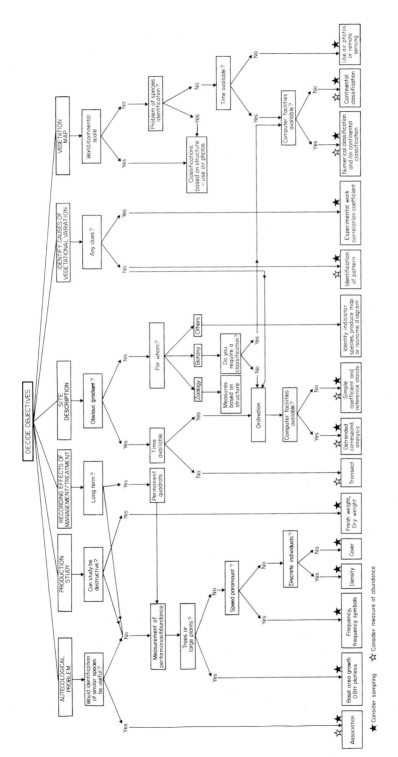

Figure 9.1 Some considerations to be used in choosing an appropriate method for describing or analysing vegetation.

1.1   The dynamics of vegetation

The subject of change in vegetation has recently received renewed attention with the publication of a small book on the subject by Miles (1979) and a special number of *Vegetatio* in 1981 (Vols 46 and 47). While the subject has interested plant ecologists from the time of the earliest serious studies it is really only methods of study that concern us here. We believe that there have been two methodological avenues which merit special attention.

One is the use of Markovian models to predict the nature of stable end-points, for example at the end of successional series. This approach has been used by Horn (e.g. 1975 and 1976) in his studies of deciduous forests on the eastern side of the United States and by Usher in his studies of wood-feeding termites and the reanalysis of Watt's classic breckland grassland data (Usher, 1981). Essentially the method involves determining the replacement probabilities of one species or seral stage by another. These values are placed in a matrix which is multiplied by a non-zero vector until a stable set of values emerge. The replacement probabilities could be the chance of different tree species colonizing a gap left by a particular tree species and then the stable end-point is the supposed climax condition for the forest. Sadly, this apparently elegant approach suffers from at least seven disadvantages identified by Usher (1981) such as the difficulty of defining the seral states, the change in transition probabilities with time, spatial patterns are ignored, the possibility of new species entering the sequence is overlooked and the history of the quadrat neglected. Mulholland (1980) however, used the method in a study of beechwoods in Epping Forest, Essex: Fig. 9.2 shows the conditions under which oak is replaced by beech, birch by oak and birch by beech which were elucidated with the help of a Markovian model. Sargent (1984) has used a similar approach with railway vegetation.

Permanent quadrats have been the backbone of long-term studies of vegetation change. A.S. Watt's classic plots in the Breckland, which cover the period 1936–57 have already been mentioned and Silvertown (1980) has recently published results from the Park grass plots at Rothamsted which represent different fertilizer regimes over a 125 year period. Austin (1980) studied the dynamics of an English garden lawn over a period of 6 years using an ordination technique (see below).

Often, these studies make use of cover values as a measure of change, but in order to determine the survival probabilities of different species under different types of management, records of individuals should be made. This kind of demographic approach has been recommended for many years by Harper and his school, and studies which have been concluded are proving very illuminating, e.g. Sarukhan & Harper (1973) and Sarukhan (1974), although it is difficult to execute in closed vegetation.

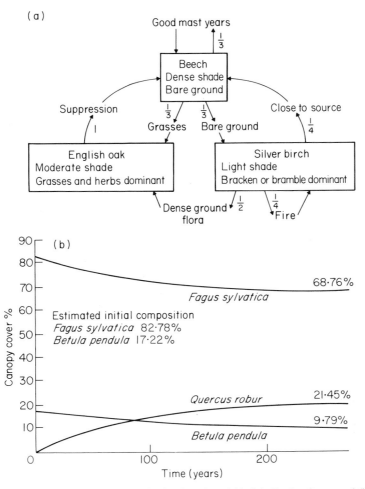

Figure 9.2 (a) Transition probabilities for beech, oak and birch in Epping forest, and (b) a Markovian model to show the probable changes from the present species composition to the stable woodland that might occur in 270 years time. (From Mulholland, 1980.)

The other apparently novel and useful approach to vegetation change is the use of time-series ordinations. By sampling vegetation in successive years by quadrats and then ordinating all of them together, it is possible to produce an ordination diagram on which the degree of change (as distance on the ordination) of a permanent quadrat or a vegetation type can be shown. Austin (1977) has done this with the lawn data, and other workers have studied changes after the establishment of grazing exclosures, e.g. Nieuwborg (pers. comm.) on dune vegetation at St Cyrus in Aberdeenshire. This appears to be a simple and useful way of extending ordination studies into the time dimension.

Swaine & Greig-Smith (1980) use an improvement of this 'ordination by

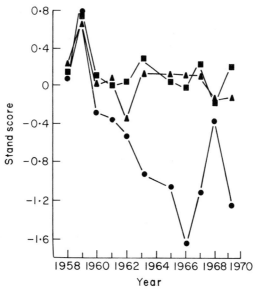

Figure 9.3 Changes in stand score on the first axis of the principal components analysis of the between-years correlation matrix for Llydaw 2. Treatments: ▲ control; ■ summer grazing; ● no grazing. (After Swaine & Greig-Smith, 1980.)

repeated enumerations' technique for permanent plots in montane grasslands from North Wales over the period 1958–70. This consists of eliminating the effect of individual initial differences between stands, and the resultant stand scores plotted against year show the difference between the control and grazed plots over time (Fig. 9.3).

2   Description of vegetation

Vegetation may be defined as an assemblage of plants growing together in a particular location and may be characterized either by its component species or by the combination of structural and functional characters that characterize the appearance, or physiognomy, of vegetation. This is an important distinction which is reflected by the range of methods available for describing vegetation. Structural or physiognomic methods do not demand species identification and are often considered more meaningful for small-scale (large-area) studies and for habitat description for scientists of other disciplines. For example, zoologists favour description of vegetation which they can interpret in terms of the niche, habitat and food resources of animals. Methods based on species composition or floristics are more useful for large-scale (small-area) studies of a more detailed, botanical, nature. They are, however, used by the Continental European school of phytosociologists for vegetation classification and mapping on an extensive scale. It may be significant that these studies are very detailed and time-consuming. Thus, they may be considered large-scale (small-area) in character, although the synthesis of several decades work covers an extensive area. It appears that classificatory methods are initially devised for large-scale (small-area) studies and are subsequently used for increasingly extensive studies. This has occurred with both the continental methods and the quantitative ones such as association-analysis (q.v.).

2.1 Measures based on physiognomy

Physiognomy is used to characterize an assemblage of plants but it is a measure which is surrounded in controversy and its usage in the literature is inexact. In general, physiognomy refers to the appearance of the vegetation: its height, colour and luxuriance, and to leaf size and shape. Although these characters may seem self-evident they tend to result from a combination of functional and structural characters (Fosberg, 1967). Functional characters are those which serve an adaptive role for survival in existing or past environments, as for example the evergreen or deciduous habit. Structural characters refer to the horizontal and vertical arrangement of components of the vegetation, as for example in the spacing between individuals and their vertical layering. Purely physiognomic characters which do not relate to either or both of these other characters are difficult to isolate. For example, leaf size may be regarded as a functional adaptation to particular climatic conditions, or as a product of the age of the individual or as a result of shading when the plant is a member of an understorey. However, physiognomic characters have proved useful for describing vegetation, and include life-form (*sensu* Du Rietz) which is essentially descriptive, life-form (*sensu* Raunkiaer) which is essentially functional, periodicity, and stratification.

2.1.1 Life-form

The definition and discussion provided by Du Rietz (1931) is perhaps the most comprehensive statement of life-form, and most other authors who utilize life-form as a character for description base their classes upon those of Du Rietz. His categories reproduced in Table 9.1 are more or less self-explanatory, as are the types of vegetation which can be recognized on this basis, e.g. forest, woodland, scrub and grassland. The types are however very superficial and generalized so that only a gross description of vegetation from a large area can be achieved.

The life-form categories recognized by Raunkiaer (1934) provide one of the most widely used functional criteria for describing vegetation. He based his approach on the position with respect to the ground surface on the plant

Table 9.1 The life-form system of Du Rietz (1931)

A   Higher plants
   I.   Ligniden (woody plants)
      (a) Magnoligniden (m)—Trees taller than 2 m
         1.   Deciduimagnoligniden (md) deciduous
         2.   Aciculimagnoligniden (ma) needleleaf evergreen
         2.   Laurimagnoligniden (ml) other evergreens
      (b) Parvoligniden (p)—Shrubs 0·8 m–2 m tall
         4.   Deciduiparvoligniden (pd)
         5.   Aciculiparvoligniden (pa)
         6.   Lauriparvoligniden (pl)
      (c) Nanoligniden (n)—Under 0·8 m tall
      (d) Lianen (li)—Climbing plants
   II. Herbiden (herbs)
      (a) Terriherbiden—Terrestrial herbs
         9.   Euherbiden (h) herbs
       10.   Graminiden (g) grasses
      (b) Aquiherbiden—Water plants
       11.   Nymphaeiden (ny) rooted with floating leaves (*Nymphaea*)
       12.   Elodeiden (e) rooted without floating leaves (*Elodea*)
       13.   Isoetiden (i) rooted, bottom rosettes (*Isoetes*)
       14.   Lemniden (le) free floating, not rooted (*Lemna*)
B   Moose (Bryophytes)
       15.   Eubryiden (b) all mosses and liverworts excluding *Sphagnum*
       16.   Sphagniden (s)—*Sphagnum* spp.
C   Fletchten
       17.   Lichens
D   Algen
       18.   Algae
E   Pilze
       19.   Fungi

of the perennating bud, that is the bud from which the next seasons growth would be made. Five, broad, life-form classes were recognized: namely phanerophytes, chamaephytes, hemicryptophytes, cryptophytes and therophytes, see Fig. 9.4. Phanerophytes bear their perennating buds freely in the air at varying heights at least 25 cm above the ground. They are mostly woody plants, trees and shrubs and are subdivided into classes according to height:

Megaphanerophytes	30 + m
Mesophanerophytes	8–30 m
Microphanerophytes	2–8 m
Nanophanerophytes	25 cm–2 m

Further subdivisions were also made on the presence or absence of protection of the bud and on whether or not the species is deciduous.

Chamaephytes are also woody or semi-woody perennials bearing their

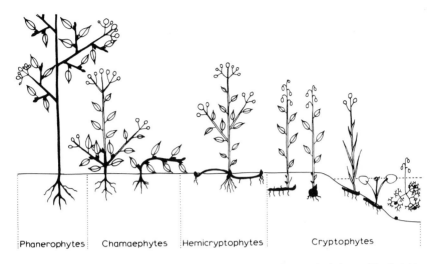

Phanerophytes Chamaephytes Hemicryptophytes   Cryptophytes

Parts of the plant which die in the unfavourable season are unshaded; persistent axes
with surviving buds are black

Figure 9.4 Diagrammatic representation of Raunkiaer's life-forms (from Raunkiaer, 1934).

buds close to the ground but less than 25 cm from the surface:

Suffrutescent or semi-shrubby forms
Passively decumbent forms
Actively creeping or stoloniferous forms
Cushion plants

Hemicryptophytes bear their renewal buds at the surface of the ground. They are a large and diverse group and include many graminaceous and herbaceous species.

Cryptophytes have their buds beneath the soil surface or in water.

Therophytes are annuals where the unfavourable season is passed as an embryo in the seed.

The contribution made by each life-form to the overall flora of an area can be expressed as a percentage of the total number of species and the resulting life-form spectrum can be depicted graphically (see Fig. 9.5). The quantitative expression of the life-form spectrum allows a more objective comparison of inter-regional floristic types. Raunkiaer also argued that since the position of the perennating bud was a functional response to climatic conditions, the relative abundance of the different life-forms could be used as a guide to climatic types. He went on to devise phyto-climatic boundaries on the basis of particular life-form spectra, and although other climatic classifications do include reference to vegetation types, for example Koppen (1928) and Thornthwaite (1948), the relationship between vegetation boundaries and climate must be extremely tenuous. The danger of a circular argument

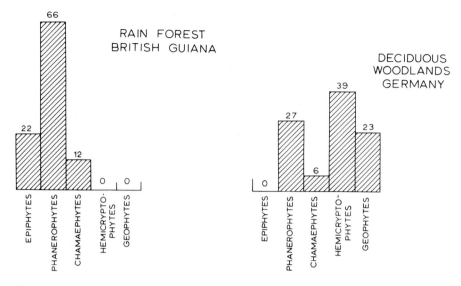

Figure 9.5 Life-form spectra for rain-forest on British Guiana and deciduous woodland in Germany.

here is such that the value of these classifications must be questioned even on a global scale. This does not negate the use of Raunkiaer's life-form classification as a tool for the description of vegetation at other scales. It has been widely used by workers in many different environments. For example, in the tropics where floristic richness, structural complexity and the paucity of up-to-date published floras make even the preliminary description of the vegetation extremely difficult, Raunkiaer's system offers a practical solution to these problems. Richards (1952) has demonstrated its usefulness as a basis for study in tropical situations as have Webb *et al.* (1970). Its value in less complex vegetation types has also been established, for example by Tansley in *The British Islands and their Vegetation* (1939). Indeed its application is made easier in those countries where hand floras include life-form information; for example, Clapham, Tutin & Warburg (1962).

2.1.2   Periodicity

A second functional feature which is often used to supplement life-form information is periodicity. Here periodicity refers to the growth phases of the vegetation or of individual species, and is an approach which has obvious attractions for the description of vegetation in seasonal climates. A record of the evergreen or deciduous character of vegetation is frequently made in preliminary descriptions, but Salisbury's (1916) approach to describing the vegetative and flowering periods of the herb flora of temperate, deciduous woodlands introduces a more dynamic element into these descriptions. He recognized three main growth phases—prevernal, vernal and aestival—and compiled phenological diagrams for the main species encountered. Other studies utilizing this approach describe the growth phases of Raunkiaer's life-forms (Preis, 1939). Periodicity records may be tedious to collect in the field, requiring a long field season for observation, but they can provide a useful basis for more detailed lines of enquiry.

2.1.3 Stratification

Periodicity is often related to the vertical and horizontal structure of the vegetation and in particular to the vertical stratification or layering of components of the vegetation. The recognition of more or less continuous layers of vegetation on the basis of height differences is a structural approach to vegetational description and is inherent in most life-form classifications. On a more local scale a structural approach can be used to simplify and describe the organization of complex vegetation types. Each layer is described in terms of height and in many instances floristic information is used to supplement this preliminary description. A technique pioneered in the tropics by Davies & Richards (1933) involves a complete visual representation of the stratification of the community in a profile diagram (see Fig. 9.6). These authors made careful measurements of the main trees felled along a route through tropical rain forest. In such an environment, characterized by species richness, luxuriance of growth and the overall complexity of the vegetation, a structural approach through stratification provides a useful means of supplementing life-form descriptions. The procedure, however, may be very destructive and expensive of time and effort. Tansley (1939) adapted this approach for use in the description of temperate deciduous woodlands but in this case estimated by visual inspection the height of the various layers *in situ*. Elton & Miller's (1954) classification of habitats, which is designed primarily for field zoologists, also adopts a structural approach through stratification characterized by differences in height.

This structural approach can also be applied to the root component of vegetation as well as to above-ground parts. In this case, a trench dug through the soil exposes the roots which can then be cleaned by spraying with a jet of water. Weaver & Clements (1938) and Coupland & Johnson (1965) provide many examples of root stratification among prairie grasses and other workers have examined the inter-relationships among heath plants (Rutter, 1955) and between calcicole and calcifuge species (Grubb *et al.*, 1969). This approach through stratification is not an end in itself but it can provide a useful preliminary guide to plant/environment relationships which might be

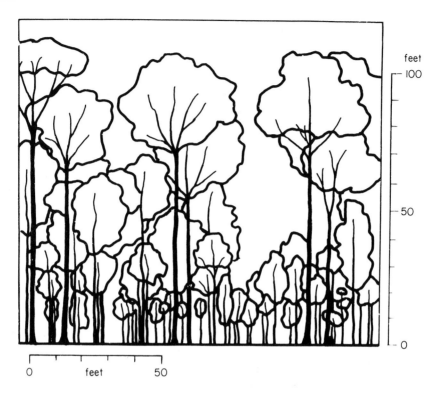

Figure 9.6 Profile diagram of forest (after Davies & Richards, 1933).

profitably examined later, and it provides a clear summary of structurally complex communities.

In summary, all descriptions of vegetation based on physiognomic characters are essentially generalized but they provide a means of arriving at informative descriptions often based on easily measurable properties of the vegetation, relatively quickly. Such characteristics can be isolated even by relatively untrained observers (see Webb *et al.*, 1970) and it is for this reason, above all others, that physiognomic descriptions retain their popularity for reconnaissance surveys among many different users, and over a wide range of scales.

2.2   Measures based on floristics

Detailed studies usually require an assessment of the species composition of an area. This may be accompanied by information about the amount or *abundance* of each species present at a site. It is useful to distinguish between abundance and *richness*, the latter being the number of species present on a particular area.

2.2.1 Destructive measures

Ideally, the ecologist requires data about the relative bulk of different species but the collection of such data requires destruction of the samples taken. This is undesirable if (i) further samples are required (if, for example, annual or long-term changes are being observed; (ii) the area is of outstanding natural beauty or biological interest; or (iii) any of the species is rare.

The most commonly used measures are fresh weight and dry weight, which are self-explanatory. Fresh weight suffers the disadvantage that it varies with the moisture content of the plant; dry weight is determined after drying to constant weight at 105°C. These measures are often also referred to as *biomass* when the total vegetation, rather than individual species is considered, or as the *yield*, if expressed on a per unit area and/or per unit time basis. (See Chapter 1.)

2.2.2 Non-destructive measures

These have the advantages of repeatability and of causing minimal damage to the vegetation. There are several methods, some of which may be subdivided, but none are Utopian in their efficacy (Brown, 1954; Kershaw, 1973; Greig-Smith, 1983). All involve the use of a sampling unit; the size and number required will be dealt with later in the section.

Density

Density is defined as the number of individuals of a particular species per unit area. Counts are usually made in a number of replicate quadrats. This measure is independent of the size of the sampling unit and is said to be 'absolute'. It is much favoured by zoologists who rarely have difficulty in defining an individual, but application to many vegetation types is impractical. The determination of the density of trees, shrubs, tussocks of grasses or sedges, arable weeds and conspicuous individual herbs such as orchids is simple, but species which spread vegetatively, such as grasses and clover, are often almost impossible to deal with in this way. An individual grass plant is impossible to define in a permanent pasture and the counting of tillers is impractical unless the quadrats are extremely small.

Cover

Cover is defined as the proportion of the ground occupied by a perpendicular projection of the aerial parts of individuals of the species under consideration (Greig-Smith, 1983) and is usually expressed as a percentage.

Table 9.2 Cover scales in general use

Class	Domin	Braun-Blanquet	Hult-Sernander	Lagerberg-Raunkiaer
+	A single individual	Less than 1%	–	–
1	1–2 individuals	1–5	0–6·25	0–10
2	Less than 1%	6–25	6·5–12·5	11–30
3	1–4	26–50	13–25	31–50
4	4–10	51–75	26–50	51–100
5	11–25	76–100	51–100	–
6	26–33	–	–	–
7	34–50	–	–	–
8	51–75	–	–	–
9	76–90	–	–	–
10	91–100	–	–	–

Because of the overlayering of different species, the total cover for an area may exceed 100%, and in the case of highly stratified forests may reach several hundred per cent. It is also an absolute measure and may be recorded (i) by visual estimate (ii) with single cover pins (often called point quadrats), or (iii) with frames of pins or frames of cross-wires.

The subjective estimation of cover simply involves a visual estimate. The percentage obtained may then be expressed as a figure indicating the range within which it falls. Such values are often called *Domin* values and are much used by phytosociologists on the continent of Europe (Bannister, 1966). The original use of the Domin scale involved information about cover and abundance, but combined scales are unpopular amongst ecologists striving for objectivity and a simple cover scale is generally preferred.

Although several cover scales have been suggested (Shimwell, 1971) there are two in regular use, namely the Domin and Braun-Blanquet scales (Table 9.2).

These measures may be criticized on the grounds that they are 'pseudo-quantitative' and give little more information than frequency symbols (Tansley & Adamson, 1913), but they are easily and quickly used in the field.

Single cover pins may be used on low vegetation such as grassland to give a quick, easy and accurate measure of abundance. The pins should theoretically have zero cross-sectional area as exaggeration occurs with increasing diameter, as shown in Fig. 9.7. The exaggeration is greater for fine-leaved species than for broad-leaved ones. Bicycle spokes are often used as cover pins and are normally held vertically in the vegetation. The pins may be held in a frame, and moved up and down through a series of holes in two parallel bars, one fixed vertically above the other (Fig. 9.8). Frames using cross-wires are similar in design (Fig. 9.8) except that the holes are larger, or tubes in which wires are inserted are used and are viewed from above (Winkworth &

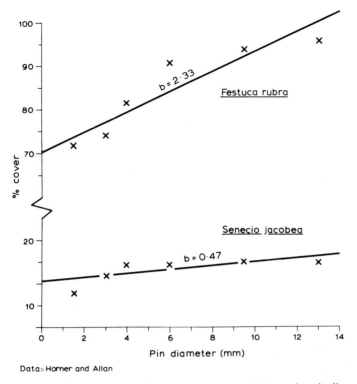

Data:- Horner and Allan

Figure 9.7 A comparison of the exaggeration of cover values with increasing pin diameter for two species.

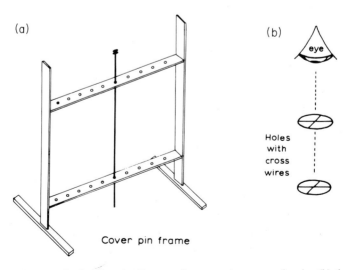

Figure 9.8 (a) Frame for holding pins for recording percentage cover showing (b) the use of cross-wires.

Goodall, 1962). Single cover pins are appropriate where an estimate of cover within a given area is required. Frames of pins are appropriate where a systematic sample (e.g. a transect) is required.

The species touched by each pin between the canopy and ground level is normally recorded although some workers note the number of times each species is touched. This is known as *cover repetition* and is likely to be more highly correlated with yield but suffers the disadvantage of being much more difficult to record. An estimate of leaf area index can be made if cover repetition is recorded using inclined pins (Warren Wilson, 1960).

The use of cover is largely restricted to low vegetation such as grassland and the herb layer of woodlands. It becomes difficult to handle the pins and/or frames in vegetation such as tall heaths and scrub. Cover had been shown to be sufficiently sensitive to record changes in vegetation resulting from change in management or biotic fluctuations. Thomas (1960) for example, successfully used this measure to record changes in chalk grassland following the decline in rabbit populations due to myxomatosis.

Frequency

Frequency is defined as the chance of finding a species in a particular area in a particular trial sample. It is obtained by using quadrats and expressed as the number of quadrats occupied by a given species per number thrown or, more often, as a percentage. It is extremely quick and easy to record and has consequently always been popular amongst ecologists. However, these advantages must be weighed against two quite serious disadvantages:

(a) The frequency is dependent upon quadrat size, and it is for this reason that the measure is considered non-absolute (Greig-Smith, 1983). Consequently care must be taken to select the optimum quadrat size (see below) and this must be stated whenever frequency values are reported.

(b) The frequency value obtained reflects the pattern of distributions of the individuals as well as their density. In other words it expresses information about both pattern and abundance and therefore confuses two basic and important features of vegetation. Figure 9.9 shows three areas each having the same density. However, each will produce a different frequency value because the individuals occur in different patterns.

It is useful when recording frequency to distinguish between root and shoot frequency (Greig-Smith, 1983), i.e. when records are based on individuals of a species which are rooted in the sample area as opposed to records based on the occurrence of any aerial part, even if it is part of a plant rooted outside the quadrat. Root frequency is generally preferred.

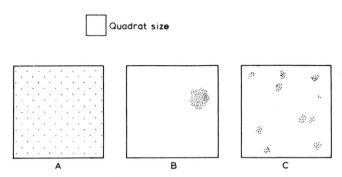

Figure 9.9 Dependence of percentage frequency on the pattern of distribution of individuals. With the same number of individuals present in each case and using the same size of quadrat, widely differing values will be obtained from the three communities. (From Kershaw, 1973.)

Some workers use quadrats subdivided by strings or wires (Archibald, 1949). These often take the form of 50 × 50 cm quadrats subdivided into 25 smaller quadrats. Local frequency may then be expressed for each of these frames and may be useful if a single value is required for a small area. This is sometimes the case when collecting data for ordination or numerical classifications.

Basal area, girth and diameter

These are all measures used for species with large individuals, e.g. tussocks, shrubs and trees. They have been extensively used by foresters and, in association with data on density, permit the prediction of timber yield. As their names imply, they involve the measurement of diameter, girth or area at various heights of the plant. Ground level, as for basal area, is not popular because of the convoluted outline of many individuals due to buttresses, prop roots, etc. Consequently an arbitrary height (4 ft 3 in or 1·3 m and called breast height) is commonly used.

Girth of trees is often measured with special tapes (quarter girth tapes) which for commercially important species permit direct reference to tables for the estimation of timber yield.

Plotless measures

The practical difficulties of using quadrats and cover pins or marking out sample areas in forest and scrub have led to the use of plotless methods to estimate density (see Shimwell, 1971; Greig-Smith, 1983). This sampling design may also be used for collecting information about species composition, growth and environmental factors. Four different procedures have

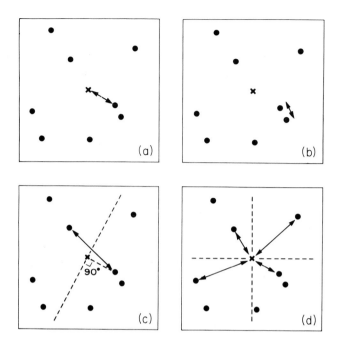

Figure 9.10 Plotless sampling methods: (a) closest individual method; (b) nearest neighbour method; (c) random pairs method, (d) point-centred quarter method.

been suggested (Cottam, 1947; Cottam & Curtis, 1949, 1955, 1956) and are based on a number of random points (Fig. 9.10).

Closest individual method. Measurements are made from each random point to the nearest individual.

Nearest neighbour method. Measurements are made from the individuals to their nearest neighbour.

Random pairs method. Measurements are made from an individual to another on the opposite side of the sample point.

Point-centred quarter method. Distances are measured from the sampling point to the nearest individual in each quadrat.

From the mean value of each of these measurements the mean area and thus the density of the species may be calculated.

$$\text{Density} = \sqrt{\frac{\text{mean area}}{2}}$$

Pielou (1959, 1962) suggests that the correction factor in the denominator should in some circumstances be calculated empirically and has discussed the use of these measures in the study of pattern and of competition.

Performance

Various measures of vigour or performance have been used by ecologists and these are described by Chapman in Chapter 1.

2.2.3 Sampling

The non-destructive and destructive measures described above require the use of some kind of sampling unit, usually a quadrat. As we have seen, the choice of method may be a problem, but even when this has been satisfactorily resolved there still remains the difficulty of deciding on the number, size and arrangement of samples. The choice of sampling method will again depend upon the nature of the problem, the morphology of the species, its pattern, and the time available. Here it is only possible to present the main alternatives and outline their theoretical and practical advantages and disadvantages. The research worker must always make the final decision within the context of the particular problem.

There are six main types of sampling arrangement from which to choose; they vary in their popularity but each is appropriate for some set of circumstances.

SELECTIVE

The quadrats are arranged subjectively to include representative areas or areas with some special feature, such as the species under study. In some circumstances practical considerations may render this the only arrangement possible; for example, where access is difficult or dangerous, as on cliffs. Data collected from such a sampling pattern are not acceptable for statistical analyses involving the assessment of significance such as *t*-tests, *F*-tests, association, correlation or regression, but are suitable for some multivariate techniques such as ordination.

RANDOM

This is often considered to be the 'ideal' method of sampling. Each sample by definition must have an equal chance of being chosen, and the samples should be positioned by using pairs of random numbers as distances along two axes positioned at right angles to each other. A variant of this type of approach, known as a random walk, may involve walking on a compass bearing for a random number of paces, sampling, changing direction and repeating the procedure. Random numbers may be taken from Statistical Tables such as those of Fisher & Yates (1963). Throwing quadrats over one's left shoulder, or any other form of haphazard sampling, does not, however, achieve a random coverage of the sample area.

Many statistical tests assume that data have been collected from a random arrangement of the sample units and, if the research worker is in any doubt, this approach should be adopted.

REGULAR OR SYSTEMATIC

One of the problems with random sampling is that the cover of the area with sample points is not regular, some areas being under-sampled and others over-sampled. Sampling using a grid, however, achieves a regular arrangement, but this does limit the statistical tests which can be applied. If a statistical analysis is not required, then the method has the advantage that fewer sample points will be required compared with random sampling.

Care must be exercised if there is a phased pattern in the area under study. The classic example usually quoted is of broad ridge-and-furrow permanent pasture. If, in this situation, a regular sampling arrangement were used and the phase of the sampling coincided with the phase of pattern in the vegetation, a very distorted result could be obtained. Similar problems could arise with dune ridges and slacks, and polygonal patterns in arctic vegetation.

RESTRICTED RANDOM (PARTIAL RANDOM)

This is a compromise between systematic and random sampling and combines some of the advantages of each. The area under study is subdivided and each subdivision then sampled at random. In this way each point in the subdivisions has an equal chance of being sampled and the data are suitable for statistical analysis. It is, however, more time-consuming than either 'parent' method because the area has to be marked out with a grid as well as necessitating the location of the random points.

TRANSECTS

Transects are a form of systematic sampling in which samples are arranged linearly and usually contiguously. They are very commonly used in studies and are appropriate to the investigation of gradients of change when they should be positioned at right angles to the zonation. For example, on salt-marshes, inter-tidal zones, successions on dunes, hydroseres, altitudinal gradients, from dry to wet heath, and across gradients of trampling intensity.

Another appropriate use of transects is in the study of pattern.

STRATIFIED

This method of sampling has been more extensively used in disciplines other than ecology. It involves subdividing the field of study into relatively homogeneous parts and then sampling each subdivision according to its area, or some other parameter. For example, a mosaic of grassland and scrub could be divided into these two components and then each sampled, or a zonation with clearly demarcated zones is suitable for this type of sampling.

UNALIGNED STRATIFIED SYSTEMATIC

As with restricted random sampling the area to be sampled is subdivided into equal plots; each plot is then sampled using x and y co-ordinates that are unique for each row and column of a two-dimensional grid, and constant for all strata in a row and column (see Smartt & Grainger, 1976). Such a strategy gives good reproduction of both large and small vegetation units and avoids the local clustering of sample points that may arise even with restricted random sampling. Most recently Smartt (1978) has proposed a flexible systematic model for sample location which is designed to distribute samples in such a way that areas of greatest vegetation heterogeneity are sampled most intensively. The two-stage strategy involves an initial skeleton of sample points arrayed systematically, within which subsequent points are interpolated on the basis of the floristic diversity of the vegetation. In comparison to other non-flexible schemes, such a strategy ensures that extensive areas of homogeneous vegetation do not absorb a large proportion of the samples but also protects against gross under-sampling of very diverse but less extensive vegetation units. As with other variants of systematic sampling this flexible approach would appear to be potentially rewarding, particularly for the sampling of large areas and where time is limited and the number of samples to be collected is restricted. However, the results should not be used for statistical analysis and it would not be possible to extrapolate from them to indicate the relative extent of each vegetation unit.

Number of quadrats

When considering the number of sampling units the research worker should adopt the general rule 'the more the better'. However, the objective of sampling, as opposed to recording everything, is to reduce the amount of labour and time involved. The actual number to be used depends on the variability between samples, and plotting the running mean or the variance against the number of quadrats (Fig. 9.11). The minimum number is the number of quadrats that correspond to the point where the oscillations damp down.

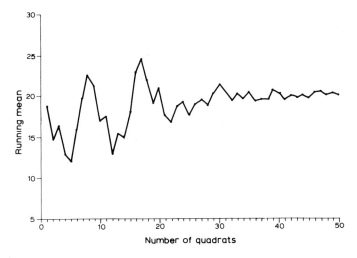

Figure 9.11 Simple test for estimating the minimum number of samples required. The running mean or variance is plotted against the number of quadrats.

Alternatively the number may be reached from a percentage basis. Assuming that the optimal area of the sampling unit has been decided (see below) the optimal number of quadrats can be chosen by deciding whether to sample 5%, 10% or 20% of the area under study. Some statistical tests require a minimum of thirty values and this may be considered a useful indication in the absence of other criteria. Another useful guide is that the variance within a sample area should be less than the variance between sample areas. If the data are relatively uniform, fewer samples are required than if values vary widely. No absolute number can be suggested, however, because the range of heterogeneity encountered in the field is so wide and research projects vary in the degree of accuracy required.

Size of quadrats

The principal considerations in choosing the size of quadrat are the morphology of the species in the vegetation to be sampled and the homogeneity of the vegetation. Small quadrats are appropriate to the study of small plants; e.g. 10 × 10 cm or 25 × 25 cm quadrats may be suitable for chalk grassland, arable weeds and fixed dune grassland, and large quadrats for scrub and woodlands. The sampling unit for the collection of data for vegetation classification or ordination should be related to the homogeneity of the vegetation and be slightly larger than the 'minimal area' of that vegetation (*sensu* Poore, 1955, see below).

Shape of quadrats

Quadrats by tradition are square although one of the causes of error, edge effect, is slightly reduced if the perimeter is reduced relative to the area, i.e. by using round quadrats. Clapham (1932) advocates the use of rectangular quadrats, orientated parallel to the principal gradient of variation in the vegetation. However, the greater the length/ breadth ratio the greater will be the edge effect. The difference that will be obtained with variously shaped quadrats is very small and not an important consideration.

Permanent quadrats, photographic recording

Long-term changes in vegetation are best studied by means of permanent quadrats or permanent transects. Watt (1962), for example, used permanent quadrats to study the long-term changes in grassland after the removal of rabbit grazing and Thomas (1960) used transects for a similar purpose. For large-scale (small-area) studies it is also possible to map individuals reasonably accurately using marked tapes from two reference pegs. Smaller scale studies require techniques of plane-tabling if accuracy is not of paramount importance or surveying using theodolite traverses. At the smallest scale (largest area) aerial photography or a series of old maps are recommended. Moore (1962) used the latter approach to study changes in the Dorset heathlands.

Direct photographic recording of vegetation with hand-held 35 mm cameras may save time in the field but the determination of species abundance subsequently becomes more difficult. Modifications of this type of approach include the use of balloons to hold the camera several feet above the ground (Duffield & Forsythe, 1972).

Marking of experimental plots is usually effected by means of posts, pegs or coloured polythene tags. Wadsworth (1970) however, has recommended a method which is vandal-proof by using magnetic markers which are relocated with a magnetometer.

Minimal area

This is both a concept and a measure, or series of measures, which aim to reflect that tantalizing property of vegetation, its relative homogeneity at different scales. Minimal area is sometimes useful as a concept and a means of characterizing vegetation but very often it produces more confusion than clarification. Shimwell (1971, p. 15) defines minimal area as the smallest area which provides enough environmental space (environmental and habitat

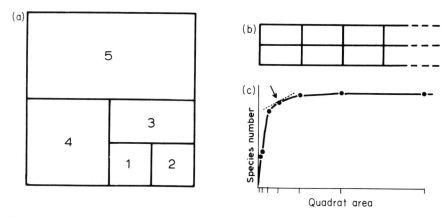

Figure 9.12 Quadrats may be conveniently arranged in the manner shown (a or b) to produce (c) a species–area curve which may be used to identify the minimal area of a plant community (From Shimwell, 1971.)

features) for a particular community type or stand to develop its true characteristics of species complement and structure.

Species number may be plotted against sample area (Fig. 9.12), to produce a species–area relation curve. A small quadrat should be used initially and subsequent ones should double the sample area using a nested arrangement. The quadrats may be conveniently arranged in the manner shown in Fig. 9.12.

Species–area curves initially rise steeply, then level off and sometimes rise again as another phase of the vegetation is entered. The point at which the curve levels off is considered to be the minimal area of the community. It may be identified visually from the graph, as by most phytosociologists on the continent of Europe, or at a predetermined gradient (Hopkins, 1955). This method is simple to use, effective and, although it has no special ecological significance, it may be used as a preliminary guide for sample size in small-scale vegetation surveys (Poore, 1955).

Alternatively, if frequency is estimated using an increasing quadrat size, the number of species with over 90% frequency at each quadrat size is referred to as the number of constants. The number of constants may then be plotted against quadrat size as described above and the minimal area is defined as that area in which the full number of constants occurs.

2.2.4   Indices of diversity

There are two major types of diversity of interest to ecologists: *habitat diversity* and *species diversity*. The former is sometimes used as a measure of ecological interest or value and the habitats of a particular area can be identified using an established list (e.g. Elton & Miller, 1954; Elton, 1966; Society for the Promotion of Nature Reserves, 1969) and can be communicated as the number of habitats per unit area.

Species diversity consists of two related components, viz, species richness and relative abundance/dominance/*equitability* (see below). Species richness is easy to measure but it is important to state the area sampled (see Krebs, 1972; Wratten & Fry, 1980). Most species have few individuals and few species have large numbers of individuals in sample data. Thus, data is not normally distributed but can be normalized by using different geometric scales for the number of individuals represented in each sample. The scales could be $\times 2$, $\times 3$ or $\times 10$, for example (Preston, 1948).

Williams (1964) recommended a measure of diversity independent of sample size where

$$S = \alpha \log_e \left(1 - \frac{N}{\alpha} \right)$$

where S = number of species in the sample, N = number of individuals in the sample, and α = index of diversity. Another index of diversity is provided by the *Shannon–Weiner function* (Krebs, 1972):

$$H = -\sum_{i=1}^{S} (p_i)(\log_2 p_i)$$

where H = index of species diversity derived from the information statistic, S = number of species and p_i = proportion of total sample belonging to the ith species.

There is a slight theoretical advantage to using $\log_2$ rather than $\log_{10}$ or $\log_e$, but any base may be used. In practice for vegetation, this is calculated from the values for the cover, fresh weight or dry weight, of each species and bare ground is usually omitted. We would recommend about 50 quadrats in normal circumstances (see Pielou (1966) for a critical review) located at random. The maximum species diversity occurs if all species are of equal abundance and can be shown to be:

$$H_{max} = \log_2 S \text{ (Krebs, 1972).}$$

The evenness of distribution of individuals between samples or equitability, E, can then be defined as H/H_{max} which is simple to calculate.

Another means of defining diversity can be derived from probability theory as recommended by Simpson (1949):

$$D = 1 - \sum_{i=1}^{S} (p_i)^2$$

Simpson's index gives more weight to common species than the Shannon–Weiner index, but there is little real difference between the methods which we have distinguished as α, H or D indices. In practice—for example, considering the effects of grazing, trampling or fertilizer application on vegetation—species richness is as useful a measure as any of the three other indices.

Foliage height diversity is used by ornithologists to estimate the vertical layering of leaves in forest, scrub and grassland. In forest, for example, the vegetation could be divided into 50 cm layers and the proportion of leaves in each layer estimated from cover values. Then either H or D as indicated above could be calculated.

3   Analysis of vegetation

The techniques discussed above have been concerned with the description of vegetation on the ground. We now consider the techniques available to the ecologist to (i) compare areas, (ii) relate vegetational variation to environmental factors, and (iii) identify controlling factors. In any research or management problem concerning vegetation, reference should be made to both parts of this chapter; the first half to select the measure of abundance, and this second half to select the appropriate method of analysis, as well as to the choice of method diagram (Fig. 9.1).

3.1   Association and correlation

The distinction between association and correlation is made here in the statistical sense. Association concerns two attributes, usually species, which may be present or absent, i.e. qualitative data, whereas correlation concerns two variables which are quantitatively related.

3.1.1   Association

Association is used in ecology in either the abstract sense to refer to a characteristic assemblage of species comparable to a community, that appears as a unit of vegetation or, in the concrete sense, as a measure of the similarity of occurrence of two species. The former is discussed below (p. 474). In the concrete sense it can be described using a statistic, such as chi-squared (χ^2). Presence of the species should be recorded in a large number of random quadrats and the data arranged in the form of a contingency table such that (a) is the number of quadrats containing both species, (b, c) the number with only one and (d) the number with neither. The formula is as shown below:

2 × 2 contingency table

Species A

		+	−	
	+	a	b	$a + b$
Species B				
	−	c	d	$c + d$
		$a + c$	$b + d$	n

$$\chi^2 = \sum \frac{(\text{obs} - \text{exp})^2}{\text{exp}}$$

or $\chi^2 = \dfrac{(ad - bc)^2 n}{(a + b)(b + d)(c + d)(a + c)}$

where the observed and expected values for each of the four cells of the contingency table are used, and n is the total number of quadrats.

The result can be examined for significance by reference to χ^2 tables. For 2 × 2 tables there is one degree of freedom. The result may indicate positive or negative association, depending on whether the observed number of joint occurrences is greater than expected or not. The result is dependent upon quadrat size because the data are of frequency type and should be interpreted with care. It is possible that association may be negative at one scale and positive at another for the same pair of species in the same area.

In most cases, and particularly when the number of samples is small it is advisable to use Yates' correction for continuity, which makes the observed values nearer to the expected by 0·5. Alternatively the formula below may be used which achieves the same result.

$$\chi^2 = \frac{n\{(|ad - bc|) - \frac{1}{2}n\}^2}{(a + b)(c + d)(a + c)(b + d)}$$

The χ^2 test of significance should not be applied when any of the expected values is less than 5. In such cases, the Fisher Exact Test should be applied (Siegel, 1956).

Two species which are positively associated occur together more often than would be expected by chance. This does not necessarily indicate that the presence of one causes the occurrence of the other but more probably that the two are responding to a similar combination of environmental factors. This information is in itself useful, but its real value is indicated when the χ^2 values

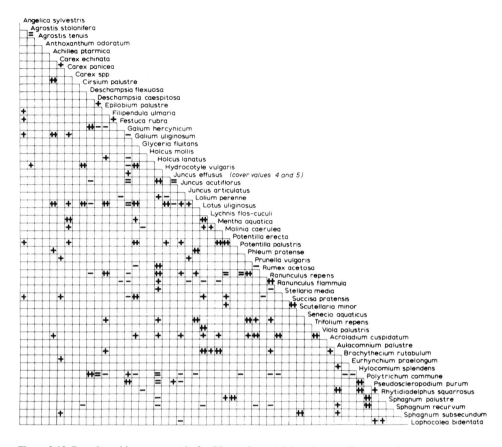

Figure 9.13 Complete chi-square matrix for 99 stands containing *Juncus effusus* showing positive and negative species relationships present. (From Agnew, 1961.)

for all the species in an area are arranged in a half-matrix (Fig. 9.13) as it permits the construction of a constellation diagram (Fig. 9.14). Agnew (1961) used this approach in his study of *Juncus effusus* in North Wales and found it useful as a means of representing the inter-relationships between species. Half-matrices of χ^2 values also form the basis for the numerical classification technique known as association-analysis and this technique is discussed further below (p. 485).

3.1.2 Correlation

The correlation coefficient can be used to relate the quantitative measures of abundance of one species to quantitative values of another and is appropriate when both species occur in most of the quadrats. The relationship between the performance of a species and the level of an environmental

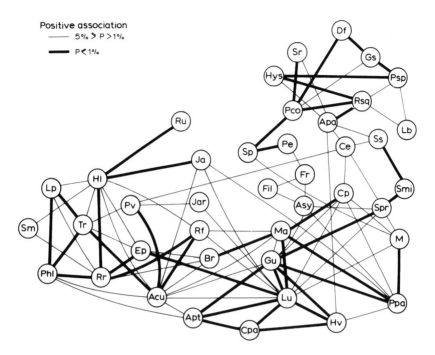

Figure 9.14 A constellation of species inter-relationships based on positive chi-squared values from 99 stands containing *Juncus effusus*. (From Agnew, 1961.)

factor may also be examined using the correlation coefficient. The formula used is:

$$r = \frac{\sum\limits_{n}^{i=1} (x - \bar{x})(y - \bar{y})}{\sqrt{\sum\limits_{n}^{i=1} (x - \bar{x})^2 \sum\limits_{n}^{i=1} (y - \bar{y})^2}}$$

Where r is the correlation coefficient, x and y are the values for each species or environmental factor and $\bar{x}$ and $\bar{y}$ are their means, n = number of quadrats. The significance of r may be obtained from tables using $n - 2$ degrees of freedom.

The correlation of a species with an environmental factor may be used to indicate possible causal factors but care should be taken when interpreting results as the environmental factor tested may itself be correlated with the causal factor. Thus, a highly significant correlation is no proof of cause-and-effect.

3.1.3   Regression

Techniques of regression are described in all statistical text books and the reader is recommended to refer to Bishop (1983) for a very simple, introductory account.

Regression is usually used as a technique to fit a line to a series of co-ordinates on a graph and as a means of assessing the degree to which two or more variables are related. Ideally, one variable should be treated as the dependent or Y variable, and the other as the independent or X variable. The approach is comparable to the calculation of the correlation coefficient, except that with the regression coefficient only the variance of the dependent variable is taken into consideration. Normally, as in the study of the effect of nitrogen level on the abundance of a species, it is easy to identify the independent and dependent variables. Sometimes however there are no *a priori* reasons why one variable should be considered dependent and the other independent, e.g. the regression of one species against another. Whilst this may not matter, it should be realized that the regression coefficient of X on Y will differ from that of Y on X.

Regression analysis is appropriate in a situation where one continuous variable, e.g. abundance of a species, is to be related to a series of levels of another variable, e.g. concentrations of a nitrogen fertilizer. This application most often arises in designed experiments and will not be discussed here (see Fisher, 1960). There are, however, circumstances in the analysis of vegetation in the field when regression is appropriate. One such example is the prediction of timber yield (Y variable) from a series of values for one or more (X) environmental variables. If several independent variables are involved methods of multiple regression are appropriate (Searle, 1966). An example of this type of approach has been used as the basis for the formulation of simple predictive models (see for example, Yarranton, 1969, 1971).

3.2   Measures of non-randomness

Non-randomness in vegetation is often referred to as *pattern*. It may take the form of an aggregation of individuals known as *contagion*, or an even distribution referred to as *regularity* (see Fig. 9.15). The former is more common than either the latter or randomly distributed individuals. The departure from randomness interests the ecologist because it is a way of characterizing the vegetation or a particular species. Also because it must have a cause and it provides an opportunity for identifying the factors that control the distribution of a species.

The causes may be either *intrinsic*, i.e. a property of the plant, or *extrinsic*, due to environmental factors, or both. The former tends to occur at a smaller

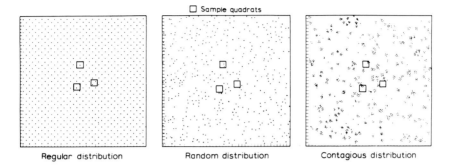

□ Sample quadrats

Regular distribution   Random distribution   Contagious distribution

Figure 9.15 Examples of (a) regular, (b) random and (c) contagious distribution of individuals.

scale than extrinsically-caused pattern and is less interesting to the ecologist. Intrinsically caused pattern may be the result of inefficient dispersal of seed referred to as reproductive pattern, or to the vegetative morphology of the species.

Pattern may be examined in terms of its occurrence, scale, intensity and degree of association between species.

3.2.1 Detection of pattern

Patterns may be detected by examining the departure of observed values from a model distribution which assumes that individuals are randomly distributed. The Poisson distribution is often used and is appropriate where there are few occurrences in relation to the total number possible, i.e. where the mean number of individuals (density) per quadrat is low. The observed number of quadrats containing 0, 1, 2, 3, 4, etc., individuals is first calculated and then the mean number of individuals per quadrat (m) permits the calculation of the corresponding expected values for each of these classes.

Number of individuals per quadrat	0	1	2	3	4	etc.
Expected number of quadrats in each class	e^{-m}	me^{-m}	$\dfrac{m^2}{2!}e^{-m}$	$\dfrac{m^3}{3!}e^{-m}$	$\dfrac{m^4}{4!}e^{-m}$	etc.

where 4! = factorial 4, i.e. 4 × 3 × 2 × 1.
   e = base to natural logarithms = 2·7183.

The departure of the observed values from the expected values may be tested by chi-squared, χ^2:

$$\text{where } \chi^2 = \sum \frac{(\text{observed} - \text{expected})^2}{\text{expected}}$$

summed for the N classes and where the number of degrees of freedom $= N - 2$. For statistical reasons the expected value for each class should be greater than 5. In order to satisfy this condition it may be necessary to pool some of the classes. If the data conform exactly to the Poisson distribution then, by definition, the variance (V) equals the mean ($\bar{x}$) or, $V/\bar{x} = 1$. If there is a tendency towards clumping or contagion, then, $V/\bar{x} > 1$ and if there is a tendency towards regularity, then, $V/\bar{x} < 1$.

$$\text{Variance } (V) = \frac{\sum x^2 - \frac{(\sum x)^2}{n}}{n - 1}$$

where $n = $ the total number of quadrats.

3.2.2   Scale of pattern

Pattern exists in vegetation at a variety of scales ranging from those reflecting different climatic zones of the earth to those reflecting local environmental factors that can only be detected by statistical methods. It is the latter which will be discussed in this section.

It is possible to examine the scale of pattern using the technique described above with different sized quadrats. However, the method generally referred to as 'pattern analysis' and developed by Greig-Smith and Kershaw (Greig-Smith, 1952, 1961; Kershaw, 1957, 1958, 1959) is more efficient. The data may be collected from a grid of nested quadrats or more usually from a transect of contiguous sampling units. Any measure of abundance can be recorded. The calculation required to analyse the pattern demonstrated by each species is a modified analysis of variance. The total sum of squares (variance) of the data is calculated and partitioned into its different component scales (block sizes).

The procedure is best explained by reference to Table 9.3 where entries in the left hand column are the data as collected in the field. The transect may be any length that is a power of 2, for example, 32, 64, 128, or 256 units. The raw data are taken as block size 1 and increasing scales of pattern are examined by adding adjacent scores together in pairs to form larger block sizes. The blocking continues until there is only one value left. Thus, for a transect of 32 basic units there will be block sizes 1, 2, 4, 8, 16 and 32.

For each block size the values are squared and then summed giving $\sum x_1^2$,

Table 9.3 The arrangement of the original data and its blocking in the calculation of pattern analysis

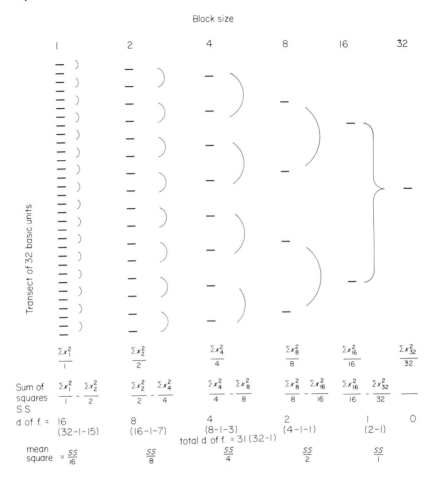

Σx_2^2, Σx_4^2, Σx_8^2, Σx_{16}^2 and Σx_{32}^2. Each of these is divided by the appropriate block size, and the sum of squares for each block size is obtained from:

$$\frac{\Sigma x_1^2}{1} - \frac{\Sigma x_2^2}{2}; \quad \frac{\Sigma x_2^2}{2} - \frac{\Sigma x_4^2}{4}; \quad \frac{\Sigma x_4^2}{4} - \frac{\Sigma x_8^2}{8}; \quad \frac{\Sigma x_8^2}{8} - \frac{\Sigma x_{16}^2}{16} \quad \text{etc.}$$

The mean square is obtained by dividing the sum of squares by the corresponding number of degrees of freedom (number of observations minus one, minus the number of degrees of freedom already accounted for).

The mean square, an estimate of variance, can then be plotted against the corresponding block size to obtain a *pattern analysis* graph (Fig. 9.16). This graph illustrates the intensity and scale of pattern for the species under study. There are problems involved in assessing the significance of the peaks

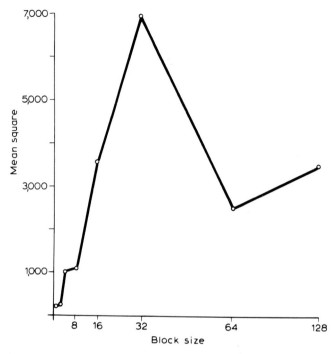

Figure 9.16 An example of a pattern analysis graph. Three scales of pattern in *Trifolium repens* are shown by three peaks, at block sizes 1, 4 and 32 on the graph of mean square/block size. (From Kershaw, 1973.)

(Thompson, 1958) and it is generally preferred to construct several graphs from several sets of data and to examine the peaks for consistency. When it is possible to assess significance (e.g. with density data), the confidence limits increase with increasing block size, so that a small peak at a small block size may be more significant than a higher peak at a larger block size. A table of confidence limits is given in Greig-Smith (1983).

Pattern analysis has proved to be a useful ecological tool, but the technique as described above has a number of weaknesses, e.g. the problem of testing significance already described. Two other problems are worth mentioning. The first arises from the hierarchical blocking procedure which gives blocks of 1, 2, 4, 8, 16, etc. units, with the result that a pattern equivalent to a block size 12, for example, may be overlooked or imprecisely estimated. The second problem is related to the starting position of the transect; the block size at which pattern is indicated being sensitive to this, and an underestimating (usually by one block size) of the scale of pattern can result. In practice, this may only be important at the larger block sizes.

A number of suggested solutions to these problems has been put forward (e.g. Usher, 1969; Goodall, 1974) and a relatively straightforward solution

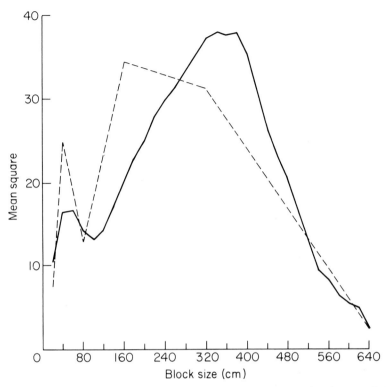

Figure 9.17 Data analysed by pattern analysis (-----) and two-term local variance analysis
(———). Data from Morton (1974).

has been suggested by Hill (1973a). Hill's analysis, referred to as two-term local variance analysis, is a modification of pattern analysis which gives mean squares for all block sizes 1, 2, 3, 4, 5, etc. up to half transect length. The transect length is not restricted to a power of 2, but can be any even number. Because of the way in which the mean squares are calculated, the results are not affected by the starting position of the transect. Fig. 9.17 shows results of pattern analysis and two-term local variance analysis applied to the same data. Note the more precise estimate of pattern scale at block size 60 cm, and the underestimate (by pattern analysis) of the 360 cm peak.

Hill's analysis is much more time-consuming to perform than Greig-Smith's analysis and a computer is necessary for reasonable sized data sets. The routine is, however, easily programmed.

Spectral analysis has been suggested for analysis of pattern in transect data (e.g. Ripley, 1978). The analysis attempts to fit a sine wave to fluctuations in abundance of a species along a transect. This is in contrast to the rectangular form of model inherent in the analysis of variance of blocked data. Spectral analysis has not been widely applied to analysis of pattern in ecological research, so its usefulness in this respect cannot be assessed at present.

3.2.3 Intensity of pattern

The height of the peaks on the pattern analysis graphs gives an impression of the 'intensity' of pattern which, in a contagious distribution, is related to the degree of aggregation. It is difficult, however, to compare the intensity of pattern from one set of data to another because the variance of a contagious distribution increases in a non-linear way with the mean. Hill (1973a) suggests the use of the statistic

$$\frac{(\text{variance} - \text{mean})}{\text{mean}^2}$$

to describe pattern intensity. This statistic is probably only appropriate to density data.

3.2.4 Pattern and correlation

Useful information can be obtained by comparing pattern analysis graphs for different species and for environmental factors. If a species has a consistent peak at the same block size as an environmental factor it may indicate a cause-and-effect relationship and this should be further studied experimentally.

The relationship between the patterns of pairs of species or a species and an environmental factor along the same transect may be further examined by means of the calculation of a covariance analysis. The procedure to obtain the mean squares at different block sizes for each of the two species (V_A and V_B) is first followed. Then the data for the two species are added together, quadrat by quadrat, and the corresponding mean squares calculated (V_{A+B}). Then,

$$V_{A+B} = V_A + V_B + 2C_{AB}$$

Where C_{AB} is the covariance between the two species. It may be interpreted directly or converted into the correlation coefficient, r: where

$$r = \frac{C_{AB}}{\sqrt{V_A V_B}}$$

This form of covariance analysis may also be applied to Hill's two-term local variance analysis.

The correlation coefficient can be plotted against the block size for

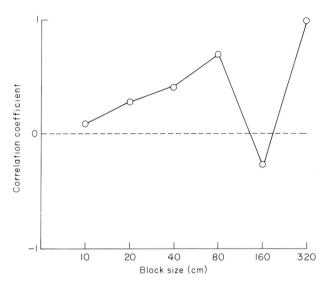

Figure 9.18 Graph to show correlation between *Festuca rubra* and *Carex flacca* in a fixed dune grassland. Both species showed pattern analysis peaks at 80 cm. Data from Morton (1974).

different species pairs. It may be positive at one block size and negative at another (Fig. 9.18) which may indicate that the two species are responding similarly at one scale but not at another. The significance of the correlation coefficient can be tested.

Pattern and covariance analyses provide the ecologist with techniques for describing the spatial variability of species distribution as well as for investigating the environmental factors that control species distribution. The analyses are applied to visually homogeneous vegetation as opposed to vegetational gradients, although the amount of 'textural' variation within a visually homogeneous area does not matter. When pattern and covariance analyses are used to identify causal factors the caution concerning hypothesis generation must be made, i.e. the detection of correlation is no proof of cause-and-effect. Experimental work is necessary to test hypotheses erected as a result of correlations. The methods have considerable potential in the study of competition, especially in the form of covariance analysis, although examples of this application are few.

Examples of the application of pattern analysis are numerous and the reader is recommended to refer to Phillips (1954) for work revealing the morphology of the rhizome system of *Eriophorum angustifolium*, and to Greig-Smith (1979, 1983) for a review of pattern caused by soil water and nutrients, animals, interactions between plants, disturbance, fire, inefficient dispersal, historical factors, and chance.

3.3   Classification and mapping

Classification is a sorting procedure which creates groups or classes of similar objects which are defined by the possession of certain prescribed attributes. Some vegetation classifications are based on classes which are arbitrarily assigned using criteria such as physiognomy, structure and floristics. On a small scale (i.e. when dealing with a large area) these criteria do not lend themselves readily to statistical treatment. Other classifications achieve a greater degree of objectivity by erecting classes within which there is a minimum of variation so that the objects, together with their attributes, satisfy more rigorous, statistically based criteria. Such classifications are based upon floristic information and include the association-analysis of Williams & Lambert (1959, 1960) and a group of methods referred to as cluster analysis (Sokal & Sneath, 1963). They are primarily used for large-scale (small-area) studies of the variability of vegetation encountered *within* particular community types.

3.3.1   Small-scale (large-area) classifications

Two basic approaches are available at this scale. The first proceeds by the subdivision of broad, inclusive classes to provide a large number of small, exclusive groups; for example, the methods suggested by Fosberg (1967) and Ellenberg & Mueller-Dombois (1967) and earlier, by Drude (1913). The second is an agglomerative method which proceeds by recognizing a large number of small exclusive units which are subsequently joined together to form large, collective groups; for example, the method advocated by Braun-Blanquet (1932) and closely followed by Tuxen (1937) and to a lesser extent by Poore (1955). In either situation, however, the central objective of the classification is to provide a description of the vegetation of large regional areas within a single, unitary framework.

Divisive methods

Methods which first identify large major vegetation classes and then proceed by subdividing these into successively smaller classes generally employ their life-form criteria as the divisive character; for example, the classifications of Rubel (1930), Du Rietz (1930, 1931) and Ellenberg (1956), or structural criteria, as for example in Fosberg (1961) and Drude (1913). In either case these main divisions correspond to the major vegetation formations of the world, e.g. forest, woodland, grassland, etc. Subsequent divisions can be made on a variety of criteria such as physiognomy, structure, periodicity, function, habitat and floristics. Fosberg (1961, 1967), for example,

Table 9.4 Fosberg extract table (from Fosberg, 1967)

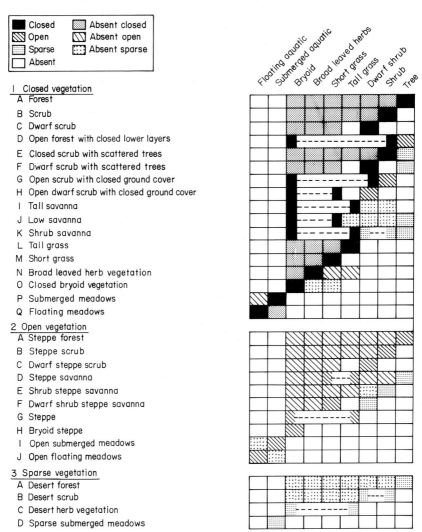

Legend:
- Closed
- Open
- Sparse
- Absent
- Absent closed
- Absent open
- Absent sparse

Column headers: Floating aquatic, Submerged aquatic, Bryoid, Broad leaved herbs, Short grass, Tall grass, Dwarf shrub, Shrub, Tree

1 Closed vegetation
A Forest
B Scrub
C Dwarf scrub
D Open forest with closed lower layers
E Closed scrub with scattered trees
F Dwarf scrub with scattered trees
G Open scrub with closed ground cover
H Open dwarf scrub with closed ground cover
I Tall savanna
J Low savanna
K Shrub savanna
L Tall grass
M Short grass
N Broad leaved herb vegetation
O Closed bryoid vegetation
P Submerged meadows
Q Floating meadows

2 Open vegetation
A Steppe forest
B Steppe scrub
C Dwarf steppe scrub
D Steppe savanna
E Shrub steppe savanna
F Dwarf shrub steppe savanna
G Steppe
H Bryoid steppe
I Open submerged meadows
J Open floating meadows

3 Sparse vegetation
A Desert forest
B Desert scrub
C Desert herb vegetation
D Sparse submerged meadows

dismisses physiognomy as being inexact, floristic composition as requiring specialist knowledge, habitat as possibly involving a circularity of argument; he recommends and adopts a system based on structure and function. His primary structural groups are based on spacing (open, closed and sparse) and a second division into formation classes (forest, scrub and grass) is based on vertical stratification (Table 9.4). The two formation groups within each formation class are based on function and are evergreen or deciduous; similarly the formations are based on function (orthophyll/sclerophyll; microphyll/mesophyll/megaphyll). This system is logically based and has

Table 9.5 Ellenberg's seven formation classes

---

   i Closed forest
  ii Woodlands
 iii Fourrés (scrub)
 iv Dwarf scrub and related communities
  v Terrestrial herbaceous communities
 vi Deserts and other sparsely vegetated areas
vii Aquatic plant formations

---

provided a useful framework for describing and processing IBP check sheets. Descriptions of the method are available in Peterken (1967) and Kuchler (1967).

An alternative system to that of Fosberg has been prepared for UNESCO by a working party (Ellenberg & Mueller-Dombois, 1967) and is based on plant life-forms or physiognomy but includes ecological (habitat) criteria. There are seven major formation classes (Table 9.5). Each formation is subdivided into formation subclasses, formation group, formation and subformation. This method has been less widely used than Fosberg's system probably because of the involvement of ecological criteria. This is unfortunate if the classification is to be used for the subsequent correlation of the vegetation classes with environmental factors because there is a risk of circular argument. However, practical trials have indicated that it is as easy to allocate vegetation in the field to one of these categories as it is any other of the numerous physiognomic systems (Goldsmith, 1974). Other workers such as Richards, Tansley & Watt (1940) suggest that 'vegetation should be primarily characterised by its own features not by habitat, indispensable as is the study of habitat for an understanding of its nature and distribution'. They argue that it is structure and composition which should form the basis of the description of plant communities. Indeed most early classifications were based upon physiognomic characters such as those reviewed above (see p. 443). Works such as those of Rubel (1930), Beard (1944), Cain & Castro (1959), Danserseau (1961) and Kuchler (1967) include examples of these physiognomic classifications. In addition a number of small-scale (large-area) classifications have been devised with particular problems in mind, e.g. Warming's classification (1923) emphasizes plant/water relations, and is useful for the description of aquatic vegetation, and Raunkiaer's life-form classification (1934) emphasizes the relationship between plant growth and the severity of the climate. In comparison to these classifications and those early classifications based upon physiognomy, the more recent classifications of Fosberg, and Ellenberg & Mueller-Dombois, are best described as synthetic physiognomic methods which bring together into one system many of the useful characteristics of other more specialized classifications. For this reason they are more flexible and likely to satisfy the needs of small-scale (large-area) studies more readily.

Agglomerative methods

In a recent review Beard (1978) suggests that an emphasis on physiognomy in the study of local vegetation, particularly in the tropics and subtropics, has lead to effective systems of formations that can be reasonably subdivided into floristic units. He also suggests that a floristic treatment of vegetation on a local level can be combined with a physiognomic treatment and hence assist in the interpretation of the broader biogeographic relationships of plant communities (see Beard, 1967).

The second approach to small-scale vegetation description is an agglomerative method which is associated with the Zurich–Montpellier school of phytosociology under the guidance of Braun-Blanquet (1932). This system has found widespread acceptance among continental workers and has a growing number of followers among British and American workers. While its initial aim was to provide a classification framework for the vegetation of the world, realistically it has been most fruitfully used on regional and national scales. For example, the *Prodrome des groupements végétaux d'Europe* aims to provide a comprehensive syntaxonomy of European communities based on the accumulated studies of phytosociologists working in several countries (Beeftink & Gehu, 1973). Such classifications seek to erect a hierarchy of plant communities arranged from the simplest unit, the association, to the most complex, the class (Table 9.6) based on a floristic analysis of vegetation samples (relevées). Plant communities are conceived as types of vegetation recognized by their floristic composition, and for practical classification the approach seeks to use certain diagnostic species to organize communities into a hierarchical classification. Proponents of the method argue that the hierarchy organizes and summarizes a vast amount of in-

Table 9.6 Taxonomic units of the Braun-Blanquet system of classification

Rank	Ending	Example
Class	-etea	Festuco–Brometea
Order	-etalia	Brometalia erecti
Alliance	-ion	Mesobromion
Association	-etum	Cirsio–Brometum
Sub-association	-etosum	Cirsio–Brometum caricetosum
Variant	–	(specific names used)

formation and is invaluable for the understanding and communication of community relationships (Westhoff & Maarel, 1978).

The Braun-Blanquet method proceeds in three phases: analytic, synthetic and syntaxonomical. The analytic phase involves reconnaissance of an area, choice of sample plots and the full description of the vegetation samples or relevées. The field procedure is well documented by Moore (1962) and Mueller-Dombois & Ellenberg (1974). Traditionally only those stands which are recognizably homogeneous in composition are selected for description. Their selection is often subjective, even though the 'minimal area' of an association is frequently advocated as providing the basis for selection (see above, p. 459). In the absence of any statistically based sampling design, the overall appearance of the vegetation is likely to dictate to the observer which stands are selected, and this in turn is influenced by the field dominants as much as it is by the experience of the observer. Sample selection is one of the most intractible problems facing the investigator wishing to work on a regional scale for, in the absence of any definite knowledge of the area and characteristics of the vegegation to be surveyed, even systematic or random sampling is difficult. At present, phytosociologists on the continent of Europe favour subjectively-chosen sample plots of varying sizes to arrive at a representative inventory of the vegetation concerned: thus, a plot 1–25 m^2 is regarded as adequate for sampling herbaceous vegetation, while in small scrub, a plot 25–100 m^2 is preferred, and in forests, plots of 200–500 m^2 are used for trees, with appropriate sizes substituted for the shrub and herb layers (Table 9.7).

The floristic record for each stand includes a cover–abundance rating for each species on a 5-point scale (see above) and each is assigned a sociability index, also derived from a 5 (or 10)-point scale (see Table 9.8). The dual value for each species, together with information on various environmental factors relevant to the site, is entered into the field book or special form or card. The second phase of synthetical research takes place in the laboratory where several relevées from similar vegetation types are used to compile a large data table referred to as the raw table (see Table 9.9). By a process of sorting and continuous refinement, similar stands are grouped together on the basis of their floristic composition. This process of sorting relies heavily upon the recognition of potential differential species. These are species which may be mutually exclusive in their occurrence in the table or species which have joint occurrences over a range of sites. Such species are used as the basis upon which several partially ordered tables are produced. A final orderly extract table (see Table 9.10) arranges the stands in groups which can be characterized by the presence or absence of certain more-or-less mutually exclusive species which are termed differential species (underlined in Table 9.10). At this stage the worker returns to the field to establish the

Table 9.7 Minimal area values for various communities

Community	Area (m^2)
Epiphytic communities	0·1–0·4
Terrestrial moss communities	1–4
Hygrophylous pioneer communities (Isoeto–Nanojuncetea)	1–4
Dune grasslands (Koelerio–Corynephoretea)	1–10
Salt marshes (Asteretea–Tripolii)	2–10
Pastures (Lolio–Cynosuretum)	5–10
Mobile coastal dune communities (Ammophiletea)	10–20
Hay meadows (Arrhenatheretalia)	10–25
Heathlands (Nardo–Callunetea)	10–50
Alpine meadow and dwarf shrubs (Elymo-Seslerietea)	10–50
Calcareous grasslands (Festuco–Brometea)	10–50
Chaparral, temperate sclerophyll shrubland	10–100
Weed communities (Secalietea)	25–100
Scrub communities (Rhamno–Prunetea)	25–100
Steppe communities	50–100
Temperate deciduous forest (Querco–Fagetea)	100–500
Mixed deciduous forest (North America)	200–800
Tropical secondary rain forest	200–1000
Tropical swamp forest	2000–4000

validity of the groups which have been erected in the laboratory. The laboratory procedure is well documented in Kuchler (1967), Shimwell (1971), Moore (1962), and Mueller-Dombois & Ellenberg (1974). The subsequent recognition of distinctive associations defined by character species can only be made after a considerable number of relevées has been described from a wide geographical area. Character species, which were originally thought to be almost exclusive to a particular association, represent a special case of the differential species used in the sorting of the data tables. Their selection for a particular association can only be made by understanding the concepts of constancy and fidelity. Here, constancy refers to the percentage occurrence of a species across a range of relevées and is established by reference to a 5-point presence scale. Fidelity refers to the selective preference exhibited by a species for particular communities. Both constancy and fidelity can only be

Table 9.8 Sociability and cover–abundance classes

Sociability	1	Growing once in a place, singly
	2	Grouped or tufted
	3	In troops, small patches or cushions
	4	In small colonies, in extensive patches or forming carpets
	5	In great crowds or pure populations

Table 9.9 Raw table of saltmarsh community from Blakeney Point, Norfolk

	1	2	3	4	5	6	7	8	9	10	11	12	13	14	15	16	17	18	19	20	21	22	23
Aster tripolium	+	1	2	2	2	3	2	1	1	1	3	3	2	+	2	1	2	2	1	1	2	3	2
Halimione portulacoides				2	1		3	1	5				+			3	+	1	1	+	+	+	+
Limonium vulgare			+			+		1	+	+	+	+		+			+	1		1	+		
Limonium humile			+	1		1	+				1	+	+				+			+			
Puccinellia maritima							2	2	2	2			4				+	3			1	+	2
Salicornia herbacea	2	2	4	4	3	4	2	1	2	+	2	4	1	2	3	1	3	2	1	1	1	2	2
Salicornia perennis							1	1				1			1			+					
Spartina anglica						1	+	1	+	+	1	+	+			+	2	1			+	3	1
Suaeda maritima	1	2	1	1			1	1	1	+	1	+	1	2	+	+	+	1	+	+	+	1	2
Triglochin maritima							+		1									1		+			
Bostrychia scorpioides			+	+	+	+	1	+	1	1	5	4	4		+	1	1	2	3		1	4	2
Enteromorphia spp.	3	+	+	+	+	+					1		+	2	+		+	+	+	+			
Enteromorpha intestinalis	+	+	+	+	+	+	+				+		+		+	+	+						
Fucus volubilis	1	2	2	1	1	+	+		+		+	1	+	2	1		+						
Pelvetia canaliculata							+						3							+			

Table 9.10 Orderly extract table of saltmarsh community from Blakeney Point, Norfolk

	46	13	24	19	45	21	36	9	7	11	44	23	33	49	50	18	22	34	35	48	12	30	37
Pucinellia maritima	4	4	3	3	3	3	2	2	2	2	1	1	1	1	1	±	±	±	±	±			
Limonium vulgare	1	+	2	1	1	1	+	1	+	+	+		+	2	1	+	+	+	1		+	+	1
Limonium humile	1	+			1	+	+		+	1	1			+	+	1	+	1	+	1	+	1	1
Triglochin maritima	4		1	1		+		+		1	2				+						+		+
Spartina anglica	1	+	1	1	1	+	+	1	1	+	+	1	1	+	+	2	3	+	+	+	1	+	+
Bostrychia scorpioides	1	4	2	3	3	1	1	1	1	5	1	2	4	3	3	2	4	1	2	2	4	1	2
Halimione portulacoides		+	+	1	+	+	+	1			+	+				1	+	+					1
Salicornia perennis	+	1	+	+			1		2										+				
Pelvetia canaliculata								+						3		1			3	+			
Fucus volubilis		+	+				+	+	+	+			+	+					+	+	1	+	
Enteromorpha spp.	±	±		±				1				±		1	±					±			
Enteromorpha intestinalis		+					+	+						+					+				
Suaeda maritima	1	1	+	+	+	+	1	1	1	1	2	+	+	+	1	1	1	+	1	+	+	+	+
Salicornia herbacea	1	1	2	1	2	1	+	2	2	2	2	2	1	2	3	2	2	1	2	1	4	2	1
Aster tripolium	2	2	1	1	3	2	1	1	2	3	2	2	1	2	2	2	3	1	1	3	3	1	2

```
24 25 26 27 28 29 30 31 32 33 34 35 36 37 38 39 40 41 42 43 44 45 46 47 48 49 50   Frequency

1  1  1  1  1  2  1  1  1  1  1  1  1  2     1  2  1  3  2  2  3  2  1  3  2  2
1           1  +  1  3        +     +                 2  3  +  +     3
2     +        1           +  +  1  +              +     +  1  1     +  2  1
            +  1  +     +  1  +  +  1        +  1  1  1  1  1  +  1  +  1
3              +     1  +  +  3                    1  3  4  1  +  1  1
2  +  2  2  2  2  +  2  1  1  2  +  1  1  3  2  1  2  2  2  2  1  1  1  2  3
+                                      2     +  1  1
1        +     +  +     1  +  +  +  +  +        +  +  1  +  1  1        +  +  +
+     1  +  +  +  +  +  1  +  1  +  +  +  +     2  +  1  +  1  +  1  1  1  +  +
1                          +     +                 +           2  +

2                 1  +     4  1  2  1  2           1  1  1  3  1  1  2  3  3
   5  +  +     +     1  +              +  1        2     1        +  +  +     1
   +           +  +  +  +  +                       1  +                       +
+  +  2  1  2  +  +              +     +  1  2  1  1  1     +           +  +  +
               +        +     1                       +     +
```

```
10 47 8   31 42 43 16 5   6   28 20 17 29 25 1   2   3   4   15 14 41 26 38 27 32 39 40

2  1  2  ±
+           +              +  +              +
   +     +  1  1        1           +           +  1        +

+     +  +  +  1           +     +                 +  +     +
1  1  +  +  1  1  1  +  +        1              +  +  +
5  3  3  3  2  3  3  2  1  1  +  +  +
   1  1              1
            +     +
+           1  1     +  +  2        +  +  1  2  2  1  1  2  1  2  1  1        2  1
±     ±  1     1     ±  ±     ±  ±  ±  5  3  ±  ±  ±  ±  2  2  ±  1  ±  ±
         +  1  +  +  +  +  +        +  +  +  +  +  +  +  +  +              +

+  1  1  +  1  +  +           +  +  +  +        1  2  1  1  +  2  +  1  +  +  1     2
+  1  1  +  2  2  1  3  4  2  1  3  2  +  2  2  4  4  3  2  1  2  1  2  2  3  2
1  1  1  1  3  2  1  2  3  1  1  2  2  1  +  1  2  2  2  +  1  1     1  1  1  2
```

determined once the full range of the species concerned and the communities in which they occur have been fully studied. Where high constancy is accompanied by a high degree of fidelity the species can be used as character species and these latter species apply at the level of the Association, Alliance and Order (see Table 9.6). But although constancy and fidelity are related, this is not always the case and, for this reason, a more flexible approach to defining character species has been adopted in recent studies.

Becking (1957), Shimwell (1971) and Szafer (1966) illustrate how the concept of character species now embraces (i) absolute and relative character species, and (ii) regional and local character species. It nevertheless remains to be said that even these concepts require a fairly thorough knowledge of the vegetational variation on a local and regional scale. The final or syntaxonomical phase involves the assignment of the provisional units identified in tabular rearrangement to previously recognized associations of an established hierarchy or to new associations. Established hierarchies exist for the vegetation of western and central Europe and Scandinavia, so that in several cases absolute character species can be assigned at both the level of the Association and the Alliance. These character species are clearly documented in the continental publications, notably in those contributions to the journal *Vegetatio*, in Oberdorfer (1957) and Braun-Blanquet (1964). Elsewhere in the world, the establishment of character species will have to wait until vegetational descriptions have progressed more widely. Nomenclature of syntaxa is detailed in Barkman *et al.* (1976).

Criticisms of the Braun-Blanquet method by British workers include that of Pearsall (1924) with respect to minimal area, and that of Tansley & Adamson (1926). Poore's reassessment and modification of the method (1955) includes a positive attempt to reconcile the British tradition of reliance on the field dominants with the need for more accurate floristic description and comparisons. Moore (1962) also reiterated Poore's attempt to consider the Braun-Blanquet method as a reconnaissance technique worthy of serious attention. The modifications which Poore suggested have probably served to confuse rather than to clarify but he recommends the use of dominants as well as constants to characterize the lowest vegetation units which he terms *noda*. *Dominants*, he defines as those species occurring with the highest cover–abundance ratings using the Domin scale (see above) and *constants* as those species occurring in more than 80% of like relevées. Where, within a nodum, the number of species occurring in Constancy Class V (> 80% of the lists) is greater than the number occurring in Constancy Class IV (60–80% of the lists) the unit is raised to the status of the association. Poore utilizes Sørensen's coefficient (see below) of community similarity as a more objective means of comparing like noda. Poore's approach is very similar to that of Dahl (1956) in his description of the mountain vegetation of Rondane, in

Norway, that of McVean & Ratcliffe (1962) and Burnett (1964) for Scottish vegetation, and that of Edgell (1969) for a survey of Welsh upland vegetation. Fundamentally all these methods are attempting to provide a realistic frame of reference within which the vegetation of a relatively large geographical area can be adequately described. Poore thus recognizes that two noda may overlap in their ranges and suggests that any new stand of vegetation described from a similar area could be located within the framework of variation provided by his reference points, the noda.

The advantages which these methods of vegetation classification provide are several. They are primarily reconnaissance methods and are particularly useful at a local to regional scale. They are easy to apply in the field and quick to execute. The arrival of sorting routines and computer programmes have removed the tedious and somewhat 'muddled' procedure required to compare hundreds of relevées. Tabular rearrangement is now achieved systematically and rapidly using the programs TABORD (Maarel *et al.*, 1978), PHYTO (Moore, 1970) and PHYTOPAK (Huntley *et al.*, 1981), the latter favouring an ability for observer manipulation, and the versatile program TWINSPAN—a two-way indicator species analysis (Hill, 1979)—allows for tabular rearrangement based on a polythetic divisive method. Many of these programs have taken advantage of advances made in the storage and retrieval of large data sets and of developments in multivariate clustering techniques. The analysis of very large data sets (1000 + relevées) is best approached through the use of initial clustering programs such as COMPCLUS (Gauch, 1982).

Application of the Braun-Blanquet method

In situations where a rapid and comprehensive inventory of vegetation is required the Braun-Blanquet method has immediate advantages. For example, the *National Vegetation Classification* commissioned by the Nature Conservancy Council in part applies this method together with numerical classification, with the objective of providing a full description and classification of all the semi-natural vegetation of Britain. The classification will provide research workers and practitioners with a comprehensive frame of reference within which studies and land-use decisions can be undertaken, and incidentally will do much to foster a better understanding among European phytosociologists. Other successful applications of the Braun-Blanquet method using computing facilities are to be found in Jensén (1979), and Moore & O'Sullivan (1978). Most practitioners of the method will agree with Huntley *et al.* (1981) who note that despite the simplicity of most of the systems for classifying vegetation from a physiognomic or structural basis and the logical approach of many of them (e.g. Fosberg's) there remains the problem of sampling. The ecologist needs to decide how to choose his sample stands and to determine their size and number. Whilst the discussion about sampling design for quadrats (see above, p. 457) is appropriate here there are special problems encountered with the methods used on an extensive scale. It is generally accepted that these techniques are subjective and many authors advise that 'representative' areas should be described. But who is to choose what is representative, and at what scale? If mapping is the ultimate objective, it may be acceptable to recognize only areas sufficiently large to be mappable at the scale selected (Kuchler, 1967). If, however, nature conservation is the primary objective, there may be extremely small areas such as vertical cliff faces, to which endemic or rare species are restricted, whose description should be subjected to extensive initial reconnaissance prior to sampling to familiarize the ecologist with the range of vegetation represented. The use of aerial photographs may speed up this stage but extensive direct observation on the ground is nevertheless essential. The range of the main vegetation types should be noted, the approximate extent of each estimated, and the number of samples determined in proportion to the area of each type. The arrangement of stands within each vegetation type ideally should approximate to randomness, and the size of each should be chosen on the basis of its minimal area (see above, Table 9.7). It is thus desirable that several minimal area curves be prepared for each vegetation type in the study area, but if time is limiting, and the worker is familiar with the vegetation, a visual estimate may be sufficient. In the final analysis however, as the phytosociologists on the continent of Europe have reported, sampling at a small-scale (large-area) must rely heavily upon the experience of the research worker both with respect to familiarity with the methods and the area under study.

3.3.2 Large-scale (small-area) classifications

The Braun-Blanquet method has been used by a number of workers as a basis for detailed investigations of vegetation/environmental relations, but it is frequently argued that the vegetation units erected by the small-scale classifications are so subjective and generalized that they do not provide useful reference points. This dissatisfaction arises from the growing awareness of the complexity of vegetational variation. But even on a large scale there is still a need for some means of simplifying this complexity, and the simplification must not be at the expense of informativeness. Physiognomic and structural differences do not adequately reflect the complexities of large-scale variation whereas both qualitative and quantitative measures of species occurrence are considered to be relevant since they are likely to reflect local differences in environmental gradients more closely. It is for these reasons that large-scale studies recourse to some numerical treatment of species occurrence (see below) and to the numerical classification of groups based on differences and/or similarities between floristic samples. These numerical classifications can either proceed by division from above as, for example, in Goodall (1953) and Lambert & Williams (1966) or by fusion from below, as in the agglomerative methods such as cluster analysis. Fundamentally, however, these two different strategies work to the same end.

In the following sections we shall assume that a classification of the stands is required and thus the word STAND will be used instead of individual, vegetation sample, quadrat or relevé, and SPECIES will be used to mean attribute or variable.

Divisive methods

Divisive procedures for classification can be either monothetic or polythetic. In monothetic procedures each division is made on the basis of a single species (e.g. its presence or absence), whereas in polythetic procedures each division is based on some or all of the species.

DIVISIVE MONOTHETIC METHODS

These methods achieve a classification of the N stands by successive divisions into smaller groups using a single species at each division. Williams & Lambert's (1959) association-analysis has been the most widely used of these methods and the normal analysis derives a hierarchical classification of stands by divisions based on the presence or absence of a single species at each division. Association-analysis is a development of Goodall's (1953) original approach, but his method produced a non-hierarchical classification which was based on positive associations only and is little used today.

The starting point for association-analysis is a stand-by-species matrix containing qualitative (presence and absence) data. A 2×2 contingency table is then calculated for each species pair and χ^2 calculated (see above). Williams & Lambert (1959) favoured the index $\sqrt{\chi^2/N}$, where N is the number of stands in the group to be divided, as a measure of association; this is calculated for each species pair and placed in a species-by-species matrix. Summing the columns or rows of this symmetric matrix gives $\Sigma \sqrt{\chi^2/N}$ for each species, and the species with the highest value of this is taken to be the species giving rise to maximum heterogeneity. The N stands are then divided into two groups according to whether or not they contain that species. These two groups are then each subdivided by the same procedure, and so on. Division is usually continued until there are no significant ($P = 0.05$,

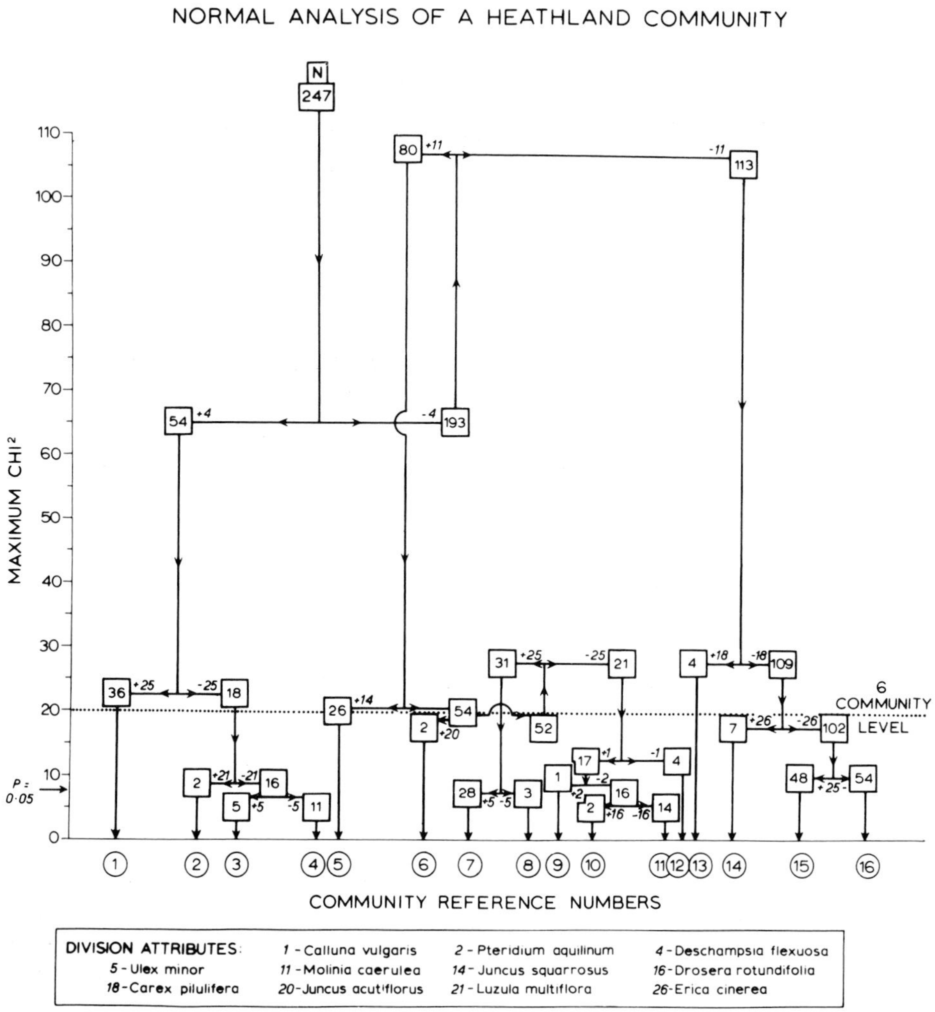

NORMAL ANALYSIS OF A HEATHLAND COMMUNITY

Figure 9.19 An example of normal association analysis hierarchy (from Harrison, 1970).

$\chi^2 = 3\cdot84$) associations remaining. An example of this normal association-analysis hierarchy is given in Fig. 9.19.

This method has been widely used in many different community types ranging from the tropical savannas (Kershaw, 1968) and tropical rain forest (Austin & Greig-Smith, 1968) to temperate grasslands (Gittins, 1965) and heaths (Harrison, 1970; 1971). Most analyses are performed on data collected from local areas or small regions, although Proctor (1967) used data collected on a national basis to examine the distribution of liverworts in Britain. It is considered to be one of the most useful classificatory procedures for the preliminary analysis and reconnaissance of vegetation and the end groups

SIX COMMUNITY LEVEL OF A HEATHLAND COMMUNITY

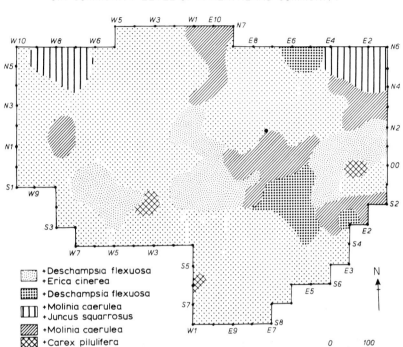

Figure 9.20 Map of normal association-analysis end groups.

can be mapped (see Fig. 9.20). The data are easy to collect, although some difficulties arise, notably in the choice of quadrat size and the influence of pattern (Kershaw, 1961), since small quadrats introduce strong negative associations (see above, p. 462). Kershaw also points out that certain frequent species may dominate the analysis and for this reason most programmes specify that species occurring in more than 98% or less than 2% of the quadrats are removed. In addition, differences of richness between quadrats remain in the data even after standardization and these also affect the hierarchy. Nevertheless, association-analysis in this form has been widely used by ecologists and has proved useful and informative particularly at both the preliminary reconnaissance stage of enquiry and as a summarizing procedure.

Association-analysis also exists in inverse (Williams & Lambert, 1961) and nodal form (Lambert & Williams, 1962). With inverse analysis the inter-relationships between samples are examined by inverting the data matrix so that species become individuals and the quadrats in which they occur, their attributes (Table 9.11). The resultant end groups reflect species

Table 9.11 Some possible combinations of 'individual' and 'attributes' in the analysis of ecological data matrices

Individual	Attribute	Comment
Stand	Species	Most common situation. Finite number of possible attributes.
Stand	Environmental factors	Infinite number of possible attributes.
Species	Stands	Species ordination.
Environmental factors	Stands	Ordination of environmental factors.
Stand	Species + Environmental factors	Not recommended. Interpretation confusing.

of similar performance. For example, in Fig. 9.21 end group A represents a mixed heath community, group B a scrub community, and end group E a distinctive community of wet hollows. The remainder represents variable grass heath units. Nodal analysis draws together the results of both normal and inverse analysis into a single, two-way table in which the normal end groups are arranged horizontally and the inverse groups, vertically. Each cell of the matrix is then subjected to an approximation of both normal and inverse analysis. The sites and species on which the divisions are made in each cell are termed coincidence parameters. Where a cell is fully defined in both directions by the presence of two coincidence parameters, the cell is termed a nodum (Fig. 9.22). Where the cell is only fully defined in one direction it is termed a subnodum (Lambert & Williams, 1962). In this way groups of species which characterize certain site groups can be revealed and vice versa. Technical difficulties in execution, the problem of species richness and abundance, together with the emergence of other techniques which offer similar facilities (Lambert & Williams, 1966) have meant that in practice inverse and nodal analysis have only limited use in ecological investigations. However, it is pertinent to point out that both the underlying aims and the mechanics of these numerical methods have much in common with those of more traditional phytosociological methods. For example, Ivimey-Cook & Proctor (1966) suggest that 'association-analysis is of value in examining the consistency of evidence and conclusions in phytosociology and in detecting directions of variation overlooked by traditional methods, and that it provided a powerful tool for the detection of faithful (character) and differential species'.

Another form of divisive monothetic classification is described by Lance & Williams (1968). This routine (DIVINF) divides on the basis of pre-

INVERSE ANALYSIS OF A HEATHLAND COMMUNITY

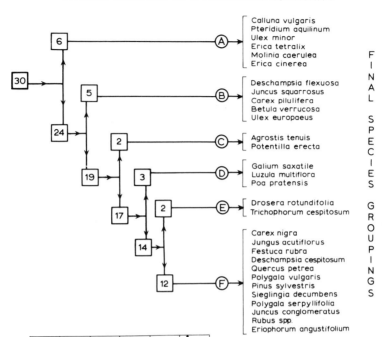

Figure 9.21 An example of an inverse association-analysis hierarchy. (From Harrison, 1970.)

sence or absence of a single species at each division, but the species is selected as that which gives the maximum 'information-fall' after division. A description of information theory and statistics is beyond the scope of this chapter and readers are referred to Greig-Smith (1983, p. 183) for a description.

Finally, it should be stated that, in theory at least, monothetic methods are prone to 'misclassification'. This refers to the fact that a stand may be similar in most respects to one group of stands, but it may be placed in another group on the basis of the presence of one particular species. The divisive polythetic techniques described in the next section are more reliable in this respect.

DIVISIVE POLYTHETIC METHODS

On theoretical grounds, divisive polythetic procedures are superior to both divisive monothetic and agglomerative procedures because a maximum amount of information is used at the major (first) divisions of the hierarchy. Until fairly recently, however, divisive polythetic procedures have not been

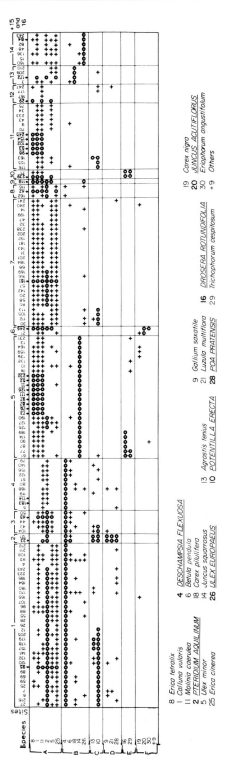

Figure 9.22 A two-way table of a heathland community (nodal analysis). (From Harrison, 1970.)

usable because of the length of the necessary computations. At each dichotomous division there is a large number ($2^{n-1} - 1$) of possible combinations of stands which could form the resulting two groups. For example, with only 15 stands to classify there are 16 383 possible solutions, and each has to be tested in order to find the optimal division into two groups. Even with fast modern computers this is only possible for a small (less than about 20) number of stands. One way of circumventing this problem, however, is to use a *directed search* strategy. This involves placing the stands in such an order that the optimal solution can be arrived at by simply finding the position in this order where the split into two groups should be made. In effect, the stands are ordinated along one axis and the 'correct' position for the split along this axis is determined. This reduces the number of trial groupings from $2^{n-1} - 1$ to $n - 1$, which means that data consisting of very large numbers of stands can be classified. The procedure is repeated at each division and can be continued down to any desired level in the hierarchy. Lambert *et al.* (1973) describe the directed search approach in more detail and in their routine AXOR they use principal components analysis to order the stands along the maximum variance axis and a coefficient ΔSS (the between-group sum of squares) which is calculated for each possible split, the maximum value of which indicates the optimal split. Other ordination techniques could be used for ordering the stands of each group to be divided; for example, Morton & Bates (unpublished) in their routine DIVRA use reciprocal averaging.

A method proposed by Noy-Meir (1973b) uses principal components analysis, but the subdivision into groups is based on successive components (axes) in a single ordination rather than a re-ordination of stands within each group formed by division. The criterion for deciding where the optimal split along the axis should occur is based on the stand scores (positions) on the axis and is not based on a recalculation of sum of squares from the original stand × species data matrix as it is in AXOR. The statistic used by Noy-Meir is the sum of variances within groups,

$$\bar{s}_p^2 = s_1^2 + s_2^2 = d_1^2/(n_1 - 1) + d_2^2/(n_2 - 1)$$

where d_1^2 and d_2^2 are the sums of the squares of the deviations within groups from the group means for the two groups being considered. This statistic is calculated for each of the $n - 1$ ordered splits and the optimal split is where this statistic is minimal. Noy-Meir suggests that a hierarchical type of division is appropriate for principal components analysis; i.e. the first split into two groups is based on the first component, then each of the two groups thus formed are examined separately on the second component, and so on for subsequent components. The attractiveness of this method is that it is a simple appendage to an ordination technique and is relatively low in its demands on computing time.

A method suggested by Gauch & Whittaker (1981), detrended correspondence analysis space partition (DCASP) is a variant of Noy-Meir's method which uses detrended correspondence analysis (Hill & Gauch, 1980) instead of PCA, and the divisions are made subjectively rather than being optimized mathematically.

Indicator species analysis (Hill *et al.*, 1975) is a method of polythetic divisive classification based on the ordination method of reciprocal averaging (Hill 1973b). It derives its name from the fact that, after an initial dichotomous division of stands is made on the ordination axis, a group of species is defined which are indicative of each of the resulting groups of stands. At each division the method proceeds as follows:

1   The first axis of a reciprocal averaging ordination is extracted and the mean of the stand scores (positions) is calculated.

2   The stands are divided into two groups; those with scores less than the mean (the 'negative' group) and those with scores greater than the mean (the 'positive' group).

3   Each species (j) is given an 'indicator value',

$$I_j = |(m_1/M_1) - (m_2/M_2)|$$

where m_1 and m_2 are the number of stands in which the species occurs on the 'negative' and 'positive' sides respectively of the dichotomy, and M_1 and M_2 are the total number of stands on each of the two sides of the dichotomy. Thus, a species which is a perfect indicator of either the 'negative' group or the 'positive' group will have a score of 1. Other species will have scores of less than 1, with those occurring equally on both sides of the dichotomy scoring 0.

4   The species are placed in order according to their indicator values and an arbitrary number, usually five, of species are selected which are the best indicators of either the 'negative' group or the 'positive' group. These are the 'indicator species'.

5   Using the indicator species only, 'indicator scores' are calculated for each stand. The indicator species of the 'negative' group are assigned a value of -1, the 'positive' indicators are assigned $+1$ and these values are summed for each stand. If, for example, three of the indicator species each score $+1$ and the other two each score -1, the indicator scores for the stands will range from $+3$ to -2, giving a simplified ordination of species based on the indicator species only. This simplified ordination can be plotted against the original first axis of the reciprocal averaging ordination (see, e.g. Fig. 9.23).

6   A division of the stands into two groups is then made. Firstly, an 'indicator threshold' is set up to divide the stands into two groups on the basis of the indicator scores. This is shown in Fig. 9.23 between the values of 0 and

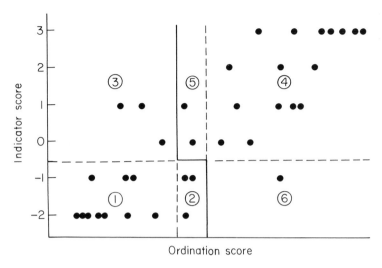

Figure 9.23 Scatter diagram showing the relation between indicator score and original ordination score (hypothetical data). The numbered regions are referred to in the text. (From Hill *et al.*, 1975).

−1, a horizontal broken line. The stands in the regions 1 and 4 are the negative and positive groups respectively, both in the indicator score ordination and in the original ordination. Stands in regions 3 and 6 are 'misclassified'; their classifications in the two ordinations are discordant. The position of the indicator threshold is adjusted to give a minimum number of misclassifications. The misclassified stands are placed in the group dictated by the original ordination. The original ordination is now divided at a point which agrees as closely as possible to the selected indicator threshold, allowing a narrow (typically, 10% of the axis length) 'zone of indifference' for borderline cases (see Fig. 9.23). The stands in the borderline regions (2 and 5) are placed in the group dictated by the indicator score ordination. Thus, referring to Fig. 9.23, stands in the regions 1, 2 and 3 are placed in one group and stands in the regions 4, 5 and 6 are placed in the other group and the whole process recommenced for each of these two groups.

The results obtained by this somewhat tortuous process have the advantage that the classification can be used as a key to vegetation stands, and because the key utilizes five species at each dichotomy it is less prone to misclassification than a key produced by a monothetic classification such as association analysis. Indicator species analysis is primarily used with qualitative data but a modification involving 'pseudo-species' makes the method usable with quantitative data. The reader is referred to Hill *et al.* (1973) for a more detailed account of the method, and to Gauch & Whittaker (1981) for a comparison with some other methods.

Pentecost (1980) used indicator species analysis in a study of the bryophytes and lichens of rock outcrops in Snowdonia. The analysis clearly identified groupings of cryptogams associated with rock-type (first division) and exposure and bird-perching sites (second division). Bunce (1982) has used the method to derive a classification of woodland ground vegetation in Britain and Goldsmith *et al.* (1982) used it to identify indicator species for different types of broadleaved forest as a basis for forest management.

TWINSPAN, two-way indicator species analysis (Hill, 1979), is a modification of indicator species analysis which gives both a classification of stands and species. These classifications are then used to produce an ordered two-way table of the original data. The results of this technique are similar in some ways to the association tables produced by phytosociological methods and TWINSPAN has been adopted as one of the main methods used in the National Vegetation Classification supported by the Nature Conservancy Council.

Agglomerative methods

A large number of agglomerative methods of classification are now available that use information on all the species present in the data, i.e. they are polythetic (see Gauch & Whittaker, 1981; Orloci, 1978). Their objective is to build up a hierarchy of successively larger and larger clusters until ultimately all clusters fuse into a single cluster that contains all the stands. Two steps are involved in the classification procedure. First the initial data matrix of stands-by-species is transformed (Greig-Smith, 1980) into a matrix that expressed 'dissimilarities' between all sample pairs. Dissimilarity is generally computed along one of the numerous distance measures that have found widespread acceptance among ecologists (see for example p. 507). Williams *et al.* (1966) review a large number of indices of similarity and dissimilarity, among them squared Euclidean distance, standardized squared Euclidean distance, several non-metric coefficients such as that of Czekanowski (see below) and Bray and Curtis and the information statistic. The second stage involves a choice of sorting strategy whereby stands, now characterized in terms of geometric dissimilarity, are fused together to form clusters. Numerous sorting strategies are available (see CLUSTAN). Single-linkage or nearest neighbour clustering is widely used in taxonomic research but Hill (1975) points out that it is seldom appropriate for most ecological situations. It produces straggly clusters which quickly agglomerate very dissimilar stands and, unlike most other sorting strategies, the information content of the cluster does not improve upon that of the first individual. More useful are complete-linkage and average-linkage strategies.

The former produces tight clusters of similar stands and two clusters fuse when their dissimilarity (distance) is equal to the greatest dissimilarity for any pair of stands with one in each cluster. It is hence sometimes called the maximum or furthest-neighbour method. Average-linkage sorting fuses clusters on the basis of the average dissimilarity, and the most common method of computing the average is the unweighted, arithmetic average. Sokal & Sneath (1973) point out that this latter method maximizes the match between distances in the original data matrix with the distances first used to fuse any given stand pair in the resulting dendrogram. Minimum-variance clustering produces a cluster that is based on the minimization of a squared distance weighted by cluster size (Orloci 1967). The output from all of these agglomerative methods takes the form of a branched hierarchy or dendrogram in which fusion levels and cluster membership are readily identifiable. Most of the polythetic agglomerative methods discussed here have been applied across a wide range of ecological data and, as with other numerical classificatory techniques, have been used to assist with the sorting of Braun-Blanquet tabulation (see Campbell, 1978). Frenkel & Harrison (1974), Moore *et al.* (1970) and Gauch & Whittaker (1981) provide comparative assessments of several methods using field and simulated data.

Perhaps the single most telling disadvantage of these methods is their reliance on a large number of initial calculations both to store the dissimilarity matrix and to execute the numerous initial calculations that are needed to form the first fusions of the hierarchy. Gauch (1982) states that computation time is non-linear and costly for large numbers of stands, and that dendrogram display is for all practical purposes restricted by the size of paper to perhaps 50–100 stands. Another weakness is that the maximum amount of information is used at the lower, less important, part of the hierarchy.

3.3.3 Vegation mapping

Many methods of vegetation classification lend themselves very readily to vegetation mapping and other methods have been designed specifically for this purpose. Kuchler (1967) has provided a valuable summary of these methods and of the kinds of decisions which have to be taken during the preparation of such maps. For example, the first consideration is the purpose for which the map is designed and the second is the scale at which the map is to be drawn. A third consideration is the method for characterizing the mappable unit, that is between physiognomic/structural characteristics and community units based on floristics. Small-scale maps with a representative fraction of between 1:1 000 000 and 1:100 000 will only allow generalized features of the vegetation to be shown, while large-scale maps on a scale between 1:10 000 and 1:50 000 allow a greater amount of detail to be shown. None of these decisions, however, can be made without reference to a fourth factor, namely the characteristics of the area to be mapped. For example, a large, flat area of uniform vegetation requires different considerations from the same area in a complex mountainous district. In effect the design of a map has therefore to be a compromise between these four considerations.

Kuchler (1967) and Dansereau (1957) have both designed methods of vegetation mapping for use on a small scale. Because each is prepared for the sole purpose of mapping there is a heavy reliance on the use of symbols and no implications are made about the universal applicability of the classificatory units. Kuchler's method relies upon preparatory work being accomplished on air photographs (see below) before field work commences, and most methods of small-scale vegetation classification such as those of Fosberg and Ellenberg (see above) require that air photographs or ground survey maps be available before mapping can proceed. These photographs and maps need to be on as large a scale as possible for use in the field even though the final map may be prepared on a small scale. In tropical countries and the more remote parts of the globe, such basic information may be lacking, in which case vegetation mapping can only proceed alongside ground survey work, or if special reconnaissance flights are undertaken (Poore & Robertson, 1964). Where floristic information is required for mapping on scales between 1:5000 and 1:50 000, Braun-Blanquet's method may prove suitable. In this case the method is based on detailed floristic inventories made in the field and reference to air photographs is only made for ease of field orientation or providing a base map upon which mapping in the field can proceed. Mapping the vegetation units generated during the laboratory analysis of selectively chosen samples (see above) proceeds using cadastral or topographical maps, which are readily available in most European countries on a range of relatively large scales. The community units recognized in the laboratory are used to design a key and the units are checked for their 'real' value in the field and the vegetation within the area assigned to one of them. Maps prepared using this system of classification are published at a variety of scales, many of which are beautifully coloured, and Kuchler's bibliography includes a full inventory of the classic maps. The community units of this system of classification were originally designed to embrace the vegetation of the world but current practice among the European phytosociologists reduces the emphasis on universal applicability and each map is designed to represent adequately the floristic composition of the vegetation for the particular area under examination (Ellenberg, 1954).

Mapping on a large-scale (small-area) presents more problems to the ecologist because large-scale cadastral or topographic maps may not be available or are out of date. In this case the student has to prepare his own base map using aerial photographs or ground survey methods such as the theodolite traverse, chain surveying and plane tabling. Suitable manuals which include methods of survey for large-scale maps are Curtin & Lane (1955) and Cain & Castro (1959). All maps should be provided with a scale and a linear grid for ease of interpretation and for the location of randomly selected or systematic samples. Such an approach is obviously useful in the preparation of maps to illustrate the end groups of the numerical classificatory methods described above. Preparing maps in open and wooded country can each present their difficulties. Taylor (1969) offers a method for the rapid reconnaissance mapping of open country on relatively large scales, but mapping in woodland can best proceed by careful ground surveying as suggested above. In either case reference should be made to all available reference material such as topographic maps, estate maps and air photographs before embarking on time-consuming and precise surveys.

The use of isonomes for the diagrammatic representation of local variability in community composition is a technique similar to the transect, in that it permits the diagrammatic representation of vegetation variation in the field (Ashby & Pidgeon, 1942). A grid of sample units is marked out in the field and information about the species and often the environmental factors are collected and transferred to graph error. Contours (isonomes) are then drawn around species showing the distribution of uniform areas of abundance and so producing a series of maps. This simple technique, however, only works satisfactorily where gradients of variation are pronounced and the study covers a small area.

3.3.4 Use of aerial photography

The use of aerial photography in ecology and especially in the preparation of vegetation maps can be traced back many decades (e.g. Stamp, 1925) and has been widely used overseas in resource survey work (Bawden, 1967). The range of methods currently available includes various types of remote sensing, i.e. infra-red imagery, multi-spectral photography and radar imagery, as well as photography *sensu stricto*, and the choice of method depends upon the purpose.

Overlapping, vertical panchromatic prints are usually viewed in pairs, to produce stereoscopic images. Simple mosaics of prints may be useful for initial interpretation and photographs taken at an oblique angle to the ground may be used for special purposes, such as identifying archaeological features. The production of photogrammetric maps is a subject in its own right and usually requires an initial ground survey to correct for vertical and horizontal aberrations in the plane's flight path. However, the transfer of ecological information to such maps can be carried out by ecologists providing they have had some introductory training. Although the preparation of vegetation maps is the most commonly encountered application there are a variety of specialist uses reported in the literature and these include recording the extent of fire, incidence of crop disease (Colwell, 1956), counting of animals (Perkins, 1971), recording of individual tree species (Howard, 1970, p. 235) and mapping of exposure (Goldsmith, 1973a). In theory, the ecologist is faced with the choice of type of imagery, scale and time of year for flying and should consult a standard text on aerial photography (Howard, 1970). In practice, however, if he is working on a limited budget, the characteristics of the imagery may be determined by the type of cover available and his problem is principally one of interpretation.

The advantages of using aerial photography are that large areas can be covered rapidly, can be viewed stereoscopically, and vegetation boundaries can be mapped very accurately. These advantages become increasingly important as the scale of the area of study decreases and its degree of inaccessibility increases.

The major disadvantages of vegetation mapping using these techniques are that characterization must be largely based on physiognomy and the problem of producing an initial classification has to be resolved. In practice a ground survey may be conducted initially and then as many vegetation types as possible characterized on the imagery using differences in tone, texture and pattern (spacing).

Examples of the application of aerial photography by the Nature Conservancy in the British Isles are discussed by Goodier (1971) and this includes papers describing the Dartmoor Ecological Survey in which both vegetation maps and maps of the distribution of grazing animals were prepared from air photographs.

Trials to compare four types of multispectral imagery (infra-red colour, infra-red panchromatic, true colour, panchromatic) have recently been conducted and some of these (Goldsmith, 1972) indicate that a combination of types of imagery may provide more ecological information than can be obtained from any one alone.

3.3.5 Satellite imagery

Earth resources satellites carry a number of imaging systems including radar and multispectral optical sensors. Radar systems have been used largely for oceanographic applications whereas multispectral scanners are particularly useful for sensing the vegetation cover of the earth's surface. The data recorded by these scanners are the intensities of reflected radiation in each of a number of wavebands. The Landsat 3 satellite, for example, records in four bands representing the reflected green, red and infra-red (2 bands) radiation. Different vegetation types have different spectral responses and can thus be distinguished. The ground resolution of satellite imagery is relatively coarse compared with that produced by airborne systems, e.g. Landsat 3 produces data for pixels which represent ground areas of 79 × 79 m, after processing. The scanner scans the earth's surface recording data for contiguous pixels and these data are transmitted to a receiving station and stored. The data are then made available in the form of magnetic tapes, each tape comprising a scene covering an area of 185 × 185 km. Each Landsat scene is recorded by the satellite in approximately 25 seconds, and the same scene is recorded every 18 days.

The digital data recorded on magnetic tape are pieced together to form an image of the scene, or part of it, by an image analyser. The image analyser is linked to a fast computer which is capable of processing the large amounts of pixel data efficiently and rapidly and the image is displayed on a visual display unit screen. A suite of computer programs is used for image representation, enhancement and classification, and a typical sequence of events in processing an image would be:

1   production of single band grey images;
2   merging of these to form a composite image;
3   assigning a different colour to each separate band to give a false colour composite;
4   production of a classified image.

The production of a classified image is akin to the production of a vegetation map and there are a number of ways of achieving this. One method is to utilize 'training areas'. This requires knowledge of the location of relatively homogeneous areas of distinct vegetation and other land cover types within the area of the image being processed. The image analyser is then used to statistically characterize the pixels in these areas and this information is used by the computer to assign each pixel in the whole scene or subscene to one of the defined classes. The individual classes can then each be assigned a different colour and the classified image displayed. The computations involved during classification are lengthy and the speed of the computer limits, to some extent, the range of methods which can be used.

One of the main possibilities for application is in the monitoring of change or loss of habitats. Vegetation mapping can be achieved as described above and, because satellite imagery is multitemporal, images can be compared at different points in time. Another important advantage of using satellite imagery in this way is that the data are in digital form and the areas covered by different vegetation classes can simply be estimated from the number of pixels in each class. The main limitations of this application are of course related to the accuracy of the classification, the ground resolution, and the availability of cloud-free scenes. The possibility of monitoring habitat change using satellite imagery is a relatively recent development, following the launch of the first Landsat satellite in 1972; but it is likely to be of increasing importance in future years, particularly with increased spectral and spatial resolution.

Multispectral data can also be used to estimate the productivity of growing vegetation. The red and infra-red reflectance is particularly important here; actively photosynthesizing vegetation absorbs red light, giving a low red reflectance but a high infra-red reflectance. This means that stressed vegetation can be detected and leads to the possibility of satellite imagery being used to detect or monitor the effects of pollutants. The imagery cannot, however, attribute the stress to a particular cause, and the resolution currently available may not be sufficient for some purposes. To some extent, pollutants themselves can be detected and monitored; for example, smoke plumes from industrial sources, and oil slicks in the marine environment.

Another possible application of satellite imagery is in the detection of areas of unimproved semi-natural vegetation; for example, the detection of unimproved grazing land within a matrix of improved grasslands. The unimproved pastures have a different spectral response due to their lower productivity, different species composition and sometimes, accumulations of leaf litter. The use of satellite imagery in this way could indicate those areas worthy of more detailed ground-based ecological survey.

The main limitation referred to above is that of spatial resolution, most of the data currently available having a 79 m resolution. Landsat 4, launched in 1982, carried a 'thematic mapper' which gave improved spectral (7 bands) and spatial (30 m) resolution. The system failed, however, after a short period and only a very limited amount of data are available. Landsat 5 was launched in March 1984 and also carries the thematic mapper. Data from this will greatly facilitate the applications described above. The launch, in 1985, of the French SPOT satellite, with its multispectral ground resolution capability of 20 m will further facilitate these applications. For a textbook on remote sensing see Curran (1985).

3.4   Vegetational and environmental gradients

Gradients in vegetation have generated considerable debate about their causes, significance, and the most appropriate methods of study. Vegetational gradients usually reflect environmental gradients and may be static or dynamic. An extreme case of a static gradient is a zonation. If dynamic gradients have been demonstrated and are moving in a directional manner they are part of a successional sequence (Clements, 1916). The principal disparity of opinion occurs between workers who emphasize the homogeneity of vegetation, play down the importance of gradients and consider classification the most appropriate way of representing vegetational variation

(Poore, 1955; Moore, 1962), and those who consideer such variation to be essentially continuous. The proponents of the *continuum concept* (McIntosh, 1967b) favour different methods for representing vegetational variation, such as gradient analysis and ordination. The nature of vegetational variation, however, hovers tantalizingly between these two extreme points of view (Webb, 1954). The situation is further complicated by the cyclical changes which take place in vegetation (see above).

In practice, it is important that the research worker be familiar with the alternative concepts of vegetation which can be achieved by reference to the key papers given above and then select a method by considering the objectives. If a classification is required a classificatory technique should be chosen and if a gradient analysis or ordination is wanted an alternative technique should be chosen. Classes of vegetation can nevertheless be ordinated.

It may be useful to distinguish between direct gradient analysis where vegetation samples are arranged using observed environmental gradients, and indirect gradient analysis, where the samples are arranged along axes generated from the vegetational data (Shimwell, 1971; Whittaker, 1967). The former include the continuum method (Curtis & McIntosh, 1951) and gradient analysis (Whittaker, 1956) and the latter the various types of ordination procedure (Goodall, 1954; Gauch, 1982; Greig-Smith, 1983; Pielou, 1984).

3.4.1 Gradient analysis

Methods of gradient analysis were first proposed by Whittaker (1956) and they offer a means of representing a continuum in a very simple manner. However, they give most satisfactory results where one or two environmental gradients have marked overall control of vegetational variation. Vegetation samples or species are arranged along axes characterized by these controlling environmental factors (Fig. 9.24).

Loucks (1962) has modified and developed this technique by constructing synthetic scalars for the axes. These integrate two or more environmental factors to position sites in the form of a two-dimensional diagram. However, the selection of factors and the construction of the scalars is highly subjective and is most likely to be successful where the principal controlling factors are obvious. Elsewhere the construction of ordinations based on more objective methods (see below) is to be recommended.

Gradient analysis may also be used to relate vegetational variation to aspect by constructing polar diagrams. This approach has also been successfully used by Perring (1959) to demonstrate variation on chalk grassland and Goldsmith (1967) to examine the pattern of deposition of salt spray on sea-cliffs (Fig. 9.25).

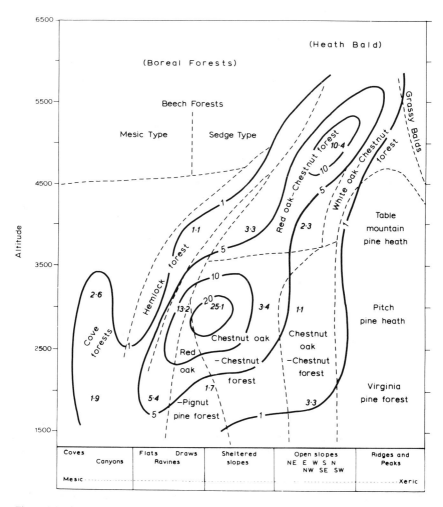

Figure 9.24 An example of gradient analysis (from Whittaker, 1956).

Austin (1972) has re-examined his chalk grassland data and calculated 'environmental dynamic scalars' from stepwise regression analysis of the performance of selected species, environmental factors and seasonal factors. Then the response of, say *Carex flacca* can be predicted on the basis of a multiple regression with the scalars and he prefers this approach to principal components analysis alone. This technique, however, assumes that the scalars are independent of each other and normally distributed.

3.4.2   Ordination

An ordination is a spatial arrangement of samples such that their position reflects their similarity. It is used as a framework upon which to

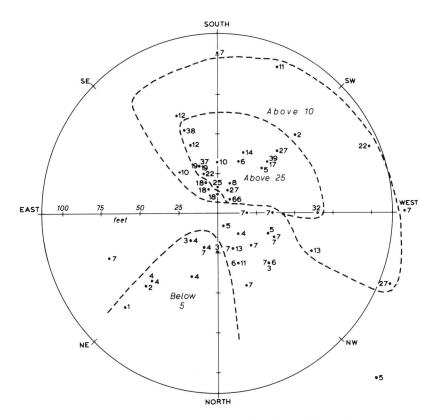

Figure 9.25 An example of a polar gradient analysis. Sea-cliff stands have been arranged using altitude on the radii and aspect on the circumference of the graph. Soil conductivity values have been plotted against the size positions to indicate the distribution of salt spray on sea-cliffs.

compare species and environmental factors as a basis for erecting hypotheses about cause-and-effect relationships. There are two basic types of ordination:

1 Environmental ordination, where the axes are constructed from environmental data.

It is used in situations with a large number of attributes, e.g. species, to summarize the variation in the data as a small number of orthogonal axes (independent, i.e. at right angles to each other). These axes can then be used to arrange the individuals (i.e. stands or vegetation samples) in a meaningful way, that is as an ordination diagram. Ordination is therefore a technique for reducing the number of dimensions, for example from eighty species to three ordination axes which can be more easily interpreted. There is a considerable amount of jargon involved but for simplicity we shall assume that a stand ordination is usually required, thus the word STAND will be used in place of individual, quadrat or vegetation sample and the word SPECIES will be used to mean attribute or variable. See Table 9.11 and the ordination glossary and synonyms on p. 511 for other terms which may be confusing.

2 Phytosociological ordination, where the axes are constructed from vegetational data.

Methods of gradient analysis may be considered types of environmental ordination *sensu lato* but normally the term is reserved for techniques that use all the variables of the data set. Phytosociological ordinations are considered more flexible since all the vegetational variables are immediately obvious and can be used. Individual environmental factors can then be compared as they become available.

The early simple ordinations used reference stands to represent the two extremes of variation with the vegetation samples (stands) described. More recent techniques, usually variants of factor analysis, extract the principal axes of variation directly from the data. The earlier techniques were developed before electronic computers were generally available and are still useful for small projects, work in areas where computers are not available and for teaching purposes.

The procedure of continuum analysis was first advocated by Curtis & McIntosh in 1951. A measure of abundance or importance value (IV) for each species is multiplied by a 'climax adaptation number' (CAN). The bell-shaped curves for each species can then be compared with those for environmental factors (Fig. 9.26). The process is highly subjective and has been largely superseded by the method proposed by Bray & Curtis (1957). These authors replaced the continuum index by a similarity coefficient for comparison between stands. The coefficient used was suggested by Sørenson

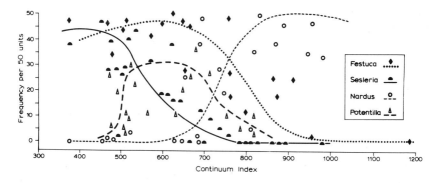

Figure 9.26 Use of the continuum index to arrange vegetation samples along an axis. The importance value of the four most abundant species has been plotted against the continuum index to show their relative abundance along a gradient from limestone grassland on acidic drift.

(1948) or possibly Czekanowski (1913):

$$C = \frac{2W}{A + B}$$

where A and B are the quantities of all the species found in each of the two stands to be compared and W is the sum of the lesser values for the species common to the two stands.

Animal ecologists have used other indices of which the best known is probably Mountford's (1962). These are discussed in Southwood (1966). Coefficients of similarity are then inverted ($100 - C$ or $Cmax - C$) to produce a distance coefficient which is then entered in a half-matrix and the two most dissimilar stands chosen as the end-points of the first axis. All other stands are located along this axis by intersecting arcs drawn proportional in length to the distances between the stand to be located and the first two reference stands. A second axis can be constructed at right angles to this first axis by using, as reference stands, that pair of stands closest to the mid-point of the original axis but having the largest interstand distance entered in the half-matrix. On the arrangement of stands in two-dimensional space, species performance and the levels of selected environmental factors may be plotted (Fig. 9.27). By visual comparison between species/environmental relationships tentative hypotheses may be erected regarding causal relationships. The same approach can be used to extract a third axis although interpretation usually becomes difficult after the third.

Various improvements on this basic approach have been made. For example, Beals (1960) suggests a formula for positioning stands along each axis with respect to the reference stands and so permits the electronic computation of the ordination diagrams (Orloci, 1966). Coefficients, such as interstand or Euclidean distance (Sokal, 1961), which are theoretically more acceptable have also been suggested and are now more widely used than the original one suggested by Sørensen

$$D_{j.h.} = \sqrt{\sum_{n}^{i=1} (X_{ij} - X_{ih})^2}$$

where j and h are the stands being compared, i is the species under consideration, and x is the score.

This formula is geometrically more acceptable as it is an extension of Pythagoras' theorem into multidimensional space. It overcomes some of the failings of Sørensen's coefficient, such as the distance AC sometimes being longer than $AB + BC$ for three stands, A, B, and C. Bannister (1968) has discussed the informativeness of different measures of species abundance for ordination and concludes that presence-or-absence data may be as infor-

(a) *Armeria maritima*

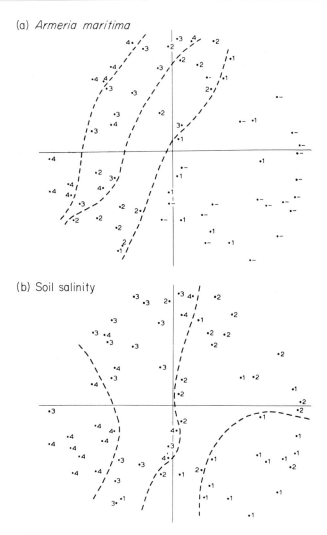

(b) Soil salinity

Figure 9.27 Examples of ordination diagrams derived from sea-cliff frequency data using the weighted similarity coefficient and principal components analysis. (a) Abundance of *Armeria maritima* (as quartiles). (b) Distribution of soil salinity (as quartiles). (From Goldsmith, 1973b.)

mative as some more detailed quantitative measures which are more time-consuming to collect in the field. However, this approach suffers from the serious limitations imposed by the use of reference stands and this can only be overcome by the application of types of factor analysis. Orloci (1966) recommended the use of the weighted similarity coefficient

$$\mathrm{WSC} \;=\; \sum_{n}^{i=1} (X_{ij} - \bar{X}_i)(X_{ih} - \bar{X}_i)$$

(where j and h are the two stands being compared, and i is the species under consideration) together with principal components analysis and this has been extensively tested (Austin & Orloci, 1966; Austin, 1968b; Bunce, 1968; Allen, 1971) and found to be extremely effective. More recently a modified form of principal components analysis, known as reciprocal averaging, has been developed which simultaneously ordinates both the stands and the species (Hill, 1973b) (see below).

Other similarity and dissimilarity (distance) measures which have proved useful are:

$$\text{Euclidean distance}_{jk} = \left[\sum_{i=1}^{n} (X_{ij} - X_{ik})^2 \right]^{1/2} = \text{interstand distance}$$
$$(\text{Sokal, 1961})$$

Internal association (IA) = similarity between replicate stands (or 100 used if they are assumed to be identical).

$$\text{Percentage similarity} = \text{PS}_{jk} = \frac{200 \sum_{i=1}^{n} \min(A_{ij}, A_{ik})}{\sum_{i=1}^{n} (A_{ij} + A_{ik})}$$

$$\text{Coefficient of dissimilarity} = \text{CD}_{jk} = \text{IA} - \text{CC}_{jk}$$

$$\text{Coefficient of community} = \text{CC}_{jk} = \frac{200 \, S_c}{S_j + S_k} \quad \begin{array}{l} (\text{Czekanowski, 1913;} \\ \text{Bray & Curtis, 1957}) \end{array}$$

n = no. of species.
A_{ij} and A_{ik} = abundances of species i in stands j and k.
S_j and S_k = no. of species in stands j and k.
S_c = no. of species in common.

See Williams *et al.* (1966), Clifford & Williams (1976), Gauch (1982).

Ordinations are not restricted to an arrangement of stands using species but the raw data matrix may be transformed to ordinate species using stands (Fig. 9.28). An environmental factor matrix may also be used to obtain either an ordination of environmental factors or an ordination of stands. In this manner, species/stand/environmental factor relationships may be thoroughly explored.

Ordination has now become one of the most popular techniques for analysing vegetational variation. However, a study of the literature indicates that it is usually used for two, sometimes complementary, purposes. These are (i) to prepare a framework for describing vegetational variation and (ii)

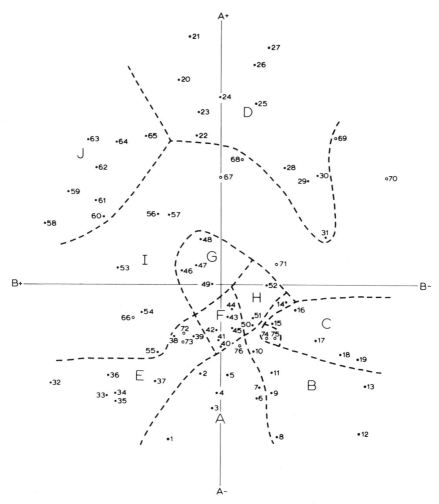

Figure 9.28 An ordination of 76 species from frequency data for sea-cliff vegetation processed along the weighted similarity coefficient and principal components analysis. (From Goldsmith, 1973b).

in the search for causal factors. In both types of application the raw data is invariably floristic and usually quantitative. Nevertheless, there are no *a priori* reasons why presence-or-absence data should not be used and some ecologists recommend their use (Bannister, 1966; Greig-Smith, 1969).

Ordination techniques are appropriate to any scale of study and published accounts range from the relatively local scale studies of Gittins (1965) and Austin (1968) on small areas of calcareous grassland, to the more extensive studies of Ashton (1964) in the Dipterocarp forests of Brunei and of Greig-Smith *et al.* (1967) in the Solomon Islands. In theory ordination is appropriate in situations where vegetational variation is continuous; in practice is is used as an alternative to classification. The choice between these two major alternative approaches to vegetation is an empirical one, depending on whether the ecologist require categories of vegetation or ordination diagrams (Anderson, 1965). It is important to note, however, that in the search for causality ordination only serves to erect hypotheses and, as in any correlation-type approach, the testing of such hypotheses must be carried out with other data or by experimentation, for example see Goldsmith (1973b,c).

Correspondence analysis or reciprocal averaging (RA)

This approach to ordination developed by Hill (1973b, 1974) operates on a traditional stand-by-species matrix, involves an iterative procedure and produces simultaneous ordinations of stands and species. It is based on a rationale similar to gradient analysis. Species are weighted along a rough initial gradient and weightings are used to calculate stand scores (see Appendix 2 of Hill (1973b) for worked example). These stand scores are then used to calculate a new and improved calibration of the species. The new species weightings then provide further improvement of the stand calibration. The process is continued until a stable, optimal solution is obtained.

Reciprocal averaging has the advantages that: (i) computing time only rises linearly with the amount of data analysed; (ii) only positive entries in the data matrix need to be stored in the memory so that large data sets present no particular difficulty; (iii) the results appear to be readily interpreted; (iv) theoretical advantages as set out in Hill (1973b); for example, there is no use of reference stands as end-points. The average (root mean square) species abundance profile on each axis has a standard deviation of 1 and thus the ordination axis length becomes a certain number of species standard deviations. Thus, this approach to ordination follows the widely held conceptual model of Gaussian bells for each species along an axis representing a gradient of vegetational variation.

Its faults are that the second axis often produces an 'arch' or 'horseshoe' shape, which is why Hill developed 'detrended correspondence analysis'—the computer program is called DECORANA (Hill, 1979)—which rather crudely breaks the axis into sections and straightens them out. However, even when using the detrended version of the method, it is advisable to remove extreme outlying stands and any discontinuities prior to analysis. Another fault of RA is that distances in the ordination space do not have a consistent meaning in terms of compositional change (especially at the ends of the axis), i.e. there is distortion as illustrated in Hill (1979) or Gauch (1982). This is largely corrected or rescaled for all axes by the detrended version of RA.

DECORANA provides the option of transforming the data, downweighting rare species, varying the number of segments from 26 for detrend-

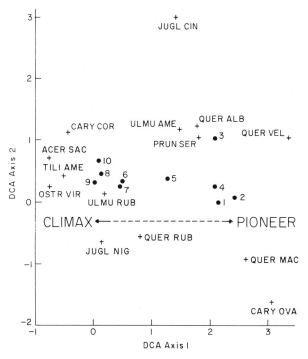

Figure 9.29 Detrended correspondence analysis ordination of southern Wisconsin upland forests. The data and species names are given in Table 4.4 of Gauch (1982). Only the first axis is ecologically meaningful, expressing a climax-to-pioneer gradient (indicated by the dashed line). Axis scales are in units of average standard deviations of species turnover. (From Gauch, 1982.)

ing or of omitting some stands. Figure 9.29 shows a species ordination of southern Wisconsin upland forests produced by this method (Gauch, 1982).

Non-metric multidimensional scaling (MDS)

The use of the rank order of the information in a dissimilarity matrix was originally developed by Shepard (1962) and Kruskal (1964a,b) and has been assessed for ecological purposes by Fasham (1977) and Prentice (1977). Although the method uses a matrix of dissimilarities between the stands the method of axis construction is very different from other ordinations and has the advantage that it avoids the curvilinear distortions between species and the effects of beta diversity which cause the horseshoe-shaped ordinations. The four main stages of the calculation can be summarized as follows (Fasham 1977):

1 a matrix of dissimilarities between stands is produced, e.g. Euclidean distances;

2 the stands are ordered (ranked) so that their dissimilarities are in ascending order;

3 a non-parametric regression is used to position samples on a smooth curve, using a measure of departure or 'stress' from the line;

4 stands are shuffled iteratively to minimize the 'stress'.

The coefficient need not be Euclidean and even a matrix of similarity values could be used. Readers should also note that only the ranking (step 2) is non-metric and that the rest of the procedure, e.g. the calculation of the dissimilarity values, can be metric or non-metric.

This method appears to have considerable promise but has not been widely taken up by plant ecologists. The most serious problem appears to occur if the initial configuration chosen at step 2 is inappropriate as it may then be difficult to obtain the optimal solution. Also the amount of computation time increases with the square or cube of the number of samples so data set size is severely limited (Gauch, 1982).

Johnson & Goodall (1979) have described the application of an alternative non-linear method of ordination involving the fitting of bell-shaped species-response curves, and then a maximum likelihood approach to obtain a 'best' approximation of the positions of the samples along a one-dimensional gradient. The search for a more appropriate non-linear ordination model is likely to continue (see also Austin, 1976).

Evaluation of ordination techniques

1 Techniques can be compared by using simulated data sets and comparing the recovered structure with the original structure (see, for example, Noy-Meir, 1973a).

2 On field data a regular spacing of stands on the ordination is sometimes considered desirable but that may result from a distortion of the true relationships of stands. In practice, consumer preferences are made subjectively and a consensus emerges amongst plant ecologists after a few years of use. At the moment (1985) detrended correspondence analysis is widely preferred on both practical and theoretical grounds.

Ordination glossary and synonyms

Here are our interpretations of a selection of confusing terms.

Catenation: a catena is usually considered to be a gradient of soil variation, often down a slope on a uniform parent material, but catenation is sometimes used as a synonym for ordination. This use should however be avoided.

Classical scaling: a synonym for principal components analysis.

Coenocline: a simulated community gradient, equivalent to an axis of vegetational variation and could be an ordination axis.

Coenoplane: a two-dimensional vegetation pattern.

Correspondence analysis: a synonym for reciprocal averaging.

Direct ordination: refers to species plotted against an environmental axis, otherwise known as gradient analysis.

Indirect ordination: based on a stand-by-species matrix and is the usual procedure. Hence, the term is somewhat confusing.

Non-linear ordination: no assumption of linear relationship between species and environmental factors, e.g. non-metric multidimensional scaling.

Principal components analysis (PCA) using the *correlation* matrix: PCA of data standardized by species to zero mean and unit variance.

PCA using the *covariance* matrix: PCA of unstandardized data.

Polar ordination: usually indicates the early Bray & Curtis (1957) ordinations with the axes crossing in the centre of the ordination which depend upon reference stands serving as poles of the axes.

Principal axes and *principal coordinates analysis*: similar to principal components analysis but can be applied to matrices other than covariance and correlation types.

Q/R type ordinations: A Q-type ordination of stands is one based on similarities between stands, whereas an R-type ordination of stands is one based on correlations/associations between species.

Simple or perpendicular ordination: Orloci (1966) term for simple ordinations calculated with Euclidean distance as the coefficient.

Transformation of data: usually the distributions of the species are centred when the coefficients are calculated (e.g. weighted similarity coefficient), but sometimes raw data is standardized before calculating coefficients (Greig-Smith, 1983; Noy-Meir, 1973a; Noy-Meir *et al.*, 1975).

3.4.3   Trend-surface analysis

Geographers have frequently mapped trends in quantitative variables by fitting polynomial surfaces to their observations (Chorley & Haggett, 1965). Many ecological variables, such as species' abundance and levels of environmental factors, are mappable in the same manner. Gittins (1968) has applied trend-surface analysis to his Anglesey limestone grassland data and to Goodall's (1954) Australian mallee. He concludes that the method should prove a useful tool for vegetation mapping and relating mathematically extracted components back to the field. However, many ecologists consider the technique to represent a further abstraction from reality which is only useful in certain circumstances (Norcliffe, 1969).

4   Choice of methods

This chapter has outlined the methods most frequently used for describing and analysing vegetation. In some circumstances the choice of method is immediately obvious, but more frequently the ecologist is faced with a selection from as many as half-a-dozen, all of which appear to have equivalent advantages and disadvantages. In this situation the final decision will have to be made on the basis of the characteristics of the particular problem. Figure 9.1 is designed to assist the research worker select a method. It is not, however, the last word on the choice of an appropriate measure or technique and should only be used after reading this chapter and then with considerable thought and caution. It deals only with that part of a research problem concerning the vegetation, and similar decision-making processes will be required for studies of fauna and soil.

Stage one must always be to decide the objectives of the study. It often helps to list these on paper and whenever possible discuss them with another ecologist. Six major types of problem frequently confronted by ecologists have been recognized. Every reader will be able to think of others and most people will realize that some studies will include two or more of the categories in Fig. 9.1. If this is the case each category should be examined and more than one approach considered. Each major category of problem is followed by a series of questions and yes/no answers. These should be traced until the recommended method is identified. For example, if a zoologist wishes to describe the vegetation on a research site 'site description' forms the starting point. Whether or not the site had an obvious gradient of topographic, edaphic or vegetational variation is then considered and if the answer is negative the researcher is asked whether the description is for a zoological, botanical or other type of study. The suggestion would then be made that a method based on vegetation structure should be used. Reference would have to be made to the variety of methods described in the text; for example, those of Du Rietz (1931), Raunkiaer (1934) and Elton & Miller (1954). Before making a decision, however, it is advisable to consult the references appropriate to that method to obtain a full description of them and examples of their application. Decision-making is a slow process but a period of deliberation spent at this stage in the project is likely to save several days or weeks of unnecessary effort during the relatively tedious process of collecting the data.

Books on experimental designs and the planning of experiments are useful even when selecting a method for vegetation analysis (see Ch. 7). Jeffers (1978) has produced a leaflet in the Statistical Check List series which are published by ITE. There are seventy-two points that he asks the reader to consider and a bibliography. Other leaflets in the series deal with sampling, modelling and plant growth analysis.

Table 9.12 Computer programs available for the analysis of vegetation data

Library or package name	Authors	Language	Brief description	Further details from:
CLUSTAN	–	FORTRAN	The CLUSTAN package contains a wide range of procedures for agglomerative and divisive monothetic classification	CLUSTAN, 16 Kingsburgh Road, Murrayfield, Edinburgh EH12 6DZ
ECOLIB	A.J.Morton & J.W. Bates	FORTRAN	*Programs for classification, ordination and analysis of pattern.* Includes: association-analysis, polythetic divisive classification, centroid sorting, optimal agglomeration, principal components analysis, reciprocal averaging, multidimensional scaling, pattern and covariance analysis, two-term local variance and covariance analysis.	Dr A.J. Morton, Imperial College Silwood Park, Ascot, Berks SL5 7PY
PHYTOPAK	B. Huntley, J.P. Huntley & H.J.B. Birks	FORTRAN	*A suite of computer programs for the handling and analysis of phytosociological data.* Can be used in conjunction with ordination and classification routines obtained elsewhere.	Dr B. Huntley, Dept of Botany, Science Laboratories, South Road, Durham DH1 3LE
–	R.A. Goldstein & D.F. Grigal	FORTRAN	*Computer programs for the ordination and classification of ecosystems.* Includes: agglomerative classifications based on information statistics, sums of squares, and distance measures. Principal components analysis, canonical analysis.	National Technical Information Service, US Dept of Commerce, 5285 Port Royal Rd, Springfield, Virginia, USA.
–	H.G. Gauch & M.O. Hill	FORTRAN	*The Cornell Ecology Program Series.* Includes: polar ordination, principal components analysis, reciprocal averaging, detrended correspondence analysis, two-way indicator species analysis.	Hugh G. Gauch, Jr Ecology & Systematics, Cornell University, Ithaca, New York 14853, USA.
–	L. Orloci	BASIC	Includes: principal components analysis, reciprocal ordering, agglomerative classifications	Orloci (1975).

5 Computer programs

Many of the methods described in the preceding sections of this chapter require access to a computer and computer programs. Computer programs for individual methods are often available from the authors of the methods, but it is sometimes useful to have a variety of programs available in the form of a package or program library. Table 9.12 gives information about some packages which are available. The list is not intended to be complete but it does give a source of programs for many of the methods mentioned in this chapter which require computer analysis.

6 References

AGNEW A.D.Q. (1961) The ecology of *Juncus effusus* in North Wales. *J. Ecol.* **49**, 83–102.
ALLAN J.A. (Ed.) (1981) Matching remote sensing technologies and their applications. Remote Sensing Society, SOAS, London.
ALLEN T.F.H. (1971) Multivariate approaches to the ecology of algae on terrestial rock surfaces in North Wales. *J. Ecol.* **59**, 803–826.
ANDERSON A.J.B. (1971) Ordination methods in ecology. *J. Ecol.* **59**, 713–26.
ANDERSON D.J. (1965) Classification and ordination in vegetation science: Controversy over a non-existent problem? *J. Ecol.* **53**, 521–526.
ANDERSON D.J. (1967) Studies on structure in plant communities. III. Data on pattern in colonizing species. *J. Ecol.* **55**, 397–404.
ARCHIBALD E.E.A. (1949) The specific character of plant communities. I. Herbaceous communities. *J. Ecol.* **37**, 260–273.
ASHBY E. and PIDGEON T.M. (1942) A new quantitative method of analysis of plant communities. *Aust. J. Sci.* **5**, 19.
ASHTON P.S. (1964) Ecological studies in the mixed Dipterocarp forests of Brunei State. *Oxford Forestry Memoirs N. 25.* Clarendon Press, Oxford.
AUSTIN M.P. (1968b) An ordination of a chalk grassland community. *J. Ecol.* **56**, 739–758.
AUSTIN M.P. and GREIG-SMITH P. (1968) The application of quantitative methods to vegetation survey. II. Some methodological problems of data from rain forest. *J. Ecol.* **56**, 827–844.
AUSTIN M.P. (1972) Models and analysis of descriptive vegetation data. *Mathematical Models in Ecology* (ed. J.N.R. Jeffers), pp. 61–84. Blackwell Scientific Publications, Oxford.
AUSTIN M.P. (1976) Performances of four ordination techniques assuming three different non-linear species response models. *Vegetatio* **33**, 43–49.
AUSTIN M.P. (1977) Use of ordination and other multivariate descriptive methods to study succession. *Vegetatio* **35**, 165–175.
AUSTIN M.P. (1980) Exploratory analysis of grassland dynamics, an example of lawn succession. *Vegetatio* **43**, 87–94.
AUSTIN M.P. and BELBIN L. (1981) An analysis of succession along an environmental gradient using data from a lawn. *Vegetatio* **46**, 19–30.
AUSTIN M.P. and GREIG-SMITH P. (1968) The application of quantitative methods to vegetation survey. II. Some methodological problems of data from rain forest. *J. Ecol.* **56**, 827–844.
AUSTIN M.P. and ORLOCI L. (1966) Geometric models in ecology. II. An evaluation of some ordination techniques. *J. Ecol.* **54**, 217–229.
BANNISTER P. (1966) The use of subjective estimates of cover-abundance as the basis for ordination. *J. Ecol.* **54**, 665–674.
BANNISTER P. (1968) An evaluation of some procedures used in simple ordinations. *J. Ecol.* **56**, 27–34.

BARCLAY-ESTRUP P. and GIMINGHAM C.H. (1969) The description and interpretation of cyclical processes in a heath community. I. Vegetational change in relation to the *Calluna* cycle. *J. Ecol.* **57**, 737–758.

BARKMAN J.J., MORAVEC J. and RAUSCHERT S. (1976) Code of phytosociological nomenclature. *Vegetatio* **32**, 131–185.

BAWDEN M.G. (1967) Applications of aerial photography in land system mapping. *Photogrammetric Record, V* **30**, 461–464.

BEALS E. (1960) Forest bird communities in the Apostle Islands of Wisconsin. *The Wilson Journal* **72**, 156–181.

BEARD J.S. (1944) Climax vegetation in tropical America. *Ecology* **25**, 127–158.

BEARD J.S. (1967) Some vegetation types of tropical Australia in relation to those of Africa and America. *J. Ecol.* **55**, 271–290.

BEARD J.S. (1978) The physiognomic approach. In *Classification of Plant Communities.* pp. 33–64 (ed. R.H. Whittaker), pp. 33–64. Junk. The Hague.

BECKING R.W. (1957) The Zurich-Montpellier school of phytosociology. *Bot. Rev.* **23**, 411–488.

BEEFTINK W.G. and GÉHU J-M. (1973) Spartinetea maritimae *Prodrome des groupements végétaux d'Europe.* Vol. I. Cramer, Lehre.

BILLINGS W.D. and MOONEY H.A. (1959) An apparent frost hummock-sorted polygon cycle in the alpine tundra of Wyoming. *Ecology* **40**, 16–20.

BIRSE E.L. and ROBERTSON J.S. (1967) Chapter 5 in *The Soils of the Country Around Haddington and Eyemarth* (ed. J.M. Ragg and D.W. Futty). Mem. Soil Survey. Gt. Br.

BISHOP O.N. (1983) *Statistics for Biology.* Longmans, London.

BOALER S.B. and HODGE C.A.H. (1962) Vegetation stripes in Somaliland. *J. Ecol.* **50**, 465–474.

BOATMAN D.J. and ARMSTRONG W.A. (1968) A bog type in north-west Sutherland. *J. Ecol.* **56**, 129–141.

BRAUN-BLANQUET J. (1932) *Plant Sociology: The Study of Plant Communities.* McGraw-Hill, New York.

BRAUN-BLANQUET J. (1964) *Pflanzensoziologie*, 3rd edn. Springer-Verlag, Berlin.

BRAY J.R. and CURTIS J.T. (1957) An ordination of the upland forest communities of southern Wisconsin. *Ecol. Monogr.* **27**, 325–349.

BROWN D. (1954) Methods of surveying and measuring vegetation. *Comm. Bur. Pastures and Field Crops. Bull.* **42**, Comm. Agric. Bureau.

BUNCE R.G.H. (1968) An ecological study of Ysgolion Duon, a mountain cliff in Snowdonia. *J. Ecol.* **56**, 59–76.

BUNCE R.G.H. (1982) *A Field Key for Classifying British Woodland Vegetation.* Part 1. ITE, Cambridge.

BURGESS A. (1960) Time and size as factors in ecology. *J. Ecol.* **48**, 273–285.

BURNETT J.H. (Ed.) (1964) *The Vegetation of Scotland.* Oliver and Boyd, Edinburgh.

CAMPBELL M. (1978) Similarity coefficients for classifying relevées. *Vegetatio* **37**, 101–109.

CAIN S.A. (1932) Concerning certain phytosociological concepts. *Ecol. Monogr.* **2**, 475–508.

CAIN S.A. and CASTRO G.M. (1959) *Manual of Vegetation Analysis.* Harper and Bros., New York.

CHORLEY R.J. and HAGGETT P. (1965) Trend-surface mapping in geographical research. *Trans. Inst. Br. Geogr.* **37**, 47–67.

CLAPHAM A.R. (1932) The form of the observational unit in quantitative ecology. *J. Ecol.* **20**, 192–197.

CLAPHAM A.R., TUTIN T.G. and WARBURG E.F. (1962) *Flora of the British Isles* (2nd edition). Cambridge University Press, Cambridge.

CLEMENTS F.E. (1916) Plant succession: an analysis of the development of vegetation. *Carnegie Inst. Washington Publ.* **242**, 1–512.

CLIFFORD H.T. and WILLIAMS W.T. (1976) Similarity measures. In *Pattern Analysis in Agricultural Science* (ed. W.T. Williams), pp. 37–46. Elsevier, New York.

COLLANTES M.B. and LEWIS J.P. (1980) Ordenamiento de las communidades vegetales herbaceas dek departamento Rosario, *ECOSUR*, Argentina, 7(14), 171–184.

COLWELL R.N. (1956) Determining the prevalence of certain cereal crop diseases by means of aerial photography. *Hilgardia* **26**, (5), 223–286.

COTTAM G. (1947) A point method for making rapid survey of woodlands. *Bull Ecol. Soc. Amer.* **28**, 60.

COTTAM G. and CURTIS J.T. (1949) A method for making rapid surveys of woodlands by means of pairs of randomly selected trees. *Ecology* **30**, 101–104.

COTTAM G. and CURTIS J.T. (1955) Correction for various exclusion angles in the random pairs method. *Ecology* **36**, 767.

COTTAM G. and CURTIS J.T. (1956) The use of distance measures in phytosociological sampling. *Ecology* **37**, 451–460.

COUPLAND R.T. and JOHNSON R.E. (1965) Rooting characteristics of native grassland species in Saskatchewan. *J. Ecol.* **53**, 475–508.

CURRAN P.J. (1985) *Principles of Remote Sensing.* Longman, London.

CURTIN W. and LANE R.F. (1955) *Concise Practical Surveying.* English Universities Press, London.

CURTIS J.T. (1959) *The Vegetation of Wisconsin: an Ordination of Plant Communities.* Wisconsin University Press, Madison.

CURTIS J.T. and McINTOSH R.P. (1951) An upland forest continuum in the prairie-forest border region of Wisconsin. *Ecology* **32**, 476–496.

CZEKANOWSKI J. (1913) *Zarys Metod Statystcznyck.* Warsaw.

DAHL E. (1956) Rondane: Mountain vegetation in south Norway and its relation to the environment. *Skr. norske Vidensk-Akad. Mat. Naturv. Kl.* **3**, 374.

DANSEREAU P. (1957) *Biogeography: An Ecological Perspective.* Ronald, New York.

DANSEREAU P. (1961) Essai de représentation cartographique des éléments structuraux de la végétation, in Gaussen, H. (ed.) *Méthodes de la Cartographie de la Végétation.* 233–255. Paris.

DAVIS T.A.W. and RICHARDS P.W. (1933) The vegetation of Moraballi Creek, British Guiana: an ecological study of a limited area of tropical rain forest. Part I. *J. Ecol.* **21**, 350–372.

DRUDE O. (1913) *Die Okologie der Pflanzen.* Vieweg. (Publ.), Braunschweig.

DU RIETZ G.E. (1930) Classification and nomenclature of vegetation. *Svensk. Bot. Tidskr.* **24**, 489–503.

DU RIETZ G.E. (1931) Life-forms of terrestrial flowering plants. *Acta Phytogeographica Suecica* **3**, 1–95.

DUFFIELD B.S. and FORSYTHE J.F. (1972) Assessing the impacts of recreation use on coastal sites in East Lothian. In *The Use of Aerial Photography in Countryside Research.* Countryside Commission, London.

EDGELL M.C.R. (1969) Vegetation of an upland ecosystem: Cader Idris, Merionethshire. *J. Ecol.* **57**, 335–359.

ELLENBERG H. (1954) Zur Entwicklung der Vegetationssystematik in Mitteleuropa *Angew, PflSoziol* (Wien) Festschr. Aichinger, **1**, 134–143.

ELLENBERG H. (1956) Aufgaben und Methoden der Vegetationskunde, in H. Walter *Einfuhrung in die Phytologie* Vol. IV, pt I. Stuttgart.

ELLENBERG H. and MUELLER-DOMBOIS D. (1967) Tentative physiognomic-ecological classification of plant formation of the Earth. *Ber. geobot. inst. Stiftg. Rubel* **37**, 21–46.

ELTON C.S. (1966) *The Pattern of Animal Communities.* Methuen, London.

ELTON C.S. and MILLER R.S. (1954) The ecological survey of animal communities with a practical system of classifying habitats by structural characters. *J. Ecol.* **42**, 460–496.

FASHAM M.J.R. (1977) A comparison of nonmetric multidimensional scaling, principal components and reciprocal averaging for the ordination of simulated coenoclines. *Ecology* **58**, 551–561.

FISHER R.A. (1960) *The Design of Experiments.* Oliver & Boyd, Edinburgh.

FISHER R.A. and YATES F. (1963) *Statistical Tables for Biological, Agricultural and Medical Research.* Oliver & Boyd, London.

FOSBERG F.R. (1961) A classification of vegetation for general purposes. *Trop. Ecol.* **2**, 1–28.

FOSBERG F.R. (1967) A classification of vegetation for general purposes. In *Guide to the Check-sheet for I.B.P. Areas* (ed. G.F. Peterken). IBP Handbook 4. Blackwell. Scientific Publications, Oxford.

FRENKEL R.E. and HARRISON C.M. (1974) An assessment of the usefulness of phytosociological and numerical classificatory methods for the community biogeographer. *Journal of Biogeography*, **1**, 27–56.

GAUCH H.G. JR (1975) *Eigenvector Ordination: Reciprocal Averaging and Principal Compounds Analysis.* Cornell University, New York.

GAUCH H.G. (1977) *ORDIFLEX*—A flexible computer program for four ordination techniques: weighted averages, polar ordination, principal components analyses, and reciprocal averaging. Cornell University, New York.

GAUCH H.G. JR (1982) *Multivariate Analysis in Community Ecology.* Cambridge University Press, Cambridge.

GAUCH H.G.J. and SCRUGGS W.M. (1979) Variants of polar ordination. *Vegetatio* **40**, 147–153.

GAUCH H.G. JR and WHITTAKER R.H. (1981) Hierarchical classification of community data. *J. Ecol.* **69**, 537–558.

GAUCH H.G. JR, WHITTAKER R.H. and WENTWORTH T.R. (1977) A comparative study of reciprocal averaging and other ordination techniques. *J. Ecol.* **65**, 157–174.

GITTINS R. (1965) Multivariate approaches to a limestone grassland community. I. A stand ordination. II. A direct species ordination. III. A comparative study of ordination and association analysis. *J. Ecol.* **53**, 385–425.

GITTINS R. (1968) Trend-surface analysis of ecological data. *J. Ecol.* **56**, 845–869.

GITTINS R. (1981) Towards the analysis of vegetation succession. *Vegetatio* **46**, 37–39.

GOLDSMITH F.B. (1967) *Some aspects of the vegetation of sea-cliffs.* Ph.D. thesis, University of Wales.

GOLDSMITH F.B. (1972) Vegetation mapping in upland areas and the development of conservation management plans. In *The Use of Aerial Photography in Countryside Research.* Countryside Commission, London.

GOLDSMITH F.B. (1973a) The ecologist's role in development for tourism: a case study in the Caribbean. *Biol. Linn. Soc.* **5**, 265–287.

GOLDSMITH F.B. (1973b) The vegetation of exposed sea cliffs at South Stack, Anglesey, I. The multivariate approach. *J. Ecol.* **61**, 787–818.

GOLDSMITH F.B. (1973c) The vegetation of exposed sea cliffs at South Stack, Anglesey, II. Experimental studies. *J. Ecol.* **61**, 819–830.

GOLDSMITH F.B. (1974) An assessment of the Fosberg and Ellenberg methods of classifying vegetation of conservation purposes. *Biol. Cons.* **6**, 3–6.

GOLDSMITH F.B., HARDING J., NEWBOLD A. and SMART N. (1982) An ecological basis for the management of broadleaved forest. *Q.J. Forestry* **76**, 237–247.

GOODALL D.W. (1953) Objective methods for the classification of vegetation. I. The use of positive interspecific correlation. *Aust. J. Bot.* **1**, 39–63.

GOODALL D.W. (1954) Vegetational classification and vegetation continua. *Angew. PflSoziol.* **1**, 168–182.

GOODALL D.W. (1974) A new method for the analysis of spatial pattern by random pairing of quadrats. *Vegetatio* **29**, 135–146.

GOODIER R. (1971) *The Application of Aerial Photography to the Work of the Nature Conservancy.* Nature Conservancy, Edinburgh.

GREIG-SMITH P. (1952) Ecological observations on degraded and secondary forest in Trinidad, British West Indies. I. General features of the vegetation. *J. Ecol.* **40**, 283–315. II. Structure of the communities. *Ibid.* 316–330.

GREIG-SMITH P. (1961) Data on pattern within plant communities. I. The analysis of pattern. *J. Ecol.* **49**, 695–708.

GREIG-SMITH P. (1979) Pattern in vegetation. Presidential Address 1979. *J. Ecol.* **67**, 755–779.

GREIG-SMITH P. (1980) The development of numerical classification and ordination. *Vegetatio* **42**, 1–9.

GREIG-SMITH P. (1983) *Quantitative Plant Ecology*, 3rd edition. Butterworths, London.

GREIG-SMITH P. (1969) Analysis of vegetation data: the user viewpoint. *Proceedings of the International Symposium on Statistical Ecology*. Pennsylvania State University Press, 149–166.

GREIG-SMITH P., AUSTIN M.P. and WHITMORE T.C. (1967) The application of quantitative methods to vegetation survey. I. Association-analysis and principal component ordination of rain forest. *J. Ecol.* **55**, 483–503.

GROENEWOLD H. VAN (1965) Ordination and classification of Swiss and Canadian forests by various biometric and other methods. *Ber. Geobot. Inst. E.T.A., Stiftg. Rubel* **35**, 28–102.

GRUBB P.J., GREEN H.E. and MERRIFIELD R.C.J. (1969) The ecology of chalk heath: its relevance to the calcicole–calcifuge and soil acidification problems. *J. Ecol.* **57**, 175–212.

HALL J.B. (1977) Forest-types in Nigeria, an analysis of pre-exploitation forest enumeration data. *J. Ecol.* **65**, 187–199.

HARPER J.L. (1967) A Darwinian approach to plant ecology. *J. Ecol.* **55**, 247–270.

HARRISON C.M. (1970) The phytosociology of certain English heathland communities. *J. Ecol.* **58**, 573–589.

HARRISON C.M. (1971) Recent advances in the description and analysis of vegetation. *Trans. Inst. Br. Geogr.* **52**, 113–127.

HARRISON C.M. and WARREN A. (1970) Conservation, stability and management. *Area 2*, 26–32.

HILL M.O. (1973a) The intensity of spatial pattern in plant communities. *J. Ecol.* **61**, 225–235.

HILL M.O. (1973b) Reciprocal averaging: an eigenvector method of ordination. *J. Ecol.* **61**, 237–250.

HILL M.O. (1974) Correspondence analysis: a neglected multivariate method. *J. Roy. Stat. Soc.*, Ser. C, **23**, 340–354.

HILL M.O. (1979a) *DECORANA: A Fortran program for detrended correspondence analysis and reciprocal averaging.* Cornell University, New York.

HILL M.O. (1979b) *TWINSPAN: A Fortran program for arranging multivariate data in an ordered two-way table by classification of the individuals and attributes.* Cornell University, New York.

HILL M.O. and GAUCH H.G.J. (1980) Detrended correspondence analysis: an improved ordination technique. *Vegetatio* **42**, 47–58.

HILL M.O., BUNCE R.G.H. and SHAW M.W. (1975) Indicator species analysis, a divisive polythetic method of classification, and its application to a survey of native pinewoods in Scotland. *J. Ecol.* **63**, 597–614.

HOPE-SIMPSON J.F. (1940) On the errors in the ordinary use of subjective frequency estimations in grassland. *J. Ecol.* **28**, 193–209.

HOPKINS B. (1955) The species area relations of plant communities. *J. Ecol.* **43**, 409–426.

HORN H.S. (1975) Markovian properties of a forest succession. In *Ecology and Evolution of Communities* (eds. M.L. Cody and J.M. Diamond), pp. 196–211.

HORN H.S. (1976) Succession. *Theoretical Ecology: Principles and Applications* (ed. R.M. May), pp. 187–204.

HOWARD J.A. (1970) *Aerial Photo-Ecology.* Faber & Faber, London.

HUNTLEY B., HUNTLEY J.P. and BIRKS H.J.B. (1981) Phytopak: a suite of computer programs designed for the handling and analysis of phytosociological data. *Vegetatio* **45**, 84–95.

IVIMEY-COOK R.B. and PROCTOR M.C.F. (1966) The application of association-analysis to phytosociology. *J. Ecol.* **54**, 179–192.

JEFFERS J.N.R. (1978) *Design of Experiments.* Statistical Checklist **1**, Institute of Terrestrial Ecology, Cambridge.

JENSÉN S. (1979) Classification of lakes in southern Sweden on the basis of their macrophyte composition by means of multivariate methods. *Vegetatio* **39**, 129–146.

JOHNSON R.W. and GOODALL D.W. (1979) A maximum likelihood approach to non-linear ordination *Vegetatio* **41**, 133–142.

KAISER H.F. (1958) The varcimax criterion for analytic rotation in factor analysis. *Psychometrika* **23**, 187–200.

KERSHAW K.A. (1957) The use of cover and frequency in the detection of pattern in plant communities. *Ecology* **38**, 291–299.

KERSHAW K.A. (1958, 1959) An investigation of the structure of a grassland community. I. The pattern of *Agrostis tenuis*. *J. Ecol.* **46**, 571–592. II. The pattern of *Dactylis glomerata, Lolium perenne and Trifolium repens. Ibid.* **47**, 31–43. III. Discussion and conclusions. *ibid.* **47**, 44–53.

KERSHAW K.A. (1961) Association and co-variance analysis of plant communities. *J. Ecol.* **49**, 643–654.

KERSHAW K.A. (1962) Quantitative ecological studies from Landmannhellir, Iceland. I. *Eriophorum angustifolium*, and II. The rhizome behaviour of *Carex bigelowii* and *Calamagrostis neglecta. J. Ecol.* **50**, 163–179.

KERSHAW K.A. (1968) Classification and ordination of Nigerian savanna vegetation. *J. Ecol.* **56**, 483–495.

KERSHAW K.A. and LOONEY J.H. (1985) *Quantitative and Dynamic Ecology*, 3rd edition. Edward Arnold, London.

KOPPEN W. (1928) *Klimakarte der Erde*. Perthes, Gotha.

KREBS C.J. (1972) *Ecology: The Experimental Analysis of Distribution and Abundance*. Harper & Row, New York.

KRUSKAL J.B. (1964a) Multidimensional scaling by optimising goodness of fit to a non-metric hypothesis. *Psychometrika* **29**, 1–27.

KRUSKAL J.B. (1964b) Non-metric dimensional scaling: a numerical method. *Psychometrika* **29**, 115:129.

KUCHLER A.W. (1967) *Vegetation Mapping*. Ronald Press, New York.

LAMBERT J.M., MEACOCK S.E., BARNS J. and SMARTT P.F.M. (1973) AXOR and MONIT: Two new polythetic divisive strategies for hierarchical classification. *Taxon* **22**, 173–176.

LAMBERT J.M. and WILLIAMS W.T. (1962) Multivariate methods in plant ecology. IV. Nodal analysis. *J. Ecol.* **50**, 775–802.

LAMBERT J.M. and WILLIAMS W.T. (1966) Multivariate methods in plant ecology. VI. Comparison of information-analysis and association-analysis. *J. Ecol.* **54**, 635–664.

LANCE G.N. and WILLIAMS W.T. (1967) A general theory of classifying sorting strategies. 1. Hierarchical systems. *Comput. J.* **9**, 373–380.

LANCE G.N. and WILLIAMS W.T. (1968) Notes on a new information-statistic classifactory program. *Comput. J.* **11**, 195.

LEEUWEN C.G. VAN (1966) A relation theoretical approach to pattern and process in vegetation. *Wentia* **15**, 25–46.

LOUCKS O.L. (1962) Ordinating forest communities by means of environmental scalars and phytosociological indices. *Ecol. Monogr.* **32**, 137–166.

LOUPPEN J.M.W. and K. VAN DER MOORES (1979) CLUSLA: A computer program for the clustering of large phytosociological data sets. *Vegetatio* **40**, 107–114.

MAAREL E. VAN DER (1971) Plant species diversity in relation to management. In *The Scientific Management of Animal and Plant Communities for Nature Conservation* (ed. E. Duffey and A.S. Watt). Blackwell Scientific Publications, Oxford.

MAAREL E. VAN DER (1979) Vegetatio 1980: the past six years and the near future. *Vegetatio* **41**, 129–132.

MAAREL E. VAN DER, JANSSEN J.G.M. and LOUPPEN J.M.W. (1978) TABORD, a program for structuring phytosociological tables. *Vegetatio* **38**, 143–156.

MACARTHUR R.H. (1964) Environmental factors affecting bird species diversity. *Amer. Nat.* **98**, 387–397.

MACARTHUR R.H. (1965) Patterns of species diversity. *Biol. Rev.* **40**, 510–533.

MARGALEF R. (1968) *Perspectives in Ecological Theory*. Chicago University Press, Chicago.

MCINTOSH R.P. (1967a) An index of diversity and the relation of certain concepts to diversity. *Ecology* **48**, 392–404.

MCINTOSH R.P. (1967b) The continuum concept of vegetation. *Bot. Rev.* **33**, 130–187.

MCVEAN D.N. and RATCLIFFE D.A. (1962) *Plant Communities of the Scottish Highlands*. Monographs of the Nature Conservancy, No. 1. H.M.S.O. London.

MILES J. (1979) *Vegetation Dynamics*. Chapman & Hall, London.

MOORE J.J. (1962) The Braun-Blanquet system: A reassessment. *J. Ecol.* **50**, 761–769.

MOORE J.J. (1970) *Phyto: a suite of programs in Fortran IV for manipulation of phytosociological tables according to the principles of Braun-Blanquet*. Dept. of Botany, Dublin, 11 p. mimeog.

MOORE J.J., FITZSIMMONS P., LAMBE E. and WHITE J. (1970) A comparison and evaluation of some phytosociological techniques. *Vegetatio* **20**, 1–20.

MOORE J.J. and O'SULLIVAN A. (1978) A phytosociological survey of the Irish Molinio-Arrhenatheretea using computer techniques. *Vegetatio* **38**, 89–93.

MOORE N.W. (1962) The heaths of Dorset and their conservation. *J. Ecol.* **50**, 369–391.

MORTON A.J. (1974) Ecological studies of fixed dune grassland at Newborough Warren, Anglesey. I. The structure of grassland, *J. Ecol.* **62**, 253–260. II. Causal factors of the grassland structure. *J. Ecol.* **62**, 261–278.

MOUNTFORD M.D. (1962) An index of similarity and its application to classificatory problems. In *Progress in Soil Zoology* (ed. P.W. Murphy), pp. 43–50. Butterworth, London.

MUELLER-DOMBOIS D. and H. ELLENBERG (1974). *Aims and Methods of Vegetation Ecology*. J. Wiley & Sons, New York.

MULHOLLAND P. (1980) *A study of the cycle of regeneration in the beech-dominated woodlands of Epping Forest, Essex*. M.Sc. Thesis, UCL, London.

NAVEH ZEV and WHITTAKER R.H. (1979) Structural and floristic diversity of shrublands and woodlands in Northern Israel and other Mediterranean areas *Vegetatio* **41**, 171–190.

NEAL M.W. and KERSHAW K.A. (1973) Studies on lichen-dominated systems. IV. The objective analysis of Cape Henrietta Maria raised-beach systems. *Can. J. Bot.* **51**, 1177–1190.

NORCLIFFE G.B. (1969) On the use and limitations of trend surface models. *Canadian Geographer* **8**, 338–348.

NOY-MEIR I. (1973a) Data transformation in ecological ordination. I. Some advantages of non-centering. *J. Ecol.* **61**, 329–41.

NOY-MEIR I. (1973b) Divisive polythetic classification of vegetation data by optimized division on ordination components. *J. Ecol.* **61**, 753–760.

NOY-MEIR I., WALKER D. and WILLIAMS W.T. (1975) Data transformation in ecological ordination. II. On the meaning of data standardization. *J. Ecol.* **63**, 779–800.

OBERDORFER E. (1957) *SudDeutsche Pflanzengesellschaften*. Fisher, Jena.

ODUM E.P. (1969) The strategy of ecosystem development. *Science* **164**, 262–270.

ORLOCI L. (1966) Geometric models in ecology. I. The theory and application of some ordination methods. *J. Ecol.* **54**, 193–216.

ORLOCI L. (1967) An agglomerative method for classification of plant communities. *J. Ecol.* **55**, 193–206.

ORLOCI L. (1975) *Multivariate Analysis in Vegetation Research*. Junk, The Hague.

ORLOCI L. (1978) *Multivariate Analysis in Vegetation Research*, 2nd edn. Junk, The Hague.

OSVALD H. (1923) Die vegetation des Hochmoores Komosse. *Akad. Abh. Uppsala.*

PEARSALL W.H. (1924) The statistical analysis of vegetation: a criticism of the concepts and methods of the Uppsala school. *J. Ecol.* **12**, 135–139.

PENTECOST A. (1980) The lichens and bryophytes of rhyolite and pumice-tuff rock outcrops in Snowdonia, and some factors affecting their distribution. *J. Ecol.* **68**, 251–267.

PERKINS D.F. (1971) Counting sheep, cattle and ponies on Dartmoor by aerial photography. In *The Application of Aerial Photography to the work of the Nature Conservancy* (ed. Goodier). The Nature Conservancy.

PERRING F. (1959) Topographical gradients of chalk grassland. *J. Ecol.* **47**, 447–481.

PETERKEN G.F. (1967) *Guide to the check sheet for I.B.P. Areas*. International Biological Programme, Handbook 4. Scientific Publications, Oxford.

PHILLIPS M.E. (1954) Studies in the quantitative morphology and ecology of *Eriophorum angustifolium* Roth. Part III. *New Phytol.* **53**, 312–342.

PIELOU E.C. (1959) The use of point-to-plant distances in the study of the pattern of plant populations. *J. Ecol.* **47**, 607–613.

PIELOU E.C. (1962) The use of plant-to-neighbour distances for the detection of competition. *J. Ecol.* **50**, 357–367.

PIELOU E.C. (1966) The measurement of diversity in different types of biological collections. *J. Theoret. Biol.* **13**, 131–144.

PIELOU E.C. (1984) *A Primer on Classification and Ordination*. Wiley, Chichester.

POISSONET P. *et al.* (1981) Vegetation dynamics in grasslands, heathlands and mediterranean liqueous formations. *Vegetatio* **46–47**, 286 pp.

POORE M.E.D. (1955) The use of phytosociological methods in ecological investigations. I. The Braun-Blanquet System. *J. Ecol.* **43**, 226–244. II. Practical issues involved in an attempt to apply the Braun-Blanquet system. *J. Ecol.* **43**, 245–269.

POORE M.E.D. and ROBERTSON V.C. (1964) *An approach to rapid description and mapping of biological habitats based on a survey of the Hashemite Kingdom of Jordan*. Sub-commission on Conservation of Terrestrial Biological Communities. I.B.P.

PRENTICE I.C. (1977) Non-metric ordination methods in ecology. *J. Ecol.* **65**, 85–94.

PRENTICE F.W. (1948) The commonness and variety of species. *Ecology* **29**, 254–283.

PREIS K. (1939) Die *Festuca vallesiaca—Erysimum crepidifolium* Assoziation aux Basalt, Glimmerschiefer und Granitgneis. *Beih Bot. Cbl.* **59B**, 478–530.

PRESTON F.W. (1948) The commonness and rarity of species. *Ecology* **29**, 254–283.

PRITCHARD N.M. and ANDERSON A.J.B. (1971) Observations on the use of cluster analysis in botany with an ecological example. *J. Ecol.* **59**, 727–748.

PROCTOR M.C.F. (1967) The distribution of British liverworts: a statistical analysis. *J. Ecol.* **55**, 119–136.

RATCLIFFE D.A. (1971) Criteria for the selection of nature reserves. *Adv. Sci.* **27**, 294–296.

RATCLIFFE D.A. and WALKER D. (1958) The Silver Flowe, Galloway, Scotland. *J. Ecol.* **46**, 407–445.

RAUNKIAER C. (1934) *The Life-Forms of Plants and Statistical Plant Geography*. Clarendon Press, Oxford.

RICHARDS P.W. (1952) *The Tropical Rain Forest*. Cambridge University Press, Cambridge.

RICHARDS P.W., TANSLEY A.G. and WATT A.S. (1940) The recording of structure, life-forms and flora of tropical forest communities as a basis for their classification. *J. Ecol.* **28**, 224–239.

RIPLEY B.D. (1978) Spectral analysis and the analysis of pattern in plant communities. *J. Ecol.* **66**, 965–981.

ROSKAM E. (1971) *Programme ORDINA: multidimensional ordination of observation vectors*. Programme Bull 16, Psychology Lab. University of Nimegen.

RUBEL E. (1930) *Die Pflanzengesellschaften der Erde*. Huber Verlag, Bern.

RUTTER A.J. (1955) The composition of wet-heath vegetation in relation to the water-table. *J. Ecol.* **43**, 507–543.

SABO S.R. (1980) Niche and habitat relations in sub-alpine bird communities of the White Mountains of New Hampshire. *Ecol. Monogr.* **50**, 241–59.

SALISBURY E.J. (1916) The oak–hornbeam woods of Hertfordshire. Parts I and II. *J. Ecol.* **4**, 83–120.

SARGENT C. (1984) *Britain's Railway Vegetation*. ITE, Huntingdon.

SARUKHAN J. (1974) Studies on plant demography: *Ranunculus repens* L., *R. bulbosus* L. and *R. acris* L. II. Reproductive strategies and seed populations dynamics. *J. Ecol.* **62**, 151–177.

SARUKHAN J. and GADGIL M. (1974) Studies on plant demography: *Ranunculus repens* L., *R. bulbosus* L. and *R. acris* L. III. A mathematical model incorporating multiple modes of reproduction. *J. Ecol.* **62**, 921–936.

SARUKHAN J. and HARPER J.C. (1973) Studies on plant demography: *Ranunculus repens* L., *R. bulbosus* L. and *R. acris* L. I. Population flux and survivorship. *J. Ecol.* **61**, 675–716.

SCHANDA E. (Ed.) (1976) *Remote Sensing for Environmental Sciences*, Ecological Studies 18, Springer-Verlag, Berlin.

SEARLE S.R. (1966) *Matrix Algebra for the Biological Sciences*. Wiley, New York.

SHEPARD R.N. (1962) The analysis of proximities. Multidimensional scaling with an unknown distance function. I and II. *Psychometrika* **27**, 125–140, 217–246.

SHIMWELL D.W. (1971) *The Description and Classification of Vegetation*. Sidgwick & Jackson, London.

SIBSON R. (1972) Order invariant methods for data analysis. *J. Roy. Statist. Soc. B* **34**, 311–49.

SIEGEL S. (1956) *Non-parametric Statistics for the Behavioural Sciences.* McGraw-Hill, New York.

SILVERTOWN J. (1980) The dynamics of a grassland ecosystem: botanical equilibrium in the Park Grass Experiment. *J. Appl. Ecol.* **27**, 491–504.

SMARTT P.F.M. (1978) Sampling for vegetation survey: a flexible systematic model for sample location. *J. Biogeogr.* **5**, 43–56.

SMARTT P.F.M. and GRAINGER J.E.A. (1974) Sampling for vegetation survey: some aspects of the behaviour of unrestricted, restricted and stratified techiques. *J. Biogeogr.* **1**, 193–206.

SIMPSON E.H. (1949) Measurement of diversity. *Nature* **163**, 688.

SNEATH P.H.A. and SOTAL R.R. (1973) *Numerical Taxonomy.* Freeman, San Francisco.

SOCIETY FOR THE PROMOTION OF NATURE RESERVES (1969) *Biological Sites Recording Scheme.* S.P.N.R. Conservation Liaison Committee Technical Publication, I.

SOKAL R.R. (1961) Distance as a measure of taxonomic similarity. *Syst. Zoology* **10**, 70–79.

SOKAL R.R. and SNEATH P.H.A. (1963) *Principles of Numerical Taxonomy.* Freeman, San Francisco.

SØRENSON T. (1948) A method of establishing groups of equal amplitude in plant sociology based on similarity of species content. *Kong. Dan Vidensk. Selsk. Biol. Skr.* **5**, 1–34.

SOUTHWOOD T.R.E. (1966) *Ecological Methods.* Methuen, London.

STAMP L.D. (1925) The aerial survey of the Irrawaddy Delta forests (Burma). *J. Ecol.* **13**, 262–276.

SWAINE M.D. and GREIG-SMITH P. (1980). An application of principal components analysis to vegetation change in permanent plots. *J. Ecol.* **68**, 33–41.

SZAFER W. (1966) *The Vegetation of Poland.* International Series of Monographs in Pure and Applied Biology **9**. Pergamon, Oxford.

TANSLEY A.G. (1935) The use and abuse of vegetational concepts and terms. *Ecology* **16**, 284–307.

TANSLEY A.G. (1939) *The British Islands and their Vegetation.* Cambridge University Press, Cambridge.

TANSLEY A.G. and ADAMSON R.S. (1913) Reconnaissance in the Cotteswolds and the Forest of Dean. *J. Ecol.* **1**, 81–89.

TANSLEY A.G. and ADAMSON R.S. (1926) A preliminary survey of the chalk grasslands of the Sussex Downs. *J. Ecol.* **14**, 1–32.

TAYLOR J.A. (1969) Reconnaissance surveys and maps. In *Geography at Aberystwyth* (eds. E.G. Bowen, H. Carter and J.A. Taylor). University of Wales Press.

THOMAS A.S. (1960) Changes in vegetation since the advent of myxomatosis. *J. Ecol.* **48**, 287–306.

THOMPSON H.R. (1958) The statistical study of plant distribution patterns using a grid of quadrats. *Aust. J. Bot.* **6**, 322–342.

THORNTHWAITE C.W. (1948) An approach towards a rational classification of climate. *Geog. Rev.* **38**, 55–94.

TOWNSHEND J.R.G. (Ed.) (1981) *Terrain Analysis and Remote Sensing.* George Allen & Unwin, London.

TÜXEN R. (1937) Die Pflanzengesellschaften nordwest-Deutschlands. *Mitt. Flor.-soz. Arbeitsg. in Niedersachsen*, **3**, 1–170.

UNWIN D. (1974) An introduction to trend surface analysis. *Concepts and techniques in modern geography* **5**, 1–40.

USHER M.B. (1969) The relation between mean squares and block size in the analysis of similar patterns. *J. Ecol.* **57**, 505–514.

USHER M.B. (1979) Markovian approaches to ecological succession. *J. Anim. Ecol.* **48**, 413–426.

USHER M.B. (1981) Modelling ecological succession, with particular reference to Markovian models. *Vegetatio* **46**, 11–18.

WADSWORTH R.M. (1970) An invisible marker for experimental plots. *J. Ecol.* **58**, 555–557.

WARD J.H. (1963) Hierarchical grouping to optimise an objective function. *J. Am. Stat. Ass.* **58**, 236–244.

WARMING E. Okologiens Grundformer, Udkast til en systemastisk Ordning. *K. Danske Vidensk. Selsk. Naturv. Math. Afd. Skr.* **8**, R4, 119–187.

WARREN WILSON J. (1960) Inclined point quadrats. *New Phytol.* **59**, 1–8.

WATT A.S. (1947) Pattern and process in the plant community. *J. Ecol.* **35**, 1–22.

WATT A.S. (1962) The effect of excluding rabbits from Grassland A (Xerobrometum) in Breckland, 1936–60. *J. Ecol.* **50**, 181–198.

WATT A.S. (1964) The community and the individual. *J. Ecol. (Suppl.)* **52**, 203–212.

WEAVER J.E. and CLEMENTS F.E. (1938) *Plant Ecology.* McGraw-Hill, London.

WEBB D.A. (1954) Is the classification of plant communities either possible or desirable? *Bot. Tiddsk.* **51**, 362–370.

WEBB L.J., TRACEY J.G., WILLIAMS W.T. and LANCE G.N. (1970) Studies in the numerical analysis of complex rain-forest communities. V. A comparison of the properties of floristic and physiognomic-structural data. *J. Ecol.* **58**, 203–232.

WESTOFF W. and MAAREL E. VAN DER (1978) The Braun-Blanquet approach. In *Classification of Plant Communities* (ed. R.H. Whittaker), pp. 287–399. Junk, The Hague.

WHITTAKER R.H. (1956) Vegetation of the Great Smoky Mountains. *Ecol. Monogr.* **26**, 1–80.

WHITTAKER R.H. (1965) Dominance and diversity in land plant communities. *Science* **147**, 250–260.

WHITTAKER R.H. (1967) Gradient analysis of vegetation. *Biol. Rev.* **49**, 207–264.

WHITTAKER R.H. (1969) Evolution of diversity in plant communities. In *Diversity and Stability in Ecological Systems. Brookhaven Symp. Biol.* **22**, 178–196.

WHITTAKER R.H. (1970) *Communities and Ecosystems.* Macmillan, London.

WHITTAKER R.H. (1973) *Handbook of Vegetation Science V Ordination and Classification of Communities.* Junk, The Hague.

WHITTAKER R.H. (Ed.) (1978) *Classification of Plant Communities.* Junk, The Hague.

WILDI O. (1979) GRID—a space density analysis for recognition of noda in vegetation samples. *Vegetatio* **41**, 95–100.

WILLIAMS C.B. (1964) *Patterns in the Balance of Nature and Related Problems in Quantitative Ecology.* Academic Press, London.

WILLIAMS W.T. and DALE M.B. (1965) Fundamental problems in numerical taxonomy. *Adv. Bot. Res.* **2**, 35–68 (ed. R.D. Preston). Academic Press, London.

WILLIAMS W.T. and LAMBERT J.M. (1959) Multivariate methods in plant ecology. I. Association analysis in plant communities. *J. Ecol.* **47**, 83–101.

WILLIAMS W.T. and LAMBERT J.M. (1960) II. The use of an electronic digital computer for assocation analysis. *J. Ecol.* **48**, 689–710.

WILLIAMS W.T. and LAMBERT J.M. (1961) III. Inverse association-analysis. *J. Ecol.* **49**, 717–729.

WILLIAMS W.T., LAMBERT J.M. and LANCE G.N. (1966) Multivariate methods in plant ecology. V. Similarity analyses and information-analysis. *J. Ecol.* **54**, 2; 427–445.

WILLIAMSON M.H. (1978) The ordination of incidence data. *J. Ecol.* **66**, 911–920.

WILLIS A.J. (1973) *Introduction to Plant Ecology.* Allen & Unwin, London.

WINKWORTH R.E. and GOODALL D.W. (1962) A crosswise sighting tube for point quadrat analysis. *Ecology* **43**, 342–343.

WISHART D. (1978) CLUSTAN IC. User Manual. 3rd edn. Inter-University Research Council Series. Report 47, 175 pp.

WRATTEN S.D. and FRY G.L.A. (1980) *Field and Laboratory Exercises in Ecology.* Edward Arnold, London.

YARRANTON G.A. (1966) A plotless method of sampling vegetation. *J. Ecol.* **54**, 229–238.

YARRANTON G.A. (1969) Plant ecology: a unifying model. *J. Ecol.* **57**, 245–250.

YARRANTON G.A. (1971) Mathematical representations and models in plant ecology: response to a note by R. Mead. *J. Ecol.* **59**, 221–224.

10 Site history

P. D. MOORE

1   Introduction

Placing current ecological data in their correct time context is an important aspect of many ecological studies. Any research involving the spatial pattern of plant populations, or population dynamics, or vegetation processes at a site must involve some consideration of site history, for the current behaviour or location of a species can often be understood only by reference to its history in the area. Otherwise detailed studies of this kind must often remain speculative in their wider interpretations, simply because precise information concerning temporal changes at a site are lacking. It is the addition of a time dimension to ecological studies which has been the prime object of the discipline palaeoecology.

Palaeoecological studies, however, vary greatly in the scale (both temporal and spatial) of the questions which they seek to answer. The time scale can vary from millions of years—where questions concerning evolutionary responses of taxa are being asked—to historic events, such as a drought or a fire. The spatial scale can vary from global proportions to that of the spacing of individual plants. In this chapter, emphasis will be placed upon the techniques available for conducting research work concerned with historical changes at the lower end of both these scales, i.e. dealing with a time scale of, at the most, centuries and a spatial scale of local habitat proportions. This is not to imply that work dealing with larger areas and longer time scales are of lesser relevance to the ecologist, any more than ancient human history is of lesser value to the sociologist than recent history, but there are already good, modern texts dealing with paleoecological methods on this scale (see, for example, Birks & Birks, 1980). Also, ecologists working on modern problems relating to current vegetation or conservation often raise questions which can be answered with the aid of information concerning recent vegetation or habitat history.

The required evidence may be available from one or more of three sources:

(a) biotic evidence, derived from the biota itself;

(b) documentary evidence, derived from written sources;

(c) stratigraphic evidence, derived from stratified sediments in the vicinity of the site under study.

2   Biotic evidence

The plants and animals present at a site may themselves provide evidence relating to the site's history. The species of plant found and the overall diversity of species may relate to the history of disturbance, or its lack. Also, the population structure of selected species and the growth history of individuals may record catastrophes in the past.

2.1   Floristic richness and indicator species

A simple list of species present in a site can often provide interesting information, particularly if the list includes species which are poor colonists of disturbed areas. Perhaps the most detailed investigations based upon this type of evidence have been conducted in woodlands, for the complete clearance of forest may cause the loss of many plant species which have a limited capacity for subsequent regeneration or invasion. The presence of such species at a site, therefore, indicates a lack of serious disturbance in the past. Plants can be indicators of ancient woodland.

The technique has been exploited with considerable success in Britain by Peterken (1974). On the basis of extensive historical and botanical surveys he was able to list the woodland plants which are confined to primary forests, and also to list those which have progressively weaker association with undisturbed habitats. Once such relationships have been established, a botanical survey can indicate fairly precisely the degree of disturbance a site is likely to have experienced in the past. Such work in Britain has now been placed upon a very firm basis by the recent surveys of Peterken (1981).

Not only the trees, shrubs and herbaceous flora are of value in this respect; epiphytic bryophytes and especially lichens have also proved of value (Rose, 1976). Not only the presence of disturbance-sensitive indicators can be used, but also the overall diversity of sites. Ancient, undisturbed forest is generally richer in species than secondary woodland. Rose & James (1974) have provided estimates of the potential richness of lichen epiphytes in forest and have come to the conclusion that primary forest may contain up to 120 species km^{-2}. In the same way, Peterken & Game (1984) have shown that ancient forest has a richer assemblage of vascular plants than secondary forest and also, most significantly, that this richness is not severely reduced by fragmentation of the habitat. So ancient forest fragments can still be recognized on the basis of their flora.

It has been realized that some hedges are relic fragments of former forest and some others are of considerable antiquity (Pollard *et al.* 1974). Hooper (1970) developed a formula for calculating the age of a hedge based upon its floristic richness. Having established a linear relationship between the number of woody plant species found in a 30-yard length of hedge and the age of that hedge (based on documentary evidence), he was able to derive a simple formula. Accumulation of further data has led to slight modifications of this, but it can be given as:

Age of hedge = (99 × number of species) − 16.

Thus, a hedge with 10 species in 30 yards is dated at 974 years old (± 175 years for 95% confidence intervals).

The history of some grassland areas may also be deduced from their botanical character. Ancient grassland is known to be floristically rich and also contains species indicative of lack of disturbance (Sheail & Wells, 1969). In southern British grasslands, much research has been conducted upon *Pulsatilla vulgaris* which is one such species (Wells, 1968). It may also be possible to deduce past grazing pressures on the basis of indicator species; light grazing often results in a build up to relatively unpalatable plants (Harper, 1969). The value of this approach, coupled with documentary evidence and historical research has been shown by Crompton & Sheail (1975) in their grassland studies in the English brecklands.

2.2 Growth history of a species

In some species of plants it is possible to determine their growth success at different times in the past. A simple example of this is shown in Fig. 10.1 which relates to the growth of *Pinus nigra* on sand dunes in eastern England. In this species a whorl of branches is produced at the commencement of each year's growth, so the distance between whorls represents the annual height increment. Taking a given cohort of trees, in this case 12–15 years old, one can measure the annual height increment and compare the succeeding years to detect any particular stress or catastrophe. In this particular case there was a severe summer drought in 1976 which greatly reduced the growth of the trees in that year.

A more widely applicable technique is the use of annual girth increments in woody species. The technique has been used extensively in archaeological research and in studies of climatic history. Some of the theoretical and practical problems are described in detail by Baillie (1982). In addition to providing information on the growth of an individual in the past, this method may also supply information concerning catastrophes. For example, LaMarche & Hirschboeck (1984) found frost-damaged zones in sections of wood dated to 1628–1626 BC from the Aegean which corresponds to a period of volcanic eruption. Cwynar (1977) has reconstructed the fire history of a study area in Canada on the basis of the incidence of fire scars in woods and trees.

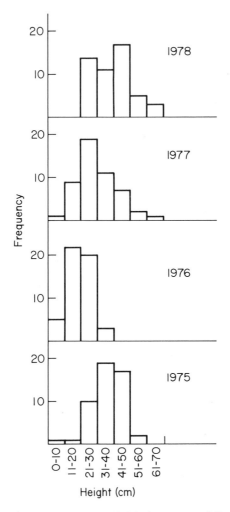

Figure 10.1 Frequency distributions of annual height increments of *Pinus nigra* on sand dunes in the Holkham National Nature Reserve, Norfolk, England (data of H.D. Moore). A drought in 1976 suppressed growth, but recovery had taken place by 1978.

Occasionally the population structure of certain plant species is informative concerning the history of a site. Crisp and Lange, in their studies of *Acacia burkittii* in Australia showed that intense grazing by sheep and rabbits in the past had reduced the rate of recruitment of young trees, resulting in a pattern of depressions in the frequency of certain age-classes. Other studies of this type among desert shrubs have not always proved as simple to interpret because of the complicating factor of water stress and intraspecific competition (Moore & Bhadresa, 1978). In the New Forest in Southern England, Peterken & Tubbs (1965) found three distinct generations of tree species which relate to periods of slackening browsing intensity in historic times (see also p. 403).

3   Documentary evidence

Detailed information on the past state of vegetation at a site is unlikely to date back more than a few decades and is of very limited geographical availability. Air photographs are similarly restricted in the time period which they cover, but are more widely available in terms of spatial coverage. Excellent examples of the type of work which can be done on this basis are given by Peterken & Harding (1974, 1975).

Earlier documentary information has rarely been assembled with subsequent ecological research in mind and may, therefore, be difficult to interpret. The value of old maps and other documents has been discussed most lucidly by Sheail (1980). There are many examples of historical ecological reconstructions based upon such documentary evidence, one of the most thorough being that of Yates (1979).

4   Stratigraphic evidence

One of the most valuable sources of evidence concerning the history of a site is any sedimentary deposit in its vicinity. Even relatively unpromising, shallow deposits of superficial humus within a wood, silt in a marsh, or buried, organic soils have proved of considerable value in reconstructing recent vegetation history. Much of the work conducted on stratified sediments has been based on large sites which may be expected to record regional vegetation events and has been aimed at reconstructing regional vegetational and climatic history, on the basis of which fossil maps or even reconstructed vegetation maps have now been produced (Birks et al., 1975; Birks & Saarnisto, 1975; Bernabo & Webb, 1977; Huntley & Birks, 1983). From such broad scale analyses, particularly of pollen grains in sediments, it has recently become possible to make climatological calibrations relating pollen assemblages to temperature and rainfall parameters and hence to reconstruct climatic history (Webb & Clark, 1977; Birks, 1981). Many studies of this sort are now available from various parts of North America, such as the northwest coast (Heusser et al., 1980), California (Adam & West, 1983) and Ontario (McAndrews, 1981).

But smaller depositional sites with a more local representation of vegetation can also be of great value. Where the questions asked by the investigator are of an essentially local nature, then the selection of small deposits for analysis may be most appropriate.

Work by Jacobson & Bradshaw (1980) has permitted the construction of

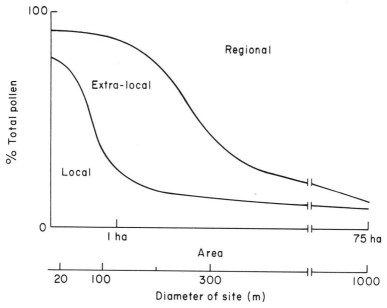

Figure 10.2 The relative proportions of pollen from different sources around a site of deposition in relation to the size of the site. It is assumed that there is no input from a stream. From Jacobson & Bradshaw (1980).

a model (Fig. 10.2) relating the diameter of the deposit to the relative proportion of local and extra-local input of microfossils, particularly pollen grains and spores. Such a model can only be a general guide, and will vary with such features as local vegetation (whether forested) and topography. It will also be modified considerably if the site is stream fed. However, it is a useful guide to the source area from which pollen will have originated.

4.1   Site selection

It may well be the case that there is little choice in the selection of sites for analysis, depending upon the habitat involved, but the following types of material may be suitable for palaeovegetational analysis (more detailed discussions are given in Moore & Webb, 1978).

(*a*) Pond or lake sediments.
(*b*) Peats.
(*c*) Soils.
(*d*) Other materials such as faecal matter, etc.

4.1.1   Pond sediments

These prove particularly useful for vegetation reconstruction of small catchments where local vegetation history is of interest. A classic study of this type in Janssen's (1967) analysis of a 114 cm core from an 80 m diameter pond in Minnesota. Care must be exercised in selecting an appropriate location for extracting a sample core since, as in many water bodies, sediment can become concentrated ('focused') in the deeper parts of the basin, leaving shallower areas with little or no sediment accumulation (Davis & Ford, 1982). The precise sampling site should be determined only after a series of trial borings has been conducted. An example of a pollen analysis of a small pond site is given in Fig. 10.3.

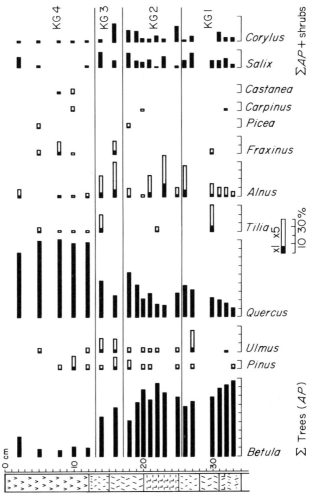

Figure 10.3 Pollen diagram from Kingswood Glen, Kent, a small pond situated in a relic fragment of oak woodland in south-east London. The pollen curves of *Betula* and *Quercus* demonstrate the development of a woodland succession interrupted by disturbance episodes at the commencement of zones KG2 and KG3. Stratigraphic symbols are given in Fig. 10.6.

4.1.2   Peats

Wet hollows in which organic peats have accumulated have often been used in palaeoecological investigations. They normally have lower rates of input of materials from surrounding catchments than do ponds, hence their record, particularly of macrofossils, is even more local. They may show periods of mineral inwash associated with severe catchment erosion as a result of disturbing influences of hydrological changes. Good examples of the kind of work which can be achieved using such sites are given by : Girling & Greig (1977) from Hampstead Heath, London; Baker *et al.* (1978) from Epping Forest, England; Birks (1982) in Cumbria; Watts (1975) from north-western Georgia. Fig. 10.4 shows a pollen diagram from a marsh (*marismas*) site in southern Spain where sandy muds and peats are eventually buried by wind-blown sand. Here the approach of the mobile sands is indicated by rising values of *Pteridium* and the falling levels of the local *Isoetes* in the wet hollow. The use of small, peat-forming sites of this kind is also proving vauable in reconstructing small-scale dynamics in Southern Hemisphere woodlands, as in the sclerophyll forests of Tasmania (MacPhail, 1984).

Peat deposits from below current sea level have also provided interesting sources of palaeoecological information, particularly regarding local conditions (e.g. Godwin & Godwin, 1960; Havinga & van den Berg, 1982), as have shallow peats from high altitudes and from tundra regions (e.g. Fredskild, 1978), though particular consideration must be given to long-

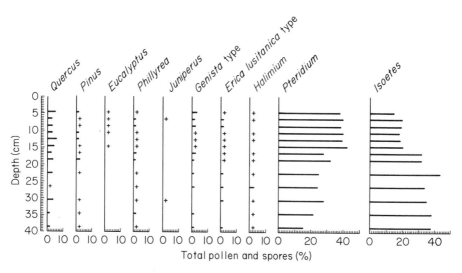

Figure 10.4 Part of a pollen diagram from an organic sand band buried beneath sands in the heathlands of the Coto Doñana National Park in southern Spain. The changing proportions of spores of *Pteridium* and *Isoetes* (probably *I. hystrix*) shows that the seasonal wetness of the site was decreasing as mobile sands covered the hollow and *Pteridium* invaded.

distance pollen dispersion and irregular air mass movements in such treeless environments.

4.1.3 Soils

The interpretation of soil microfossil profiles is notoriously difficult (see, for example, Havinga, 1974), but such sites are often the only ones available, particularly in sandy regions (Dimbleby, 1962; Vanhoorne, 1979). A major problem is the reliability of stratigraphic sequences in soils as a consequence of bioturbation; Dimbleby regards a pH of 5·5 as maximal for pollen preservation. Some studies have concentrated on buried soil profiles where conditions have essentially been preserved in the state when the burial occurred (e.g. Valentine & Dalrymple, 1976) and such sites have proved particularly valuable in archaeological contexts. General reviews of the subject are to be found in Dimbleby (1978, 1985) and in Greig (1982). The latter account is of particular relevance to archaeological sites in an urban context.

The organic, mor humus which accumulates on the surface of acid soils has been the subject of considerable attention in recent years. It was Iversen (1958) who first recognized the potential of such materials for pollen analysis in his studies in the Draved Forest, Denmark. Since then such analyses have continued to provide a wealth of data concerning the local development of woodland conditions (see, for example, O'Sullivan, 1973; Stockmarr, 1975). Microfossil studies of such soil profiles have also provided much evidence concerning the process of podzolization in soils which began as brown earths (Anderson, 1979). Fig. 10.5 (from Stockmarr, 1975) shows the development of a stratified podsol profile and the accumulation of mor humus.

In drier parts of the world, alluvial soils may supply one of the few sources of stratified materials for palaeoecological research, as shown in the work of Freeman (1972) in New Mexico and Moore & Stevenson (1982) in Iran. Some analyses on the transport of microfossils in alluvial sediments and its importance in the interpretation of such materials has been carried out by Solomon et al. (1982).

Not only fossil stratigraphy has been used in palaeoecological work on soil profiles, but also chemical stratigraphy. An excellent example of the results which can be achieved by such methods is that of Konrad et al. (1983), who were able to pinpoint the location of palaeo-indian settlement sites and hearths using phosphorus and calcium analyses in soil profiles. The value of phosphate analysis has been further discussed by Ottaway (1984).

4.1.4 Other materials

Plant fossils survive in a variety of other materials which may be of

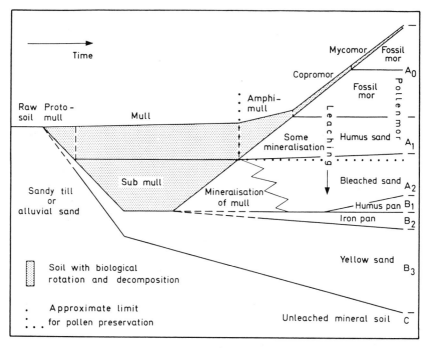

Figure 10.5 Jens Stockmarr's model of the natural development of forest soils in north-west Europe, showing the change from mull to mor humus types during podzolization. From Stockmarr (1975).

local value for palaeoecological studies. In high latitudes, for example, pollen may be retrieved from stratified ice deposits, as in glaciers. The survival of pollen in such ice was first described by Vareschi (1935) and the significance of his work was reviewed by Godwin (1949). Since then work has proceeded, particularly on the Greenland ice cap where not only pollen but other plant fragments have been described (Fredskild & Wagner, 1974). Stratification at such sites is so well maintained that summer and winter accumulation can be distinguished on the basis of pollen content (Ambach et al., 1966).

Analysis of faecal material for plant fragments as a guide to a herbivore's diet is described elsewhere in this volume (Bhadresa, p. 61). The technique has been applied extensively in archaeological work both in the United States (e.g. Martin & Sharrock, 1964) and in Britain (Greig, 1981, 1982). Microfossils in faecal material, however, are difficult to interpret (Moe, 1983).

Other stratified detritus, such as the middens of packrats have proved of value in recording past vegetation. In the Grand Canyon, Cole (1982) was able to reconstruct the dominant vegetation over the past 24 000 years using evidence from this source.

Analysis of honey and honey products, such as mead, is mainly the province of the apiculturist (see, for example, Adam & Smith, 1981), but such materials have been located and analysed in an archaeological context, such as the Bronze Age deposits described from a Scottish burial site by Dickson (1978).

4.2   Field sampling

Critical accounts of the methods used for both field and laboratory sampling of sediments are available in the literature, e.g. Faegri & Iversen (1975), Moore & Webb (1978), Birks & Birks (1980). Perhaps the most detailed set of practical manuals available is the International Geological Correlation Programme Project 158 B Project Guide, edited by Björn Berglund (1979–82, 1985), and this should certainly be consulted by all involved in detailed palaeoecological investigation. Here, an outline only can be given of the more basic techniques used in the field and the laboratory.

4.2.1   Sections

Exposed sections of sediments are ideal for palaeoecological work in many respects, for they display the extent and degree of lateral variation in sediment stratigraphy. Their exposure, however, is often dependent upon engineering construction, agricultural drainage, or extreme natural erosion as in the case of peat hags.

As Barber (1976) has pointed out, detailed examination and mapping of an exposed face is often best carried out before any cleaning of the surface takes place. In this way, weathered features, such as plant tussocks, may be more evident. The use of quadrat grids upon such faces is often helpful in assisting detailed mapping, and photographic records, particularly in colour, may also be valuable for subsequent checking. Carefully aligned colour transparencies, taken from a tripod, may even be projected onto paper for mapping.

Cleaning of the surface for sampling should be carried out with a knife or trowel, using lateral movements to minimize the vertical mixing of materials. Cleaning is also necessary to expose fresh, unoxidized organic samples, which darken quickly on exposure to air. The colour of this material can be recorded against standard charts (see Ball, p. 228) or simplified scales of a subjective type can be used (Aaby, in Berglund, 1979–82). Humification is also often recorded on subjective scales. Perhaps the most usable in the field is that of Troels-Smith (1955), which simplifies the older von Post scale (1924), given here in brackets.

Hum. O. Well preserved plant structure. When squeezed in the hand, yields clear water (= von Post 1 and 2).

Hum. 1. Some homogeneous ground material present, though generally well preserved. Squeezing yields turbid water with consequent loss of about a quarter of the solids (= von Post 3 and 4).

Hum. 2. Plant structure partly distinct, up to half of solid material lost by squeezing (= von Post 5 and 6).

Hum. 3. Plant structure in advanced decay and indistinct. Squeezing involves loss of up to three-quarters of the solid material (= von Post 7 and 8).

Hum. 4. Plant structure hardly discernible. Whole sample may be lost on squeezing (= von Post 9 and 10).

Some macrofossil remains may be discernible in the field, such as larger fruits, leaves or cones, but these are best examined in the laboratory. Even in the field, however, it is usually possible to come to certain conclusions about the nature of the sediments being examined. One can distinguish between limnic (formed underwater), telmatic (formed approximately at the water table) and terrestric (formed above the water table) deposits, and it may be possible to subdivide these according to major constituents, either inorganic or organic.

Traditionally, there have been two main approaches to the description of sediment types, both in the field and the laboratory. One seeks to describe the sediments in purely geological terms, considering the deposit as an assemblage of various elements either geologically or biologically derived. Such has been the approach of Troels-Smith (1955). The other approach attempts to combine the geological, sedimentary aspect of the description with an account of the major features of the plant community involved in the biological input (first put forward by von Post (1924) and later used and modified by many workers, e.g. Godwin (1975) and Faegri & Gams (1937), set out in Faegri & Iversen (1975)). This latter system has attractions from the ecological point of view, since it permits a high degree of resolution in terms of ecological relationships, but it can also lead to a proliferation of minor adaptations in symbolic representation as further plant species need to be catered for, especially when used in geographical areas far removed from the orginal sites.

The Troels-Smith system is probably of greater general value, since it can be applied to any sediment type, regardless of geographical or biological origin. It has been slightly simplified recently by Aaby (in Berglund, 1979) both by a reduction in the number of sediment classes and also by altering certain symbols in such a way that they are easier to draw. These are shown in Fig. 10.6.

The classification is sufficiently simple to permit field description of sediments. it copes with mixed sediments and can be used to designate the frequency of the material on a three-point scale (Troels-Smith in fact uses a four-point scale), which can be represented diagrammatically in terms of symbol density, and also humification on the scale already described, which

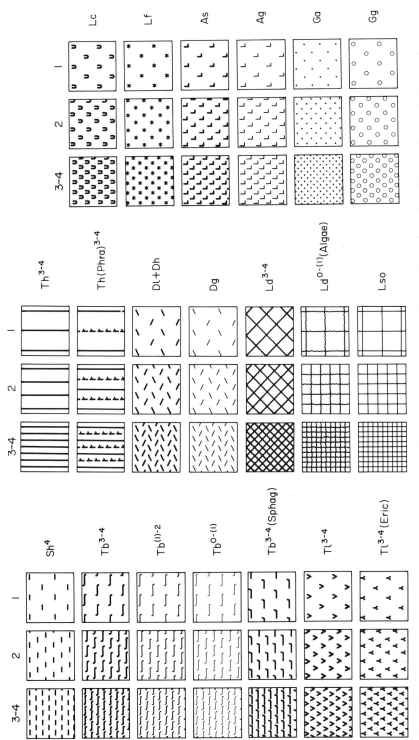

Figure 10.6 Symbols used for described sediment types as modified by Aaby from the original suggestions of Troels-Smith. Numbers at the heads of columns refer to density (greatest on the left) and the letters and figures in the right hand column refer to the nature of the material and the humification respectively. These are explained in the text (p. 538). From Aaby, in Berglund (1979–82).

can be indicated on diagrams by the line thickness of the symbol concerned (see Fig. 10.6).

The major sediment elements recognized in Aaby's modified Troels-Smith scheme are as follows:

Substantia humosa (Sh). Disintegrated organic material, structureless. Often the origin of such material becomes evident upon microscopic examination in the laboratory.

Turfa bryophytica (Tb). Moss peat. Often this is determinable to genus, e.g. *Sphagnum, Polytrichum, Rhacomitrium,* and sometimes to species, e.g. *S. imbricatum.* The conventional symbol can be adapted to a more precise indication of the species involved by adding letters. Aaby suggests the use of a distinctive symbol for the very commonly encountered *Sphagnum* peat (Fig. 10.6).

Turfa lignosa (Tl). Wood peat. Aaby proposes that the ecological distinction between wood from tree species (such as *Betula, Salix* or *Alnus*) needs to be distinguished in a stratigraphic diagram from that of dwarf shrubs, such as the Ericales. He suggests the symbol shown in Fig. 10.6.

Turfa herbacea (Th). Herbaceous peat. It may be necessary to add extra symbols to differentiate between some ecologically important herbaceous elements (e.g., *Phragmites, Carex, Cladium, Eriophorum*). Aaby suggests that these can be inserted between vertical lines, as in the case of *Phragmites,* shown in Fig. 10.6.

Detritus (D). Detritus muds, having coarse fragments of organic materials greater than 2 mm (*Detritus humosus,* Dh, and *Detritus lignosus,* Dl) or granular organic particles less than 2 mm (*Detritus granosus*), deposited under shallow water conditions.

Limus detritosus (Ld). Fine fragments of organic material, derived from a variety of biological sources and deposited under lake conditions. Specific symbols have been proposed for algal detritus and for diatomite (*Limus siliceus organogenes,* Lso).

Limus calcareus (Lc). Marl, derived from lakes containing high levels of calcium carbonate.

Limus ferrugineus (Lf). Produced in iron-rich lakes.

Argilla steatodes (As). Clay. Particle size < 0·002 mm.

Argilla granosa (Ag). Silt. Particle size 0·02–0·06 mm.

Grana arenosa (Ga). Sand. Particle size 0·06–2·0 mm.

Grana glareosa (Gg). Gravel. Particle size 2·0–20·0 mm.

Sampling of an exposed face can be conducted in the field, but there are many advantages in collecting an entire monolith.

(*a*) Laboratory sampling is cleaner and more comfortable and can therefore be tackled with greater accuracy.

(*b*) Larger samples are usually transported in this way, thus permitting more extensive macrofossil examination and making possible the extraction of samples for radiocarbon dating.

(*c*) It may be necessary to extract intermediate samples at a later stage in analysis and this is possible if a monolith has been taken.

Removal of a monolith is best carried out by using stainless steel monolith holders. A useful size is 50 × 15 × 15 cm; the holder can be constructed from a sheet of 1 mm thick stainless steel 45 × 50 cm which is folded at right angles along two lines to produce a three-sided column. This can be hammered into the peat face and then dug away with a trowel. These can be transported and stored in polythene bags; a low temperature, a little above freezing point is best since this reduces drying and microbial attack.

More details of studies based upon analyses of peat sections are given by Casparie (1972), Barber (1981) and Aaby (in Berglund, 1979). Such work can provide valuable ecological data about the changes in plant community patterns in time (e.g. Stewart & Durno, 1969) and also may permit certain general deductions about climatic changes in the past (e.g. Barber, 1982).

4.2.2 Coring

Where sections of sediment are not exposed, it is necessary to use coring equipment and there are several types available. These have been discussed in detail by West (1977), and Wright *et al.* (1965) to whom the reader is referred for detailed discussion. The construction and the relative merits of the various borers are shown in Table 10.1 and Fig. 10.7 which are taken from West's account.

Corers can be divided into four main types: (i) augers, (ii) rotating core samplers, (iii) piston corers and (iv) free fall samplers.

Augers are commonly used for soil sampling, but have considerable disadvantages from the point of view of palaeoecological sampling, the chief of which is the disturbance they cause in sediments as they penetrate them. The samples which are recovered from them are often highly contaminated by passing through upper layers during withdrawal and are not suitable for any work beyond that of preliminary survey.

Rotating samplers include the Hiller corer (Faegri & Iversen, 1975) and the Russian corer (Jowsey, 1966). Both have chambers which are 50 cm in length and about 3 cm in diameter, though a larger version of the Hiller type sampler has been described by Smith *et al.* (1968).

The Hiller sampler has an auger tip to the sampling chamber, hence can be rotated clockwise as it penetrates the sediments and this is of considerable value when coring stiff materials. On the other hand, it disturbs the sediments prior to sampling, cores cannot be extruded for return to the laboratory and

Table 10.1 Comparisons of the structure and efficiency of various hand-operated samplers and an indication of their suitability for different sediments. From West (1977)

Type	Sample Length (cm)	Width (cm)	Deformation of sediment	Compression of sediment	Suitability for sediments Good for	Bad for
Hiller	50	3	Much	None	Peat, compact mud	Loose organic sediment, inorganic sediment
Russian	50	5	None	None	Peat, mud	Loose organic sediment, inorganic sediment
Dachnowski	30	5	Little	Little	Non-fibrous peat, mud	Fibrous peat, loose organic sediment
Livingstone	50	4	Little	Some	Non-fibrous peat, mud	Fibrous peat, loose organic sediment, inorganic sediment
Punch	50	6	Little	Some	Compact organic sediment, clay silt, fine sand	Fibrous peat, loose organic sediment
Screw auger	25	4	Much	Much	Compact organic sediment, clay and silt	Soft organic and inorganic sediment
Other augers	10–30	5–10	Much	Little	Compact organic sediment, clay, silt, sand, gravel	Soft organic and inorganic sediment

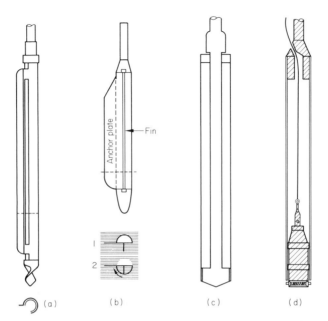

Figure 10.7 Peat samplers in common use. (a) Hiller, (b) Russian, (c) Dachnowski, (d) Livingstone. From West (1977).

cleaning is very difficult. Samples are usually taken directly from the chamber and returned to the laboratory in tubes. The problems of core removal and cleaning have been partially solved in the modified Hiller sampler described by Thomas (1964). In this model, the auger head screws off and the inner rotating chamber can be removed for core extrusion and sampling. It is best to use zinc liners wedged within the inner chamber. Both of these samplers, however, have the disadvantage that the chambers may fail to open and take a sample in very soft sediments.

The Russian corer is now widely used for peat survey work, being clean to operate, simple and reliable in its mechanical operation, relatively free from contamination problems, providing a core which can be removed intact to the laboratory and avoiding the disturbance of sediments prior to sampling. Its one disadvantage is that it cannot be rotated during penetration, hence it may prove difficult to operate in very stiff sediments, especially at depth.

Piston corers may be of the Dachnowski or the Livingstone type, depending upon whether the piston is operated by a central rod or by wire (see Fig. 10.7). During operation, the piston remains stationary while the tube descends to enclose the sample core. This type of corer provides uncontaminated cores which can be collected intact, but is not appropriate for the sampling of soft, loose sediments, which tend to drain from the sample chamber during withdrawal. There are many modifications of this type of sampler, including large-capacity samplers used for radiocarbon samples or for macrofossil analysis (e.g. Cushing & Wright, 1965).

Free fall samplers are simple lengths of tubing, attached to a wire, which can be used for lake sediments simply by permitting them to drop into the superficial layers and penetrate by its own momentum. An interesting modification has been developed in which the tube itself is seated at one end and filled with dry ice. Prior to release, trichloroethylene is added to depress the freezing point. When this sampler penetrates the sediments, they freeze in a block around its outside and can be recovered in a frozen state. The technique is particularly useful for sampling very loose, easily disturbed layers, when even delicate laminations can be preserved in the frozen mass. The idea was first put forward by Shapiro (1958) and has since been used for studies involving detailed analysis of, for example, charcoal layers (e.g. Swain, 1973) and annual laminations (e.g. Saarnisto *et al.*, 1977).

It is always a good idea to make as detailed as possible a description of cores in the field, even if they are to be transported intact to the laboratory. Very often the colour and textural distinctions which can be made by observing fresh, unoxidized samples, are lost once the cores have been exposed even for a few minutes to the atmosphere. The description and nomenclature of sediment types has been outlined in the previous section.

Cores should be wrapped in polythene to prevent desiccation and transported in rigid containers. They are best stored prior to analysis at about 2°C, which reduces microbial decay but does not produce the distortion which may result from freezing.

4.2.3   Soils

Soils can generally be investigated stratigraphically by means of a soil pit (see Ball, p. 221). Sampling for pollen and microfossil analysis can be difficult if the mor humus layers are friable. The use of monolith holders, as described for open peat sections, is to be recommended where possible, but extra care needs to be taken in the transport of such samples. Dimbleby (1985) recommends taking samples from soil profiles in the field and considers that contamination is minimized if basal samples are taken first.

Techniques used in the examination and description of soil profiles are provided by Ball (p. 222).

4.3   Macrofossils

4.3.1   Plant macrofossils

If macrofossils are to be determined quantitatively, then a known volume of sediment must be used and preferably the sample should be fairly large, to reduce sampling errors. Aaby (in Berglund, 1979) recommends a sample size of at least 50 cm^3, but often the diameter of the core precludes such large samples. Storage of the samples is best achieved by saturating with ethanol, or in a 1:1:1 mixture of glycerol, ethanol and water.

Dispersants are usually needed to break up samples, and alkalis such as 10% KOH or Na$_2$CO$_3$ are often used. The samples can then be sieved and many workers use a 0·4 mm mesh sieve, which is capable of retaining even the tiny seeds of Ericaceae and Juncaceae. Wasylikova (in Berglund, 1979) suggests 0·5 and 0·2 mm.

For calcareous samples dilute HCl may be used (Watts & Winter, 1966; Dickson, 1970) and some recommend soaking in dilute HNO$_3$ when macrofossils, especially seeds, often float to the top and can be recovered in the surface froth (Watts, 1959).

The cleaning of fruits and seeds embedded in mineral matter can be achieved by soaking for 24 h in 30% HF, followed by 30 min in 30% HCl and subsequent washing (Wasylikova, in Berglund, 1979).

Impressions of macrofossils especially leaves, in compacted muds, are particularly difficult to deal with. They are best photographed or drawn *in situ*.

Details of macrofossil extraction and examination techniques are given by Wasylikova (quoted above), Birks (1980), Dickson (1970), Cutler (1982) and Scott & Collinson (1978).

The identification of plant macrofossils is best achieved by comparison with recent material, which can be treated in a similar way to the fossil material for the construction of a type collection. The following works may prove of value as an aid to identification.

Fruits and seeds: Berggren (1969, 1981), Aalto (1970), Körber-Grohne (1967), Katz *et al.* (1965), Grosse-Brauckmann (1974), Beijerink (1974), Swarbrick (1971) and Bertsch (1941).

Plant cuticles: Grosse-Brauckmann (1972).

Wood: Barefoot & Hankins (1982) microfiche keys.

Bryophytes: Dickson (1973), Smith (1978), Grosse-Brauckmann (1974), Arnell (1981) and Nyholm (1981).

It may be possible to make estimations of the density of certain types of macrofossil in the peat, such as fruits and seeds, and these can be expressed in terms of number of fossils per unit volume of sediment (see Fig. 10.8). Often, however, such precise counts cannot be achieved, as in the case of bryophyte or root remains, or charcoal fragments. In this case it may still be possible to express relative abundance on a fine point scale, as has been done by Barber (1976). Janssen (1983) has developed a technique for assessing

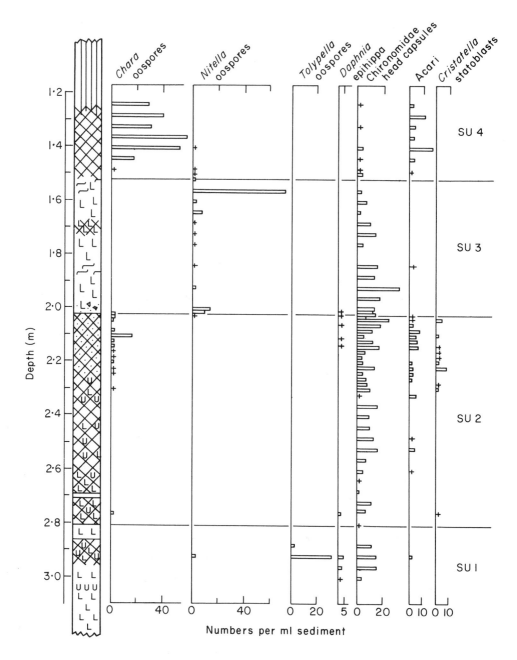

Figure 10.8 Macrofossil diagram and macrofossil zones from late-glacial sediments in south-east Scotland. From Webb & Moore (1982).

the quantitative abundance of bryophyte leaves in peat in terms of fragment number per unit volume, which could prove valuable for detailed reconstruction of the changes of plant assemblages in time. This improves on the semi-quantitative techniques of Stewart & Durno (1969).

The reconstruction of local plant communities on the basis of fossil assemblages will be greatly assisted by studies of the present-day relationship between the species composition and density in different habitats and the accumulation of fossil fragments. At present there are relatively few such studies; for example, Birks (1973) in North America and GreatRex (1983) and Collinson (1983) in Britain.

4.3.2 Animal macrofossils

Sediments are often rich in remains of invertebrate animals and these may often play an important part in aiding the reconstruction of both local and regional environments. Beetle carapaces, head capsules of Chironomid midges, Cladocera, mites, rhizopods and molluscs are all of frequent occurrence in various types of samples.

For small fossils in peats, such as rhizopods, the following extraction process is effective (see Tolonen in Berglund, 1979).

1 Take a sample of about $3 \, cm^3$.
2 Boil in water for 10 min.
3 Sieve through a 0·6 nm sieve.
4 Centrifuge suspension.
5 Mount in glycerol jelly.

Larger items, such as Chironomid head capsules are best removed by sieving (Hofmann in Berglund, 1979).

1 Take a sample of about 5–10 g fresh weight.
2 Boil in 10% KOH.
3 Sieve at 200 μm aperture and examine sievings.
4 Material which passes through this sieve size should be sieved at 100 μm aperture.
5 If sediments are calcareous, the collected sievings at 100 μm aperture should be treated with HCl.
6 Mount in polyvinyl lactophenol.

For removal of Ostracoda a 50 μm sieve aperture is required; and for Cladocera a 37 μm aperture.

Larger animal macrofossils, such as beetles and molluscs, require larger samples to provide adequate counts. For molluscs, Evans (1972) recommends 0·5–2·0 kg samples and the following treatment, which is also suitable for vertebrate bones and wood charcoal.

1 Form a slurry with water.
2 Pass through 0·5 mm sieve and oven dry the sievings.
3 Rewash until all humic material is removed.
4 Resistant organic fragments can be dissociated using 100 volume H_2O_2.

For beetles, the removal of calcareous matter is less of a problem, and the following technique is an outline of that recommended by Coope (in Berglund, 1979).

1 Form a slurry with water.
2 If calcareous, dissolve with CH_3COOH.
3 Dissociate if necessary with:
   (a) warm solution of Na_2CO_3;
   (b) 5% cold NaOH depending on the requirement of the sediment.
4 Sieve through 300 μm aperture sieve, retain sievings.
5 If necessary, concentrate insect material by flotation in kerosene (use protective gloves). Wash in detergent and hot water.
6 If necessary, dry for storage on cards.

4.4   Microfossils

4.4.1 Pollen and spores

Perhaps the most widely used evidence for environmental change has been pollen and spores, which are efficiently preserved in large quantities in a variety of sediments. The extraction process required varies according to the sediments concerned and also varies between different laboratories. The techniques are summarized by Faegri & Iversen (1975), Moore & Webb (1978) and Berglund (1979). The method outlined here is a hybrid derived from a number of sources.

1 Approximately 1 cm^3 of peat or limnic sediment is usually required. The precise volume does not need to be known unless absolute densities are to be calculated, in which case it can be determined by displacement of water in a 5 cm^3 measuring cylinder.
2 Add about 10 ml of 10% KOH and place in a boiling water bath for 10 min. Add water as evaporation takes place.
3 Sieve with 200 μm sieve, discard coarse fraction unless needed for macro-fossil examination.
4 Wash twice, centrifuging at 3000 rpm for 3 min.
5 If mineral material is present, add about 5 ml of 60% HF to the pellet in a polypropylene tube and heat in a boiling water bath until all silica is dissolved (use protective clothing and an efficient fume cupboard). If silica persists, stand for 24 hours or more in HF.

6 Centrifuge and decant into a suspension of $CaSO_4$ in a fume cupboard.

7 Add dilute HCl and place in a water bath for 2 min if silicofluorides have formed.

8 Wash twice.

9 Wash in glacial acetic acid.

10 Add a mixture of acetic anhydride and concentrated H_2SO_4 in a ratio of 9 : 1. Heat in boiling waterbath for 2 min. Centrifuge and discard supernatant carefully into running water.

11 Wash twice.

12 Wash in 2% KOH.

13 Stain with two drops of 5% aqueous safranine.

14 Add glycerol jelly to the tube, mix thoroughly and mount on slides.

Certain modifications may be made to the above procedure.

(*a*) If calcareous material is present, the sample should be treated with cold 10% HCl before commencing the KOH digestions.

(*b*) If siliceous material is absent, omit the HF treatment (steps 5–8).

(*c*) Mounting in silicone oil is preferred by many research workers. It has the advantage that pollen grains do not swell when thus treated (see Andersen, 1978) but it has the disadvantage that it remains fluid at room temperature, so mounted pollen grains are mobile. The mounting procedure is as follows:

Stages 1–12 as above.

13 Add absolute ethanol with 2 drops 5% aqueous safranine. Centrifuge and decant.

14 Wash in absolute ethanol.

15 Suspend in tertiary butyl alcohol, centrifuge and decant.

16 Add the amount of silicon oil (refractive index 1·4) needed to provide an optimal concentration of pollen, mix and allow excess tertiary butyl alcohol to evaporate.

17 Mount and seal.

(*d*) When a sample is very rich in clay, some fine colloidal material may survive HF treatment and obscure the final samples. If this is the case, a deflocculation procedure can be inserted after KOH digestion and washing and before HF treatment (i.e. between steps 4 and 5). The sample must be washed clean of KOH before treatment, tested by pH measurement. 0·1 M sodium pyrophosphate ($Na_4P_2O_7$) is added to the pellet, which is then stirred thoroughly and heated in a water bath for 10–20 min. The liquid becomes saturated with dispersed clay particles which remain in suspension when the sample is centrifuged for 5 min at 3000 rpm. The process can be repeated if necessary prior to continuing with HF treatment. Further details of this technique are given by Bates *et al.* (1978).

(e) An additional technique for the extraction of pollen grains from very fine grained materials has been suggested by Cwynar et al. (1979). They propose sieving the sample through a 7 μm aperture nylon sieve after acetolysis and washing (between steps 11 and 12). This permits the very fine clays which have survived HF treatment to pass through, but retains most pollen and spores (they record only a 0·4% loss).

(f) If facilities are not available for HF treatment, a reasonable separation of pollen can be effected by discontinuous density gradient centrifugation. A 15 cm³ capacity centrifuge is carefully loaded with 4 cm³ of sucrose with specific gravity 1·35, followed by a second layer of sucrose SG 1·2. 1 cm³ of suspension which has been treated with KOH and washed is carefully added to the tube, forming a third layer and the tube is centrifuged at 3000 rpm for 30 seconds.

After this treatment, large mineral particles will have been transported to the base of the tube, fine particles remain in suspension at the top and the junction between the lower two sucrose layers has an accumulation of pollen and plant debris, together with some mineral material of small particle size. This layer can be removed using a Pasteur pipette and acetolysed.

The method still needs some refinement, but it offers an opportunity to produce a countable preparation of microfossils from a siliceous material without HF treatment.

If the precise concentration of pollen within a sediment is required a variety of techniques can be used, which have been described and compared by Peck (1974). The most widely used involves the addition of a known number of exotic pollen grains to a known volume of sediment at an early stage in processing. The standard pellets described by Stockmarr (1971) have been extensively used, but the preparation of standard density suspension of an exotic pollen type is not difficult and can be checked by haemocytometer counts.

Mention was made earlier of extraction of pollen from honey and honey products by archaeologists (and apiarists). The technique for this recommended by Louveaux et al. (1978) is as follows:

1 Dissolve 10 g honey in 20 ml water at about 40°C.
2 Centrifuge for 10 min at 2500 rpm.
3 Wash twice in water.
4 If rich in colloidal organic matter, dissolve in 10% KOH or dilute H_2SO_4.
5 Wash again.
6 Acetolyse as above, stain in safranine and mount.

Apiarists often neglect the latter stages and mount material after stage 3 or stage 5. Acetolysis, however, produces cleaner pollen samples which are more easily compared with reference material and standard pollen keys.

Several keys to pollen grains are now easily available, such as Faegri & Iversen (1975), Moore & Webb (1978), Punt (1976) and Punt & Clark (1980, 1981) for northern Europe, and Kapp (1969) and McAndrews *et al.* (1973) for North America. These are extremely useful as an aid to pollen identification, but there is no substitute for the possession of a reference collection of modern, well determined pollen types. These can easily be prepared from fresh or herbarium plant material, using the same methods as those described for fossil material. Some commercial suppliers also provide a selection of prepared pollen types.

4.4.2 Diatoms

The siliceous frustules of diatoms survive well in lake sediments and they provide useful indicators of local conditions in the water body, such as lake chemistry. A review of their ecology has been published by Patrick (1977), and a detailed account of laboratory techniques involved in extraction has been given by Battarbee in Berglund (1979–82).

1 A sample of about $0 \cdot 1$ g is usually adequate.

2 If the sediment is calcareous, add hot 10% HCl for about 15 min.

3 To remove organic matter, add cold 30% H_2O_2 and heat gently.

4 Wash and centrifuge three times.

5 Pass through a $0 \cdot 5$ mm sieve to remove coarse material.

6 Further separation of coarse and very fine clay materials can be achieved by careful differential sedimentation from suspension. The use of a dilute ammonia solution is recommended to maintain the clays in suspension.

7 If this technique fails to remove all clay particles, flotation on heavy liquids, such as a solution of zinc bromide of SG $2 \cdot 4$ can be used. Centrifuge for 5 min at 3000 rpm. The supernatant should contain most of the diatoms but it is worth checking the pellet and repeating if necessary.

8 Water wash.

9 Evaporate on a slide and mount in Naphrax (RI $1 \cdot 74$).

Diatom taxonomy is complex and identification is often made difficult by intra-taxon variability. Drawings or photographs of each type should be made.

4.4.3 Other microfossils

Rhizopoda can be extracted from a sediment by boiling in water for 10 min, sieving through a $0 \cdot 6$ mm mesh sieve, centrifuging and mounting in glycerol jelly.

An illustrated account of the testate rhizopods of *Sphagnum* bogs is given by Corbet (1973).

Analyses of fungi in peats and mor humus layers are described by van Geel (1972) and Aaby (1983) respectively. Various other fungal and algal microfossils and their palaeoecological value are described by van Geel & van der Hammen (1978), van Geel *et al.* (1981) and Smol *et al.* (1984).

The siliceous particles found in grass leaves, phytoliths, may be preserved in sediments, but their extraction must be conducted without the use of HF, as in the case of diatoms. An example of their use in palaeoecology is the study of the cultivation of maize in ancient Ecuador by Pearsall (1978).

4.5   Chemical stratigraphy

The chemical analysis of stratified sediments can often provide information about previous conditions at the site. This is true of both the inorganic and the organic components of the sediment chemistry.

4.5.1   Inorganic chemistry

Inorganic analysis of sediments has often been used with some effect in palaeoenvironmental reconstruction. For example, Chapman (1964) has used Ca : Mg ratios in peat profiles as an index of climatic continentality or oceanicity. In a similar way, Pennington & Lishman (1971) used iodine levels in lake sediments as a guide to past oceanicity.

Chemicals may also reveal the impact of man upon a region as in the study of silica depletion in the sediments of the Great Lakes resulting from cultural eutrophication between 1820 and 1850 and the consequent increase in diatom growth (Schelske *et al.*, 1983). The history of atmospheric pollution has also been traced by chemical analysis of peat profiles. Magnetic particles in British bogs become more abundant after the build up of the Industrial Revolution, at about 1800 AD (Oldfield *et al.*, 1978) and lead has been shown to increase since about 1700 AD (Lee & Tallis, 1973).

4.5.2   Organic chemistry

The organic compounds derived from the degradation of plant materials are incorporated into stratified sediments and may provide useful information. For example, the phenolic compounds derived from lignin oxidation are relatively stable because of their resistance to microbial attack and these may indicate the extent to which woody plants occupy a catchment (Ertel *et al.*, 1984). It may even be possible to distinguish angiosperm and gymnosperm wood products on the basis of the types of phenols involved (Leopold *et al.*, 1982).

Chlorophyll degradation products are not difficult to extract from peats and lake sediments (Sanger & Gorham, 1973), and they can provide an index to the productivity of a site in the past (Sanger & Gorham, 1972; Whitehead *et al.*, 1973).

Recently there have been some very interesting developments in the use of $^{13}C : {}^{12}C$ isotope ratios in fossil materials (van der Merwe, 1982). This ratio varies according to the biochemical pathway by which it has been fixed; some plants (called C_4 plants) have a specialized photosynthetic mechanism whereby the $^{13}C : {}^{12}C$ ratio is less negative in the resulting products of photosynthesis. Since these C_4 plants include some interesting cultivated species, such as maize, the index can be used by archaeologists to demonstrate a shift in diet onto such C_4 species.

5 References

AABY B. (1983) Forest development, soil genesis and human activity illustrated by pollen and hypha analysis of two neighbouring podzols in Draved Forest, Denmark. *Danm. Geol. Unders* IIR No. 114 1–114.

AALTO M. (1970) Potamogetonaceae fruits. I. Recent and subfossil endocarps of Fennoscandian species. *Acta. Bot. Fenn.* **88**, 1–85.

ADAM D.P. and WEST G.J. (1983) Temperature and precipitation estimates through the last glacial cycle from Clear Lake, California pollen data *Science* **219**, 168–170.

ADAM R.J. and SMITH M.V. (1981) Seasonal pollen analysis of nectar from the hive and of extracted honey. *J. Apicultural Research* **20**, 243–248.

AMBACH W., BORTENSCHLAGER S. and EISNER H. (1966) Pollen analysis investigation of a 20 m firn pit on the Kessel wandferner (Ötztal Alps). *J. Glaziol (Innsbruck)* **6** (44) 233–236.

ANDERSON S.TH. (1978) On the size of *Corylus avellana* L. pollen mounted in silicone oil. *Grana* **17**, 5–13.

ANDERSON S.TH. (1979) Brown earth and podzol: soil genesis illuminated by microfossil analysis. *Boreas* **8**, 59–73.

ARNELL S. (1981) *Illustrated Moss Flora of Fennoscandia.* I. *Hepaticae,* 2nd edn. Swedish Science Research Council, Stockholm.

BAILLIE M.G.L. (1982) *Tree-Ring Dating and Archaeology* Croom Helm, London.

BAKER C.A., MOXEY P.A. and OXFORD P.M. (1978) Woodland continuity and change in Epping Forest. *Field Studies* **4**, 645–669.

BARBER K.E. (1976) History of vegetation. In *Methods in Plant Ecology*, 1st edn (ed. S.B. Chapman), pp. 5–83. Blackwell Scientific Publications, Oxford.

BARBER K.E. (1981) *Peat Stratigraphy and Climatic Change.* A.A. Balkema, Rotterdam.

BARBER K.E. (1982) Peat bog stratigraphy as a proxy climate record. In *Climatic Change in Later Prehistory* (ed. A.F. Harding), pp. 103–113. Edinburgh University Press, Edinburgh.

BAREFOOT A.C. and HANKINS F.W. (1982) *Identification of Modern and Tertiary Woods.* Clarendon Press, Oxford.

BATES C.D., COXON P. and GIBBARD P.L. (1978) A new method for the preparation of clay-rich sediment samples for palynological investigation. *New Phytol* **81**, 459–463.

BEIJERINK W. (1974) *Zadenatlas der Nederlandsche Flora.* N. Veenman and Zoren, Wageningen.

BERGGREN G. (1981) *Atlas of Seeds. Part 2 Cyperaceae.* Swedish National Science Research Council, Stockholm.

BERGGREN G. (1981) *Atlas of Seeds. Part 3 Salicaceae–Cruciferae.* Swedish Museum of Natural History, Stockholm.

BERGLUND B.E. (Ed.) (1979–82) International Geological Correlation Programme Project 158

B. Palaeohydrological changes in the temperate zone in the last 15000 years. Lake and mire environments. Project Guide, 3 volumes, Lund, Sweden. See also below.

BERGLUND B.E. (Ed.) (1985) *Handbook of Quaternary Palaeoecology.* Wiley, Chichester (a revised version of the IGCP guidebooks).

BERNABO J.C. and WEBB T. (1977) Changing patterns in the Holocene pollen record of northeastern North America: a mapped summary. *Quat. Res.* **8**, 64–96.

BERTSCH K. (1941) *Früchte und Samen. Handbuch der Praktischen Vorgeschichtesforschung.* Ferdinand Enke, Stuttgart.

BIRKS H.H. (1973) Modern macrofossil assemblages in lake sediments in Minnesota. In *Quaternary Plant Ecology* (ed. H.J.B. Birks and R.G. West), pp. 173–189. Blackwell Scientific Publications, Oxford.

BIRKS H.H. (1980) Plant microfossils in Quaternary lake sediments. *Arch Hydrobiol. Beih. Ergebn. Limnol* **15**, 1–60.

BIRKS H.J.B. (1981) The use of pollen analysis in the reconstruction of post climates: a review. In *Climate and History* (ed. T.M.L. Wigley, M.J. Ingram & G. Farmer), pp. 111–138. Cambridge University Press, Cambridge.

BIRKS H.J.B. (1982) Mid-Flandrian forest history of Roundsea Wood National Nature Reserve, Cumbria *New Phytol.* **90**, 339–354.

BIRKS H.J.B. and BIRKS H.H. (1980) *Quaternary Palaeoecology*, Edward Arnold, London.

BIRKS H.J.B., DEACON J. and PEGLAR S. (1975) Pollen maps for the British Isles 5000 years ago. *Proc. Roy. Soc. Lond.* B. **189**, 87–105.

BIRKS H.J.B. and SAARNISTO M. (1975) Isopollen maps and principal components analysis of Finnish pollen data for 4000, 6000 and 8000 years ago. *Boreas* **4**, 77–96.

CASPARIE W.A. (1972) *Bog Development in South-Eastern Drenthe (The Netherlands).* Junk, The Hague.

CHAPMAN S.B. (1964) The ecology of Coom Rigg Moss, Northumberland. II. The chemistry of peat profiles and the development of the bog system. *J. Ecol.* **52**, 315–321.

COLE K. (1982) Late Quaternary zonation of vegetation in the eastern Grand Canyon. *Science* **217**, 1142–1145.

COLLINSON M.E. (1983) Accumulations of fruits and seeds in three small sedimentary environments in southern England and their palaeoecological implications. *Ann. Bot.* 583–592.

CORBET S.A. (1973) An illustrated introduction to the testate rhizopods in *Sphagnum* with special reference to the area round Malham Tarn, Yorkshire. *Field Studies* **3**, 801–838.

CROMPTON G. and SHEAIL J. (1975) The historical ecology of Lakenheath Warren in Suffolk, England: a cse study. *Biol. Conserv.* **8**, 299–313.

CUSHING E.J. and WRIGHT H.E. (1965) Hand operated piston corers for lake sediments. *Ecology* **46**, 380–384.

CUTLER D.F. (1982) Anatomical methods used in the identification of plant remains from archaeological sites. In *Microscopy in Archaeological Conservation* (ed. M. Corfield and K. Foley), pp. 10–13. Occassional Paper No. 2. UK Institute for Conservation.

CWYNAR L.C. (1977) The recent fire history of Barron Township, Algonquin Park. *Can. J. Bot.* **55**, 1524–1538.

CWYNAR L.C., BURDEN E. and McANDREWS J.H. (1979) An inexpensive sieving method for concentrating pollen and spores from fine-grained sediments. *Can. J. Earth Sci.* **16**, 1115–1120.

DAVIS M.B. and FORD M.S. (1982) Sediment focussing in Mirror Lake New Hampshire. *Limnol. Oceanogr.* **27**, 137–150.

DICKSON C.A. (1970) The study of plant macrofossils in British Quaternary deposits. In *Studies in the Vegetational History of the British Isles* (ed. D. Walker and R.G. West), pp. 233–254. Cambridge University Press, Cambridge.

DICKSON J.H. (1973) *Bryophytes of the Pleistocene.* Cambridge University Press, Cambridge.

DICKSON J.H. (1978) Bronze Age mead. *Antiquity* **52**, 108–113.

DIMBLEBY G.W. (1962) The development of British heathlands and their sites. *Oxford Forestry Memoirs No. 23.*

DIMBLEBY G.W. (1978) *Plants and Archaeology.* Granada, London.

DIMBLEBY G.W. (1985) *The Palynology of Archaeological Sites*. Academic Press, London.

ERTEL J.R., HEDGES J.I. and PERDUE E.M. (1984) Lignin signature of aquatic humic substances *Science* **223**, 485–487.

EVANS J.G. (1972) *Land Snails in Archaeology*. Seminar Press, London.

FAEGRI K. and GAMS H. (1937) Entwicklung und Vereinheitlichung der Signaturen für Sediment —und Torfarten. *Geol. Foren. Stockh. Förh* **59**, 273–284.

FAEGRI K. and IVERSEN J. (1975) *Textbook of Pollen Analysis* (3rd edn). Blackwell Scientific Publications, Oxford.

FREDSKILD B. (1978) Palaeobotanical investigations of some peat deposits of Norse age at Qagssiarssuk, South Greenland. *Meddelelser om Grønland* **204** (5), 1–41.

FREDSKILD B and WAGNER P. (1974) Pollen and fragments of plant tissue in core samples from the Geenland Ice Cap. *Boreas* **3**, 105–108.

FREEMAN C.E. (1972) Pollen study of some Holocene alluvial deposits in Dona Ana County, Southern New Mexico *Texas. J. Sci.* **24**, 203–220.

GIRLING M. and GREIG J.R.A. (1977) Palaeoecological investigations of a site at Hampstead Heath, London. *Nature* **268**, 45–47.

GODWIN H. (1949) Pollen analysis of glaciers in special relation to the formation of various types of glacier bands. *J. Glaciol* **1**, 325–333.

GODWIN H. (1975) *The History of the British Flora*. Cambridge University Press, Cambridge.

GODWIN H. and GODWIN M.E. (1960) Pollen analysis of peats at Scolt Head Island, Norfolk. In *Scolt Head Island* (ed. J.A. Steers), pp. 73–84. Heffer & Sons, Cambridge.

GREATREX P.A. (1983) Interpretation of macrofossil assemblages from surface sampling of macroscopic plant remains in mire communities *J. Ecol.* **71**, 773–791.

GREIG J.R.A. (1981) The investigation of a medieval barrel-latrine from Worcester. *J. Archaeol. Sci.* **8**, 265–282.

GREIG J.R.A. (1982) The interpretation of pollen spectra from urban archaeological deposits. In *Environmental Archaeology in the Urban Context* (ed. A.R. Hall and H.K. Kenward), pp. 47–65. CBA Research Rpt, No. 43, London.

GROSSE-BRAUCKMANN G. (1972) Über pflanzliche Makrofossilien Mitteleuropäischer Torfe. 1 Gewebereste Krautiger Pflanzen und ihre Merkmale. *Telma* **2**, 19–55.

GROSSE-BRAUCKMANN G. (1974) Über pflanzliche Makrofossilien mittel europäischer Torfe. 2 Weitere Reste (Früchte und Samen, Moose u.a.) und ihre Bestimmungsmöglichkeiten *Telma* **4**, 51–117.

HARPER J.L. (1969) The role of predation in vegetational diversity. *Brookhaven Symp. in Biol.* **22**, 48–62.

HAVINGA A.J. (1974) Problems in the interpretation of pollen diagrams of mineral soils. *Geologie Mijnb.* **53**, 449–453.

HAVINGA A.J. and VAN DEN BERG R.M. (1982) Vegetational development of a wood peat deposit, as read from its pollen content. In *Proceedings of the Symposium on Peat Lands below Sea Level* (ed. H. de Bakker and M.W. van den Berg), pp. 275–281. International Institute for Land Reclamation and Improvement, Wageningen, Netherlands.

HEUSSER C.J., HEUSSER L.E. and STREETER S.S. (1980) Quaternary temperatures and precipitation for the north-west coast of North America. *Nature* **286**, 702–704.

HOOPER M.D. (1970) Dating hedges. *Area* **4**, 63–65.

HUNTLEY B. and BIRKS H.J.B. (1983) *An Atlas of Past and Present Pollen Maps for Europe: 0–13000 Years Ago*. Cambridge University Press, Cambridge.

IVERSEN J. (1958) Pollenanalytischer Nachweis des Reliktencharakters eines jüstischen Linden-Mischwaldes. *Veröff. Geobot. Inst. Rubel.* **33**, 137–144.

JACOBSON G.L. and BRADSHAW R.H.W. (1980) The selection of sites for palaeovegetational studies. *Quat. Res.* **16**, 80–96.

JANSSEN C.R. (1967) A post-glacial pollen diagram from a small *Typha* swamp in northwestern Minnesota, interpreted from pollen indicators and surface samples. *Ecol. Monogr.* **37**, 145–172.

JANSSEN J.A. (1983) A quantitative method for stratigraphic analysis of bryophytes in Holocene peat. *J. Ecol.* **71**, 189–196.

JOWSEY P.C. (1966) An improved peat sampler. *New Phytol.* **65**, 245–248.

KAPP (1969) *How to Know Pollen and Spores.* W.C. Brown, Dubuque, Iowa.

KATZ N.J., KATZ S.V. and KIPANI M.G. (1965) *Atlas of Keys of Fruits and Seeds Occuring in the Quaternary Deposits of USSR.* Nanka, Moscow (in Russian, illustrated).

KONRAD V.A., BONNICHSEN R. and CLAY V. (1983) Soil chemical identification of ten thousand years of prehistoric human activity areas at the Munsungun Lake Thoroughfare, Maine. *J. Archaeol. Sci.* **10**, 13–28.

KÖRBER-GROHNE U. (1967) *Geobotanische Untersuchungen der Feddersen Wierde*, Franz Steiner, Wiesbaden (2 vols).

LAMARCHE V.C. and HIRSCHBOECK K.K. (1984) Frost rings in trees as records of major volcanic eruptions. *Nature* **307**, 121–126.

LEE J.A. and TALLIS J.H. (1973) Regional and historical aspects of lead pollution in Britain. *Nature* **245**, 216–218.

LEOPOLD E.B., NICKMANN R., HEDGES J.I. and ERTEL J.R. (1982) Pollen and lignin records of Late Quaternary vegetation, Lake Washington. *Science* **218**, 1305–1307.

LOUVEAUX J., MAURIZIO A. and VORWOHL G. (1978). Methods of melissopalynology. *Bee World* **59**, 139–157.

MCANDREWS J.H. (1981) Late Quaternary climate of Ontario: temperature trends from the fossil pollen record. In *Quaternary Paleoclimate* (ed. W.C. Mahaney), pp. 319–333. Geo Abstracts Ltd, Norwich.

MCANDREWS J.H., BERTI A.A. and NORRIS G. (1973) *Key to the Quaternary Pollen and Spores of the Great Lakes Region.* Life Sci. Misc. Publ. Royal Ontario Museum, Toronto, Canada.

MACPHAIL K.K. (1984). Small-scale dynamics in an early Holocene wet sclerophyll forest in Tasmania. *New Phytol.* **96**, 131–147.

MARKGRAF V. (1980) Pollen dispersal in a mountain area. *Grana* **19**, 127–146.

MARTIN P.S. and SHARROCK F.W. (1964) Pollen analysis of prehistoric human faeces; a new approach to ethnobotany. *Amer. Antiquity* **30**, 168–180.

MOE D. (1983) Palynology of sheep's faeces: relationship between pollen content, diet and local pollen rain. *Grana* **22**, 105–113.

MOORE P.D. and BHADRESA R. (1978) Population structure, biomass and pattern in a semi-desert shrub, *Zygophyllum eurypterum* Bois & Buhse, in the Turan Biosphere Reserve of north-eastern Iran. *J. appl. Ecol.* **15**, 837–845.

MOORE P.D. and STEVENSON A.C. (1982) Pollen studies in dry environments. In *Desertification and Development: Dry Land Ecology in Social Perspective* (ed. B. Spooner and H.S. Mani), pp. 249–267. Academic Press, London.

MOORE P.D. and WEBB J.A. (1978) *An Illustrated Guide to Pollen Analysis.* Hodder and Stoughton, London.

NYHOLM E. (1981) *Illustrated Moss Flora of Fennoscandia.* II. *Musci* 2nd edn (6 vols). Swedish Natural Science Research Council, Stockholm.

OLDFIELD F., THOMPSON R. and BARBER K.E. (1978) Changing atmospheric fallout of magnetic particles recorded in recent ombrotrophic peat sections. *Science* **199**, 679–680.

O'SULLIVAN P.E. (1973) Pollen analysis of mor humus layers from a native Scots pine ecosystem, interpreted with surface samples. *Oikos* **24**, 259–272.

OTTAWAY J.H. (1984) Persistence of organic phosphates in buried soils. *Nature* **307**, 257–259.

PATRICK R. (1977) Ecology of freshwater diatoms—diatom communities. In *The Biology of Diatoms* (ed. D. Werner), Botanical Monographs No. 13. Blackwell Scientific Publications, Oxford.

PEARSALL D.M. (1978) Phytolith analysis of archaeological soils: evidence for maize cultivation in formative Ecuador. *Science* **199**, 177–178.

PECK R.M. (1974) A comparison of four absolute pollen preparation techniques. *New Phytol.* **73**, 567–587.

PENNINGTON W. and LISHMAN J.P. (1971) Iodine in lake sediments in Northern England and Scotland. *Biol. Rev.* **46**, 279–313.

PETERKEN G.F. (1974) A method of assessing woodland flora for conservation using indicator species. *Biol. Conserv.* **6**, 239–245.

PETERKEN G.F. (1981) *Woodland Conservation and Management*, Chapman and Hall, London.

PETERKEN G.F. and GAME M. (1984) Historical factors affecting the number and distribution of vascular plant species in the woodlands of central Lincolnshire. *J. Ecol.* **72**, 155–182.

PETERKEN G.F. and HARDING P.T. (1974) Recent changes in the conservation value of woodlands in Rockingham Forest. *Forestry* **57**, 109–128.

PETERKEN G.F. and HARDING P.T. (1975) Woodland conservation in eastern England: comparing the effects of changes in three study areas since 1946. *Biol. Conserv.* **8**, 279–298.

PETERKEN G.F. and TUBBS C.R. (1965) Woodland regeneration in the New Forest, Hampshire, since 1650. *J. appl. Ecol.* **2**, 159–170.

POLLARD E., HOOPER M.D. and MOORE N.W. (1974) *Hedges*. Collins New Naturalist, London.

PUNT W. (1976) *The Northwest European Pollen Flora, I*. Elsevier, Amsterdam.

PUNT W. and CLARKE G.C.S. (1980) *The Northwest European Pollen Flora, II*. Elsevier, Amsterdam.

PUNT W. and CLARKE G.C.S. (1981) *The Northwest European Pollen Flora, III*. Elsevier, Amsterdam.

ROSE F. (1976) Lichenological indicators of age and environmental continuity in woodlands. In *Lichenology: Progress and Problems* (ed. D.H. Brown, D.L. Hawksworth and R.H. Bailey), pp. 279–307. Academic Press, London.

ROSE F. and JAMES P.W. (1974) Regional studies on the British lichen flora. I. The corticolous and lignicolous species of the New Forest, Hampshire. *Lichenologist* **6**, 1–72.

SAARNISTO M., HUTTENEN P. and TOLONEN K. (1977) Annual lamination of sediments in Lake Lovojärvi, southern Finland, during the past 600 years. *Ann. Bot. Fenn.* **14**, 35–45.

SANGER J.E. and GORHAM E. (1972) Stratigraphy of fossil pigments as a guide to the postglacial history of Kirchner Marsh, Minnesota. *Limnol. Oceanogr.* **15**, 59–69.

SANGER J.E. and GORHAM E. (1973) A comparison of the abundance of fossil pigments in wetland peats and woodland humus layers. *Ecology* **54**, 605–611.

SCHELSKE C.L., STOERMER E.F., CONLEY D.J., ROBBINS J.A. and GLOVER R.M. (1983) Early eutrophication in the lower Great Lakes: new evidence from biogenic silica in sediments. *Science* **222**, 320–322.

SCOTT A.C. and COLLINSON M.E. (1978) Organic sedimentary particles: results from scanning electron microscope studies of fragmentary plant material. In *Scanning Electron Microscopy in the Study of Sediments* (ed. W.B. Whalley), pp. 137–167. Geo.Abstracts, Norwich.

SHAPIRO J. (1958) The core-freezer—a new sampler for lake sediments. *Ecology* **39**, 748.

SHEAIL J. and WELLS T.C.E. (1969) *Old Grassland, its Archaeological and Ecological Importance*. Monks Wood Experimental Station Symposium No. 5, Nature Conservancy.

SHEAIL J. (1980) *Historical Ecology: The Documentary Evidence*, NERC, Institute of Terrestrial Ecology, Cambridge.

SMITH A.G., PILCHER J.R. and SINGH G. (1968) A large capacity hand-operated peat sampler. *New Phytol.* **67**, 119–124.

SMITH A.J.E. (1978) *The Moss Flora of Britain and Ireland*. Cambridge University Press, Cambridge.

SMOL J.P., CHARLES D.F. and WHITEHEAD D.R. (1984) Mallomonadacean microfossils provide evidence of recent lake acidification. *Nature* **307**, 628–630.

SOLOMON A.M., BLASING T.J. and SOLOMON J.A. (1982) Interpretation of flood-plain pollen in alluvial sediments from an arid region. *Quat. Res.* **18**, 52–71.

STEWART J.M. and DURNO S.E. (1969) Structural variations in peat. *New Phytol.* **68**, 167–182.

STOCKMARR J. (1971) Tablets with spores used in absolute pollen analysis. *Pollen Spores* **13**, 615–621.

STOCKMARR J. (1975) Retrogressive forest development as reflected in a mor pollen diagram from Natingerbos, Drenthe, The Netherlands. *Palaeohistoria* **17**, 37–48.

SWAIN A.M. (1973) A history of fire and vegetation in north-eastern Minnesota as recorded in lake sediments. *Quat. Res.* **3**, 383–396.

SWARBRICK J.T. (1971) The identification of the seeds and indehiscent fruits of common Scottish aquatic herbs. *Trans. Bot. Soc. Edinb.* **41**, 9–25.

THOMAS K.W. (1964) A new design for a peat sampler. *New Phytol.* **63**, 422–425.

TROELS-SMITH J. (1955) Characterization of unconsolidated sediments. *Danm. Geol. Unders. IV R* **3**(10), 1–73.

VALENTINE K.W.G. and DALRYMPLE J.B. (1976) The identification of a buried paleosol developed in place at Pitstone, Buckinghamshire. *J. Soil Sci.* **27**, 541–553.

VAN DER MERWE N.J. (1982) Carbon isotopes, photosynthesis and archaeology. *Am. Scient.* **70**, 596–606.

VAN GEEL B. (1972) Palynology of a section from the raised peat bog 'Weitma scher Moor' with special reference to fungal remains. *Acta Bot. Neerl.* **21**, 261–284.

VAN GEEL B., BOHNCKE S.J.P. and DEE H. (1981) A palaeoecological study of an upper late-glacial and Holocene sequence from 'De Borchert', the Netherlands. *Rev. Palaeobotan. Palynol.* **31**, 367–448.

VAN GEEL B. and VAN DER HAMMEN T. (1978) Zygnemataceae in Quaternary Colombian sediments. *Rev. Palaeobotan. Palynol.* **25**, 377–392.

VANHOORNE R. (1979) The stratigraphical importance of the pollen analysis of fossil soils in sandy regions. *Acta Univ. Ouluensis A 82, Geol.* **3**, 143–154.

VARESCHI V. (1935) Blütenpollen in Gletschereis. *Zeitschr. für Gletscherkunde* **23**, 255–276.

VON POST L. (1924) Das genetische system der organogenen Bildungen Schwedens. *Comité Internat. de Pedologie IV. Commission* No. 22.

WATTS W.A. (1959) Interglacial deposits at Kilbeg and Newtown, Co. Waterford. *Proc. R. Irish Acad.* **60** B2, 79–134.

WATTS W.A. (1975) Vegetation record for the last 20 000 years from a small marsh on Lookout Mountain, northwestern Georgia. *Geol. Soc. Amer. Bull.* **86**, 287–291.

WATTS W.A. and WINTER T.C. (1966) Plant macrofossils from Kirchner Marsh, Minnesota: a palaeoecological study. *Geol. Soc. Amer. Bull.* **77**, 1339–1360.

WEBB J.A. and MOORE P.D. (1982) The late Devensian vegetation history of the Whitlaw Mosses, southeast Scotland. *New Phytol.* **91**, 341–398.

WEBB T. and CLARK D.R. (1977) Calibrating micropaleontological data in climatic terms: a critical review. *Annals N.Y. Acad. Sci.* **288**, 93–118.

WELLS T.C.E. (1968) Land use change affecting *Pulsatilla vulgaris* in England. *Biol. Conserv.* **1**, 37–43.

WEST R.G. (1977) *Pleistocene Geology and Biology*, 2nd edn. Longman, London.

WHITEHAD D.R., ROCHESTER H., RISSING S.W., DOUGLASS C.B. and SHEEHAN M.C. (1973) Late glacial and postglacial productivity changes in a New England pond. *Science* **181**, 744–747.

WRIGHT H.E., LIVINGSTONE D.A. and CUSHING E.J. (1965) Coring devices for lake sediments. In *Handbook of Palaeontological Techniques* (ed. B. Kummel and D.M. Raup), pp. 494–520. Freeman, San Francisco.

YATES E.M. (1979) The historical ecology of the royal forest of Wolmer, Hampshire. *Biogeographica* **16**, 93–112.

Author index

Where more than two authors are responsible for works referred to in the text, only the first named is indexed. Page where references appear in full shown in italics

557

Subject index

First page only cited when topic extends continuously over several pages